KU-641-655

DEVELOPMENT 1994 SUPPLEMENT

THE EVOLUTION OF DEVELOPMENTAL MECHANISMS

EDITED BY

MICHAEL AKAM, PETER HOLLAND, PHILIP INGHAM
AND GREG WRAY

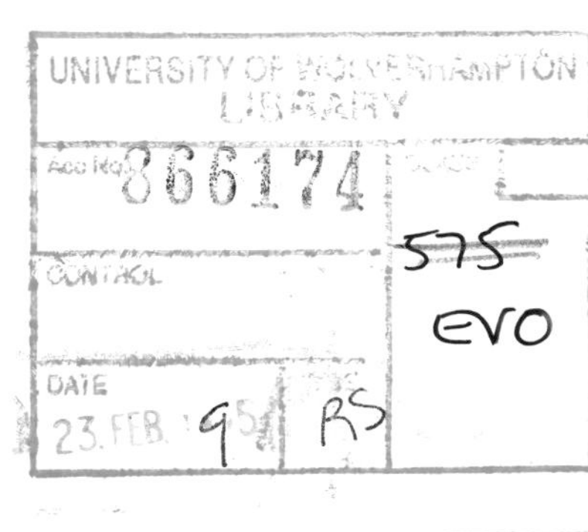

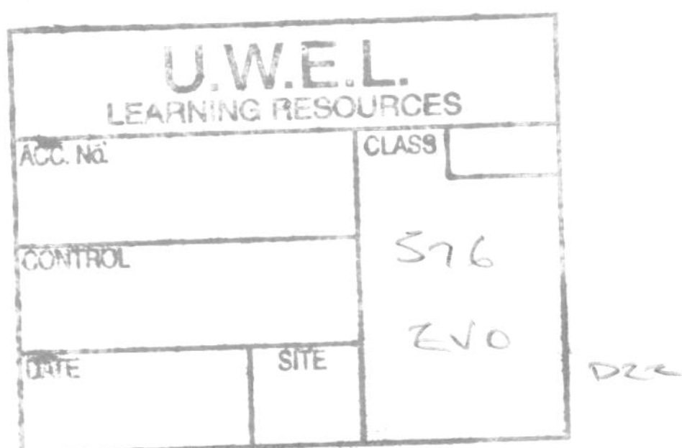

THE COMPANY OF BIOLOGISTS LIMITED
CAMBRIDGE

Typeset, Printed and Published by
The Company of Biologists Limited
Department of Zoology, University of Cambridge, Downing Street,
Cambridge CB2 3EJ

©The Company of Biologists Limited 1994

Cover picture
Acanthostega, Triops and Hox by John Rodford, 1994

Contents

1994 Supplement

Preface

"If facts of the old kind will not help us, let us seek facts of a new kind." (Bateson, W. 1894. 'Materials for the Study of Variation' Preface page vi)

Relationships between the processes of development and evolution were central to biological thought in the latter half of the nineteenth century. Accordingly, comparative morphology and embryology were at the pinnacle of the biological sciences. By the turn of the century, though, comparative morphology had been pushed to its limits, and perhaps beyond. Frustration set in. William Bateson described it well. Trained in the old school, he had been investigating what light the anatomy and development of acorn worms might throw on the origin of the chordates. Reflecting on the outcome of this study, he wrote "From the same facts, opposite conclusions are drawn; facts of the same kind will take us no further. Need we waste more effort in these vain and sophistical disputes" (Bateson, 1894). From then on, the old comparative biology was pushed to the sidelines. Any relationship between development and evolution remained peripheral to the major triumphs of twentieth century biology – to the emergence of molecular cell biology on the one hand, and to the 'modern synthesis' in evolutionary biology, with all its subsequent ramifications and revisions, on the other.

Yet even Wilhelm Roux, champion of the experimental approach and polemicist against the old biology, envisaged a time when his 'developmental mechanics' might encompass phylogenetic studies. He recognised that this endeavour would have to be adjourned until "*Entwickelungsmechanik* has been developed so far that we have gained deep insights, not only into the mechanisms forming the individual from the germ plasm, but also into the mechanisms that vary the germ plasm" (Roux, 1892; by germ plasm Roux meant something close to our 'genome'). Would Roux agree that his criteria are, at least in some measure, fulfilled? We believe he would. This volume charts some of the routes that are bringing new facts to bear on these old problems.

Underpinning all discussion is the need for a secure phylogenetic framework. Unlike Bateson, we do not seek to establish relationships by the comparative analysis of development. We have reason to hope that, as the database for molecular phylogenetic analysis expands, the domain where phylogeny equals mythology will shrink. We believe this, not because molecular data are intrinsically 'better' than morphological data, but simply because the extent of the potential database is so great. Already, the systematic analysis of a very few molecular species, notably large and small subunit ribosomal RNA, has confirmed that these molecules conserve phylogenetic information that is useful at many taxonomic levels. However, in their survey of these data, Philippe et al. (pp. 15-25) point out just how extensive sequence data must be, even under ideal conditions, to resolve relationships within explosive radiation events. It will be a long while before molecular taxonomy can approach the temporal resolution of a good stratigraphic series.

Molecular phylogenetics may resolve the relationships between taxa – but they can never tell us what sort of beast a stem group species was. Hypothetical archetypes are no substitute for a few good fossils - a point emphasised by Conway-Morris (pp. 1-13), and by Coates (pp. 169-180) in his reconsiderations of tetrapod origins. Palaeontology may not often give us a developmental series (juvenile trilobites notwithstanding) but it can constrain speculation about the end products of ancestral developmental processes – and provide clues to ancestral states that are no longer evident in extant species.

There would be no need of such clues if ontogeny did recapitulate ancestral developmental stages faithfully, but even Haeckel admitted that it does not. Modern species reflect their phylogenetic history selectively, and even then, the reflections are strangely distorted. In the context of vertebrate gastrulation, De Robertis et al. (pp. 117-124) show how molecular probes can reveal the unity of pattern and process beneath the distortions of morphology. Haeckel would be delighted.

One thing that has not been over-written or replaced during the diversification of the metazoa is the basic tool-kit of development. Since the 1940s, we have been used to the idea that metabolic enzymes and pathways are universal – but only recently has it become clear how extensively conserved are the molecules that regulate development: transcription factors, receptors and extracellular matrix molecules. As little as ten years ago, the idea of searching for insect homologues of such 'vertebrate' molecules as fibronectin or Myc seemed almost an irrelevance. (All credit to those few who persevered). Now the very idea of 'vertebrate' molecules seems a nonsense, and the relevance of functional studies in tractable model systems is established beyond dispute. Chothia (pp. 27-33) utilises data accumulating in the protein and nucleic databases to survey this metazoan tool-kit. He concludes that life uses rather few of the possible protein structures to generate its diversity – the great majority of proteins may be referred to perhaps no more than a thousand structural families.

Within a few years, genome projects will provide a complete inventory of these proteins for a usefully diverse set of reference species. Such sequences alone will be no sure guide to structure or function, but, without doubt, one striking lesson from these inventories will be the extent to which we are indeed one flesh, mite and man and lowly worm. Yes, there will be new genes – mostly made by recombining old domains (Engel et al., pp. 35-42); duplication of old genes will be rife (Holland et al., pp. 125-133; Ruddle et al., pp. 155-161); some old molecules will be seen to have acquired wholly new functions (the lens crystallins are a good case in point (Piatogorsky and Wistow, 1989)). Even so, as we see it at present, the history of life since the Cambrian has been dominated by the elaboration of regulatory mechanisms that exploit a common set of genes.

How have these regulatory mechanisms evolved? We cannot yet see 'the big picture'. Are the same cell types specified by homologous regulatory molecules in different phyla? How conserved is the molecular basis for induction, lateral inhibition or neurogenesis? Several papers in this volume provide glimpses of conserved developmental mechanisms - the hedgehog and TGFβ family signalling molecules used in analogous ways in vertebrates and invertebrates (Fietz et al., pp. 43-51; Hogan et al., pp. 53-60); transcription factors that seem to specify the same organs in vertebrates and invertebrates – heart, eyes, – despite the most diverse morphology (Manak and Scott, pp. 61-77). Are we seeing homologous mechanisms? If so, at what level does the homology lie? Will downstream and upstream regulatory networks be conserved, or are there constrained steps in cellular differentiation (cytotypic stages?) just as there are during embryogenesis, above and below which regulatory networks are more fluid? There is a new world of evolutionary biology here.

The 'new facts' of molecular biology pertain not just to molecular phylogeny and cell biology, but to the questions of organismal form – Bauplan; zootype. The Hox genes provide the outstanding example. The linear deployment of Hox genes along the anteroposterior axis of nematodes, insects and chordates provides a strong argument to establish the primitive homology of this axis in all bilateria (Ruddle et al., pp. 155-161; Manak and Scott, pp. 61-77). The expression of the same genes in echinoderms, in molluscs, even in Cnidaria, now provides a criterion to assess how the body axes of these groups relate to those of other triploblasts. Analogous data may yet place Geoffrey St Hilaire's classic conjecture (1822) concerning the relation of vertebrate and insect dorsoventral axes in the realm of testable hypothesis.

More immediately, the same Hox genes are being used as molecular labels to indicate homology between specific body regions – between insects and crustaceans (Akam et al., pp. 209-215); between cephalochordates and vertebrates (Holland et al., 1992), even between phyla (Morgan and Tabin, pp. 181-186). It remains to be seen whether this 'internal representation' of the genes will reveal relationships where comparative morphology has failed. Whether or not it succeeds, the comparison must provide some indication of the mechanisms underlying morphological change, for the Hox genes and their like are not just passive labels, but tools that sculpt morphology.

This direct link with mechanism is perhaps the most important characteristic of the 'new facts'. In the past, evolutionary change has been analysed by comparing, not the processes of development, but the static forms generated by these processes. Increasingly, it is becoming possible to compare the processes themselves, at the cellular level (Wray and Bely, pp. 97-106; Sommer et al., pp. 85-95) as well as the molecular (Patel, pp. 201-207; Tautz et al., pp. 193-199; Morgan and Tabin, pp. 181-186). Only when we understand the process of development can we begin to map the relationship between genetic change and morphological effect. It is a commonplace of developmental genetics that minimal genetic change can lead to the most dramatic morphological effect (a single base substitution in the *bicoid* gene of *Drosophila* can reverse the axes and symmetry of the embryo (Frohnhöfer and Nüsslein-Volhard, 1986; Struhl et al., 1989)). What we do not yet know is the genetic complexity of observed transitions in evolutionary history – of heterochronic changes in rates of growth, of duplications or suppression in segmentation, or inventions of morphological novelty. Papers in this volume provide glimpses of understanding. How did the complex and beautiful patterns on the wings of a butterfly arise – and how have evolutionary pressures moulded them for immediate adaptation? Nijhout (pp. 225-233) sketches, in a formal model, the outlines of a common mechanism that can generate the apparent complexity; Carroll (pp. 217-223) raises the hope that genes we already know, identified in *Drosophila*, may provide the material basis for part of this complexity.

We do not fully understand butterfly wings, insect segments or vertebrate limbs. Far from it. But we can now pose questions that address the diversity of life in geological time and species space, with some hope of finding answers that are neither trivial nor obvious: answers that go some way towards illuminating that obscure sector on the Venn diagram where genetics, evolution and development intersect.

M. A., P. H., G. W., August 1994

REFERENCES

Bateson, W. (1894). *Materials for the Study of Variation*. London: McMillan & Co.

Frohnhöfer, H. G. and Nüsslein-Volhard, C. (1986). Organization of anterior pattern in the *Drosophila* embryo by the maternal gene *bicoid*. *Nature* **324**, 120-125.

Geoffrey St. Hilaire, E. (1822). *Philosophie Anatomique*. Paris: J.-B. Baillière.

Holland, P. W. H., Holland, L. Z., Williams, N. A. and Holland, N. D. (1992). An Amphioxus homeobox gene: sequence conservation, spatial expression during development and insights into vertebrate evolution. *Development* **116**, 653-661.

Piatogorsky, J. and Wistow, G. J. (1989). Enzyme/Crystallins: Gene sharing as an evolutionary strategy. *Cell* **57**, 197-199.

Roux, W. (1892). Ziele und Wege der Entwickelungsmechanik. *Erg. Anat. Entwickelungsgesch* **2**, 415-445. (p 423, translated by K. Sander, personal communication).

Struhl, G., Struhl, K. and Macdonald, P. M. (1989). The gradient morphogen bicoid is a concentration dependent transcriptional activator. *Cell* **57**, 1259-1273.

Development 1994 Supplement, 1-13 (1994)
Printed in Great Britain © The Company of Biologists Limited 1994

Why molecular biology needs palaeontology

S. Conway Morris

Department of Earth Sciences, University of Cambridge, Downing Street, Cambridge CB2 3EQ, UK

SUMMARY

Molecular biology has re-opened the debate on metazoan diversification, including the vexing question of the origin of the major body plans (phyla). In particular, sequence analyses of *rRNA* have reconfigured significantly metazoan phylogeny, while homeobox genes suggest there could be an underlying similarity of developmental instructions in nominally disparate phyla. Despite this dramatic progress I argue that this renaissance of activity is lop-sided, but can be redressed by palaeontological data, especially from the Cambrian and immediately preceding Vendian. The fossil record complements and amplifies the conclusions derived from molecular biology, notably in the early radiation of cnidarians (Ediacaran faunas) and key steps in the diversification of the protostomes.

Key words: Cambrian, Ediacaran, evolution, fossil, Metazoa, phylogeny

INTRODUCTION

Until very recently the fundamental problem of metazoan evolution, that of the origins and relationships of the thirty to forty phyla that are generally recognized, remained effectively intractable. This was despite more than 200 years of research and controversy, and the production of a literature that now almost defies synthesis (e.g. Hyman, 1940-1961; Willmer, 1990). This did not prevent, of course, a plethora of proposals and hypotheses. Some tackled supposedly fundamental problems: is the coelom a primitive structure (the archicoelomate hypothesis, much favoured by the German school; see Willmer, 1990) or can larval anatomy, e.g. the trochophore, provide unique clues to the ancestry and relationships of otherwise disparate phyla (e.g. Nielsen and Nørrevang, 1985)? Other proposals concerning metazoan relationships were more specific but ranged from the frankly controversial, such as Løvtrup's (1977) analysis that proposed a relationship between arthropods, specifically arachnids, and chordates, to the apparent consensus that places brachiopods and the other lophophorates firmly within the deuterostomes. As we will see below, however, this latter idea may be less secure than popularly imagined.

For each and every proposal concerning metazoan relationships there was almost invariably a counter-suggestion. The existing literature is like the Sargasso Sea of mythology, dotted with hulks of varying decrepitude, each manned by a crazed crew or more likely ghosts. This morass persisted because whatever phylogenetic scheme was proposed had to rely on a handful of features being chosen as axiomatically suitable for identification of monophyly, e.g., segmentation, coelom, setae, blood-pigments, and the tacit admission that other characters accordingly had to be homoplasic. For various reasons, features such as segmentation and type of body cavity came to occupy effectively inviolate positions in terms of phylogenetic reliability, although in their more candid moments most zoologists will admit that there is little *a priori* evidence that metameric segmentation, for example, could not have evolved independently several times. The net result is that any scheme of metazoan phylogeny will inevitably involve characters that to one set of workers are crystal-clear guides to relationships, but to another group are examples of rampant convergence.

The irruption of molecular biology and specifically sequence analyses, however, has revitalized this hitherto moribund area of biology (Conway Morris, 1993a). It promises a release from the deadlock of mutually incompatible phylogenetic hypotheses by offering an independent source of evidence, most obviously to date in the reading of sequences within molecules of ribosomal *RNA* where there is no apparent connection between the molecule and either anatomical expression or environmental preference. These molecular methods, of course, are not foolproof. Different parts of a sequence evolve at different rates, discrepancies exist if comparisons are drawn between the 3′ and 5′ ends of the molecule (e.g. Patterson, 1989), and some taxa (e.g. dipteran insects (see e.g. Carmean et al., 1992), sipunculan worms) appear to evolve very rapidly. Further problems arise where the morphological distinctiveness of a group is echoed in its molecular sequence and so it remains phylogenetically isolated, as has been suggested for the nematodes by Wolstenholme et al. (1987; but see Brandl et al., 1992). Other disagreements arise about the most suitable molecule. For example, 5S rRNA is now generally regarded as too small to be appropriate in this context (Halanych, 1991), and differences also arise concerning methods of tree-building and their most appropriate interpretation.

Nevertheless, some degree of consistency seems to be emerging. As an increasing number of genes and their products are scrutinized the long-standing problems of metazoan phylogeny, such as the placement of the brachiopods, or the monophyly of arthropods, will be resolved. Should then whole-organism zoologists, not to mention palaeontologists, wait patiently for the laboratories of molecular biology to issue a series of phylogenetic dicta and then hasten to provide unquestioning assent? No, and for four reasons:

(1) Even if it is shown unequivocally, in terms of molecular biology, that two phyla are closely related, say molluscs and annelids (e.g. Ghiselin, 1988), this will tell us nothing about how the anatomical, ecological and behavioural transitions that led to these now distinct phyla were achieved. Nor will this information explicate the evolutionary processes that were involved in the origin of what we now regard as body plans. For example, if as indeed now seems likely molluscs and annelids are closely related, then how might we explain features held to be fundamental in one or both phyla such as metamery, the coelom, and chaetae? Below I will argue that fossil data are an essential component in helping to answer this question. In many cases new discoveries will provide a crucial key, but earlier material will be subject to continuing reinterpretation (e.g. Ramsköld and Hou, 1991).

In this paper I will concentrate almost entirely on the relationships between molecular biology and palaeontology so far as they concern early metazoan evolution. It would be remiss, however, if attention was not drawn to the flourishing interactions in tetrapod biology (e.g. Coates and Clack, 1990; Eernisse and Kluge, 1993). Ultimately sequence and morphological data will be complimentary, but we need to be reminded of their respective advantages and limitations. The great majority of readers will be aware of these in the context of molecular biology, but perhaps less so with respect to palaeontology. The principal advantage of fossils is that they provide taxa across the fourth dimension, many of which possess character states that otherwise have been entirely lost and so provide crucial bridges in reconstructing evolutionary trees.

(2) Molecular biologists (e.g. Christen et al., 1991) have been careful to emphasize that during times of rapid divergence, such as appears to typify the protostomes (e.g. Field et al., 1988), the precise order of branching may be very difficult to resolve. Nobody pretends the fossil record is perfect, not least in terms of relative timing of appearance. Nevertheless, a cautious analysis of metazoan evolution during the Vendian and Cambrian suggests that branching orders may yet be discernible. The difficulty, however, is not an over-abundance of characters but recognizing homologous structures at deep levels of metazoan branching and especially anatomical transformations from one state to another. From our perspective, the end-results of these transformations may look very different, but in the Cambrian they may have evolved by a series of rather trivial alterations.

(3) There is more to life than molecules. If we can generate a sensible framework for metazoan phylogeny then this reopens a whole series of questions on the nature of evolutionary convergence. Whatever scheme of evolution of the phyla is ultimately accepted it inevitably must involve the independent acquisition of major features such as body cavities, respiratory pigments, skeletal hard-parts, osmoregulatory organs, and arrangements of the nervous system that include optical sensors. The forty-odd bodyplans and the myriad of lower taxa represent a massive experiment in biological occupation and the constraints that govern the evolutionary process.

(4) Finally, old problems might receive new insights. For example, are body plans really as conservative as is usually thought? Indeed, should we abandon the essentialist concept of the body plan and the related taxon of phylum? Why do some phyla have almost invariant body plans, and why in such examples is a phylum sometimes of low diversity (e.g. sipunculans) but in others of high diversity (e.g. nematodes, see Burglin and Ruvkun, 1993, p. 619). Alternatively, why do other phyla demonstrate a much greater plasticity of form?

THE PRESENT STATE OF PLAY

At first sight the field of metazoan relationships is strewn with over-turned apple carts, and beside each one stands a molecular biologist. Certainly within various groups there have been some major surprises. Avise (1994, Table 8.2), for example, lists a significant number of new insights into the inter-relationships of birds. In terms of the relationships between the phyla, there is, however, less sign of truly radical reorganizations. Thus, molecular biologists agree with many zoologists that the cnidarians are among the most primitive of the metazoans (e.g. Adoutte and Philippe, 1993; Christen et al., 1991; Field et al., 1988; Lake, 1991; Telford and Holland, 1993). Recent evidence also suggests that within the Cnidaria the anthozoans arose first (Bridge et al., 1992; see also Schuchert, 1993) a proposal that may be consistent with new evidence from Ediacaran-like fossils (see below). Shostak (1993) has reiterated the notion that the stinging cells (or cnidae) in cnidarians are symbiont acquisitions from protistan microsporidians, an idea that in principle should be easy to test by comparing molecular sequences of proposed host and cnidae.

Even more primitive, perhaps, are the sponges, although there is also some evidence that this group is polyphyletic (Lafay et al., 1992). Concerning the triploblasts, the platyhelminthes retain a rather primitive position and appear to be monophyletic (e.g. Ruitort et al., 1993; see also Telford and Holland, 1993; Adoutte and Philippe, 1993), and investigations into the inter-relationships within this phylum also continue (e.g. Riutort et al., 1992). Within the so-called higher triploblasts a clear-cut division remains between the deuterostomes (echinoderms, hemichordates, chordates) and the protostomes (including annelids, arthropods, molluscs and sipunculans; e.g. Lake, 1990). Nevertheless, despite this overall consensus there are plenty of adjustments in sight concerning existing hypotheses of inter-relationships, and most will prove controversial.

Within the protostomes it may be that the arthropods are an early branch, which is paraphyletic (e.g. Lake, 1990; see also Adoutte and Philippe, 1993). Inter-relationships within the arthropods are also in some state of flux. Recently, Ballard et al. (1992) have argued that while onychophorans are definitely arthropods, contrary to received opinion they are highly derived, whereas the myriapods are primitive. Some support for this comes from other areas of developmental biology, including Whitington et al.'s (1993) examination of neural development in arthropods. They find evidence for homologous expression in insects and crustaceans (see also Averof and Akam, 1993 for similar conclusions), but conclude that the myriapods are less close. More in accordance with established thinking is molecular evidence for the placement of the parasitic pentastomids in the crustaceans (Abele et al., 1989).

Several other phyla now appear to be close to, if not within, the protostomes. One highly significant achievement is the recognition by Telford and Holland (1993; see also Wada and Satoh, 1994; Wright, 1993) of the protostomous nature of the chaetognaths. Their almost invariant body plan without clear similarities to any other phylum has puzzled systematists for decades, who have had an ill-defined preference for placing them near to the deuterostomes (e.g. Hyman, 1959). Admittedly, this uncertainty has been contested recently by a minority view arguing for a close relationship between chaetognaths and molluscs, specifically the opisthobranch gastropods (Casanova, 1987). The new molecular view does not explicitly support a link with molluscs, but placing chaetognaths in the protostomes has some important ramifications. In particular, the characteristic radial cleavage of chaetognath embryos strongly suggests that this feature *per se* cannot be employed usefully to distinguish protostomes, in which cell cleavage is often spiral, from the supposedly diagnostic radial cleavage of deuterostomes (Telford and Holland, 1993).

Almost as surprising as the reassignment of the chaetognaths is the recent molecular evidence concerning the nemerteans. This phylum, long imagined as a separate offshoot of the platyhelminthes that paralleled coelomate development, is now also placed firmly in the protostomes (Turbeville et al., 1992), a point that accords with Turbeville's (1986) earlier discussion of the similarity between the nemertean coelom and that of the polychaete *Magelona*. The final example of probable recruitment of a phylum to the protostomes concerns the brachiopods, and by implication the related ectoprocts (bryozoans) and phoronids. Prior to the molecular evidence for a protostomous position (Field et al., 1988), the zoological consensus had hovered between a place close to the base of the deuterostomes, or possibly intermediate between this superphylum and the protostomes (see Willmer, 1990). However, the fate of the blastopore, radial cleavage, and the trimerous coelom with the lophophore arising from the second segment, persuaded most investigators that a deuterostomous placement was appropriate, with specific similarities being drawn with the rhabdopleurid hemichordates. If the molecular evidence (Field et al., 1988; Lake, 1990) is correct, then despite the similarity, for example, of feeding mechanisms in the lophophore of hemichordates and ectoprocts (Halanych, 1993) this and other features are convergent.

The next few years should, therefore, see some interesting developments. Some progress may be made in unravelling the orders of branching. For example, amongst arthropods are onychophorans really primitive (cf. Ballard et al., 1992), but amongst lophophorates could phoronids transpire to be derived (e.g. Emig, 1982), rather than basal to this radiation, as is generally thought? For too many phyla we still lack crucial information. Various lines of evidence indicate priapulid worms to be protostomes, and comparisons between haemerythrin pigments argue for a close relationship to brachiopods (Runnegar and Curry, 1992). Will new molecular data support this relationship? Concerning the possibly polyphyletic aschelminthes, little is yet known, apart from the nematodes, which appear to be quite primitive but perhaps fairly close to the arthropods (Brandl et al., 1992). One clue comes from some preliminary evidence that the acanthocephalans may also occupy a relatively primitive position with respect to the main group of protostomes (Telford and Holland, 1993).

PALAEONTOLOGY: CRUCIAL OR MARGINAL?

What then is the role of palaeontology in unravelling the interrelationships of metazoan phyla? Employment of fossils in this specific area involves a paradox. On the one hand the geologically abrupt appearance of a wide variety of fossils near to the base of the Cambrian is consistent with a major radiation (e.g. Lipps and Signor, 1992). This is most notable in the geologically abrupt appearance of skeletal hard-parts (Fig. 1), but is also evident in a parallel diversification of trace fossils, many of which would have been made by animals with a very low preservation potential (Frey and Seilacher, 1980). On the other hand, it is received wisdom that a phylum maintains its integrity as far back as it can be traced, which in many cases is to the Cambrian, and that the fossil record provides no evidence for recognizing intermediates between phyla (e.g. Bergström, 1989, 1990). In the Cambrian, although they are by no means restricted to sediments of this age, there is a plethora of so-called "bizarre", "enigmatic", "problematic" or simply "weird" animals. These fossils have attracted quite widespread attention, and have even earned the cognomen of extinct phyla. Such additions to the Cambrian bestiary have not only reinforced the perception of the magnitude of this evolutionary explosion, but have been used to imply the necessity to search for new evolutionary mechanisms to explain this seeming disparity of anatomies (e.g. Gould, 1989). The recent demonstration that celebrated members of this bestiary, such as *Hallucigenia* and *Wiwaxia*, are not as peculiar as once thought (Butterfield, 1990; Ramsköld and Hou, 1991) has provided satisfaction to those who view zoology as largely an exercise in the correct filling of taxonomic pigeonholes. More significant is the placement of these taxa in schemes of phylogeny (Conway Morris, 1993a). For example, it is argued below that the assertion that *Wiwaxia* may be regarded as a "true polychaete" (Butterfield, 1990) is a less useful statement than treating it as part of a protostome stem group.

A further complication to the above discussion is a widespread perception that fossils are ultimately irrelevant to this type of phylogenetic discussion, in as much as "instances of fossils overturning theories of relationship based on Recent organisms are very rare, and may be non-existent" (Patterson, 1981, p. 218). This critique has received wide attention and as stated can hardly be faulted: fossils are relatively poor in characters when compared with the richness of data obtainable from living organisms. Nevertheless, the utility of fossils in deciding between phylogenies has been underestimated. In some cases fossil data make a decisive difference, a point made forcibly by a number of palaeontologists (e.g. Doyle and Donoghue, 1987; Gauthier et al., 1988; Novacek, 1992; Eernisse and Kluge, 1993). Note, however, that this new-found enthusiasm for fossil data has more to do with the evolutionary significance of character states of morphology and their placement in a cladistic framework, than their antiquity or the elusive search for ancestors. In this latter context a particularly useful concept is ghost lineages, which correspond to the, as yet, unobserved predictions in a phylogenetic branching (Norrell, 1992). These entities can be tested against the fossil record.

What sources of palaeontological data are most relevant for understanding metazoan inter-relationships? Recent marine communities are predominantly composed of soft-bodied animals or at least ones with such delicate skeletons as to have a minimal preservation potential. If the same situation applied to ancient communities, and there is evidence that it did (Conway Morris, 1986), then it is these stratigraphic horizons where soft-tissues are fossilized that in principle will be the most valuable because they will preserve the most representative cross-section of former life. As discussed below, the Vendian assemblages of Ediacaran fossils (Fig. 1) are highly controversial in that some workers (e.g. Seilacher, 1989, 1992; see also Bergström, 1989, 1990) interpret them as separate "experiments" in multicellularity, independent of the Metazoa, which they term vendobionts. While this conceivably applies to some soft-bodied Ediacaran fossils, it is my contention that many are metazoans and so could provide crucial information on the early stages of diversification. Either way, interpretation of Ediacaran fossils remains highly controversial. Recently a compromise solution has emerged whereby the vendobionts are now interpreted as cnidarian-like metazoans, but lacking the stinging cells (cnidae) that were subsequently acquired by symbiosis with microsporidians (Buss and Seilacher, 1994; see also Shostak, 1993). The only point of general agreement on Ediacaran assemblages is that the co-existing trace fossils (Glaessner, 1969) are definitely the product of metazoan activity, albeit as unspecified "worms".

Burgess Shale-type faunas are more straightforward. Shelly animals with a high fossilization potential form a small proportion of the total assemblage: in the Burgess Shale itself probably less than 5 per cent of individuals in the standing crop had robust hard-parts (Conway Morris, 1986). Significantly, soft-part preservation in the Lower and Middle Cambrian is quite widespread (e.g. Butterfield, 1994; Chen et al., 1991; Conway Morris et al., 1987). Burgess Shale-type faunas are dominated by arthropods (which here are taken to include the lobopods, anomalocarids and opabinids, but trilobites are relatively unimportant), but also contain a significant proportion of sponges, priapulid worms (including the palaeoscolecidans), and sometimes polychaete annelids (Conway Morris, 1989). Particularly striking is the general similarity between the Burgess Shale and Chengjiang faunas, although they are separated by perhaps as much as 10 Myr (the faunas are mid-Middle Cambrian and mid-Lower Cambrian respectively) and occupy separate tectonic plates (Laurentia and South China) whose geographical separation in the Cambrian was thousands of kilometers (Conway Morris, 1989; see also Shu and Chen 1994).

It is worth stressing that simply because skeletal remains are relatively unrepresentative of the original nature of Cambrian life, they are by no means a redundant source of information. A serious problem in their study, however, is the tendency for many of the skeletons to disaggregate into dozens, if not hundreds, of component parts. In the case of the trilobites, this is seldom a problem because the group has such a rich fossil record, which often includes articulated specimens, and in any event a large part of trilobite palaeontology concentrates on the head-shield. In many other cases, however, we still have only the vaguest notion of the original skeletal configurations. In the Cambroclavida, a class of uncertain phyletic position, the

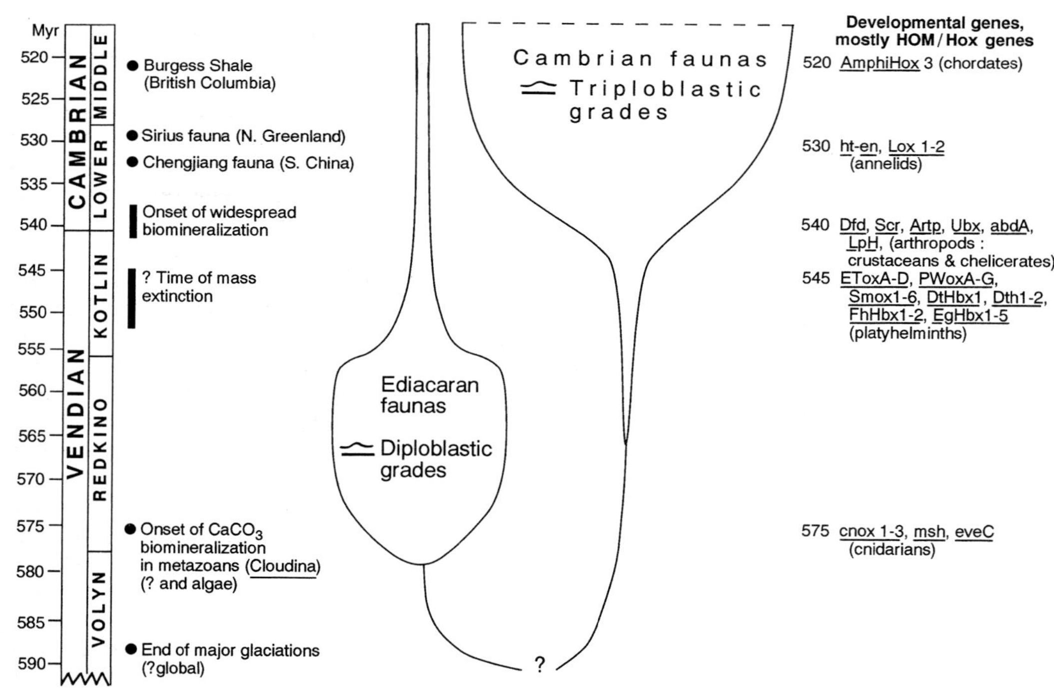

Fig. 1. Outline of the principal stratigraphic divisions and geological time scale (in Myr) relevant to the early diversification of the Metazoa, specifically an Ediacaran fauna dominated by cnidarians (or animals of a similar grade) and the subsequent Cambrian diversification that marks the rise of the numerous deuterostome and protostome phyla. The division between diploblasts and triploblasts is deep (Christen et al., 1991) and the unspecified ancestor of all metazoans probably evolved at least 600 Myr ago. Other features of this figure are various palaeontological data, including the three most important soft-bodied faunas in the Cambrian. On the right-hand side is a schematic indication of the possible time of appearance of some of the homologues presently recognized among the developmental genes of certain major phyla that appeared in the Cambrian. Sources of data are: cnidarians (Miles and Miller, 1992; Miller and Miles, 1993; Schierwater et al., 1991; Schummer et al., 1992; Shenk et al., 1993a,b); platyhelminths (Bartels et al., 1993; Garcia-Fernandez et al., 1993; Oliver et al., 1992; Webster and Mansour, 1992); arthropods (Averof and Akam 1993; Cartwright et al., 1993); annelids (Aisenberg and Macagno, 1994; Nardelli-Haefliger and Shankland, 1992; Shankland et al., 1991; Wedeen et al., 1991; Wedeen and Weisblat, 1991; Wysocka-Diller et al., 1989); chordates (Holland et al., 1992).

sclerites are usually found isolated, but specimens are known that demonstrate unequivocally articulated rows of sclerites, juxtaposed in staggered arrays and sometimes back-to-back to produce arm-like structures (Conway Morris and Chen, 1991; Yue, 1991) vaguely reminiscent of the echinoderms (Conway Morris, 1993a). But the overall appearance of the cambroclave animal remains totally unknown. In the Tommotiida (an order of uncertain phyletic position) it is again reasonable to infer a skeletal arrangement of numerous, juxtaposed sclerites. Evans and Rowell (1990) have proposed that the tommotiid was broadly similar to an armoured slug, but some sclerites are strikingly similar to certain brachiopods. Perhaps this is simply convergent, as is, more probably, the general similarity of some tommotiid plates such as those of *Dailyatia* to the plates of barnacles (Bischoff, 1976; see also Bengtson, 1977). Nevertheless, the suspicion remains that tommotiids might have been sessile rather than a vagrant crawler. In yet other Cambrian groups, such as the Mobergellidae (a family of uncertain systematic position) (Bengtson, 1968; Missarzhevsky, 1989), it is not even clear if the animal carried a single phosphatic plate, two such or perhaps even hundreds of such elements.

These outstanding problems need to be set against some recent successes. Halkieriids, long known only from isolated sclerites, are now seen to have been provided with a dorsal coating of such sclerites mantling a slug-like animal, but against all expectation the scleritome also houses a prominent shell at both anterior and posterior ends (Conway Morris and Peel, 1990; see Fig. 2D). This articulated material was collected from the Lower Cambrian Sirius Passet fauna, a Burgess Shale-like assemblage exposed in Peary Land, North Greenland (Conway Morris et al., 1987). A broadly similar disposition of associated sclerites is now claimed for the hitherto enigmatic *Volborthella*, which is otherwise known from conical sclerites built largely by accretion of sediment grains (Signor and Ryan, 1993).

The interpretation of some of these sclerite-bearing animals is hindered because of different body parts showing variable preservation potential as fossils. Thus, disarticulated calcareous sclerites of halkieriids retain a relatively high preservation potential, especially if subject to secondary diagenetic phosphatization. In contrast, the related wiwaxiids (Conway Morris, 1985), the sclerites of which are unmineralized, have a much lower chance of becoming fossilized. The conspicuous exception in this category of variable preservation potential are the brachiopods, which appear to have been invariably equipped with either a calcareous or phosphatic shell. The brachiopod radiation in the Cambrian exemplifies many of the aspects of the general metazoan divergence, including a variety of geologically short-lived "enigmatic" taxa. Properly understood Cambrian brachiopods, with their high preservation potential, could provide one of the leading exemplars of the Cambrian explosion.

How then might palaeontological data contribute to understanding early metazoan evolution and complement the insights gained from molecular biology? Below, I discuss three examples where some progress has been made.

(1) Ediacaran fossils: Metazoa or Vendobionta?

Ediacaran fossils (Fig. 1) lack obvious skeletal parts, yet they may reach substantial sizes: some of the frond-like organisms approach a metre in length. Seilacher (1989, 1992) was puzzled by the ubiquity of soft-part preservation in generally shallow-water, turbulent and presumably well-aerated environments which represent the very antithesis of the anoxic muds, so typical of exceptional tissue-preservation in younger sediments. Seilacher proposed a novel solution and interpreted the Vendobionta as a separate branch of eukaryote multicellularity. He argued that a unique composition of a tough exterior and an internal anatomy with a mattress-like construction, that lacked the metazoan features such as digestive, muscular and nervous tissue, could explain their high fossilization potential. This hypothesis was in dramatic contrast to the previous consensus that identified within the Ediacaran assemblages a preponderance of cnidarians, with anthozoans, cubozoans, hydrozoans and scyphozoans all identified with varying degrees of confidence (Glaessner, 1984; Jenkins, 1992). To this roster were added, again with fluctuating degrees of certainty, representatives of the so-called articulates, i.e. annelids and/or arthropods, as well as a possible echinoderm (Gehling, 1987).

The Vendobionta hypothesis has won both attention and adherents, but is it correct? A major problem in the interpretation of the Ediacaran fossils is their preservation in typically fairly coarse-grained sediments, making the resolution of some fine anatomical features controversial or even impossible. Even proponents of these fossils as metazoans will admit comparisons with known phyla are seldom straightforward. Nevertheless, circumstantial evidence for such features as muscular contraction and a circulatory system (Runnegar, 1982) cannot be ignored. In the Burgess Shale, moreover, a frond-like fossil (*Thaumaptilon walcotti*, Fig. 2C) closely approaches a number of Ediacaran taxa, most notably *Charniodiscus* (Conway Morris, 1993b). These latter frond-like fossils resemble anthozoan sea-pens (pennatulaceans), although Seilacher (1989) chose to emphasize the difference between some living examples, where the branches arising from the central rachis are separated, and those from the Ediacaran in which the branches are fused to a common blade (as is the case for *Thaumaptilon*). When the overall disparity of living sea-pens is considered, however, this difference seems relatively unimportant. Most significant is the recognition in *Thaumaptilon* of possible zooids and internal canals (Conway Morris, 1993b). Similar canals have been identified in Ediacaran fronds and the apparent absence of zooids can be reasonably attributed to entombment in sediments of a substantially coarser grain than the Burgess Shale. It cannot be disproved that *Thaumaptilon* is convergent with the Ediacaran fronds, but the onus of proof has shifted to supporters of the Vendobionta hypothesis. Moreover, if the *Thaumaptilon-Charniodiscus* connection is accepted, then given that the Ediacaran fronds are preserved in the same way as the other co-occurring fossils, this suggests that although the preservational circumstances of Ediacaran assemblages still require an explanation, it is unlikely to be the consequence of a unique body-organization.

Molecular evidence also appears to be consistent with an early radiation of the cnidarians, and although the division between diploblasts and triploblasts remains deep (Christen et al., 1991; Wainwright et al., 1993) there is now little support for a diphyletic origin of cnidarians and other metazoans as originally proposed (Field et al., 1988). Evidence for metazoan monophyly comes from a variety of sources (e.g. Degnan et al., 1993; Erwin, 1993; Lake, 1990; Morris, 1993). Further

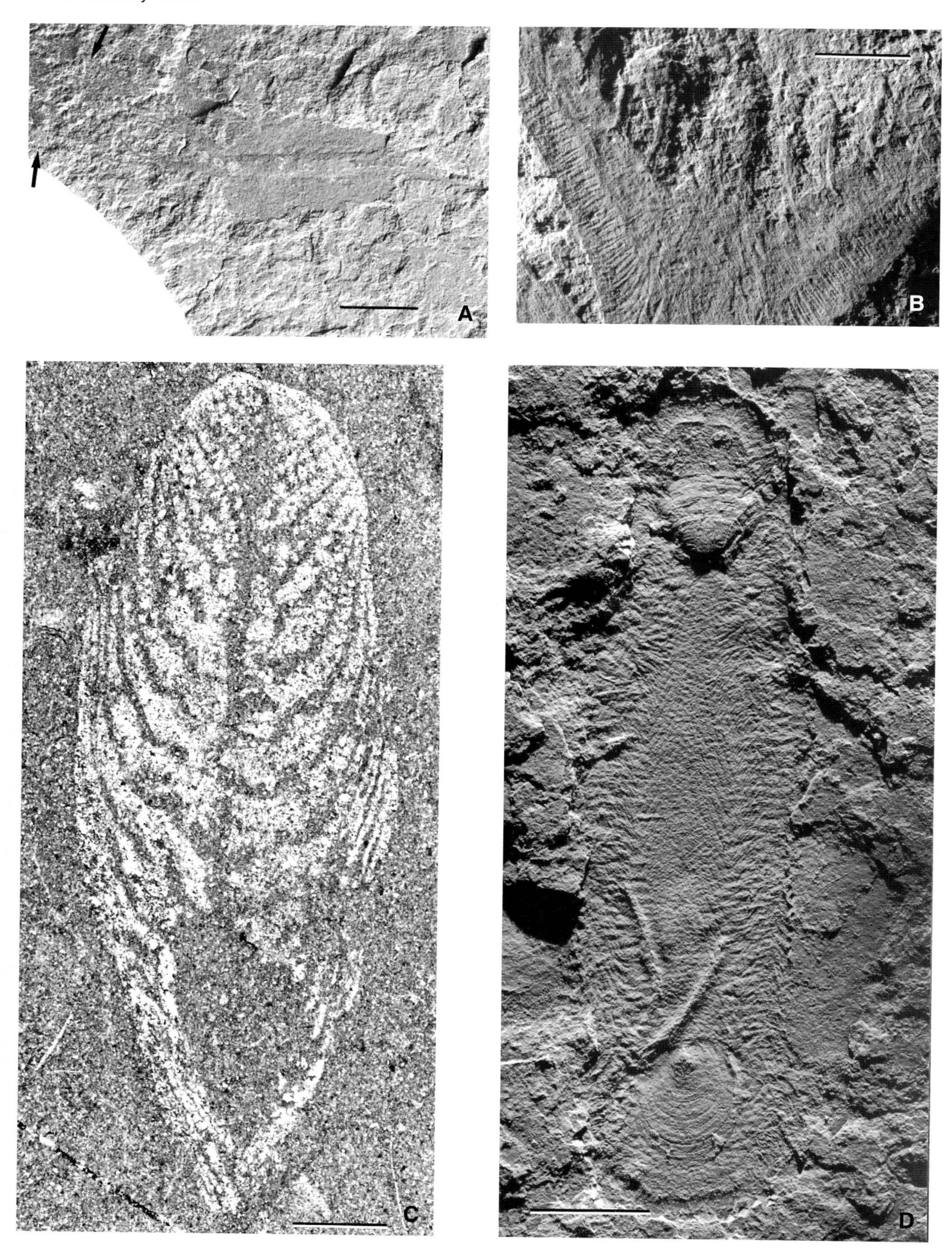
A
B
C
D

molecular data indicate a primitive status for the anthozoans (Bridge et al., 1992) and this would be consistent with the abundance of frond-like fossils in Ediacaran sediments (and the later *Thaumaptilon*). The evidence for hydrozoans, in the form of chondrophorines, is moderately compelling, and *Kimberella* may be an early cubozoan (Jenkins, 1992). Despite the abundance of discoidal fossils, comparisons with the scyphozoans seem distinctly more tenuous. More implausible, however, is the recent resurrection by Valentine (1992) of an older idea that *Dickinsonia* is a polyploid cnidarian.

Molecular evidence also places cnidarians fairly close to the ctenophores (e.g. Christen et al., 1991), and it is possible that some of the bag-like Ediacaran fossils deserve consideration as early ctenophores. Nevertheless, the chances of identifiable comb-rows of cilia surviving seems remote, and again it is only the exceptional quality of preservation of these structures in *Fasciculus* from the Burgess Shale that demonstrates ctenophores extend back to at least the Cambrian (Conway Morris, 1993a; Conway Morris and Collins, unpublished observations). The precise relationships of ctenophores with early metazoan phyla nevertheless remain enigmatic and in some ways they approach more closely the triploblasts (e.g. Ehlers, 1993; Willmer, 1990).

(2) Early arthropods

Putative arthropods, such as *Spriggina*, and more uncertainly taxa such as *Onega*, *Vendomia*, *Praecambridium* and an as-yet-undescribed "soft-bodied trilobite" (Jenkins, 1992, fig. 15) have been identified in Ediacaran assemblages, although proponents of the Vendobionta hypothesis (Bergström, 1989; Seilacher, 1989) have presented radically different hypotheses. Valentine (1988) has stressed how animals such as *Spriggina* are consistent with the paraphyletic and early branching of the arthropods as seen in some molecular trees (e.g. Lake, 1990). Unfortunately, despite its almost iconographic status as an Ediacaran representative, *Spriggina* remains remarkably poorly known. Another animal, *Vendia*, has also been persistently promoted as a primitive arthropod (e.g. Jenkins, 1992, p.168). Self-evident displacement in the only known specimen of left and right segments casts doubt on *Vendia* being an arthropod (Bergström, 1989; Fedonkin, 1985). Nevertheless, Jenkins (1992, p.168) notes that such displacement "could have occurred on burial", and Runnegar (1995) has now scotched the reiterated proposal that *Spriggina* (Bergström, 1989) and *Dickinsonia* (Bergström, 1990) showed a comparable left-right asymmetry. The bilateral symmetry of their flexible bodies was evidently distorted during burial.

Attention, moreover, should be given to a number of other Ediacaran fossils. *Bomakellia kelleri*, for example, is only known from the Ediacaran localities of the White Sea, north-east Russia. Identified as an arthropod by Fedonkin (1985), surprisingly this fossil appears to have escaped almost any mention (Bergström, 1990) in recent discussions of arthropod phylogeny. *Bomakellia* appears to possess an anterior shield and segmented trunk that bears lateral structures and a more axial series of tubercles. Fedonkin (1985) interpreted the lateral extensions as appendages, but they may be better considered as pleurae. The only known specimen of *Bomakellia* is incomplete, but it appears to have been bilaterally symmetrical, as does the possibly related taxon *Mialsemia* (Fedonkin, 1985).

The Cambrian record of arthropods has been revitalized by the analysis of the Burgess Shale and subsequent supplements addressing the Chengjiang (e.g. Bergström, 1993; Hou and Bergström, 1991; Hou et al., 1991) and Sirius Passet faunas (Conway Morris et al., 1987) (see Fig. 1). At a relatively early stage of the investigation into these early arthropods it emerged that representatives of all four major clades (chelicerates, crustaceans, trilobites, uniramians) were present, but were greatly outnumbered by a seemingly bewildering array of forms whose precise relationship to any of the above groups was enigmatic. This uncertainty was graphically expressed by Whittington (1979) in a "phylogenetic lawn" that depicted a myriad of lineages arising from an unspecified Precambrian ancestor. More recently in a series of papers Briggs and Fortey (1989; Briggs, 1990; see also Wills et al., 1994) have formulated a cladistic analysis of arthropods that is being regularly updated and from which is emerging a reasonably consistent pattern (Fig. 3).

Our understanding of early arthropod evolution, however, is far from complete as is apparent from two notable developments. First, the diversity of lobopod animals is now known to be considerable, especially on the basis of the Chengjiang finds (Hou et al., 1991). Recent recruits to this regiment include the hitherto enigmatic *Hallucigenia* from the Burgess Shale (Ramsköld and Hou, 1991). The consensus opinion would link, in some way, this lobopod radiation to the surviving terrestrial onchyophorans, although this may be in conflict with the derived position inferred from molecular biology (Ballard et al., 1990). A related and important development is the description of *Kerygmachela kierkegaardi* from the Sirius Passet fauna (Budd, 1993). In this arthropod (Fig. 2A,B) the lobopods are also associated with dorsal flaps, comparable to gills. At the anterior is a spectacular grasping apparatus. A full description of *Kerygmachela* is not yet available, but the biramous arrangement of lobopod and gill suggests a route by which the biramy of the chelicerates, crustaceans and trilobites may have arisen, by transformation of the flexible lobopods into cuticular, jointed appendages. Budd (1993) also emphasizes that *Kerygmachela* may help to constrain the systematic position of the hitherto enigmatic Burgess Shale animals *Anomalocaris* and *Opabinia*. The former is known to be equipped at its anterior with jointed appendages, and it remains possible that the prominent flaps arising from the trunk region now conceal legs. Such are depicted in a popular review of the

Fig. 2. Some key early metazoan fossils from the Cambrian. (A,B) The primitive gilled lobopod (Arthropoda) *Kerygmachela kierkegaardi* Budd. Sirius Passet fauna, Lower Cambrian, Greenland. (A) Holotype, anterior with prominent grasping apparatus to left. Note elongate spines (arrowed) extending far to anterior. Tail spines are also very elongate, but only the proximal section is shown here. (B) Detail of grasping apparatus from another specimen. (C) *Thaumaptilon walcotti* Conway Morris, an anthozoan sea-pen (Pennatulacea: Cnidaria), juvenile specimen. This fossil is strikingly similar to the Ediacaran frond-like fossils, such as *Charniodiscus*. Burgess Shale fauna, Middle Cambrian, British Columbia. (D) *Halkieria* sp. An articulated specimen showing the covering of sclerites and the prominent anterior and posterior shells, the former of which appears to be retracted from the anterior to expose an area of soft-tissue. The elongate structures towards the posterior are superimposed burrows and are not integral to the fossil. Sirius Passet fauna, Lower Cambrian, Greenland. Scale bars are equivalent to 2 cm (A,C); 4 mm (C), 1 cm (D).

Chengjiang fauna (Chen et al., 1991), although Chen et al. (1994) were unable to recognize legs or lobopods in newly discovered anomalocarids from this fauna. *Opabinia* is also known to bear an array of broad flaps, and the possibility of lobopods concealed beneath them needs re-investigation (G. Budd, personal communication).

(3) Halkieriids: stem protostomes?

The third example concerns the halkieriids (Fig. 2D), best known from the articulated specimens of the Sirius Passet (Fig. 1) fauna (Conway Morris and Peel, 1990). Research into these specimens is continuing in collaboration with Professor J.S. Peel (Uppsala), and our principal conclusions will only be touched upon. The slug-like appearance together with a dorsal coating of calcareous sclerites and two shells that grew by marginal accretion recalls in outline the appearance of primitive molluscs such as the chitons and aplacophorans. Halkieriids may indeed throw significant light on the derivation of early molluscs from a turbellarian ancestor (Bengtson, 1992; Conway Morris and Peel, 1990) and so support a recurrent proposal of invertebrate zoologists such as Vagvolgyi (1967; see also Stasek, 1972). Halkieriids appear to show homologies of sclerite arrangement with the somewhat younger wiwaxiids, best known from the Burgess

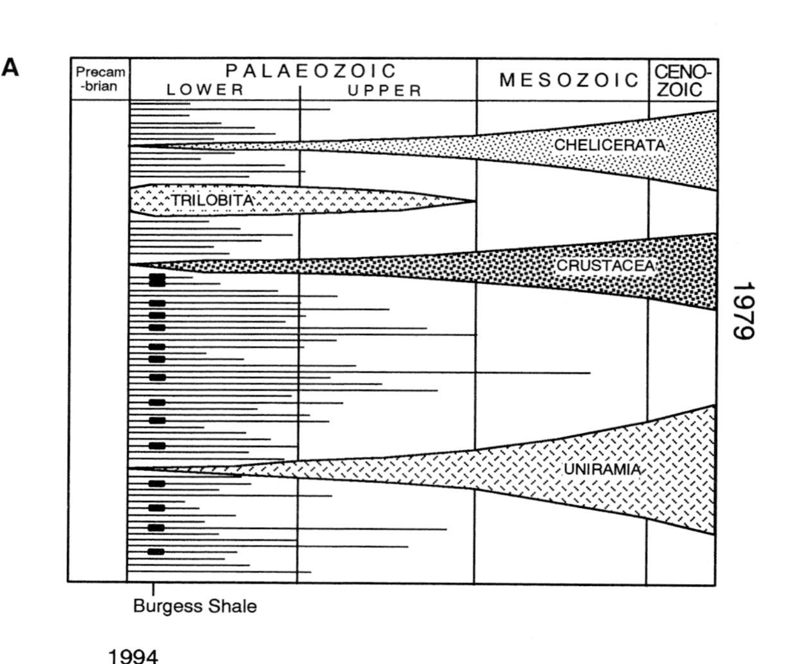

Fig. 3. Comparisons in our understanding of arthropod phylogeny. A is a modified copy of Fig. 2 of Whittington (1979), showing the early appearance of the four principal clades (chelicerates, crustaceans, trilobites (which went extinct at the end of the Palaeozoic), and uniramians (mostly insects)) and a large number of more enigmatic taxa, to produce a "phylogenetic lawn". B is a simplified version of the cladogram published by Wills et al. (1994).

B

1994

Schizoramia

Atelocerata

Arachnomorpha

Crustaceanomorpha

Lobo-
poda

Trilobita

Insecta

Shale (Conway Morris, 1985). These animals were also compared to primitive molluscs (Conway Morris, 1985), although Butterfield (1990) made the significant discovery that the ultrastructure of the wiwaxiid sclerites is closely comparable to that of the chaetae of polychaetes, including those from the Burgess Shale (see Conway Morris, 1979). Butterfield's (1990) claim, however, that wiwaxiids are true polychaetes is much more questionable, not least because of the absence of both parapodia, and the inter-ramal space (the latter is occupied by sclerites), as well as a feeding apparatus that is more like a molluscan radula than any comparable jaw in the polychaetes. Interpreting halkieriids as part of the stem group that led to polychaetes appears to be a distinctly more informative exercise. Finally, Conway Morris and Peel (1990) noted that the shells of the Sirius Passet halkieriid, especially that of the posterior (Fig. 2D), are remarkably brachiopod-like. This was regarded by us as a superficial convergence, but further research suggests that the once unpopular idea of a close link between annelids and molluscs (Field et al., 1988; Ghiselin, 1988) also needs to be supplemented by the long-overlooked proposal (e.g. Morse, 1873) of a near-relationship between annelids and brachiopods. Such a proposal is consistent with evidence from molecular biology, but is highly controversial amongst whole-organism zoologists (e.g. Willmer, 1990) who persist in allying brachiopods with other deuterostomes.

WHERE DO WE GO FROM HERE?

Palaeontological data suggest that significant new insights into metazoan evolution are already available. First, in at least two areas of protostome evolution (the annelid-brachiopod-mollusc connection and the early divergence of arthropods), it seems reasonable to invoke evolutionary transitions from turbellarians, albeit by two rather different routes. If correct, then this suggests that the metameric segmentation of annelids and arthropods arose separately, although they may still share homologous coding instructions that control the so-called pseudometamery of turbellarians (see also Holland, 1990; Newman, 1993). Second, fossil evidence of what may be termed Ediacaran survivors, principally *Thaumaptilon* from the Burgess Shale (Fig. 2C), appear to undermine the Vendobionta hypothesis and reaffirm an early origin and radiation of cnidarians. Problems also remain. Despite a rich record of Cambrian deuterostomes, including primitive echinoderms, rhabdopleurid hemichordates (e.g. Bengtson and Urbanek, 1985; Durman and Sennikov, 1993) and the cephalochordate-like *Pikaia*, their origins are still poorly understood (although see Jefferies, 1986, for a review of his controversial discussion of the calcichordates). Other outstanding problems include the origin and early divergence of the aschelminthes and ctenophores.

Here are five topics of mutual and reciprocal interest to palaeontologists and molecular biologists:

(1) There is an urgent need to expand the roster of examined metazoans. For example, in terms of molecular sequencing, next to nothing is known of groups such as priapulid worms, polychaete annelids, rotifers, articulate brachiopods, and chiton molluscs. If the comments given above concerning halkieriids, for example, win assent then we predict further molecular evidence for a close affinity not only between annelids and molluscs (Ghiselin, 1988), but also with brachiopods. Nearly all the available information on annelids refers to the highly derived leeches (e.g. Wedeen et al., 1991; Wedeen and Weisblat, 1991; Wysocka-Diller et al., 1989). Amongst the molluscs almost nothing is published on chitons, aplacophorans or monoplacophorans. The so-called "living fossil" status of *Lingula*, together with its availability, explains the interest in this supposedly primitive brachiopod, but forms such as *Crania* certainly deserve investigation.

The expansion of the data-base across all known phyla, however, must be accompanied by more extensive investigations among smaller clades. Examination of molecular trees often shows, unsurprisingly, that closely related taxa are separated by very short branch-lengths. But there seem to be some puzzling exceptions: Rosenberg et al. (1992) found remarkable variability in 28S *rRNA* (in the D6 domain) of seven species of truncatellid gastropod.

(2) The wide employment of *rRNA* is now being supplemented by other gene sequences or products, although their applicability to unravelling deep relationships within the metazoans, of course, will vary (see Kumazawa and Nishida, 1993). Nevertheless, at present there are relatively few examples (e.g. Kojima et al., 1993; Miller et al., 1993; Suzuki et al., 1993; Raff et al., 1984) where the congruence of molecular trees can be tested against unrelated sequences.

(3) There is a rich literature on biochemistry and physiology, much of which has escaped being placed in an evolutionary context. In part this is because many metabolic pathways are indeed fundamental to cellular processes. However, a comparison of more specific bioproducts and their utilization may reveal either unexpected examples of convergence or instances of supposed independent biochemical innovation actually reflecting shared ancestry. One example might be represented by the respiratory pigment haemocyanin, which occurs only in arthropods and molluscs. Accordingly, this molecule has been regarded as an important phylogenetic indicator. Evidence from both molecular biology and palaeontology, however, does not support a particularly close relationship between arthropods and molluscs. More critical analysis of haemocyanin now reveals that its higher order structure is fundamentally different in these two phyla (Mangum, 1990).

(4) Another fundamental problem is the way and extent by which developmental mechanisms have evolved. The central paradox at present is that flies and mice, for example, differ phenotypically in self-evident ways, but they share homologues in their developmental machinery, including homeotic genes (Holland, 1990). Some of these genes are remarkably widespread, including antennapedia-like sequences in cnidarians (Miles and Miller, 1992; Miller and Miles, 1993; Murtha et al., 1991; Schierwater, 1991; Schummer et al., 1992; Shenk et al., 1993a,b). Particular interest also lies in the recent detection of homeoboxes in the platyhelminthes (Bartels et al., 1993; Garcia-Fernandez, 1993; Oliver et al., 1992; Webster and Mansour, 1992), given that their primitive position relative to all other triploblasts may now be close to general acceptance. To date detailed proposals of how developmental mechanisms might have evolved are rather limited (e.g. Averof and Akam, 1993; Holland, 1990, 1992; Kappen and Ruddle, 1993; Raff, 1992), although gene duplication may

have been an important, albeit fortuitous, factor in allowing the co-opting of genes for new instructions (see Holland, 1990, 1992). Jacobs (1990) has proposed an ingenious model to link arrangement of developmental instructions, by what he labels as "selector genes", and perceived contraints of morphological expression as indicated by rates of ordinal origination. He argues that arthropods and annelids with clear serial construction underwent initial exuberance of morphological experimentation in the Cambrian, but were subsequently constrained by a regulatory system that had to remain simple owing to dispersal of the "selector" genes. Apart from general problems of testability, one difficulty with Jacobs' hypothesis is the bias introduced by the large number of supposedly enigmatic taxa (orders) of Burgess Shale arthropods, whose high-level taxonomic status is probably exaggerated (see Briggs, 1990). What is emerging from these preliminary discussions of the evolution of developmental mechanisms, however, is that, not withstanding the antiquity of such structures as the antennapedia-class (Fig. 1), the complexity of higher metazoans in part is founded on gene duplications and subsequent co-option.

To date little is known about the genome of putative ancestors, notably in either the protistan ciliates or the fungi (e.g. Baldauf and Palmer, 1993; Wainwright et al., 1993) in terms of possible homologues of developmental genes. Such information will help to constrain the nature of the ur-metazoan (see Shenk and Steele, 1993). Equally intriguing is whether a single function for primitive homeobox genes will be identified. Two items come to mind. Some evidence exists for certain genes having a primary neurogenic role, especially in the central nervous system (Patel et al., 1989; Wedeen and Weisblat, 1991). In this context perhaps we should recall that Stanley (1992) has proposed that the initiation of the metazoan radiations can be traced to the invention of the neuron. Perhaps equally fundamental, or more so, is the role of some homeoboxes in determining axis and orientation, crudely head and tail (e.g. Bartels et al., 1993). Directionality and nervous control may be the hallmarks of metazoans, and the latter might be the key to the initial diversification which was then fuelled by feedbacks, including more complex ecologies marked by the spread of predation (Vermeij, 1990) and grazing (Butterfield, 1994).

(5) The subsequent history of metazoan evolution has often been depicted as the establishment of a remarkable stability of body designs, perhaps best exemplified in the insects. This has been linked to a vague notion that somehow the genome becomes "congealed", thereby precluding the morphological experimentation that is said to characterize the Cambrian faunas. This stability may be more apparent than real. As a whole, the genome is recognized to be highly dynamic, and this appears to apply with equal force to rates of evolution of development (Wray, 1992). It is also a simplification to identify early stages of the embryology as conservative and later ones as more flexible. Wray (1992; see also Wray and Raff, 1991; Raff, 1992) presents a cogent rebuttal of this simplification, and stresses both the degree of developmental variation in some closely related taxa as well as evidence for rapid and geologically recent changes in early development. Most important, however, is Wray's (1992, p.131) emphasis that there is no evidence that the "developmental differences that distinguish phyla and classes" are materially different from those that divide lower taxa, and neither is there any reason to accept the popular notion that "developmental programmes have become too constrained by interaction since the early radiation of metazoans to allow the origin of new body plans". One could argue that the barnacle is a new body plan, but because its relationship to other arthropods, specifically the crustaceans, is clear, such a manoeuvre serves no useful purpose. But I would argue that Wray's (1992) prescient observations apply with equal force to understanding the Cambrian diversifications, although the reader may wish to consult Erwin (1994) in support of opposite views.

It may be naive to imagine that the fossil record will reveal transitions between all the phyla, but this matters little if key examples such as the role of the halkieriids in protostome diversification continue to provide useful insights. What applies to the origin of groups as disparate as annelids, brachiopods and molluscs should be equally applicable in principle across the Metazoa. Such data, combined with an understanding of the evolution of developmental mechanisms and co-option of pre-existing genes, suggests that one of the central problems of biology is close to solution. Is it now time to consider what questions will arise from this advance?

Jeff Levinton and another anonymous referee provided exceptionally helpful reviews. I thank Sandra Last for typing several versions of this paper, Hilary Alberti for assistance with drafting, and Dudley Simons for help with photography. Graham Budd kindly made available Fig 2A,B, while Derek Briggs made available the cladogram which appears in simplified fashion in Fig. 3. Earth Sciences Publication 3817.

REFERENCES

Abele, L. G., Kim, W. and Felgenhauer, B. E. (1989). Molecular evidence for inclusion of the phylum Pentastomida in the Crustacea. *Mol. Biol. Evol.* **6,** 685-691.

Adoutte, A. and Philippe, H. (1993). The major lines of metazoan evolution: Summary of traditional evidence and lessons from ribosomal RNA sequence analysis. In *Comparative Molecular Neurobiology* (ed. Y. Pichon), pp. 1-30. Basel: Birkhäuser.

Aisemberg, G. O. and Macagno, F. R. (1994). *Loxl*, an *Antennapedia*-class homeobox gene, is expressed during leech gangliogenesis in both transient and stable neurons. *Dev. Biol.* **101,** 455-465.

Averof, M. and Akam, M. (1993). *HOM/Hox* genes of *Artemia*: implications for the origin of insect and crustacean body plans. *Curr. Biol.* **3,** 73-78.

Avise, J. C. (1994). *Molecular Markers. Natural History and Evolution.* New York: Chapman and Hall.

Baldauf, S. L. and Palmer, J. D. (1993). Animal and fungi are each other's closest relatives: Congruent evidence from multiple proteins. *Proc. Natl. Acad. Sci. USA* **90,** 11558-11562.

Ballard, J. W. O., Olsen, G. J., Faith, D. P., Odgers, W. A., Rowell, D. M. and Atkinson, P. W. (1992). Evidence from 12S ribosomal RNA sequences that onychophorans are modified arthropods. *Science* **258,** 1345-1348.

Bartels, J. L., Murtha, M. T. and Ruddle, F. H. (1993). Multiple Hox/HOM-class homeoboxes in Platyhelminthes. *Mol. Phyl. Evol.* **2,** 143-151.

Bengtson, S. (1968). The problematic genus *Mobergella* from the Lower Cambrian of the Baltic area. *Lethaia* **1,** 325-351.

Bengtson, S. (1977). Aspects of problematic fossils in the early Palaeozoic. *Acta Univ. Upsaliensis* **415,** 1-71.

Bengtson, S. (1992). The cap-shaped Cambrian fossil *Maikhanella* and the relationship between coeloscleritophorans and molluscs. *Lethaia* **25,** 401-420.

Bengtson, S. and Urbanek, A. (1985). *Rhabdotubus*, a Middle Cambrian rhabdopleurid hemichordate. *Lethaia* **19,** 293-308.

Bergström, J. (1989). The origin of animal phyla and the new phylum Procoelomata. *Lethaia* **22,** 259-269.

Bergström, J. (1990). Precambrian trace fossils and the rise of bilaterian animals. *Ichnos* **1,** 3-13.

Bergström, J. (1993). *Fuxianhuia* - possible implications for the origination and early evolution of arthropods. *Lund Publ. Geol.* **109** (Abstr: Lundadayarna, Hist. Geol. Paleont. III).

Bischoff, G. C. O. (1976). *Dailyatia*, a new genus of the Tommotiidae from Cambrian strata of SE Australia (Crustacea, Cirripedia). *Senckenberg. leth.* **57,** 1-33.

Brandl, R., Mann, W. and Sprinzl, M. (1992). Mitochondrial tRNA and the phylogenetic position of Nematoda. *Biochem. Syst. Ecol.* **20,** 325-330.

Bridge, D., Cunningham, C. W., Schierwater, B., DeSalle, R. and Buss, L. W. (1992). Class-level relationships in the phylum Cnidaria: evidence from mitochondrial genome structure. *Proc. Natl. Acad. Sci. USA* **89,** 8750-8753.

Briggs, D. E. G. (1990). Early arthropods: Dampening the Cambrian explosion. In *Arthropod Paleobiology* (ed. S. J. Culver), pp. 24-43. Short Course. Paleont. Vol. 3. Knoxville: Paleontological Society.

Briggs, D. E. G. and Fortey, R. A. (1989). The early radiation and relationships of the major arthropod groups. *Science* **246,** 241-243.

Budd, G. (1993). A Cambrian gilled lobopod from Greenland. *Nature* **364,** 709-711.

Burglin, T. R. and Ruvkun, G. (1993). The *Caenorhabitis elegans* homeobox gene cluster. *Curr. Opin. Genet. Dev.* **3,** 615-620.

Buss, L. W. and Seilacher, A. (1994). The phylum Vendobionta: a sister group of the Eumetazoa? *Paleobiology* **20,** 1-4.

Butterfield, N. J. (1990). A reassessment of the enigmatic Burgess Shale fossil *Wiwaxia corrugata* (Matthew) and its relationship to the polychaete *Canadia spinosa* Walcott. *Paleobiology* **16,** 287-303.

Butterfield, N. J. (1994). Burgess Shale-type fossils from a Lower Cambrian shallow-shelf sequence in northwestern Canada. *Nature* **369,** 477-479.

Carmean, D., Kimsey, L. S. and Berbee, M. L. (1992). 18S rDNA sequences and the holometabolous insects. *Mol. Phyl. Evol.* **1,** 270-278.

Cartwright, P., Dick, M. and Buss, L. W. (1993). HOM/Hox type homeoboxes in the chelicerate *Limulus polyphemus*. *Mol. Phyl. Evol.* **2,** 185-192.

Casanova, J-P. (1987). Deux chaetognathes benthiques nouveaux du genre *Spadella* des parages de Gibraltar. Remarques phylogénétiques. *Bull. Mus. natn. Hist. nat., Paris. Sect A.* **9** (4th ser.), 375-390.

Chen J., Bergström, J., Lindström, M., and Hou X. (1991). Fossilized soft-bodied fauna. *Natl. Geograph. Res. Explorat.* **7,** 8-19.

Chen J., Ramsköld, L. and Zhou G. (1994). Evidence for monophyly and arthropod affinity of Cambrian giant predators. *Science* **264,** 1304-1308.

Christen, R., Ratto, A., Baroin, A., Perasso, A., Grell, K. G. and Adoutte, A. (1991). An analysis of the origin of metazoans, using comparisons of partial sequences of the 28S RNA reveals an early emergence of triploblasts. *EMBO J.* **10,** 499-503.

Coates, M. I. and Clack, J. A. (1990). Polydactyly in the earliest known tetrapod limbs. *Nature* **347,** 66-69.

Conway Morris, S. (1979). Middle Cambrian polychaetes from the Burgess Shale of British Columbia. *Phil. Trans. R. Soc. Lond.* **B285,** 227-274.

Conway Morris, S. (1985). The Middle Cambrian metazoan *Wiwaxia corrugata* (Matthew) from the Burgess Shale and *Ogygopsis* Shale, British Columbia, Canada. *Phil. Trans. R. Soc. Lond.* **B307,** 507-586.

Conway Morris, S. (1986). The community structure of the Middle Cambrian Phyllopod bed (Burgess Shale). *Palaeontology* **29,** 423-467.

Conway Morris, S. (1989). The persistence of Burgess Shale-type faunas: implications for the evolution of deeper-water faunas. *Trans. R. Soc. Edinb.: Earth Sci.* **80,** 271-283.

Conway Morris, S. (1993a). The fossil record and the early evolution of the Metazoa. *Nature* **361,** 219-225.

Conway Morris, S. (1993b). Ediacaran-like fossils in Cambrian Burgess Shale-type faunas of North America. *Palaeontology* **36,** 593-635.

Conway Morris, S. and Chen M. (1991). Cambroclaves and para-carinachitids, early skeletal problematica from the Lower Cambrian of South China. *Palaeontology* **34,** 357-397.

Conway Morris, S. and Peel, J. S. (1990). Articulated halkieriids from the Lower Cambrian of North Greenland. *Nature* **345,** 802-805.

Conway Morris, S., Peel, J. S., Higgins, A. K., Soper, N. J. and Davis, N. C. (1987). A Burgess Shale-like fauna from the Lower Cambrian of North Greenland. *Nature* **362,** 181-183.

Degnan, B. M., Degnan, S. M., Naganuma, R. and Morse, D. E. (1993). The *ets* multigene family is conserved throughout the Metazoa. *Nucleic Acids Res.* **21,** 3479-3484.

Doyle, J. and Donoghue, M. (1987). The importance of fossils in elucidating seed plant phylogeny and macroevolution. *Rev. Paleobot. Palynol.* **50,** 63-95.

Durman, P. N. and Sennikov, N. V. (1993). A new rhabdopleurid hemichordate from the Middle Cambrian of Siberia. *Palaeontology* **36,** 283-296.

Eernisse, P. J. and Kluge, A. G. (1993). Taxonomic congruence versus total evidence, and amniote phylogeny inferred from fossils, molecules, and morphology. *Mol. Biol. Evol.* **10,** 1170-1195.

Ehlers, U. (1993). Ultrastructure of the spermatozoa of *Halammohydra schulzei* (Cnidaria, Hydrozoa): the significance of acrosomal structures for the systematization of the Eumetazoa. *Microfauna Mar.* **8,** 115-130.

Emig, C. (1982). The biology of Phoronida. *Adv. Mar. Biol.* **19,** 1-89.

Erwin, D. H. (1993). The origin of metazoan development: a palaeobiological perspective. *Biol. J. Linn. Soc.* **50,** 255-274.

Erwin, D. H. (1994). Early introduction of major morphological innovations. *Acta Palaeont. Polonica* **38,** 281-294.

Evans, K. R. and Rowell, A. J. (1990). Small shelly fossils from Antarctica: an early Cambrian faunal connection with Australia. *J. Paleont.* **64,** 692-700.

Fedonkin, M. A. (1985). Systematic description of Vendian Metazoa. In *The Vendian System.* Vol. 1. Palaeontology (eds. B.S. Sokolov and A.B. Iwanowski), pp. 70-106. Moscow: Nauka [In Russian. English translation published by Springer-Verlag, Berlin; 1990].

Field, K. G., Olsen, G. J., Lane, D. J., Giovannoni, S. J., Ghiselin, M. T., Raff, E. C., Pace, N. R. and Raff, R. A. (1988). Molecular phylogeny of the animal kingdom. *Science* **239,** 748-753.

Frey, R. W. and Seilacher, A. (1980). Uniformity in marine invertebrate ichnology. *Lethaia* **13,** 183-207.

Garcia-Fernandez, J., Baguna, J. and Salo, E. (1993). Genomic organization and expression of the planarian homeobox genes *Dth-1* and *Dth-2*. *Development* **118,** 241-253.

Gauthier, J., Kluge, A. and Rowe, T. (1988). Amniote phylogeny and the importance of fossils. *Cladistics* **4,** 105-209.

Gehling, J. G. (1987). Earliest known echinoderm - a new Ediacaran fossil from the Pound Subgroup of South Australia. *Alcheringa* **11,** 337-345.

Ghiselin, M. T. (1988). The origin of molluscs in the light of molecular evidence. *Oxford Surv. Evol. Biol.* **5,** 66-95.

Glaessner, M. F. (1969). Trace fossils from the Precambrian and basal Cambrian. *Lethaia* **2,** 369-393.

Glaessner, M. F. (1984). *The Dawn of Animal Life. A Biohistorical Approach.* Cambridge: University Press.

Gould, S. J. (1989). *Wonderful Life. The Burgess Shale and the Nature of History.* New York: Norton.

Halanych, K. M. (1991). 5S ribosomal RNA sequences inappropriate for phylogenetic reconstructions. *Mol. Biol. Evol.* **8,** 249-253.

Halanych, K. M. (1993). Suspension feeding by the lophophore-like apparatus of the pterobranch hemichordate *Rhabdopleura normani*. *Biol. Bull.* **185,** 417-427.

Holland, P. W. H. (1990). Homeobox genes and segmentation: co-option, co-evolution and convergence. *Sem. Dev. Biol.* **1,** 135-145.

Holland, P. (1992). Homeobox genes in vertebrate evolution. *BioEssays* **14,** 267-273.

Holland, P. W. H., Holland, L. Z., Williams, N. A. and Holland, N. D. (1992). An amphioxus homeobox gene: sequence conservation, spatial expression during development and insights into vertebrate evolution. *Development* **116,** 653-661.

Hou X. and Bergström, J. (1991). The arthropods of the Lower Cambrian Chengjiang fauna, with relationships and evolutionary significance. In *The Early Evolution of Metazoa and the Significance of Problematic Taxa* (eds. A. Simonetta and S. Conway Morris), pp. 179-187. Cambridge: University Press.

Hou X., Ramsköld, L. and Bergström, J. (1991). Composition and preservation of the Chengjiang fauna - a Lower Cambrian soft-bodied biota. *Zool. Scripta* **20,** 395-411.

Hyman, L. H. (1940-1961). *The Invertebrates.* vols 1-6. New York: McGraw-Hill.

Jacobs, D. K. (1990). Selector genes and the Cambrian radiation of Bilateria. *Proc. Natl. Acad. Sci. USA* **87,** 4406-4410.

Jefferies, R. P. S. (1986). *The Ancestry of Vertebrates.* London: British Museum (Natural History).

Jenkins, R. J. F. (1992). Functional and ecological aspects of Ediacaran assemblages. In *Origin and Evolution of the Metazoa* (eds. J. H. Lipps and P. W. Signor), pp. 131-176. New York: Plenum.

Kappen, C. and Ruddle, F. H. (1993). Evolution of a regulatory gene family: HOM/Hox genes. *Curr. Opin. Genet. Rev.* **3,** 931-938.

Kojima, S., Hashimoto, T., Hasegawa, M., Murata, S., Ohta, S., Seki, H. and Okada, N. (1993). Close phylogenetic relationship between Vestimentifera (tube worms) and Annelida revealed by the amino acid sequence of Elongation Factor - 1a. *J. Mol. Evol.* **37**, 66-70.

Kumazawa, Y. and Nishida, M. (1993). Sequence evolution of mitochondrial, tRNA genes and deep-branch animal phylogenetics. *J. Mol. Evol.* **37**, 380-398.

Lafay, B., Boury-Esnault, N., Vacelet, J. and Christen, R. (1992). An analysis of partial 28S ribosomal RNA sequences suggests early radiations of sponges. *BioSystems* **28**, 139-151.

Lake, J. A. (1990). Origin of the Metazoa. *Proc. Natl. Acad. Sci. USA* **87**, 763-766.

Lipps, J. H. and Signor, P. W. (eds). (1992). *Origin and Evolution of the Metazoa*. New York: Plenum.

Løvtrup, S. (1977). *The Phylogeny of Vertebrata*. London: John Wiley.

Mangum, C. P. (1990). The fourth annual Riser lecture: The role of physiology and biochemistry in understanding animal phylogeny. *Proc. Biol. Soc. Washington* **103**, 235-247.

Miles, A. and Miller, D. J. (1992). Genomes of diploblastic organisms contain homeoboxes: sequence of *eveC*, an *even-skipped* homolog from the cnidarian *Acropora formosa*. *Proc. R. Soc. Lond.* **B248**, 159-161.

Miller, D. J. and Miles, A. (1993). Homeobox genes and the zootype. *Nature* **365**, 215-216.

Miller, D. J., Harrison, P. L., Mahony, T. J., McMillan, J. P., Miles, A., Odorico, D. M. and Lohuis, M. R. ten. (1993). Nucleotide sequence of the histone gene cluster in the coral *Acropora formosa* (Cnidaria; Scleractinia): Features of histone gene structure and organization are common to diploblastic and triploblastic metazoans. *J. Mol. Evol.* **37**, 245-253.

Missarzhevksy, V. V. (1989). Oldest skeletal fossils and stratigraphy of Precambrian and Cambrian boundary beds. *Trudy. Geol. Instit. AN SSSR* **443**, 1-237 [in Russian].

Morris, P. J. (1993). The developmental role of the extracellular matrix suggests a monophyletic origin of the kingdom Animalia. *Evolution* **47**, 152-165.

Morse, E. S. (1873). On the systematic position of the Brachiopoda. *Proc. Boston Soc. nat. Hist.* **15**, 315-372.

Murtha, M., Leckman, J. and Ruddle, J. (1991). Detection of homeobox genes in development and evolution. *Proc. Natl. Acad. Sci. USA* **88**, 10711-10715.

Nardelli-Haefliger, D. and Shankland, M. (1992). *Lox2*, a putative leech segment identity gene, is expressed in the same segmental domain in different stem cell lineages. *Development* **116**, 697-710.

Newman, S. A. (1993). Is segmentation generic? *BioEssays* **15**, 277-283.

Nielsen, C. and Nørrevang, A. (1985). The trochaea theory: an example of life cycle phylogeny. In *The Origins and Relationships of Lower Invertebrates* (eds. S. Conway Morris, J. D. George, R. Gibson, H. M. Platt), pp. 28-41. Oxford: University Press.

Norrell, M. A. (1992). Taxic origin and temporal diversity: The effect of phylogeny. In *Extinction and Phylogeny* (eds. M. J. Novacek, Q. D. Wheeler), pp. 89-118. New York: Columbia University Press.

Novacek, M. J. (1992). Fossils as critical data for phylogeny. In *Extinction and Phylogeny* (eds. M. J. Novacek and Q. D. Wheeler), pp. 46-87. New York: Columbia.

Oliver, G., Vispo, M., Mailhos, A., Martinez, C., Sosa-Pineda, B., Fielitz, W. and Ehrlich, R. (1992). Homeoboxes in flatworms. *Gene* **121**, 337-342.

Patel, N. H., Martin-Blanco, E., Coleman, K. G., Poole, S. J., Ellis, M. C., Kornberg, T. B. and Goodman, C. S. (1989). Expression of *engrailed* proteins in arthropods, annelids and chordates. *Cell* **58**, 955-968.

Patterson, C. (1981). Significance of fossils in determining evolutionary relationships. *Ann. Rev. Ecol. Syst.* **12**, 195-223.

Patterson, C. (1989). Phylogenetic relations of major groups: conclusions and prospects. In *The Hierarchy of Life. Molecules and Morphology in Phylogenetic Analysis* (ed. B. Fernholm, K. Bremer and H. Jörnvall), pp. 471-488. Nobel Symposium 70. Amsterdam: Elsevier Biomedical.

Raff, R. A. (1992). Direct-developing sea urchins and the evolutionary reorganization of early development. *BioEssays* **14**, 211-218.

Raff, R. A., Anstrom, J. A., Haffman, C. J., Leaf, D. S., Loo, J-H., Showman, R. M. and Wells, D. E. (1984). Origin of a gene regulatory mechanism in the evolution of echinoderms. *Nature* **310**, 312-314.

Ramsköld, L. and Hou X. (1991). New early Cambrian animal and onychophoran affinities of enigmatic metazoans. *Nature* **351**, 225-228.

Riutort, M., Field, K. G., Raff, R. A. and Baguna, J. (1993). 18S rRNA sequences and phylogeny of Platyhelminthes. *Biochem. Syst. Ecol.* **21**, 71-77.

Riutort, M., Field, K. G., Turbeville, J. M., Raff, R. A. and Baguna, J. (1992). Enzyme electrophoresis, 18S rRNA sequences, and levels of phylogenetic resolution among several species of freshwater planarians (Platyhelminthes, Tridadida, Paludicola). *Can. J. Zool.* **70**, 1425-1439.

Rosenberg, G., Davis, G. M. and Kuncio, G. S. (1992). Extraordinary variation in conservation of D6 28S ribosomal RNA sequences in mollusks: Implications for phylogenetic analysis. In *Abstracts of the 11th International Congress of Malacology, Siena 1992* (eds. F. Giusti and G. Manganelli), pp. 221-222. Siena: University Press.

Runnegar, B. (1982). Oxygen requirements, biology and phylogenetic significance of the late Precambrian worm *Dickinsonia*, and the evolution of the burrowing habit. *Alcheringa* **6**, 223-239.

Runnegar, B. (1995). Vendobionta or Metazoa? Developments in understanding the Ediacara "fauna". *N. Jb. Geol. Paläont. Abh.* (in press).

Runnegar, B. and Curry, G. B. (1992). Amino acid sequences of hemerythrins from *Lingula* and a priapulid worm and the evolution of oxygen transport in the Metazoa. *Int. Geol. Congr. Kyoto* (1992), Vol. 2, 346.

Schierwater, B., Murtha, M., Dick, M., Ruddle, F. H. and Buss, L. W. (1991). Homeoboxes in cnidarians. *J. Exp. Zool.* **260**, 413-416.

Schuchert, P. (1993). Phylogenetic analysis of the Cnidaria. *Z. zool. Syst. Evolut.-forsch.* **31**, 161-173.

Schummer, M., Scheurlen, I., Schaller, C. and Galliot, B. (1992). HOM/Hox homeobox genes are present in hydra (*Chlorohydra viridissima*) and are differentially expressed during regeneration. *EMBO J.* **11**, 1815-1823.

Seilacher, A. (1989). Vendozoa: Organismic construction in the Proterozoic biosphere. *Lethaia* **22**, 229-239.

Seilacher, A. (1992). Vendobionta and Psammocorallia: Lost constructions of Precambrian evolution. *J. geol. Soc. Lond.* **149**, 607-613.

Shankland, M., Martindale, M. Q., Nardelli-Haefliger, D., Baxter, R. and Price, D. J. (1991). Origin of segmental identity in the development of the leech nervous system. *Development Suppl.* **2**, 29-38.

Shenk, M. A., Bode, H. R. and Steele, R. E. (1993a). Expression of *Cnox-2*, a HOM/Hox homeobox gene in hydra, is correlated with axial pattern formation. *Development* **117**, 657-667.

Shenk, M. A., Gee, L., Steele, R. E. and Bode, H. R. (1993b). Expression of *Cnox-2*, a HOM/Hox gene is suppressed during head formation in hydra. *Dev. Biol.* **160**, 108-118.

Shenk, M. A. and Steele, R. E. (1993). A molecular snapshot of the metazoan 'Eve'. *Trends Biochem. Sci.* **18**, 459-463.

Shostak, S. (1993). A symbiogenetic theory for the origins of cnidocysts in Cnidaria. *BioSystems* **29**, 49-58.

Shu D. and Chen L. (1994). Cambrian palaeobiogeography of Bradoriida. *J. Southeast Asian Earth Sci.* **9**, 289-299.

Signor, P. W. and Ryan, D. A. (1993). Lower Cambrian fossil *Volborthella*: the whole truth or just a piece of the beast? *Geology* **21**, 805-808.

Stanley, S. M. (1992). Can neurons explain the Cambrian explosion? *Geol. Soc. Amer. Abstr. Progms.* **27(7)**, A45.

Stasek, C. R. (1972). The molluscan framework. In *Chemical Zoology* (ed. M. Florkin, B.T. Scheer), vol. 7, pp. 1-43. New York: Academic Press.

Suzuki, T., Takayi, T. and Ohta, S. (1993). N-terminal amino acid sequences of 440 kDa hemoglobins of the deep-sea tube worms, *Lamellibrachia* sp. 1, *Lamellibrachia* sp. 2 and slender Vestimentifera gen. sp. 1. Evolutionary relationships with annelid hemoglobins. *Zool. Sci.* **10**, 141-146.

Telford, M. J. and Holland, P. W. H. (1993). The phylogenetic affinities of the chaetognaths: A molecular analysis. *Mol. Biol. Evol.* **10**, 660-676.

Turbeville, J. M. (1986). An ultrastructural analysis of coelomogenesis in the hoplonemertine *Prosorhochmus americanus* and the polychaete *Magelona* sp. *J. Morph.* **187**, 51-60.

Turbeville, J. M., Field, K. G. and Raff, R. A. (1992). Phylogenetic position of Phylum Nemertini, inferred from 18S rRNA sequences: Molecular data as a test of morphological character homology. *Mol. Biol. Evol.* **9**, 235-249.

Vagvolgyi, J. (1967). On the origin of molluscs, the coelom, and coelomic segmentation. *Syst. Zool.* **16**, 153-168.

Valentine, J. W. (1988). Bilaterians of the Precambrian-Cambrian transition and the annelid-arthropod relationship. *Proc. Natl. Acad. Sci. USA* **86**, 2272-2275.

Valentine, J. W. (1992). *Dickinsonia* as a polyploid organism. *Paleobiology* **18**, 378-382.

Vermeij, G. J. (1990). The origin of skeletons. *Palaios* **4**, 585-589.

Wada, H. and Satoh, N. (1994). Details of the evolutionary history from invertebrates to vertebrates, as deduced from sequences of 18S rDNA. *Proc. Natl. Acad. Sci. USA* **91**, 1801-1804.

Wainwright, P. O., Hinkle, G., Sogin, M. L. and Stickel, S. K. (1993). Monophyletic origins of the Metazoa: An evolutionary link with Fungi. *Science* **260,** 340-342.

Webster, P. J. and Mansour, T. E. (1992). Conserved classes of homeodomains in *Schistosoma mansoni*, an early bilateral metazoan. *Mechanism. Dev.* **38,** 25-32.

Wedeen, C. J., Price, D. J. and Weisblat, D. A. (1991). Cloning and sequencing of a leech homolog to the *Drosophila engrailed* gene. *FEBS Lett.* **279,** 300-302.

Wedeen, C. J. and Weisblat, D. A. (1991). Segmental expression of an *engrailed*-class gene during early development and neurogenesis in an annelid. *Development* **113,** 805-814.

Whitington, P. M., Leach, D. and Sandeman, R. (1993). Evolutionary change in neural development within the arthropods: axonogenesis in the embryos of two crustaceans. *Development* **118**, 449-461.

Whittington, H. B. (1979). Early arthropods, their appendages and relationships. In *The Origin of Major Invertebrate Groups* (ed. M.R. House), pp. 253-268. Syst. Ass. Spec. Vol. 12. London: Academic.

Willmer, P. (1990). *Invertebrate Relationships. Patterns in Animal Evolution.* Cambridge: University Press.

Wills, M. A., Briggs, D. E. G. and Fortey, R. A. (1994). Disparity as an evolutionary index: a comparison of Cambrian and Recent arthropods. *Paleobiology* **20**, 93-130.

Wolstenholme, D. R., MacFarlane, J. L., Okimoto, R., Clary, D. O. and Wahleitner, J. A. (1987). Bizarre tRNAs inferred from DNA sequences of mitochondrial genomes of nematode worms. *Proc. Natl. Acad. Sci. USA* **84**, 1324-1328.

Wray, G. A. (1992). Rates of evolution in developmental processes. *Amer. Zool.* 32, 123-134.

Wray, G. A. and Raff, R. A. (1991). Rapid evolution of gastrulation mechanisms in a sea urchin with lecithotrophic larvae. *Evolution* **45**, 1741-1750.

Wright, J. C. (1993). Some comments on the inter-phyletic relationships of chaetognaths. In *Proceedings of the II International Workshop of Chaetognatha* (ed. I. Moreno), pp. 51-61. Palma: Universitat de les Illes Balears.

Wysocka-Diller, J. W., Aisemberg, G. O., Baumgarten, M., Levine, M. and Macagno, E. R. (1989). Characterization of a homologue of bithorax-complex genes in the leech *Hirudo medicinalis*. *Nature* **341**, 760-763.

Yue Z. (1991). Discovery of fused sclerites of early Cambrian *Phyllochiton* and its relation with zhijinitids. *Kexue Tongbao* (1991), 47-50. [In Chinese, with English abstract].

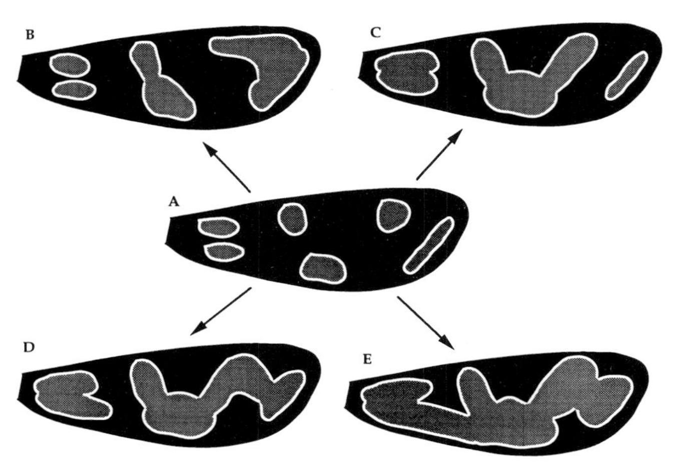

Fig. 4. Wing pattern diversity in the Zygaenidae (B-E) arises from different degrees of enlargement and fusion among roughly circular pattern elements that arise from some six centers of origin scattered across the wing (A).

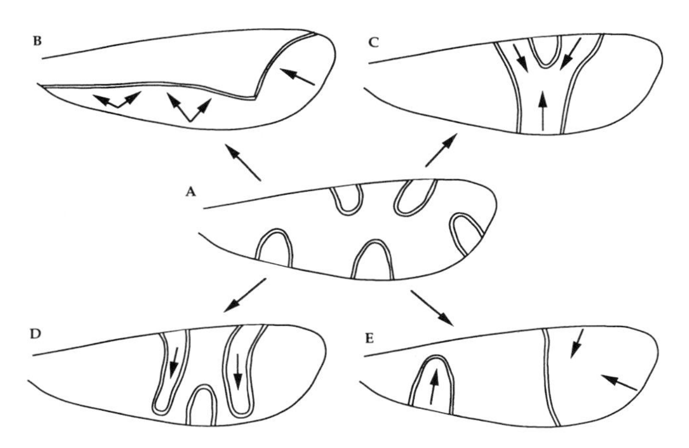

Fig. 5. A sampler of wing pattern diversity found in the 'primitive' moth families. Various banding patterns arise from differential expansion and fusion among patterns emerging from a limited number of origins (arrows).

neighboring bands (Fig. 7C), further demonstrating that all the bands are developmentally equivalent.

In the remaining Ditrysia, the number of symmetry systems is reduced. In most families there is only a single symmetry system in the middle of the wing. More seldom there is a second symmetry system near the margin (as in the Saturniidae), and at the base of the wing (as in some Geometridae). In some members of these families all three systems are present. In almost all families of moths that have multiple symmetry systems, the pigmentation of all symmetry systems is identical. The Saturniidae are a notable exception as they have a border symmetry system that is distinctively different in pigmentation from the central symmetry system. Distinctively colored borders also occur in some Notodontidae, Noctuidae and Geometridae. The wing pattern of the butterflies evolved from these ancestral systems by stabilizing the number of symmetry systems to three, and by a significant further differentiation of the individual symmetry systems. In the butterflies each symmetry systems have become individu-

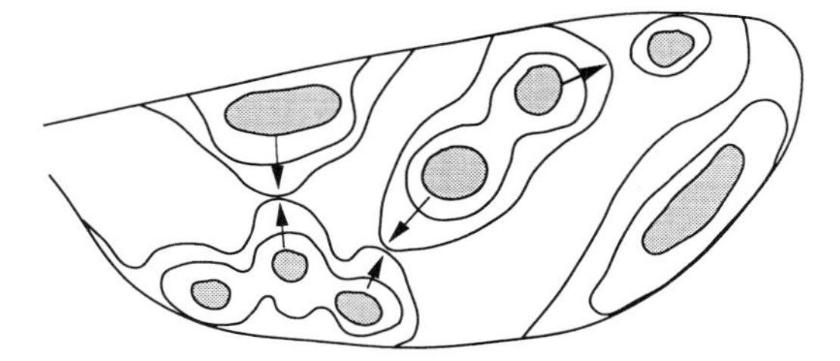

Fig. 6. Groundplan of the wing pattern of the Hepialidae. Species differ from each other in differential expansion of patterns from scattered centers of origin (gray); several possible contours are shown.

ated, as evidenced by the fact that each system develops a distinctive pigmentation, the bands of adjacent systems can no longer fuse with each other, and variation in each system is completely uncorrelated with that of adjacent systems (Nijhout, 1978, 1991; Paulsen and Nijhout, 1993).

The evolution of symmetry systems, then, appears to have occurred in the following five stages. The first stage is found in the most primitive Lepidoptera where a small number of sources for pattern determination on the wing produce a pattern of spots and bars. The pattern-determining system of these moths is doubtless derived from that of their pre-lepidopteran ancestors and may be related to the systems that also produce banding and spotting patterns on the wings of other insects. In the second stage, the number of these sources increases, dependent at least in part on the evolution of large body size; although the arrangement of these sources is constant and characteristic for each family, they are not aligned in any systematic or regular pattern. In the third stage, the sources become arranged in parallel rows and now produce a number of parallel symmetry systems. In the fourth stage, the number of rows is reduced and stabilized to three. The fifth stage is the differentiation and individuation of the symmetry systems produced by each row of sources, so that each system develops a distinctive form and pigmentation. If there was no evolution in the signal sources (the more parsimonious assumption), then the last step presumably involved the evolution of a distinctive environment in the regions of the wing where each symmetry system develops, so that in each region the signal interacts differently with its environment and induces a different pigment.

Evolution of pattern compartmentalization by the wing venation system

In butterflies, compartmentalization of the color pattern by wing veins uncouples pattern development in adjoining wing-cells and, as a consequence of this developmental uncoupling, allows evolution of the color pattern in different wing cells to become uncoupled as well. One of the consequences of compartmentalization, then, is that pattern elements in adjoining wing-cells can diverge morphologically. To discover when during lepidopteran phylogeny compartmentalization first appeared, we can search for species whose wings exhibit sharp differences of the pattern in adjacent wing cells. Such differences could be the dislocation of a band where it crosses a wing vein, the elimination of segments of a band in one or more wing-cells, or an abrupt change in the morphology or pigmentation of a band segment from one wing-cell to the next.

Dislocations of, or gaps in, symmetry system bands are

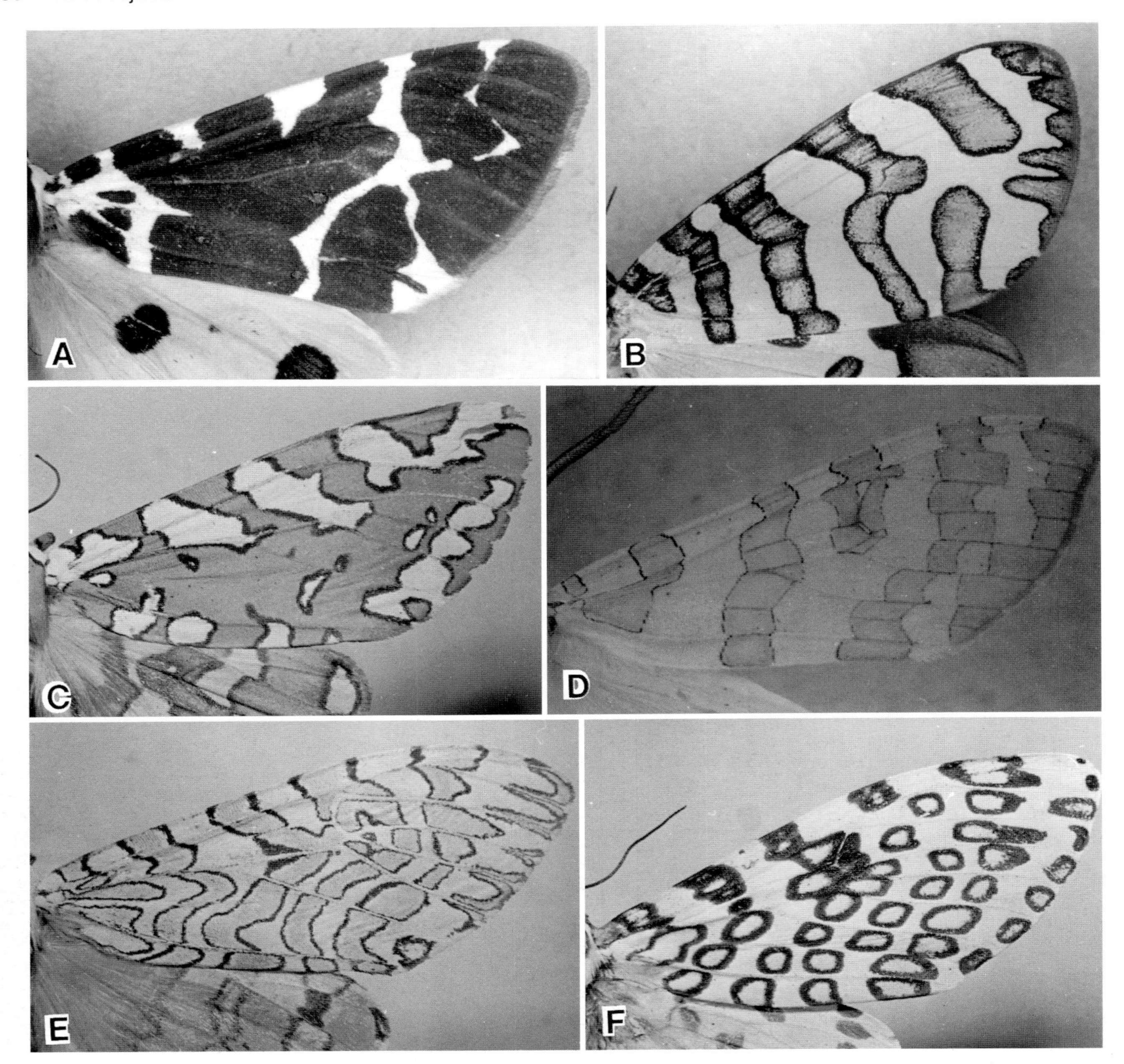

Fig. 7. The wing patterns of various Arctiidae illustrating broadly fused patterns (A), well-defined bands (B, D), lateral fusions among bands (C), dislocated bands (D), and bands broken by wing veins to various degrees (E, F). (A) *Arctia caja*; (B) *Callimorpha thelwalli*; (C) *Arachnis aulea*; (D) *Halisidota tesselaris*; (E) *Arachnis picta*; (F) *Ecpantheria scribonia*.

found rarely outside of the butterflies. There are a few cases in the Arctiidae (*Halisidota*; Fig. 7D) and the Geometridae (where the genus *Dysphanis* has a very butterfly-like wing pattern), and the occurrence in these two families suggests that the potential for dislocation must exist in most of the Ditrysia, although this potential is seldom expressed. The wing veins do, however, play important roles in other aspects of pattern formation in many families of moths, in ways that reveal much about how the compartmentalization of butterfly wing patterns may have evolved. In the moths, two pattern systems are affected by the wing veins. First, symmetry system bands are often deflected proximally or distally where they cross a wing vein (Fig. 8). Second, ripple patterns, very primitive patterns that are not part of the nymphalid groundplan, are almost always dislocated at the wing veins (Fig. 9), even in the most basal families of moths. Below I will attempt to interpret the significance of each of these morphological features for pattern development and evolution.

Distortions of symmetry system bands near a wing vein can be found in all families of the Ditrysia and in the Pyralidae and Hepialidae. In most cases, the bands are continuous across the wing but are deflected into proximally or distally pointing peaks where they cross the wing veins, giving the band a scalloped appearance (Figs 3, 8C). If each band represents a threshold on a gradient (Toussaint and French, 1988), then one can obtain scalloped bands if the threshold changes in the vicinity of a wing vein, or if the gradient has a different shape near a wing vein than it does in the middle of a wing-cell. If patterns are produced as thresholds on gradients of a diffusible signalling substance that moves from cell to cell through gap junctions (Nijhout, 1990), then a change in the density of gap junctions (which would affect the local rate of diffusion) near the wing veins could account for the observed deviations in the positions of the bands. Since developmental uncoupling of color pattern formation in adjoining wing-cells could be accomplished by abolishing communication between them, then a mechanism for modulating the density of gap junctions at the wing veins of these moths would provide the necessary preconditions for the evolution of compartmentalization of the pattern in their descendants, the butterflies.

There are also a number of cases, particularly common in the Arctiidae, where the band is constricted and fully interrupted at a wing vein. In species such as *Ecpantheria scribonia* (Arctiidae), each band forms rings that are squared off or flattened on the sides that face the wing veins (Fig. 7F). If there is a simple blockage of intercellular diffusional coupling at the wing veins then it must extend some distance away from the wing veins. If there is an organizing center at the middle of each spot (it must be remembered that each band in the Arctiidae is a symmetry system) then computer simulation suggests that the shapes of patterns found in Arctiidae could be obtained if the wing veins either absorbed or destroyed the diffusing signal (Nijhout, 1991). Such a mechanism would also inhibit communication across wing veins and could, therefore, also provide an adequate precondition for the evolution of compartmentalization.

Ripple patterns are a different case. Ripples are believed to be patterns of the 'background' unrelated to the symmetry systems of the nymphalid groundplan (Nijhout, 1991). They are composed of short irregular lines that run across the wing-cell perpendicular to the wing veins. These ripples are connected in random branching patterns. A consequence of this stochastic structure is that no two wings (even those of a single specimen) have an identical ripple pattern. The scale of this branching pattern varies from species to species; large scale ripples give the overall wing pattern a striated appearance, small scale ripples give a reticulated appearance (Fig. 9). At all scales, the ripples are sharply interrupted at the wing veins; there is no bending near the veins, as we see in the bands of symmetry systems.

Ripple patterns on the wing are older than the Lepidoptera; they are found also in the Trichoptera, Homoptera and Orthoptera, which suggests that they may well represent one of the oldest color patterns in the insects. In the Lepidoptera they are either the exclusive wing pattern, as in most Cossidae, Neopseustidae

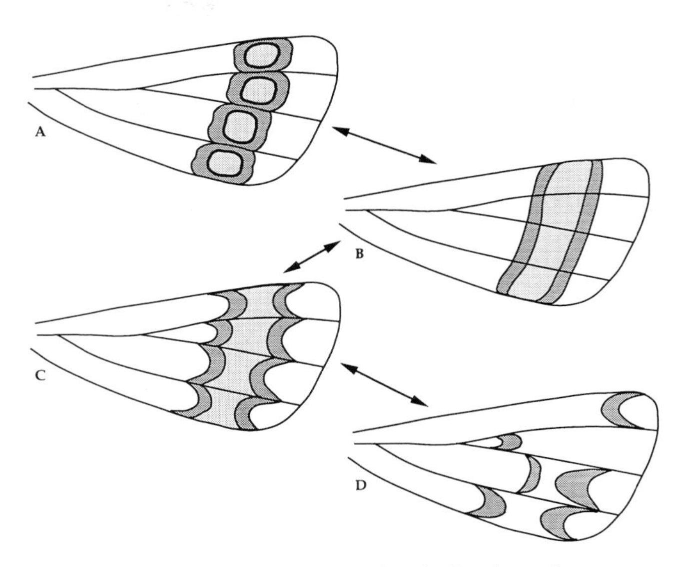

Fig. 8. Evolution of symmetry system bands. Bands can be constricted to circles (A), or deflected at wing veins (C). In butterflies the bands become dislocated and may even be lost in certain wing-cells (D). Double-headed arrows indicate possible evolutionary interconversions.

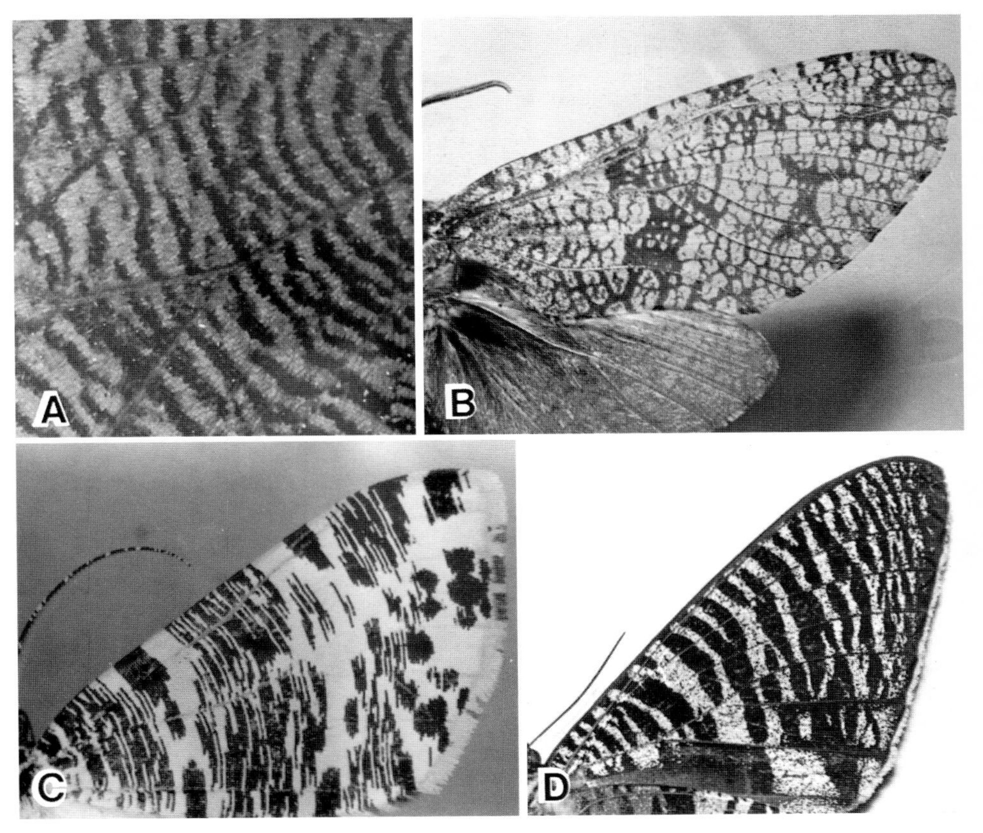

Fig. 9. Various examples of ripple patterns. (A) *Cercyonis pegala* (Satyridae); (B) *Prionoxystus robiniae* (Cossidae); (C) *Durbana tricoloraria* (Geometridae); (D) *Chrysiridia madagascarensis* (Uraniidae).

and many other non-ditrysians, or they co-occur with the symmetry systems in many Ditrysia. In the butterflies and many Ditrysia, the ripple patterns are clearly part of the background and the elements of the nymphalid groundplan develop 'on top' of the ripples without interacting with them (illustrations in Nijhout, 1991). On the basis of comparative and experimental studies, Nijhout (1991) suggested that during color pattern development ripple pattern determination precedes that of the elements of the nymphalid groundplan.

Sharp discontinuities of ripple patterns at the wing veins suggest that barriers to pattern determination exist at the veins. But in moth species that have *both* symmetry bands and ripple patterns, the former are almost always smoothly continuous across wing veins while the latter are not. Clearly the veins have a different function during the determination of these two pattern systems. The simplest interpretation of this difference is that the veins change their properties during development, so that they act as barriers during early stages of development when the ripple patterns are determined, and become 'transparent' to intercellular signals later in development when the banding pattern is determined. Presumably this could happen by having epidermal cells coupled through gap junctions early in development, but not later. Gap junctions can be controlled dynamically and are known to change during development in other systems (Fraser, 1985). If this is a correct interpretation, then this implies that compartmentalization of the wing is actually a very ancient character that is expressed early in development in many and perhaps all Lepidoptera. The butterflies (and probably some arctiids and geometrids) then differ from the moths only in the evolution of a delay in the time at which the venous barrier is lost during development, so that the compartmentalization mechanism that formerly was only active during ripple pattern determination remains active and now also affects symmetry system determination.

Intermediate stages in the transformation from a completely uncompartmentalized to a completely compartmentalized wing can actually be found sporadically throughout the moths. In the Zygaenidae, for instance, there are species in which the spots, particularly those near the wing margin, are indented where they cross a wing vein. Similar indentations and discontinuities of the pattern at or near the distal wing margin are found in other moth families as well. These observations suggests that at least the distal portions of the wing may have venous compartments in some species. A proximodistally graded 'activity' of the wing veins, such as is found in some butterflies (Nijhout, 1991), may thus have have its origin fairly early in lepidopteran phylogeny.

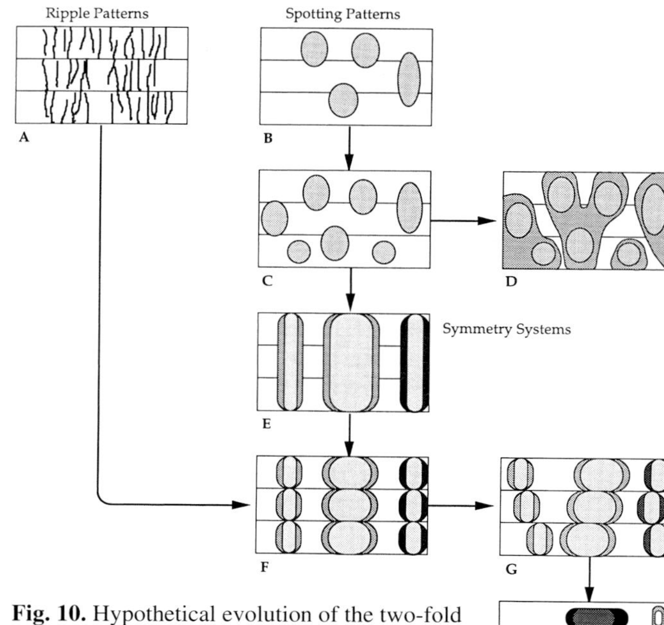

Fig. 10. Hypothetical evolution of the two-fold compartmentalization of butterfly wing patterns. Primitively, the Lepidoptera had two independent patterning systems: ripple patterns that are compartmentalized by wing veins (A), and spotting patterns (B). Spotting pattern evolved through multiplication of centers of origin (C), and diversification in the patterns of fusion between the spots (D). In the Ditrysia the centers of origin became well aligned to form discrete symmetry systems, and the symmetry systems became differentiated from each other (E). In the immediate ancestors of the butterflies the two patterning systems combined to produce a compartmentalization of symmetry systems (F). This compartmentalization enables bands of butterfly symmetry systems to become displaced independently in each wing-cell (G) and the elements in each wing-cell to evolve unique differentiation (H).

Putting it all together

The patterning system of butterflies with its two-fold compartmentalization, appears to have evolved through the convergence of two very ancient pattern-determining mechanisms, illustrated diagrammatically in Fig. 10. My hypothesis is that there were two independent primitive patterning mechanisms which the Lepidoptera inherited from their phylogenetic ancestors.

The first mechanism is one that generates spot patterns such as those illustrated in Figs 4, 5 and 6. This mechanism may be related to those that generate spotting patterns on the wings of flies and beetles, and possibly to those that generate spotted patterns in the Neuroptera (e.g. Fulgoridae) and Orthoptera. In the primitive moths, all spots are generated by the same physiological mechanism, so that any two spots can fuse smoothly. These spots are not aligned in parallel rows, though they are often arranged in regular arrays that are characteristic for a genus of a family. In the Ditrysia there evolved a mechanism that aligns the spots into parallel rows. In the lineage that gave rise to the butterflies the number of these rows became limited to three. Primitively each symmetry system on a wing is composed of identical pigmented bands. It is not until the evolution of the butterflies (or their immediate ancestors) that we see a divergence in the morphology and pigmentation of different symmetry systems.

The second mechanism is that of compartmentalization by the wing veins. This too is an ancient mechanism the Lepidoptera inherited from their ancestors. In nearly all moths, compartmentalization is restricted to the ripple patterns and does not affect the spot patterns and symmetry systems. Compartmentalized ripple patterns and uncompartmentalized symmetry systems occurred side-by-side during most of the phylogenetic evolution of the Lepidoptera. During development, ripple pattern determination precedes symmetry system determination, and this suggest that in moths compartmentalization with respect to color pattern determination may be restricted to certain times of development. If this is correct, then a simple way of compartmentalizing symmetry systems would be to extend the time during which compartmentalization is in effect into the period of symmetry system determination. This is what appears to have happened in the butterflies and their immediate ancestors.

In the butterflies we then see the coincidence of a mechanism that has evolved to produce a regular array of three symmetry systems with a mechanism that compartmentalizes the wing with respect to color pattern formation. The result is an uncoupling of development of the symmetry systems in different wing cells. The butterflies also inherited from their immediate ancestors the capacity to develop differences in the pigmentation of each of their symmetry systems, and further evolved these differences to the degree that adjacent symmetry systems no longer share a significant number of developmental determinants. Now, small portions of the bands of each symmetry system can move around within a wing-cell and be molded by natural selection into new alignments and new forms, without affecting the position, size, shape or color of any of the other segments.

It is interesting to ask at this point whether the evolution of a diurnal habit and the enormous radiation of wing color patterns as instruments for visual communication were cause or consequence of the evolution of this versatile patterning mechanism. The preconditions for the evolution of a highly compartmentalized and flexible pattern development system appear to be present in all the Ditrysia, but probably in few if any of the non-Ditrysia. A diurnal habit has clearly evolved several times in the Lepidoptera (for instance in the Castniidae, Sesiidae, Zygaenidae and Arctiidae). The comparative morphological evidence shows that indentation of distal patterns by the wing veins, and dislocation of symmetry bands occur occasionally in collateral lineages such as the Zygaenidae and Arctiidae, and this suggests that the conjunction of symmetry systems and compartmentalization of the wing long preceded the evolution of a diurnal habit by the butterflies' ancestors. Neither the Zygaenidae, nor the Arctiidae or Geometridae, however, appear to express a system that allows their symmetry systems to differentiate independently. Such a system probably evolved relatively early in the Ditrysian phylogeny, however, since it is expressed in the Saturniidae and a few Notodontidae. Thus all the individual elements necessary for the evolution of a highly flexible and adaptable developmental system for color pattern formation (symmetry systems, their individuation and their compartmentalization) existed long before the butterflies evolved. Yet, while some families of moths express one or two of these elements, only in the butterflies have all three converged. With the evolution of a diurnal habit one might expect that the needs and opportunities for visual communication would favor the developmental system that can provide the broadest vocabulary of visual signals. This may have been the selective environment that favored the evolution of an interaction among the three existing patterning systems in the butterflies. The resulting integrated developmental system enabled their color patterns to diversify to a degree unprecedented in the phylogenetic history of the Lepidoptera.

This work was supported by grant IBN-9220211 from the National Science Foundation.

REFERENCES

Brock, J. P. (1971). A contribution towards an understanding of the morphology and phylogeny of the ditrysian Lepidoptera. *J. Nat. Hist.* **5**, 29-102.

Carroll, S. B., Gates, J., Keys, D. N., Paddock, S. W., Panganiban, G. E. F., Selegue, J. E. and Williams, J. A. (1994). Pattern formation and eyespot determination in butterfly wings. *Science* **265**, 109-114.

Common, I. B. F. (1975). Evolution and classification of the Lepidoptera. *Ann. Rev. Entomol.* **20**, 183-203

Fraser, S. E. (1985). Gap junctions and cell interactions during development. *Trends Neurosci.* **8**, 3-4.

Henke, K. (1933). Untersuchungen am *Philosamia cynthia* Drury zur Entwicklungsphysiologie des Zeichnungsmusters auf dem Schmetterlingsflügel. *Wilhelm Roux' Arch. EntwMech. Org.* **128**, 15-107.

Hennig, W. (1965). Phylogenetic systematics. *Ann. Rev. Entomol.* **10,** 97-116.

Kristensen, N. P. (1984). Studies on the morphology and systematics of primitive lepidoptera. *Steenstrupia* **10**, 141-191.

Kühn, A. and Von Engelhardt, M. (1933). Über die Determination des Symmetriesystems auf dem Vorderflügel von *Ephestia kühniella. Wilhelm Roux' Arch. EntwMech. Org.* **130**, 660-703.

Minet, J. (1991). Tentative reconstruction of the ditrysian phylogeny (Lepidoptera: Glossata). *Entomol. Scand.* **22**, 69-95.

Nielsen, E. S. (1989). Phylogeny of major lepidopteran groups. In *The Hierarchy of Life*, (ed. B. Fernholm, K. Bremer and H. Jörnvall), pp. 281-294. Amsterdam: Elsevier.

Nijhout, H. F. (1978). Wing pattern formation in Lepidoptera: A model. *J. Exp. Zool.* **206**, 119-136.

Nijhout, H. F. (1981). Pattern formation on lepidopteran wings: Determination of an eyespot. *Dev. Biol. 80*, 367-274.

Nijhout, H. F. (1990). A comprehensive model for color pattern formation in butterflies. *Proc. Roy. Soc. London B*, **239**, 81-113.

Nijhout, H. F. (1991). *The Development and Evolution of Butterfly Wing Patterns*. Washington, DC, USA: Smithsonian Institution Press.

Nijhout, H. F. (1994a). Developmental perspectives on the evolution of butterfly mimicry. *BioScience* **44**, 148-157.

Nijhout, H. F. (1994b). Genes on the wing. *Science* **265**, 44-45.

Paulsen, S. M. (1994). Quantitative genetics of butterfly wing color patterns. *Devel. Genet.* **15**, 79-91

Paulsen, S. M. and Nijhout, H. F. (1993). Phenotypic correlation structure among the elements of the color pattern in Precis coenia (Lepidoptera: Nymphalidae). *Evolution* **47**, 593-618.

Schwanwitsch, B. N. (1924). On the groundplan of the wing pattern in nymphalids and certain other families of rhopalocerous Lepidoptera. *Proc. Zool. Soc. London B*, **34**, 509-528.

Schwartz, V. (1962). Neue Versuche zur Determination des zentralen Symmetriesystems bei *Plodia interpunctella. Biol. Zblt.* **81**, 19-44.

Scoble, M. J. (1992). *The Lepidoptera: Form, Function and Diversity*. Oxford: Oxford Univ. Press.

Süffert, F. (1927). Zur vergleichende Analyse der Schmetterlingszeichnung. *Biol. Zblt.* **47**, 385-413.

Süffert, F. (1929). Morphologische Erscheinungsgruppen in der Flügelzeichnung der Schmetterlinge, insbesondere die Querbindenzeichnung. *Wilhelm Roux' Arch. EntwMech. Org.* **120**, 229-383.

Toussaint, N. and French, V. (1988). The formation of pattern on the wing of a moth, Ephestia kühniella. *Development* **103**, 707-718.

Wehrmaker, A. (1959). Modifikabilität und Morphogenese des Zeichnungsmusters von *Plodia interpunctella* (Lepidoptera: Pyralidae). *Zool. Jahrb. Zool. Physiol.* **68**, 425-496.

Wiley, E. O. (1981). *Phylogenetics: The Theory and Practice of Phylogenetic Systematics*. New York: Wiley.

Williams, J. A. and Carroll, S. B. (1993). The origin, patterning and evolution of insect appendages. *BioEssays* **15**, 567-577.

Index of Authors

Subject Index

Development 1994 Supplement, 15-25 (1994)
Printed in Great Britain © The Company of Biologists Limited 1994

Can the Cambrian explosion be inferred through molecular phylogeny?

Hervé Philippe, Anne Chenuil* and André Adoutte†

Laboratoire de Biologie Cellulaire 4, URA 1134 CNRS-Bâtiment 444, Université Paris-Sud, 91405 ORSAY-CEDEX, France

†Author for correspondence
*Present address: Laboratoire Génome et Populations, Université Montpellier 2, Case 063, 34095 Montpellier cedex 05, France

SUMMARY

Most of the major invertebrate phyla appear in the fossil record during a relatively short time interval, not exceeding 20 million years (Myr), 540-520 Myr ago. This rapid diversification is known as the 'Cambrian explosion'. In the present paper, we ask whether molecular phylogenetic reconstruction provides confirmation for such an evolutionary burst. The expectation is that the molecular phylogenetic trees should take the form of a large unresolved multifurcation of the various animal lineages. Complete 18S rRNA sequences of 69 extant representatives of 15 animal phyla were obtained from data banks. After eliminating a major source of artefact leading to lack of resolution in phylogenetic trees (mutational saturation of sequences), we indeed observe that the major lines of triploblast coelomates (arthropods, molluscs, echinoderms, chordates...) are very poorly resolved i.e. the nodes defining the various clades are not supported by high bootstrap values. Using a previously developed procedure consisting of calculating bootstrap proportions of each node of the tree as a function of increasing amount of nucleotides (Lecointre, G., Philippe, H. Le, H. L. V. and Le Guyader, H. (1994) *Mol. Phyl. Evol.*, in press) we obtain a more informative indication of the robustness of each node. In addition, this procedure allows us to estimate the number of additional nucleotides that would be required to resolve confidently the currently uncertain nodes; this number turns out to be extremely high and experimentally unfeasible. We then take this approach one step further: using parameters derived from the above analysis, assuming a molecular clock and using palaeontological dates for calibration, we establish a relationship between the number of sites contained in a given data set and the time interval that this data set can confidently resolve (with 95% bootstrap support). Under these assumptions, the presently available 18S rRNA database cannot confidently resolve cladogenetic events separated by less than about 40 Myr. Thus, at the present time, the potential resolution by the palaeontological approach is higher than that by the molecular one.

key words: evolution, metazoa, rRNA

INTRODUCTION

The notion that most lines of metazoans appear rather abruptly in the fossil record, during the Cambrian, unpreceeded by identifiable forerunners, dates back to the 19th century. In fact, this observation was a point of serious concern to Darwin who devoted several pages in 'The origin of species' (1859) to discuss the possible reasons for the absence of identifiable Precambrian fossils. The notion of a 'Cambrian explosion' of animal life is now amply documented and is particularly striking in the richness of the Burgess Shale fauna, which dates from the mid-Cambrian (approx. 520 Myr ago) but also in its slightly earlier (530-535 Myr ago) close strata of Sirius Passet (Greenland) and Chengjiang (China). Briefly, the present view of the palaeontological evidence, as summarised by Conway Morris (1993 and this volume; see also Bowring et al., 1993), recognises three early episodes of animal life: (i) the pre-Cambrian Ediacara fauna (570-555 Myr ago), probably mostly of diploblastic 'grade' of organisation, and which may be unrelated to (ii) the later Cambrian fauna, largely dominated by triploblasts, which appears at the base of the Cambrian (approx. 540 Myr ago) and then radiates explosively during (iii) the third episode, yielding representatives of most of the 35 major metazoan phyla within an interval of probably less than 20 million years. Thus, the major types of body plans of metazoans may have originated during a relatively short time interval.

These observations are of considerable interest in understanding the mechanisms of large scale evolution and the role that developmental innovations may have had in shaping animal diversity. It is therefore important that they be substantiated by independent lines of evidence. The purpose of the present paper is to inquire as to whether phylogenetic analysis of gene sequences from extant metazoans might provide confirmation for the occurrence of such an evolutionary burst of triploblasts. The central argument runs as follows: if the split between the various animal phyla took place within a short time interval as compared to the length of time elapsed since its occurrence (say 10 Myr as compared to 500 Myr), one expects that determination of the order of emergence of the various lineages using sequence data will be almost impossible, i.e. that the various animal phyla will emerge as an unresolved 'bush' in the molecular phylogeny. The basic reason for this expectation is that the molecular events allowing one to establish the order of emergence of the various clades on a tree are the mutations that occur on the 'internal branches' of the tree, in between the points of emergence of the clades under analysis: these are the synapomorphies (shared derived characters) uniting the successive clades into a series of nested groups. The longer the time interval between two cladogenetic events the higher the probability that mutations will have accumulated within the corresponding branch in the tree and therefore the clearer the kinship of the taxa located after the branch will be.

The idea, then, is to reverse the argument and to assume that if we cannot satisfactorily resolve a multifurcation in a molecular phylogeny, it is because the time interval separating the emergence of the various clades involved has been too short with respect to the time elapsed since the event; thus, unresolved nodes in a molecular phylogeny would be interpreted as corresponding to an evolutionary radiation. This is indeed what has repeatedly been observed in the molecular phylogeny of Metazoa. However, it should be realised at the onset that the use of this argument rests on two essential parameters, a true historical one, the duration of time separating two cladogenesis events, and a methodological one, namely our ability, through our tree construction methods, to discriminate 'well resolved' nodes from 'unresolved' ones. From the start, then, we see that this approach to the question of the Cambrian radiation is intimately linked to the problem of evaluating the reliability of nodes in molecular phylogenies. The short history of the molecular phylogeny of Metazoa provides a good illustration of how these evaluations started in a rather intuitive and non-quantitative way to become increasingly rigorous.

Starting with the pioneering study of Field et al. (1988) for example, which was based on partial 18S ribosomal RNA sequences analysed by a distance method, the authors stressed that the order of emergence of the four major groups of coelomates they analysed, Chordata, Echinodermata, Arthropoda and a set of 'eucoelomate protostomes' could not be confidently resolved and suggested that this reflected a rapid phyletic splitting, i.e. a rapid radiation of all coelomate phyla. Their arguments were that the internal branches separating the points of emergence of the various taxa were short and, more importantly that the topology was unstable i.e. that it changed depending on the actual species sampled. These data were reanalysed by Patterson (1989) using a variety of methods, in particular parsimony, and by Lake (1990) using his evolutionary parsimony method (see Erwin, 1991 for a review of these papers). Contradictions and uncertainties in the results suggested both that a rapid radiation of eucoelomates may indeed have occurred and that the data were 'noisy'. 28S rRNA partial sequences of a broad sample of 'invertebrate' species covering ten triploblastic coelomate phyla, three pseudo-coelomates ones and one acoelomate were also obtained and analysed by one of us (Chenuil, 1993) with very similar results.

Later, two of us (Adoutte and Philippe, 1993) reanalysed a broader 18S rRNA partial sequence data set, using parsimony methods and bootstrap testing and clearly confirmed three points: (1) diploblasts were deeply split from triploblasts (as had been inferred by Raff et al. (1989) and by Christen et al. (1991); (2) platyhelminths (acoelomate triploblasts) were the sister group of coelomates and (3) the major coelomate phyla were very poorly resolved. A 'giant' multifurcation, comprising annelids, arthropods, molluscs, echinoderms, chordates and many more minor groups provided a fair representation of the results. Even the separation into the two major lineages of coelomates, protostomes and deuterostomes, although apparent in the tree, was not supported by significant bootstrap values. We suggested at that time that the latter point could be a reflection of the Cambrian explosion.

In the past ten years, much use has been made of the bootstrap value (or bootstrap proportion, BP) to estimate the reliability of nodes in phylogenetic trees (Felsenstein, 1985). The BP (the number of times a given node is obtained over the total number of nucleotide resamplings carried out) is indeed a convenient value to estimate the strength of the phylogenetic signal, within the framework of a given tree reconstruction method: values above 95% indicate that the data contain a strong signal in favour of this node while low values indicate that the node is poorly supported and thus in fact may not exist (see Zharkikh and Li, 1992a,b; Hillis and Bull, 1993 and Felsenstein and Kishino, 1993 for recent detailed analyses of the significance and statistical properties of the bootstrap). Thus, an expectation of a radiation process is that bootstrap proportions should be low in all the internal branches surrounding the radiation point in the tree.

A corollary of this criterion is the instability of the node: when poorly supported by the bootstrap, nodes often display instability in the face of variations in the length of sequence analysed and, more significantly, in the face of modifications of species sampling (as systematically analysed in Lecointre et al., 1993). That point is strikingly illustrated in the paper by Adoutte and Philippe (1993) where the addition of a single new species to the tree transforms the Metazoa from monophyletic (diploblasts + triploblasts) to biphyletic (diploblasts on one branch and triploblasts on another), in agreement with the low BP (60%) of the corresponding node. This provided an indication of the difficulty of solving this question and was interpreted as confirming the depth of the split between diploblasts and triploblasts.

In the present work, we have carried out a new analysis of an 'edited' database of 18S rRNA, eliminating fast evolving species and analysing the significance of the nodes in the trees by a procedure recently developed within our group. This involves calculating not only the bootstrap proportion at each node on a tree, but also determining how this value changes when different lengths of sequence are included in the analysis (Lecointre et al., 1994). This allows us to obtain a curve of the BP as a function of the number of nucleotides used, which is more informative than the mere BP value based simply on a single sequence length. In particular, this procedure allows one to estimate the number of additional nucleotides that would be required to transform a 'moderately supported' node into a strongly supported one. This is then combined with palaeontological data to establish a rough relationship between the length of time separating two cladogenesis events, the length of the corresponding branch and the amount of sequence information required to support it in a statistically significant way. When applied to the rRNA dataset of Metazoa, this provides an estimate of the sequencing effort that would be required to resolve closely spaced branching points in the phylogeny, an effort that turns out to be enormous and unrealistic in most cases.

MATERIAL AND METHODS

Sequences used

Only species for which the full 18S rRNA sequences is available have been used in the present study. As of December 1993, this corresponded to 69 species of Metazoa in the EMBL and GenBank data banks. All the sequences were handled and further analysed through the MUST package (Philippe, 1993), and are available upon request.

Alignment and tree construction

The sequences were aligned manually using the editing functions of the MUST package (Philippe, 1993). Only confidently aligned domains were used, using stringent criteria to eliminate all doubtful portions. The boundaries of the domains thus selected are as follows, using mouse 18S rRNA nucleotide numbering as a reference: 82-125, 137-180, 187-194, 208-243, 289-307, 311-539, 548-689, 798-834, 841-1112, 1122-1407, 1440-1551, 1558-1737.

For the 69 species, this yielded a total of 1615 aligned sites of which 1010 were variable and 690 informative under the parsimony criterion. When fast-evolving species are eliminated (see Results), 55 species remain, yielding 1474 aligned sites of which 708 are variable and 486 informative.

Trees were constructed using the Neighbour Joining method (Saitou and Nei, 1987) and were submitted to bootstrapping (Felsenstein, 1985) using the NJBOOT program of the MUST package set at 1000 resamplings. All the bootstrap calculations were carried out on a Sun-Sparc 10 computer.

Calculation and display of the 'pattern of resolved nodes' (PRN)

The method described by Lecointre et al. (1994) was used throughout, under the following conditions. The full aligned sequences of the 55 species were each submitted to random sampling of a given number of sites (a 'jack-knifing of sites') through the use of a new program, PRN, running on UNIX platforms. Thirteen different sequence lengths were chosen (25, 50, 75, 100, 150, 200, 250, 300, 350, 400, 450, 500 and 600 sites) and for each, 200 samples were drawn. Thus, a total of 2600 subsets of sequence alignments were obtained each including all 55 species. Each of these subsets was used to construct an NJ tree, which was submitted to 1000 bootstrap replicates. All the combinations of species appearing in more than 1% of the replicates were 'stored' in a file (of about 60 Mb). This yields several tens of thousands of nodes. Selection of the nodes was then carried out using the new program AFT-PRN according to the following criteria: the node should correspond to a BP with an ascending tendency and it should be present in more than 2000 of the subsets of sequences. This is a rather stringent criterion allowing us to keep only nodes that appear frequently. At a given node, one could therefore display graphically the evolution of BP as a function of the number of nucleotides that were used to generate the tree (COMPBOO program of the MUST package or DISPLAY-PRN under UNIX).

Lecointre et al. (1994) have shown that the mean of BP can be related to the number of nucleotides, x, through the function $BP = 100\,(1-e^{-b\,(x-x')})$. The parameters b and x′ are estimated by non-linear regression using the GENSTAT package.

Relationship of 'b' to branch length and to time

Branch lengths were directly provided by the NJ program and displayed using the TREEPLOT program. These lengths were plotted as a function of the value of the b parameter for the corresponding branch.

The following palaeontological dates were used to establish the relationship between the time elapsed between two cladogeneses and the b value: 300 Myr between the point of origin of Pectinidae (as represented by *Placopecten*) and that of Mactridae (as represented by *Spisula*; Rice et al., 1993), 50 Myr between the point of origin of gnathostomes and that of all vertebrates and 300 Myr between the point of origin of amniotes and that of eutherians (see Benton, 1990).

Table 1. Results of the relative rate test for a selected sample of species

Moliniformis moliniformis	1.58	19.67
Herdmania momus	3.48	19.71
Drosophila melanogaster	1.69	20.36
Oedignathus inermis	3.52	20.36
Eptatretus stoutii	1.26	21.21
Myxine glutinosa	1.26	21.21
Pugettia quadridens	3.99	21.64
Sagitta elegans	1.97	22.99
Haemonchus contortus	2.23	23.47
Haemonchus placei	2.23	23.47
Haemonchus similis	2.23	23.47
Strongyloides stercoralis	2.11	23.89
Aedes albopictus	3.10	24.50
Caenorhabditis elegans	2.10	25.74

The relative rate was calculated by establishing the mean number of either the gaps (1st column) or all types of substitutions (2nd column) between the sequences of the species considered and the sequences of 8 diploblast species taken as an outgroup.

The values for the complete set of species ranged from 14.6 to 25.74 when all types of substitutions are computed. The species displayed in this table are those at the highest extreme of the distribution, which have all been eliminated, the limit for inclusion in the dataset having been set at 18.3.

In the column corresponding to gaps, there is much greater homogeneity, even for fast evolving species except for a few species that display a number of gaps much higher than that of others (*Herdmania momus*, *Oedignathis inermis* and *Pugettia quadridens*) and which correspond to species whose sequence determination appears to have been problematic.

Relative rate test

The mean values of the distances between each of the triploblast species and the 8 diploblast ones was computed using either all types of nucleotide differences or only the gaps (Table 1).

Saturation curve

The number of inferred substitutions was calculated using the program PAUP 3 (Swofford, 1991), courtesy of H. Récipon. The corresponding matrix between all pairs of species was obtained through the TREEPLOT program, and the COMP-MAT program allowed the visualisation of the results.

RESULTS

Global 18S rRNA Neighbour-Joining tree of Metazoa

Fig. 1 shows the Neighbour-Joining (NJ; Saitou and Nei, 1987) tree of all the metazoan species for which complete 18S rRNA sequences were available in data banks as of December 1993. Partial sequences were excluded in order to maximise the amount of information for each species. This database contains representatives of 13 metazoan phyla including the most numerically important ones, with only one unfortunate omission, that of annelids, for which complete sequences are not available. The tree is arbitrarily rooted between diploblasts (Porifera, Ctenophora, Placozoa and Cnidaria) and triploblasts to avoid the use of a non-metazoan outgroup (which would decrease the length of alignable sequences) and because diploblasts constitute a clear outgroup to triploblasts on the basis of multiple previous evidence. The tree displayed is that directly obtained by the NJ method, a distance method that does not assume equality of evolutionary rates among branches and whose efficiency at recovering the actual tree has been shown to be reasonably good (Tateno et al., 1994). It is used here because of its great rapidity in terms of computer time, a critical advantage for the extensive calculations carried out in this paper.

The tree displays a number of interesting features and also immediately illustrates one major source of artefact: several species or groups of species display much longer branches, indicating either a two- or three-fold higher rate of evolution in those sequences or inaccurately determined sequences. Such inequalities are known to generate topological errors in the positioning of the corresponding taxa (Felsenstein, 1978; Hendy and Penny, 1989; Swofford and Olsen, 1990). A clear example is provided by two insects, *Drosophila* and *Aedes,* which emerge as a sister group to nematodes (the latter also all having long branches) and separate from the five other arthropods, which have smaller branches and are more traditionally positioned in the tree. Both for *Drosophila* and for *Caenorhabditis* it is known that the problem lies not in the quality of the sequences but in their rapid rate of evolution, as has been pointed out in several previous papers.

In spite of these sources of errors, Fig. 1 displays a number of features that deserve a brief comment.

(1) The distance between diploblasts and triploblasts is indeed the largest measured in the tree between high level taxa supporting the validity of the rooting.

(2) Except for the nematodes, and an acanthocephalan (*Moliniformis*), all the other major triploblast lineages emerge very close to each other, i.e. separated by very short interval branches and correspondingly low BPs.

(3) Nematodes and the acanthocephalan, traditionally grouped within the pseudocoelomates, seem to emerge between the diploblasts and the platyhelminths, i.e. acoelomate triploblasts. This is contrary to the usual view which places pseudocoelomate emergence between acoelomates and eucoelomates. However, because of the great inequalities in rates of evolution and the very poor BPs in this portion of the tree, this result should not be over emphasised.

(4) Contrary to our previous study, based on a parsimony analysis of partial 18S rRNA sequences (Adoutte and Philippe, 1993), the monophyly of coelomates is not strongly supported (33% BP) and the positioning of the platyhelminths as a sister group to coelomates is correspondingly weakened.

In summary, the view emerging from this initial tree is one of a large burst of all triploblastic metazoans, including platyhelminths. But this view should be considered with caution because of the inequalities in evolutionary rates observed within the tree.

In a second step, the relative-rate test (Sarich and Wilson, 1973) was systematically carried out on all the species of the tree (Table 1). The distribution of distances to the outgroup for the 55 remaining species is roughly gaussian and ranges from 14.6 to 18.3 whereas the distances for the fourteen discarded species vary from 19.7 to 25.7, noticeably far from gaussian. All fourteen taxa with too high a rate were discarded, yielding the more restricted set appearing in Fig. 2, which was used in all further calculations.

When the rapidly evolving species are eliminated from the database, according to this criterion, several discrepancies of Fig. 1 disappear, and the BPs rise (Fig. 2). However, some

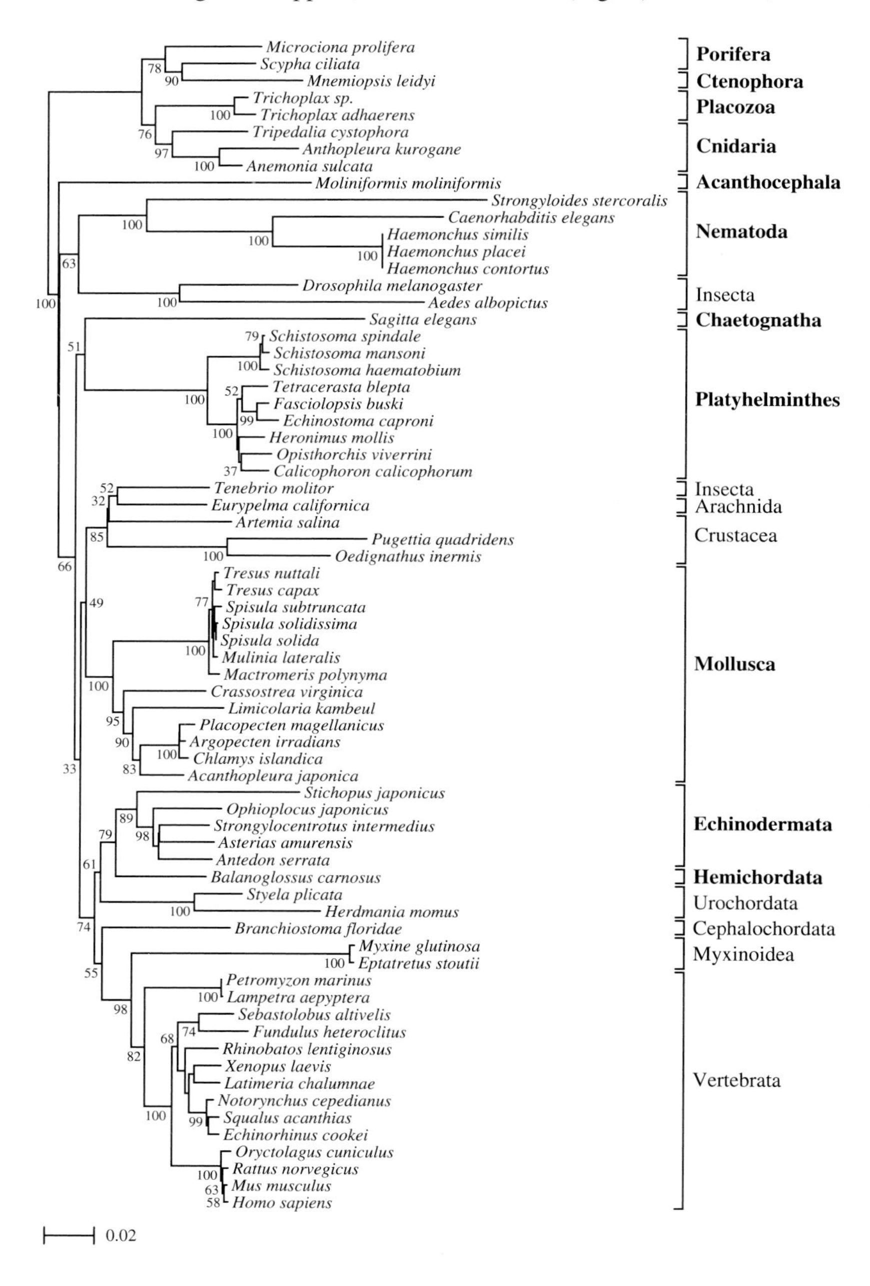

Fig. 1. Phylogeny of Metazoa based on complete 18S rRNA sequences treated by the Neighbour-Joining method. Numbers below internal branches indicate the bootstrap proportion of the corresponding node (1000 resamplings). Note the length of the branches of all the species of Nematoda and of two Insecta (*Drosophila melanogaster* and *Aedes albopictus*).

interesting taxa are discarded, most notably nematodes. Among the salient points now emerging are:

(i) A confirmation of the outgroup status of Platyhelminthes with respect to coelomates (but still with a weak support for the monophyly of coelomates, i.e. 49% BP).

(ii) Higher BPs for the monophyly of coelomate protostomes and deuterostomes (75% and 69% respectively).

(iii) High BPs for several taxa known from independent analyses to be monophyletic, such as arthropods, molluscs, echinoderms and vertebrates. This result should, again, not be over-emphasised because the sampling is limited and biased in some of these taxa (for example there is a large over-representation of bivalves within the molluscs). All these conclusions are very similar to those of Winnepennickx et al. (1992) who used a similar data set and similar methods.

(iv) In contrast, a few inconsistencies within these monophyletic groups are observed, such as the position of *Limicolaria*, a gastropod, amongst the bivalves within the molluscs. Several incongruities are also observed within the vertebrates such as the point of emergence of mammals and the position of chondrichthyans. This may well be due to the fact that more rapidly evolving portions of the rRNA were discarded from this analysis (to enable analysis of very distant taxa), with a resultant loss of information appropriate for more closely related ones such as vertebrates. The topology observed within deuterostomes, with echinoderms, hemichordates and urochordates as a sister group to cephalochordates and vertebrates, is also contrary to traditional zoological views. However, the very same topology was very recently reached by Wada and Satoh (1994) using a variety of tree construction methods.

Thus the view changes slightly from one that suggested complete lack of resolution of the order of emergence of the various clades to one starting to display some significant pattern but still with low resolution. Can this lack of resolution be the result of yet another type of bias in the data?

We have shown previously that mutational saturation of sequences can strikingly decrease the resolution of several nodes in a tree by bringing the corresponding branching points artefactually much closer to each other (Philippe et al., 1994; Philippe and Adoutte, 1994). We therefore verified that the relatively low resolving power of 18S rRNA was not due to saturation problems. This is easily achieved by plotting the number of inferred substitutions between all pairs of species as deduced from a parsimony algorithm, against the actual number of differences recorded between the extant sequences. When saturation is present, the molecular distances measured between extant sequences level off while the distances deduced from the tree continue to rise. This procedure was applied to the rRNA data set used in the present study and it can be seen in Fig. 3 that a slight saturation is discernible but that it is weak and we think it is not likely to perturb severely the interpretations. The data are therefore suggestive of a true radiation.

A closer look at critical nodes in the tree: the PRN method

Instead of simply examining the values of the BP at important nodes as a criterion of robustness of the corresponding nodes, we have recently introduced a procedure of BP analysis which involves following the value of BP as a function of an increasing number of nucleotides taken into account (Lecointre et al., 1994; see Materials and Methods). Various types of curves are thus obtained. Furthermore it has been shown that the shape of the curve in the case of resolved

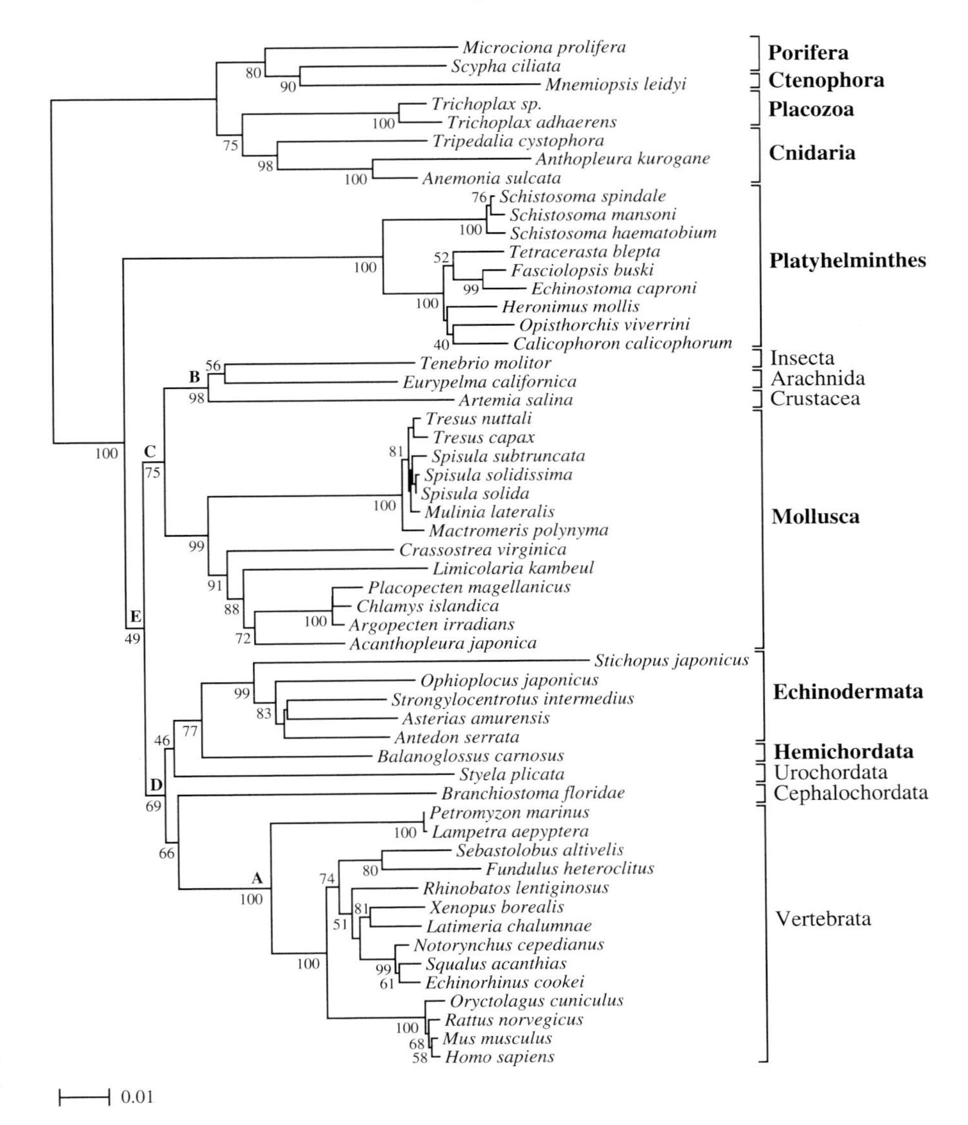

Fig. 2. Phylogeny of Metazoa derived in the same fashion as that of Fig. 1 except for the removal of fast evolving lineages; the nodes designated by the letters A, B, C, D and E are those whose PRNs are displayed in Fig. 4.

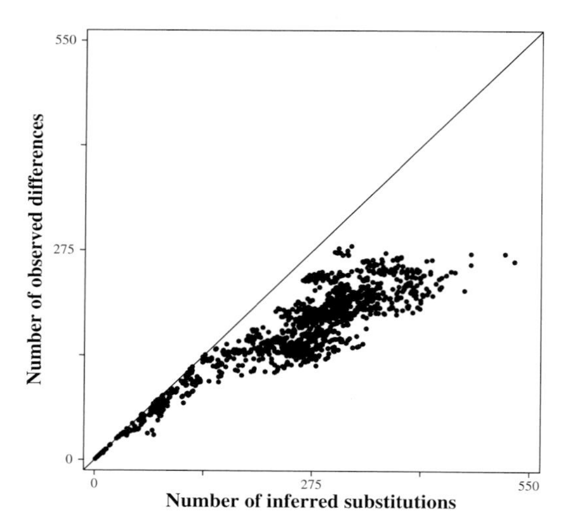

Fig. 3. Comparison of patristic distances (inferred numbers of substitutions) versus number of observed nucleotides differences between the sequences used to construct the tree of Fig. 2 (see text).

nodes (PRN) fits closely a simple monomolecular function of the form:

$$BP = 100\,(1-e^{-b\,(x-x')}) \qquad (1)$$

where x is the number of informative sites, and where b and x′ are parameters specific to each node, which can be estimated directly from the data through non-linear regression. In fact, it was shown that the b value from this equation provides an accurate descriptor of the curve because x′ is always close to 0. b is much more informative than a single BP value since it is correlated at the same time to BP and to the general shape of the curve.

The most interesting part of this procedure lies at the high extremes of BP: for these values of BPs, it discriminates between cases in which this high value is reached very quickly (i.e. using a small number of nucleotides) and those in which it is reached more slowly (i.e. after using a much larger number of nucleotides); thus, even within the category of fully resolved nodes, a discrimination can be made between those that are extremely strong and those that are moderately strong.

Let us take some examples from nodes displayed in Fig. 2. Node A and node B correspond respectively to the monophyly of vertebrates and arthropods; both are highly supported by the BP: 100% and 98% respectively. Their PRN, displayed in Fig. 4A,B, are clearly of the fully resolved type, with BPs reaching 98-100% with the 18S rRNA data already available. The shape of the curve is substantially different however in these two cases: the slope is extremely steep in the case of vertebrates, the plateau being reached after using only a very small portion of the available nucleotides, while in the case of arthropods the plateau is reached much more gradually and the full complement of nucleotides is required to reach the highest BP. In fact, the monophyly of arthropods is known, from extensive previous work, to be difficult to establish using rRNA sequence data (Turbeville et al., 1991), while that of vertebrates is a robust feature of phylogenies based on very diverse characters, both molecular and anatomical. One possible interpretation of the difference between these two clades is that diversification into the various classes occurred earlier in the history of the phylum Arthropoda than in the vertebrates.

Nodes C and D are especially interesting in the context of the present paper since they correspond to the monophyly of coelomate protostomes and deuterostomes respectively. Their BPs are 75% and 69% and they yield PRNs (Fig. 4C,D) that are clearly different from the two just discussed in that, although the curve rises steadily, they do not reach 100% BP with the available amount of data. In such a case it is interesting to compute, on the basis of these two curves, the number of nucleotides that would be needed to yield a value of 95% BP. This is respectively 1300 and 2100 variable nucleotides (and thus about a threefold higher total number of nucleotides to be sequenced) in each of the 55 species displayed in Fig. 2. Since the complete 18S rRNA was used already, one could turn to 28S rRNA, assuming that its rate of evolution is roughly similar. However, even complete 28S rRNA would not be sufficient in the present case. Thus, although both the BP and overall shape of the PRNs for these two nodes are intuitively in favour of the monophyly of the two corresponding groups, it is seen that in order to establish the point definitively using a stringent criterion would require considerable experimental effort. Such a situation is even more drastically illustrated in the case of the node which suggests the monophyly of a group composed of the echinoderms, urochordates and hemichordates on Fig. 2. This node is supported by a BP of only 46% and, as indicated above, is contrary to the traditional zoological assumption which groups this set of taxa paraphyletically with the other deuterostomes (cephalochordates and vertebrates) into a single large monophyletic unit. In this case, reaching a value of BP of 95% would require 3500 variable nucleotides.

Node E, supporting the monophyly of triploblastic coelomates again has a medium BP (49%) and a slowly ascending PRN (Fig. 3E). Confirmation of the monophyly of coelomates would require 2600 variable nucleotides. This is in contrast to our previous work, based on shorter sequences of 18S rRNA and a different taxonomic sampling which more clearly supported the monophyly of coelomates (75% BP) with platyhelminths as a sister group (Adoutte and Philippe, 1993). This is probably a reflection of the strong impact of the particular species sampling under analysis as analysed in detail in Lecointre et al. (1993). One further result illustrating the difficulty in solving this node can be found in the fact that an alternative topology, that which groups platyhelminths with coelomate protostomes, displays a PRN very similar to that obtained in the case of the monophyly of coelomates (Fig. 4F); in fact, the hypothetical node displayed in Fig. 4F requires the same number of nucleotides to be solved as that of Fig. 4E (2600), indicating nearly identical support for two opposing topologies. An ascending shape therefore does not necessarily indicate that the BP will continue to grow with increasing sequence length, but that the actual sampling of sites and species lead to this pattern by chance only. One must keep in mind that our new approach only allows the number of nucleotides to be sequenced to be estimated. It is necessary to obtain more data to establish the 'good' phylogenetic pattern (here the monophyly of coelomates).

At any rate, both the data for the monophyly of coelomates and for that of coelomate protostomes and deuterostomes conform to the idea that these various branchings are difficult to resolve and may therefore correspond to a rapid radiation.

As stated in the Introduction, our ability to resolve nodes in a phylogenetic tree critically depends on the differences that accumulate within the internal branches of the tree. Instead of simply trying to define resolved versus unresolved nodes, could it now be possible to approach the relationship between the length of time elapsed between two cladogenesis events and the length of the corresponding internal branches of the tree? That is, is it possible to estimate the smallest time interval between two events that we are able to identify with confidence using given molecular tools? If such a relationship could be established for a given molecule such as rRNA, then we would be in a position to define its resolving power in terms of millions of years.

The resolving power of rRNA

To answer the questions just raised, we need to go through three successive steps and make two assumptions. We start from the PRN function described above which relates the BP of a given node to b and to the number of nucleotides; this function simplifies to

$$BP = 100\,(1-e^{-b\,x}) \qquad (2)$$

by assuming x′ is negligible. Since we can calculate the value of b independently from x and if we set the BP at a required value (such as 95%), then we can calculate x as a function of b. If, furthermore, we have related b to time, then we can calculate x as a function of time, that is, for a given time interval expressed in millions of years (Myr), we can calculate the number of nucleotides that a given molecule requires to be discriminative.

We will therefore first establish that branch lengths and the b value defined above for the PRN formula are correlated, as could be expected; then we will show that absolute time is correlated to b using values derived from the fossil record that enables us to carry out calibrations by extrapolating values derived from palaeontology. At that stage we make the two assumptions, namely that internal branch lengths, as determined through the NJ method, reflect 'true' branch lengths, that is they are proportional to the real number of mutational events that have been fixed between two cladogeneses, and second, that branch lengths are directly proportional to real time, that is we assume a molecular clock. Having now a relationship between b and time, we can reincorporate this information into the initial formula relating b to the number of nucleotides and

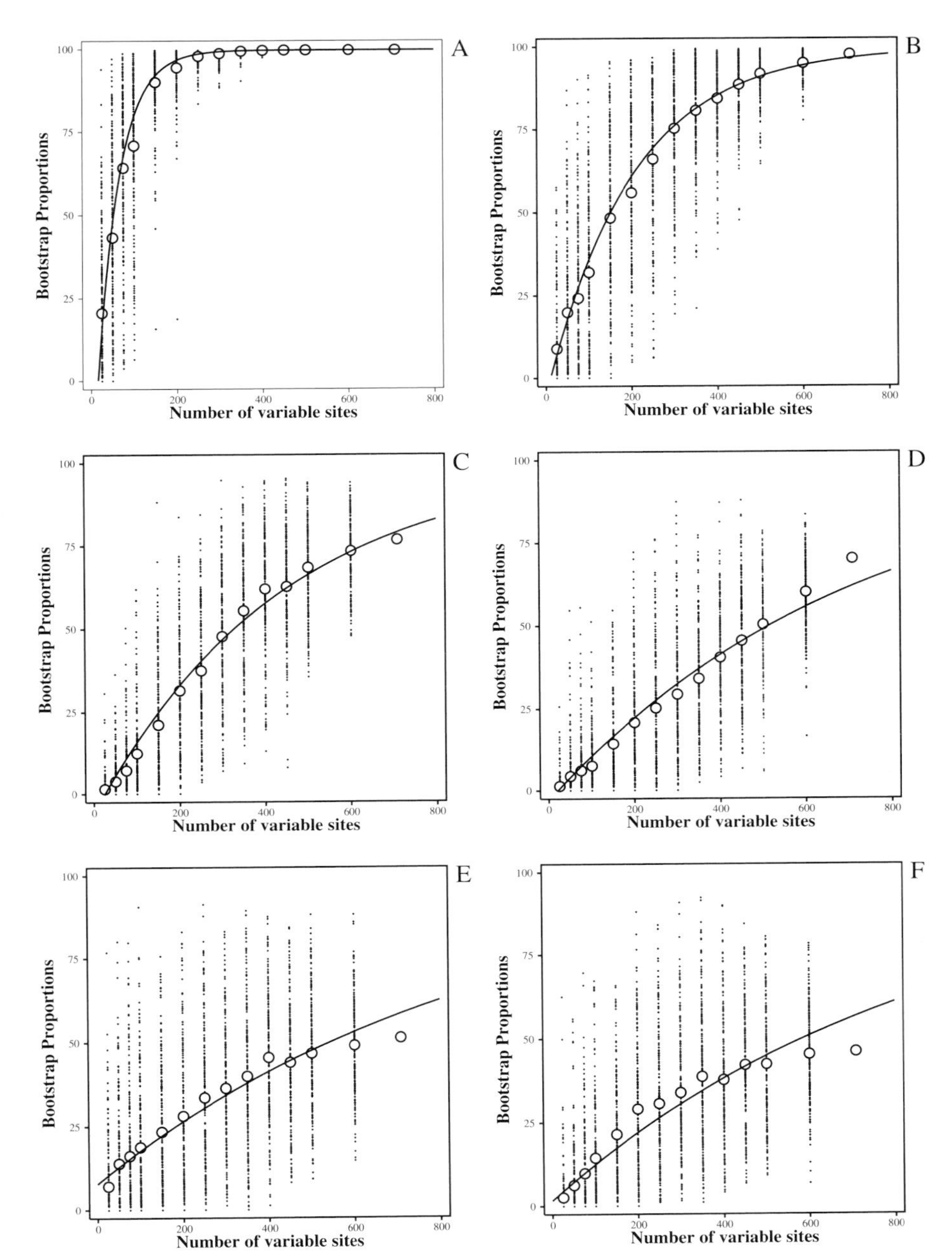

Fig. 4. The patterns of resolved nodes (PRN) for 5 nodes taken from the phylogeny of Fig. 2. In each graph, the 200 bootstrap proportions (ordinate) obtained for a given number of nucleotides are plotted vertically as a function of increased number of nucleotides taken into account (abscissa) with increments of 50 nucleotides at each step until the full number of variable sites (about 700) of the 18S rRNA data set is included. The graphs correspond respectively to the monophyly of vertebrates (A), that of arthropods (B), that of coelomate protostomes (C), and that of deuterostomes (D). Graphs E and F correspond to two possible contradictory topologies at the base of triploblasts: one which places Platyhelminthes as the outgroup to all coelomates (as shown on Fig. 2, node E) and that which incorporates Platyhelminthes within triploblasts (node F, not shown). The open circles on each set of vertical points correspond to the average value of all the bootstrap proportions. The general shape of the curve joining these open circles has been previously found to be described by a function of the type $BP = 100\,(1-e^{-b\,(x-x')})$. Parameters b and x′ can be obtained from the experimental data through non-linear regression and introduced back into the function to draw the curves as shown.

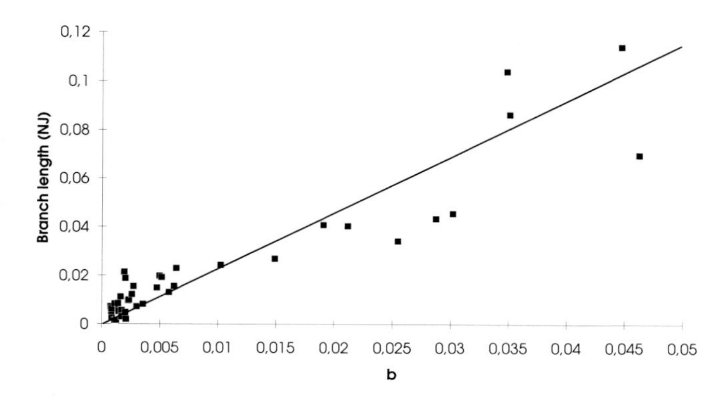

Fig. 5. Correlation between the length of internal branches as determined by the Neighbour-Joining method and the value of parameter b (see text).

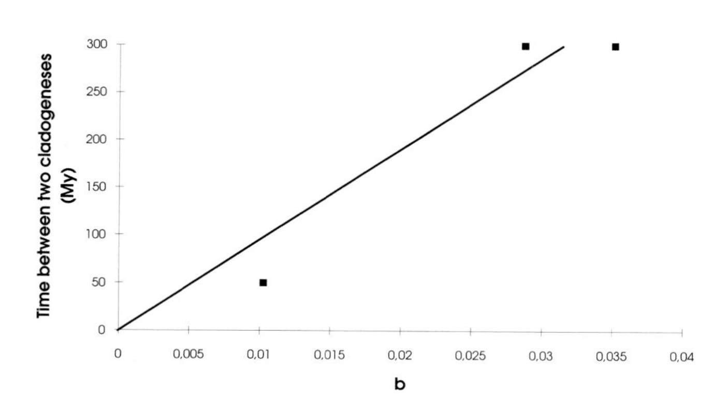

Fig. 6. Relationship between the time elapsed between two cladogeneses and the b value (see text).

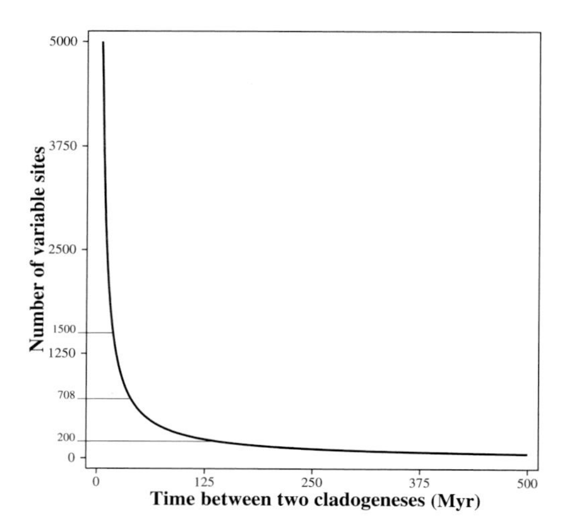

Fig. 7. Relationship between the time elapsed between two cladogeneses and the number of variable sites needed to resolve a given node with a 95% BP. The curve is drawn using the function x = 28000/ΔTc (see text). The three intersects on the ordinate correspond to the number of informative sites, respectively, in partial 28S rRNA sequences (200), full 18S rRNA (708) and full 18S + 28S rRNA (1500).

deduce the shape of the curve relating the number of variable sites to time.

The relationship of b to branch length is shown in Fig. 5. It can be seen that the experimental points fit a regression line rather well; in addition, it was observed that the quality of the correlation did not depend on the depth of the node in the tree. Thus, the first requirement is satisfied.

The relationship of b to absolute time is shown in Fig. 6. Only three time points, listed in Material and Methods, were available. The regression is therefore based on only three points. This nonetheless allows us to calculate a rough estimate of the value of k in the formula

$$b = k\Delta T_c$$

were k is a proportionality constant and ΔT_c is the time interval separating two cladogeneses. k was thus calculated to be equal to 0.000108.

We then substitute this expression of b in equation (1) and take BP to be equal to 95%. We obtain

$$95 = 100\,(1 - e^{-k\Delta T_c\, x})$$

from which we extract

$$x = \frac{-\log 0.05}{k\Delta T_c} \approx \frac{28000}{\Delta T_c}.$$

We thus see that the number of nucleotides required to resolve a node at the 95% BP level is simply related to time through a constant. We can therefore plot the curve of x as a function of the duration of time separating two cladogeneses having established the value of k above in the case of rRNA. As could be expected from the initial formula the curve obtained (Fig. 7) is a hyperbola.

From this curve, it can be calculated that with 200 variable sites (which is a common number when partial 28S rRNA sequences are used) the discrimination power of the dataset is not better than 140 Myr. With 708 variable sites, which is the case with the complete 18S rRNA dataset used in the present study, the resolving power is of 40 Myr. If we assume that 18S + 28S rRNA roughly yield to 1500 nucleotides, the resolving power with the complete rRNA unitary cluster improves slightly to 19 Myr but, due to the shape of the curve, improvement in resolution is clearly not linear with respect to sequence length. For example, to reach a resolving power of 1 Myr, 28000 **variable** nucleotides would be required so that a total of at least 80,000 homologous nucleotides should be sequenced, which is the equivalent of 40 times the 18S rRNA.

DISCUSSION

The question raised at the beginning of this paper as to whether molecular phylogenies of extant metazoans allow the visualisation of the Cambrian explosion in the form of an unresolved multifurcation has been reformulated in the course of the analysis: through the set of tools developed in this and previous work, we have been able to evaluate better the significance of nodes displaying low bootstrap values i.e. 'poorly resolved nodes' intuitively interpreted so far as corresponding to a radiation; more importantly, we have suggested a procedure allowing us to assign a time limit to such nodes (as a function of the specific data set under analysis). We are thus in a position to evaluate the resolving power of rRNA sequence data in terms of the minimum time span between two cladogenetic events required for it to identify safely the corresponding internal branch.

The most striking result obtained is that the resolving power of the presently available complete 18S rRNA database is only of about 40 Myr: this is the duration of time required for the molecule to identify a distinct cladogenesis with a bootstrap proportion of 95%. Such a time interval is much longer than those that can be resolved with the presently available radiometric geochronological methods. For example, using uranium-lead zircon dating, Bowring et al. (1993) have recently calibrated early and middle Cambrian rocks with a 5-10 Myr precision. If the fossil record is of good quality i.e. if it is reasonably continuous, without too many gaps, palaeontology performs better than 18S rRNA and better even than 18 + 28S rRNA in the sense that it can narrow the time period during which a biological group has diversified better than the molecules can. Obviously, palaeontological dating does not resolve phylogenetic patterns; it simply allows one to estimate the lapse of time during which biological groups have diversified. Because methods are now available to evaluate the quality of the fossil record (see Benton, 1994), the comparison of molecular and palaeontological data will become increasingly rigorous.

Is this conclusion confounded by methodological problems? Several parameters other than the relative shortness of the time interval can contribute to obscuring a 'true' node in a phylogenetic tree. We are presently studying these parameters systematically both on real and on simulated data sets. Of those parameters for which some data are already available, the most prominent are mutational saturation of sequences and inequalities in evolutionary rates of the different taxa under analysis. We suggest from Fig. 3 that saturation is not prominent in the dataset used here and we have reduced the problem of unequal rates by eliminating fast evolving lineages from the tree. We are aware that several additional pitfalls are conceivable such as biases in species sampling (Lecointre et al., 1993), inequalities in the density of the topology in certain areas of the tree and level of homoplasy as well as other factors. However, our present simulations suggest that the most important parameter affecting resolution of nodes remains the length of internal branch, i.e. the time interval separating the cladogenesis events (Philippe, unpublished data). To minimise the risks resulting from these various uncertainties, we have used, throughout the present work, a stringent criterion for the bootstrap value, 95%, although several recent studies conclude that 70% is a reasonable value in view of the conservative aspect of the bootstrap (Zharkikh and Li, 1992a,b; Hillis and Bull, 1993). Indeed, the impact of species sampling on BP is very strong (as illustrated by substituting a BP of 60% for monophyly of Metazoa to a BP of 60% for polyphyly of Metazoa by the addition of a single species, see Introduction), especially when the internal branch length is short and few species are used (Philippe and Douzery, 1994). Thus, since this phenomenon is not controlled, we believe our conservative criterion for BP of 95% is justified. The conclusion thus holds that the molecular data do point to a radiation of triploblastic coelomate animals if we are willing to qualify a 40 Myr interval as such.

These results have two broad implications in terms of developmental biology. Firstly, superposition of the developmental characteristics of the various lineages over the phylogeny pattern can indicate the order of emergence of developmental traits. Secondly, and more significantly in the context of the present paper, confirmation of the notion of a relatively rapid Cambrian radiation raises a number of interesting questions concerning the mechanisms of diversification of the body plan of Metazoa.

With respect to the first point, we will limit ourselves to four inferences based on the phylogeny.

(1) The confirmation of the deep split between diploblasts and triploblasts suggests that diversification into these two major types of metazoan occurred very early in animal evolution. The ancestral state in diploblasts is probably only two tissue layers. The development of an interstitial type of cell in some lineages such as some cnidarians and some ctenophores is probably derived and independent of that which occurred in triploblasts which developed the mesoderm from the outset.

(2) The sister group status of the platyhelminths with respect to the coelomates, reasonably well supported in the present study, confirms that an acoelomate triploblastic stage has preceded invention of the coelom in metazoans. Furthermore, the molecular phylogeny is compatible with a single origin of the coelom at a point located between the emergence of platyhelminths and the coelomate radiation.

(3) The position of platyhelminths also suggests that, within the triploblasts, spiral development is more primitive than radial, if spiral development within coelomate protostomes is taken to be homologous to that of platyhelminths. If such is the case, then radial cleavage is a derived feature.

(4) The existence of two major lines of coelomates, the traditional coelomate protostomes and deuterostomes, is apparent on our trees and reasonably well supported by the bootstrap. Significantly, the two corresponding nodes yield PRNs suggesting that the monophyly of each group may ultimately be confirmed if more data is available. This suggests that the embryological character traditionally used to separate these two lineages i.e. blastopore fate is, despite various troublesome exceptions, of real phylogenetic significance with protostomy being the ancestral state. Secondary mouth formation is correlated with the origin of the deuterostomes. It should be remembered that blastopore fate is also correlated with mode of origin of mesoderm and coelom formation.

As for the second point, how to account for a rapid diversification of body plans during the Cambrian, the following comments can be made. It is now well established that the common ancestor of protostomes and deuterostomes already possessed a diversified complement of HOX class homeobox-containing genes belonging to the HOM/Hox complex (Akam, 1989; Schubert et al., 1993). In fact, comparison of insect and vertebrate HOM/Hox complexes indicates that their common ancestor already possessed at least 6 or 7 of the major genes. In addition, it is now clear that the nematode, *C. elegans*, a 'pseudocoelomate' also possesses a HOM/Hox complex of at least four genes (Kenyon and Wang, 1991) and work from several laboratories (Oliver et al., 1992; Webster and Mansour, 1992; Bartels et al., 1993) including ours (Balavoine and Telford, unpublished data) indicates that platyhelminths have a large diversity of Antp-type HOX genes. Finally, such genes have also been identified in diploblasts (Schierwater et al., 1991; Shenk et al., 1993). In view of these data, Slack et al.

(1993) have proposed that possession of a basic HOM/Hox complex expressed at the phylotypic stage was a synapomorphy of all Metazoa. It can therefore be hypothesised that the ancestor of coelomates and, in fact, the ancestor of triploblasts already possessed a diversified array of HOM/Hox genes involved in defining broad domains along the anteroposterior body axis. In addition, there is now substantial evidence to indicate that genes involved in major developmental processes such as cell to cell signalling, cell adhesion, cell migration, gene regulation, etc. are of very ancient evolutionary origin (reviewed by Shenk and Steele, 1993). Our suggestion, then, is that the major genetic tools used for carrying our embryological development, defining axes and polarities, establishing cell determination and cell differentiation, were already present within Metazoa, prior to the Cambrian radiation. This radiation can therefore be viewed essentially as a tinkering process (Jacob, 1981), variously combining and regulating an already available basic set of genes; because no major genetic innovation was required but rather a shuffling of already available elements and because the various marine ecological niches were essentially empty, one can readily accept that an 'explosive' process might have taken place.

The striking conclusion reached in this paper is the shape of the curve relating the number of nucleotides required to resolve a node significantly, and the corresponding time interval: because of the hyperbolic shape of this curve, it can be seen that the number of nucleotides required and hence the sequencing effort needed do not rise directly proportionately to the decrease in time but as an inverse relationship. Nodes separated by a long time interval will be consistently resolved with very short sequences while those separated by short intervals will require a disproportionate amount of primary sequence data. This simple rule explains many of the results that have appeared through the years in the phylogenetic literature. For example, it is striking that from the early days of molecular phylogeny to the present time, the improvement in the resolution of some points of the phylogeny of vertebrates, as seen through the globin genes, has been modest (see for example Goodman et al., 1987). These, in fact, are usually the same nodes that resisted analysis using 18S or 28S rRNA (Stock et al., 1991; Lê et al., 1993) even when relatively large species samples were used such as in the study of Lê et al. (1993). In all these cases, one is probably in the ascending portion of the hyperbola where a considerable sequencing effort is required for a small improvement in resolution. Probably the clearest example of such a situation is in the case of the man-chimpanzee-gorilla tritomy where a considerable increase in data only modestly improved the quality of the resolution (Holmquist et al., 1988a,b; Felsenstein, 1988).

The resolution of difficult nodes using the same type of sequence information will therefore, in most cases, require a disproportionate experimental effort, usually out of reach. There is an alternative to this strategy, however, which displays a great similarity to that used in traditional comparative anatomy. This involves carrying out comparative anatomy of genomes. The principle is that rare genomic events such as duplications, transpositions, rearrangements, grouping of genes in the form of operons, mitochondrial genetic code, etc. constitute, in some cases, excellent phylogenetic markers. For example, the physical organisation of the genes of the HOM/Hox complexes (see Kappen and Ruddle, 1993 for review) provides several characters that can be used as strong phylogenetic indicators. These rare events, however, are difficult to exploit for two reasons: the methods for using them in a phylogenetic framework are only starting to be developed and the amount of information (number of species) is scarce. Major breakthroughs in collecting this type of data and in the treatment of the data have, nevertheless, recently appeared. For example, Sankoff et al. (1992) have used gene order in mitochondrial DNA to carry out broad scale phylogenies. The authors propose a measure of gene order rearrangement as well as an algorithm and a software to compute it. Such a method could now be applied to a variety of gene complexes for which qualitative information on gene order and organisation is available (many bacterial operons for example). A similar type of approach can be expected to be applicable to the extensive gene order and physical organisation type of information now emerging from the big genome projects. Indeed, the very recent determination of a long portion of chromosome of *C. elegans* is already yielding much information of this type (Wilson et al., 1994).

We especially thank our colleagues Hervé Le Guyader and Guillaume Lecointre for many discussions and contributions to the ideas expressed in this paper. We thank the laboratory of D. Dacunha-Castelle (Statistiques appliquées, Université Paris-Sud, Orsay) for the GenStat program, and are grateful to Max Telford for his critical reading of the text. This work was supported by grants from the CNRS, the Université Paris-Sud and DRED (Direction de la Recherche et des Etudes Doctorales) for computing equipment. Cécile Couanon is thanked for expert assistance in the preparation of the manuscript.

REFERENCES

Adoutte, A. and Philippe, H. (1993). The major lines of metazoan evolution: Summary of traditional evidence and lessons from ribosomal RNA sequence analysis. In *Comparative Molecular Neurobiology* (ed. Y. Pichon), pp. 1-30. Basel:Birkhäuser.

Akam, M. (1989). Hox and HOM: homologous gene clusters in insects and vertebrates. *Cell* **57**, 347-349.

Bartels, J. L., Murtha, M. T. and Ruddle, F. H. (1993). Multiple *Hox/HOM*-class homeoboxes in platyhelminthes. *Mol. Phyl. Evol.* **2**, 143-151.

Benton, M. J. (1990). *Vertebrate Palaeontology.* London: Harper Collins Academic.

Benton, M. J. (1994). Palaeontological data and identifying mass extinctions. *Trends Ecol. Evol.* **9**, 181-185.

Bowring, S. A., Grotzinger, J. P., Isachsen, C. E., Knoll, A. H., Pelechaty, S. M. and Kolosov, P. (1993). Calibrating rates of early Cambrian evolution. *Science* **261**, 1293-1298.

Chenuil, A. (1993). Etude des relations de parenté entre les principaux groupes d'invertébrés protostomiens par amplification, séquençage et comparaison de portions du gène de l'ARN 28S. Thesis, Université Sci. Tech. Languedoc, Montpellier, France.

Christen, R., Ratto, A., Baroin, A., Perasso, R., Grell, K. G. and Adoutte, A. (1991). An analysis of the origin of metazoans, using comparisons of partial sequences of the 28S rRNA, reveals an early emergence of triploblasts. *EMBO J.* **10**, 499-503.

Conway Morris, S. (1993). The fossil record and the early evolution of the Metazoa. *Nature* **361**, 219-225.

Conway Morris, S. (1994). Why molecular biology needs palaeontology. *Development* **120**, **Supplement** 1-13.

Darwin, C. (1859). *The Origin of Species* (6th edition, 1872) London: John Murray.

Erwin, D. H. (1991). Metazoan phylogeny and the Cambrian explosion. *Trends Ecol. Evol.* **6**, 131-134

Felsenstein, J. (1978). Cases in which parsimony or compatibility methods will be positively misleading. *Syst. Zool.* **27**, 401-410.

Felsenstein, J. (1985). Confidence limits on phylogenies: An approach using the bootstrap. *Evolution* **39**, 783-791.

Felsenstein, J. (1988). Perils of molecular introspection. *Nature* **335**, 118.

Felsenstein, J. and Kishino, H. (1993). Is there something wrong with the bootstrap on phylogenies? A reply to Hillis and Bull. *Syst. Biol.* **42**, 193-200.

Field, K. G., Olsen, G. J., Lane, D. J., Giovannoni, S. J., Ghiselin, M. T., Raff, E. C., Pace, N. R. and Raff, R. A. (1988). Molecular phylogeny of the animal kingdom. *Science* **239**, 748-753.

Goodman, M., Miyamoto, M. M. and Czelusniak, J. (1987). Pattern and process in vertebrate phylogeny revealed by coevolution of molecules and morphologies. *In* Molecules and Morphology in Evolution: Conflict or Compromise? (ed. C. Patterson) pp. 141-176. Cambridge: Cambridge University Press.

Hendy, M. D. and Penny, D. (1989). A framework for the quantitative study of evolutionary trees. *Syst. Zool.* **38**, 297-309.

Hillis, D. M. and Bull, J. J. (1993). An empirical test of bootstrapping as a method for assessing confidence in phylogenetic analysis. *Syst. Biol.* **42**, 182-192.

Holmquist, R., Miyamoto, M. M. and Goodman, M. (1988a). Higher-primate phylogeny – Why can't we decide? *Mol. Biol. Evol.* **5**, 201-216.

Holmquist, R., Miyamoto, M. M. and Goodman, M. (1988b). Analysis of Higher-Primate phylogeny from transversion differences in nuclear and mitochondrial DNA by Lake's methods of evolutionary parsimony and operator metrics. *Mol. Biol. Evol.* **5**, 217-236.

Jacob, F. (1981). Le jeu des possibles. Fayard editor.

Kappen, C. and Ruddle, F. H. (1993). Evolution of a regularoty gene family: HOM/HOX genes. *Cur. Opin. Genet. Dev.* **3**, 931-938.

Kenyon, C. and Wang, B. (1991). A cluster of *Antennapedia*-class homeobox genes in a nonsegmented animal. *Science* **253**, 516-517.

Lake, J. A. (1990). Origin of the metazoa. *Proc. Natl. Acad. Sci. USA* **87**, 763-766.

Lê, H. L. V., Lecointre, G. and Perasso R. (1993). A 28S rRNA-based phylogeny of the Gnathostomes: first steps in the analysis of conflict and congruence with morphologically based cladograms. *Mol. Phyl. Evol.* **2**, 31-51.

Lecointre, G., Philippe, H., Lê, H. L. V. and Le Guyader, H. (1993). Species sampling has a major impact on phylogenetic inference. *Mol. Phyl. Evol.* **2**, 205-224.

Lecointre, G., Philippe, H., Lê, H. L. V. and Le Guyader, H. (1994). How many nucleotides are required to resolve a phylogenetic problem? The use of a new statistical method applicable to available sequences. *Mol. Phyl. Evol.* (in press).

Oliver, G., Vispo, M., Mailhos, A., Martinez, C., Sosa-Pineda, B., Fielitz, W. and Ehrlich, R. (1992). Homeoboxes in flatworms. *Gene* **121**, 337-342.

Patterson, C. (1989). Phylogenetic relations of major groups: conclusions and prospects. In *Hierarchy of Life. Molecules and Morphology in Phylogenetic analysis.* (ed. B. Fernholm, K. Bremer and Jörnvall), pp. 471-488. Amsterdam: Excerpta Medica.

Philippe, H. (1993). MUST, a computer package of Management Utilities for Sequences and Trees. *Nucl. Acids Res.* **21**, 5264-5272.

Philippe, H. and Adoutte, A. (1994). What can phylogenetic patterns tell us about the evolutionary processes generating biodiversity ? In *Aspects of the Genesis and Maintenance of Biological Diversity* (ed. M. Hochberg, J. Clobert and R. Barbault), Oxford: Oxford University Press (in press).

Philippe, H. and Douzery, E. (1994). Quartet approach in molecular phylogeny: a note of caution as examplified by the Cetacea/Artiodactyla relationships. *J. Mam. Evol.* (in press).

Philippe, H., Sörhannus, U., Baroin, A., Perasso R., Gasse F. and Adoutte A. (1994). Comparison of molecular and paleontological data in diatoms suggests a major gap in the fossil record. *J. Evol. Biol.* **7**, 247-265.

Raff, R. A., Field, K. G., Olsen, G. J., Giovannoni, S. J., Lane, D. J., Ghiselin, M. T., Pace, N. R. and Raff, E. C. (1989). Metazoan phylogeny based on analysis of 18S ribosomal RNA. In *Hierarchy of Life. Molecules and Morphology in Phylogenetic Analysis.* (eds. B. Fernholm, K.) pp. 247-260.Amsterdam: Bremer and Jörnvall, Excerpta Medica.

Rice, E. L., Roddick, D. and Singh, R. K. (1993). A comparison of molluscan (Bivalvia) phylogenies based on palaeontological and molecular data. *Mol. Marine Biol. Biotechnol.* **2**, 137-146.

Saitou, N. and Nei, M. (1987). The Neighbor-Joining method: A new method for reconstructing phylogenetic trees. *Mol. Biol. Evol.* **4**, 406-425.

Sankoff, D., Leduc, G., Antoine, N., Paguin, B., Lang, B. F. and Cedergren, R. (1992). Gene order comparisons for phylogenetic inference: Evolution of the mitochondrial genome. *Proc. Natl. Acad. Sci. USA* **89**, 6575-6579.

Sarich, V. M. and Wilson, A. C. (1973). Generation time and genomic evolution in primates. *Science* **179**, 1144-1147

Schierwater, B., Murtha, M., Dick, M., Ruddle, F. H. and Buss, L. W. (1991). Homeoboxes in cnidarians. *J. Exp. Zool.* **260**, 413-416.

Schubert, F. R., Nieselt-Struwe, K. and Gruss, P. (1993). The Antennapedia-type homeobox genes have evolved from three precursors separated early in metazoan evolution. *Proc. Natl. Acad. Sci. USA* **90**, 143-147

Shenk, M. A., Bode, H. R. and Steele R. E. (1993). Expression of Cnox-2, a HOM/HOX homeobox gene in hydra, is correlated with axial pattern formation. *Development* **117**, 657-667.

Shenk, M. A. and Steele R. E. (1993). A molecular snapshot of the metazoan 'Eve'. *Trends Biol. Sci.* **18**, 459-463.

Slack, J. M. W., Holland, P. W. H. and Graham, C. F. (1993). The zootype and the phylotypic stage. *Nature* **361**, 490-492.

Stock, D. W., Gibbons, J. K. and Whitt, G. S. (1991). Strengths and limitations of molecular sequence comparisons for inferring the phylogeny of the major groups of fishes. *J. Fish. Biol.* **39** (suppl. A), 225-236.

Swofford, D. L. (1991). PAUP: Phylogenetic Analysis Using Parsimony (Illinois Natural History Survey, Champaign, IL), Version 3.0 s.

Swofford, D. L. and Olsen, G. J. (1990). Phylogeny reconstruction. In *Molecular Systematics.* (ed. Hillis, D. M. and Moritz, C.), pp. 411-501.

Tateno, Y., Takazaki, N. and Nei, M. (1994). Relative efficiencies of the maximum-likelihood, neighbor-joining and maximum-parsimony methods when substitution rate varies with site. *Mol. Biol. Evol.* **11**, 261-277.

Turbeville, J. M., Pfeifer, D. M., Field, K. G. and Raff, R. A. (1991). The phylogenetic status of arthropods, as inferred from 18S rRNA sequences. *Mol. Biol. Evol.* **8**, 669-686.

Wada, H. and Satoh, N. (1994). Details of the evolutionary history from invertebrates to vertebrates, as deduced from the sequences of 18S rDNA. *Proc. Natl. Acad. Sci. USA*, **91**, 1801-1804.

Webster, P. J. and Mansour, T. E. (1992). Conserved classes of homeodomains in *Schistosoma mansoni*, an early bilateral metazoan. *Mech. Dev.* **38**, 25-32.

Wilson, R., Ainscough, R., Anderson, K. et al., and Wohldman, P. (1994). 2.2 Mb of contiguous nucleotide sequence from chromosome III of *C. elegans. Nature* **368**, 32-38.

Winnepennickx, B., Backeljau, T., van de Peer, Y. and De Wachter, R. (1992). Structure of the small ribosomal subunit RNA of the pulmonate snail. *Limicolaria kambeul*, and phylogenetic analysis of the Metazoa. *FEBS lett.* **309**, 123-126.

Zharkikh, A. and Li, W. H. (1992a). Statistical properties of bootstrap estimation of phylogenetic variability from nucleotide sequences: I. Four taxa with a molecular clock. *Mol. Biol. Evol.* **9**, 1119-1147.

Zharkikh, A. and Li, W. H. (1992b). Statistical properties of bootstrap estimation of phylogenetic variability from nucleotide sequences: II. Four taxa without a molecular clock. *J. Mol. Evol.* **35**, 356-366.

Development 1994 Supplement, 27-33 (1994)
Printed in Great Britain © The Company of Biologists Limited 1994

Protein families in the metazoan genome

Cyrus Chothia

MRC Laboratory of Molecular Biology and the Cambridge Centre for Protein Engineering, Hills Road, Cambridge, CB2 2QH, UK.

SUMMARY

The evolution of development involves the development of new proteins. Estimates based on the initial results of the genome projects, and on the data banks of protein sequences and structures, suggest that the large majority of proteins come from no more than one thousand families. Members of a family are descended from a common ancestor.

Protein families evolve by gene duplication and mutation. Mutations change the conformation of the peripheral regions of proteins; i.e. the regions that are involved, at least in part, in their function. If mutations proceed until only 20% of the residues in related proteins are identical, it is common for the conformational changes to affect half the structure.

Most of the proteins involved in the interactions of cells, and in their assembly to form multicellular organisms, are mosaic proteins. These are large and have a modular structure, in that they are built of sets of homologous domains that are drawn from a relatively small number of protein families. Patthy's model for the evolution of mosaic proteins describes how they arose through the insertion of introns into genes, gene duplications and intronic recombination.

The rates of progress in the genome sequencing projects, and in protein structure analyses, means that in a few years we will have a fairly complete outline description of the molecules responsible for the structure and function of organisms at several different levels of developmental complexity. This should make a major contribution to our understanding of the evolution of development.

Key words: protein evolution, exons, introns, mosaic proteins

INTRODUCTION

An examination of the number of genes found in different organisms shows clearly how the evolution of development involves the development of new proteins:

Bacteria: *Escherichia coli*	4,000
Yeast: *Saccharomyces cerevisae*	6,000
Nematode: *Caenorhabditis elegans*	18,000
Humans	65,000

(the numbers for *C. elegans* and humans are from Wilson et al. (1994) and Fields et al. (1994) respectively). Thus, an understanding of the process by which the protein repertoire has grown and acquired new properties during the course of evolution is an essential component of a general understanding of the evolution of development.

Here I review three recent advances in our understanding of protein evolution. First, I discuss the evidence that suggests that most, or all, proteins belong to a relatively small number of families whose members are descended from a common ancestor. Second, I show that members of a family can diverge to a point where they have very few sequence identities and half their structures have different folds. Third, I describe Patthy's model for the formation of the mosaic proteins, which have played a central role the evolution of multicellular organisms. Lastly, the implications of these discoveries are discussed.

NUMBER OF PROTEIN FAMILIES

An analysis of the protein sequences and structures that were available in 1992 gave a rough estimate of the number of protein families (Chothia, 1992). The data used to make this estimate comprised (i) the initial results of the genome projects; (ii) the data bank of known protein sequences and (iii) the data bank of known protein structures. Examination of these data showed that:

(1) Of the sequences produced by each of the different genome projects, close to one third had a clear homology to an entry that was already present in the data bank of protein sequences (see Table 1). It is reasonable to assume that sequences produced by the genome project represent a random sample and so this result suggests that one third of all protein families had a representative in the sequence data bank.

(2) Of the entries in the sequence data bank, 28% matched, with a residue identity of at least 25%, the sequence of one of the entries in the protein structure data bank. (These figures come from Sander and Schneider, 1991 and personal communication quoted in Chothia, 1992, who determined the proportion of the sequences in the EMBL/SwissProt data bank that are homologous to the sequences of the proteins in the Brookhaven protein structure data bank.) The figures suggest that about one quarter of protein sequences belong to a family for which there is a known structure.

(3) The Brookhaven protein structure data bank contained

Table 1. The sequences from genome projects that are related to other previously known sequences

Source	Total number of genes	Genes related to those previously known	Reference
A. Genome Projects			
Caenorhabditis elegans			
chromosome III (part)	32	14 (44%)	Sulston et al (1992)
Yeast			
chromosome III	182	52-66 (29-36%)	Oliver et al. (1992)
chromosome IX (part)	46	15 (33%)	Barrell, Smith and Brown*
B. Large libraries of expressed genes			
Human brain			
I	~1,400	406 (~30%)	Adams et al. (1992)
II	1,531	725 (47%)	Adams et al. (1993)
Caenorhabditis elegans			
St Louis	1,517	512 (34%)	Waterson et al. (1992)
NIH	585	210 (36%)	McCombie et al. (1992)

*Unpublished results quoted in Chothia (1992).

one or more representatives of some 120 different protein families (Pascarella and Argos, 1992; Chothia, 1992).

Put together these results imply that there are some 1,500 different protein families (Table 2).

Now this calculation assumes that sequence comparisons can give a complete picture of family membership. This is not the case. It has become clear, from the structures determined by X-ray crystallography and NMR, that proteins can evolve to the point where, though they continue to share the same fold, their sequence identities are no greater than that of two randomly selected sequences (a review of recent cases is found in Murzin and Chothia, 1992). This means that sequence comparisons with reasonable thresholds underestimate the extent to which proteins are related.

If we assume that the current methods of sequence comparisons can find 80% of the proteins in one family and adjust our calculations accordingly, the estimated number of protein families drops to 1000 (see Table 2). In fact, anecdotal evidence from recently determined protein structures suggest that sequence comparisons are not this efficient. Thus a conservative view of the current evidence is that the large majority of proteins come from no more than 1000 families.

Since the initial results of the genome projects, five further reports have appeared which are in accord with this estimate. Glaser et al. (1993) sequenced a 97 kb region of the *Bacillus subtilis* genome. They found 92 open reading frames; 42 of these coded for proteins homologous to entries already in data banks. Adams et al. (1993) partially sequenced the genes for about 1500 proteins from human brain. Half of these were found to have sequences homologous to those previously known. Wilson et al. (1994) reported on 2.2 megabases of contiguous nucleotide sequence from chromosome III of the nematode *Caenorhabditis elegans*. This section of the chromosome contains 483 genes of which 40% are related to previously known sequences or code for tRNAs. These groups used the current standard sequence matching techniques.

Using matching techniques that are more sensitive to the significance of low homologies than those used by previous workers and a larger data base of sequences and structures gave a higher proportion of matches in the two more recent reports. Koonin et al. (1994) re-examined the yeast chromosome III sequences. In

Table 2. Calculation of the number of protein families

Proportion of genome sequences that have a related sequence in the sequence data bank:		one third*
Proportion of sequences in the data bank related to a protein of known structure:		one quarter*
Number of families represented by the known structures:		120*
Number of protein familes		
1. Assuming current sequences comparisons detect all members of protein families:	$3 \times 4 \times 120$:	1500
2. Assuming sequence comparisons detect 80% of family members:	$3 \times 0.8 \times 4 \times 0.8 \times 120$:	1000

* Values given here are for 1992 data; see text for details and references.

the original report, 29-36% of the 182 sequences were reported to be similar to those previously known (Table 1; Oliver et al., 1992). Koonin et al. (1994) showed that the products of 61% of these genes have a significant similarity to an entry in the sequence data banks and 19% are similar to a protein of known three-dimensional structure. Dujon et al. (1994) reported the complete sequence of yeast chromosome XI. Of its 331 open reading frames, 67% coded for either known yeast proteins or were homologous to other proteins in the data banks.

The concept of a protein family has a straight forward application to small and medium sized proteins that are built of just one domain. Large proteins, however, are usually built of several domains that are usually not homologous to one another but are often homologous to domains found in other quite different proteins. Two examples are the extracellular neural cell adhesion molecule and the intracellular muscle protein titin, which are built mostly from different combinations of immunoglobulin superfamily and fibronectin type III domains (see Fig. 1). Proteins of this kind can best be described as being built of components that come from different families of protein domains.

It should be emphasised that the number of 1000 families is very much smaller than the number of folds that are possible within the limitations of the physics and chemistry of proteins. It is difficult to calculate the number of possible folds but this point becomes apparent if we realise that taking just the 200 folds that are currently known, and changing the connections between their secondary structures (within the limits of the protein chain topology rules), would give tens of thousands of different folds (A.G. Murzin, unpublished calculation).

THE EVOLUTION OF PROTEINS

In this section I discuss the general effects of mutations on the structure of proteins and how new proteins are made from combinations of old proteins.

Two particular proteins are used to illustrate certain general points. Both are members of the immunoglobulin superfamily (Fig. 1). In prokaryotes single domains that belong to this family are attached to glycosyltransferases and cellulase where they may be involved in carbohydrate recognition (Klein and Schultz, 1991; Juy et al., 1992). The eukaryote branch of the family is very large and has members that are involved in cell-cell adhesion, in the structural organisation and regulation of muscle and in the immune system. These members of the family are usually built from several domains of roughly 100 residues each and with homologous sequences and structures. In some cases these domains combine with domains from other families.

The two members of the superfamily used below to illustrate various points are the neural cell adhesion molecule (NCAM), which was mentioned above and which is involved in the development of the nervous system, and the antibody molecule (Fig. 1).

Evolution of structural diversity

The development of specialised or new functions, through the duplication of genes and their subsequent mutation, became apparent from the comparisons of the amino acid sequences of the globins (Ingram, 1961), of trypsin and chymotrypsin (Walsh and Neurath, 1964; Hartley et al., 1965) and of lysozyme and α-lactalbumin (Brew et al., 1967).

Later it was found that this process can go beyond the point where the diverged proteins have sequences similarities that are easily recognisable. In such cases, the common evolutionary origin only becomes clear from the similarities in the details of the three dimensional structures of the proteins. A recent striking example is the discovery that the muscle protein actin, hexokinase and the ATPase domain of the heat shock cognate protein, HSC70, belong to the same protein family. Actin and the HSC70 domain have, respectively, 375 and 386 residues. Of these, 240 have the same conformation though only 39 have identical side chains (Flaherty et al., 1991).

In general, proteins respond to changes in sequence by changes in structure (Lesk and Chothia, 1980; Chothia and Lesk, 1987). As the stability of proteins is low: the folded forms are 5-15 Kcals more stable than the denatured forms, the structural changes produced by mutations can involve only small changes in energy. This means, in turn, that structural changes in evolution occur through a series of small incremental steps. Though the individual changes are small, their cumulative effects can be large. An example of this can be seen when we examine the structural differences of the variable and constant domains in antibodies.

Antibodies are built from two types of domains: the variable (V), which form the antigen binding site and the constant (C) which are involved in various effector functions (Fig. 1). V and C domains evolved from a common ancestor, which probably had a structure intermediate between the two (Harpaz and Chothia, 1994). They have structures in which the polypeptide chain runs back and forth to form two β-sheets which pack face to face (Fig. 2). Although different V domains have small differences in structure, the two β-sheets that form the core of their structures is very largely conserved (Fig. 3A). Similarly the two β-sheets that form the core of different C domains are also conserved (Fig. 3B).

A comparison of the conserved core of V domains with that of C domains shows that there is a central region common to both but around this they have quite different structures. This is shown graphically in Fig. 3C where the conserved cores of the V and C domains are shown superimposed.

The structural differences can be measured quantitatively. Typically, V and C domains have 110 and 100 residues respec-

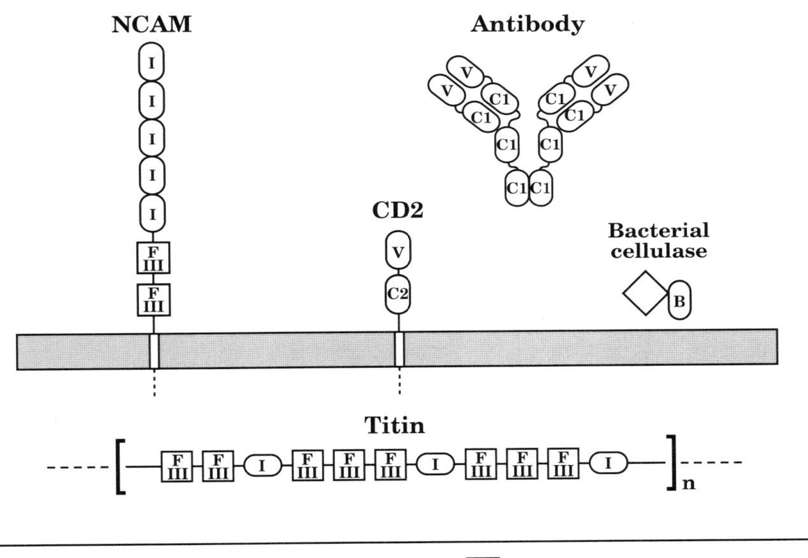

Fig. 1. Structures of five members of the immunoglobulin superfamily. Immunoglobulin superfamily domains contain approximately one hundred residues and have homologous sequences and structures. Groups of domains, that have structures more similar to one another, than they are to other members of the family, are grouped into sets. Antibodies are built of V and C1 set domains (Williams and Barclay, 1988). The CD2 adhesion molecule is built from V and C2 set domains (Jones et al., 1992) whereas the cell surface neural cell adhesion molecule (NCAM) contains five I set domains (Harpaz and Chothia, 1994) and two type III fibronectin domains that are not members of the immunoglobulin superfamily. (Some forms of the protein also contain a muscle specific domain (MSD, see Fig. 6) a transmembrane helix and a cytoplasmic domain) The muscle protein titin is also built from I set and fibronectin domains but in a different arrangement to that seen in the NCAMs. Bacterial cellulase has one B set domain linked to a non-homologous catalytic domain (Juy et al., 1992).

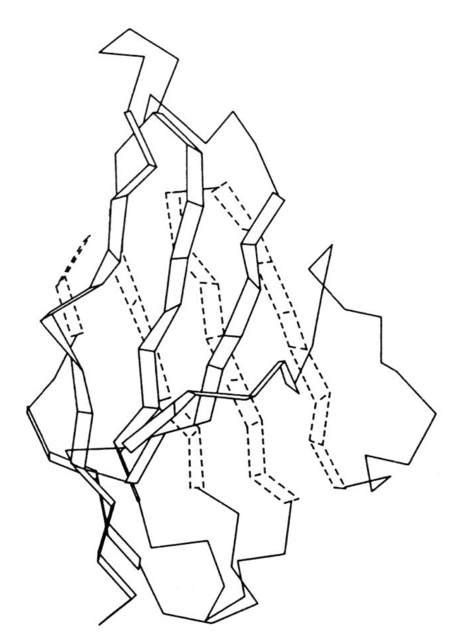

Fig. 2. The fold of the polypeptide chain in an immunoglobulin variable domain. The regions drawn as ribbons hydrogen bond to each other to form two β-sheets: one drawn in broken lines and the other in continuous lines. See also Fig. 3.

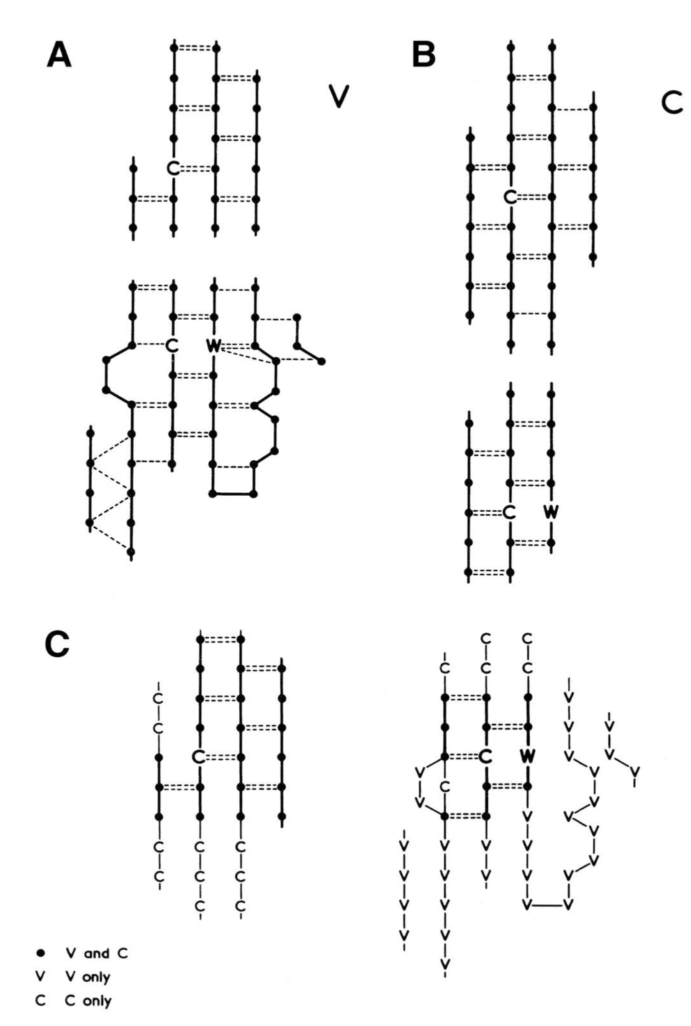

Fig. 3. The β-sheets structures in variable (V) and constant (C) domains. Small filled circles represent residues; thick lines, the polypeptide chain, and broken lines, the interchain hydrogen bonds. The sites of conserved cysteine and tryptophan residues are indicated by C and W. A shows the β-sheets structures common to all V domains and (B) shows the same for C domains. The superimposition of these structures (C) shows that only the central regions are common to both types of domains; outside this region the structures tend to be different.

tively. Superposing individual V and C structures shows that typically, some 55 of the residues in each structure have the same conformation. That is 50% of the V domain (55/110) and 55% of the constant domain (55/100) have the same conformation but the other regions in the two proteins differ in conformations. In regions with the same conformation the proportion of residues that are identical in the V and C domains is about 15%.

The same calculation was carried out on 32 pairs of homologous proteins that come from eight protein families (Chothia and Lesk, 1986). The results are shown in Fig. 4. They demonstrate that when related proteins have 40% or more identical residues the extent of the structural changes tend to be small. But, when the sequence changes have progressed so that only 20% of the residues are identical, it is common for the structural changes to be so extensive that a core comprising only about half of each protein retains the same structure (Fig. 4).

The regions that undergo changes in conformation are minor elements of secondary structure that are on the surface, and the loops of peptide that link the major elements of secondary structure. In most proteins it is these regions that form, at least in part, the structures responsible for activity and specificity. Changes in these regions change functional properties. In the evolution of enzymes they often modify specificity; changes in the catalytic mechanisms, however, are rare but do occur occasionally (Murzin, 1993).

Evolution of mosaic proteins

The discovery that eukaryotic genes, and coding regions within genes (exons), are often separated by non-coding regions (introns) led to the proposal that new proteins can be created by shuffling exons using the intron regions for recombination (Gilbert, 1978; Blake, 1978). The current evidence is strongly against the proposal that entirely new proteins are made by shuffling individual exons (see Patthy, 1991a,b, 1994 and references therein). It is clear, however, that whole protein domains (which may well have introns within their genes) are shuffled to create new structures.

Intronic recombination has been involved in the formation of cell adhesion molecules, cell surface receptors, proteins of the body fluids, extracellular matrix proteins and proteins involved in the structural organisation and regulation of muscle. These have been described as mosaic proteins because they are built of several discrete domains or modules that are the same type or a few different types. (Modules of the same type have homologous amino acid sequences and structures). The neural cell adhesion molecules (NCAMs) are an example of such proteins: they have different forms but all have an extracellular portion that contains seven domains (modules): five are members of the immunoglobulin superfamily and two are fibronectin type III domains (Fig. 1). Some forms also contain the small muscle specific domain (MSD).

A model for the evolution of mosaic proteins has been

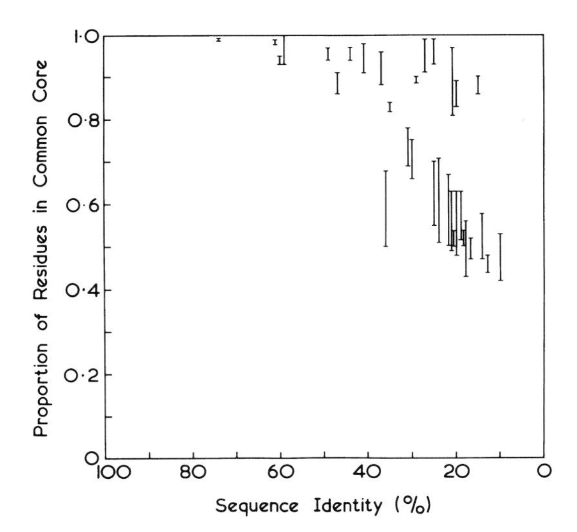

Fig. 4. For 32 pairs of homologous proteins from eight protein families, this plot shows the proportion of their structures that have the same fold, as a function of their sequence identity. If two proteins with n_1 and n_2 residues have c residues with the same fold (see Fig. 3), the proportions with the same fold are c/n_1 and c/n_2. If, of the c residues that have the same fold, b are identical, the sequence identity is 100b/c %. (This figure is adapted from Chothia and Lesk (1986) where details are given for each of the structural comparisons.)

proposed by Patthy (1991b, 1994) and is shown in Fig. 5. Starting from a gene for a cytoplasmic protein, the model proposes the following series of steps that lead to the formation of extra-cellular mosaic proteins: (i) the gain of a secretory peptide; (ii) insertions of introns at the beginning and end of the gene to form a proto-module; (iii) its duplication, and (iv) intronic recombination with other types of modules.

An intron has a "phase" that is defined by the position of the break it makes in a codon:

One implication of Patthy's model is that the introns at the

Intron phase	0	1	2
Intron position	|	|	|
Codons	T C A	G G A	
Residues	Ser	Gly	

beginning and end of modules must have the same phase: if they did not, duplicated genes, and other genes gained by recombination, would read out of phase. For reasons to do with the mechanisms of intron insertion and splicing, it is expected that the introns used to create mosaic proteins will be phase 1 (see Patthy, 1994).

In Fig. 6 the position and phase of the introns in the gene that codes for the extra-cellular domains of chicken NCAM (Owens et al., 1987; Prediger et al., 1988) are shown. As expected from Patthy's model, each module is separated at its exact boundary by a phase 1 intron.

Introns also occur within modules and in the NCAM gene there are one, two or three within each of the extra-cellular modules (Fig. 6). They occur at non-homologous positions and have different phases: 0, 1 or 2. The introns in domains 1 - 3 and 5 - 6 seem to be insertions that have not been useful in the evolution or function of the protein: they are probably just evolutionary "noise".

Multi-domain proteins can be formed by recombination without the help of introns: this occurs in bacteria. Inspection of the gene structures of the mosaic proteins involved in the formation of the metazoa, however, shows that all, or almost all, of those that are currently known arose through intronic recombination (Patthy, 1991b, 1994). Indeed Patthy (1994) has argued that the development of spliceosomal introns and the accumulation of a critical mass of module types made a crucial contribution to the formation of metazoa and their radiation.

Proteins with developmental isoforms

The product of a gene that is built of several exons can be modified by having alternative patterns of splicing that result in certain exons, and therefore peptides or domains, being expressed in some forms of the protein but not in others. These changes in sequence produce changes in functional properties, They produce versions of a protein that are specific to different types of cell types or to different stages of development (Smith et al., 1989; Maniatis, 1991).

Alternative patterns of splicing occur in the expression of the NCAM gene. For the extracellular domains it involves a ten residue exon bracketed by the two introns within domain 4 and the exons that form the small muscle specific domain (MSD) domain (Fig. 6). (Outside this region of the NCAM gene there are additional exons that code for a trans-membrane peptide and a cytoplasmic domain. These also have alternative splicing patterns.)

Alternative patterns of splicing are often involved in the specialisation or in modification of function. Thus the NCAM forms that express the MSD domain are specific for muscle cells (Dickson et al. 1987). It has been proposed that the expression of the peptide in domain 4 changes the function of NCAM from a molecule that promotes morphological plasticity to one that maintains stable cell-cell contacts (Doherty et al., 1992).

Expression of the additional exons do not affect the integrity of the NCAM domain structure. The extra peptide in domain

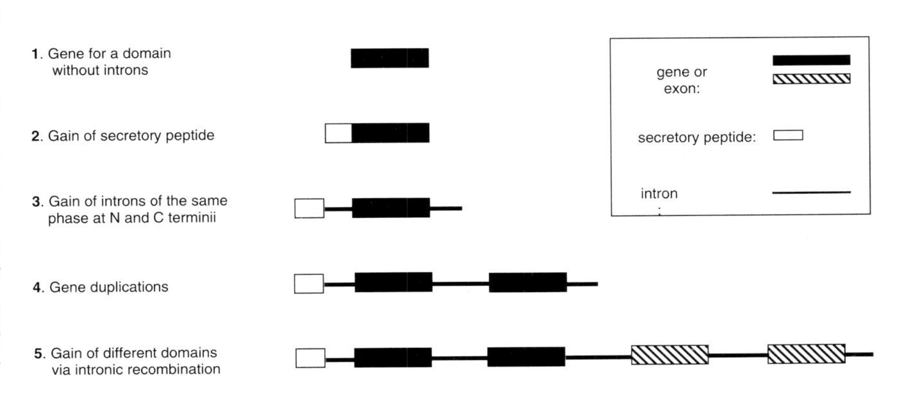

Fig. 5. The model put forward by Patthy (1991b, 1994) for the evolution of mosaic proteins (see text).

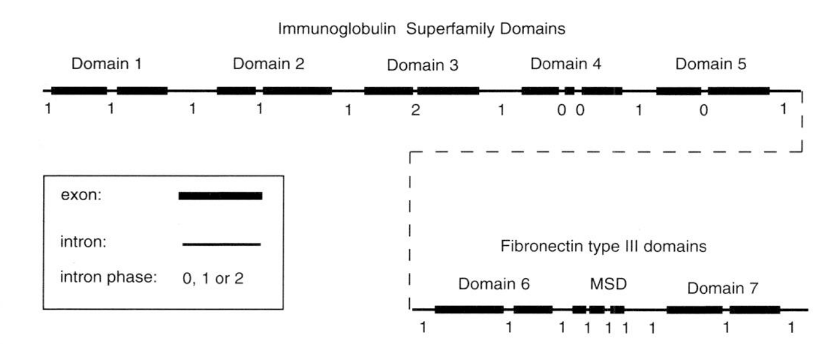

Exons and introns in the region of the gene that codes for the extra-cellular domains of neural cell adhesion molecules (NCAMs)

Fig. 6. The structure of the part of the chicken neural cell adhesion molecule (NCAM) gene that codes for the extra-cellular domains (Owens et al., 1987; Prediger et al., 1988). Note that the exon and intron regions are NOT drawn to scale. This region of the gene codes for five immunoglobulin superfamily modules or domains (numbered 1 to 5) and two type III fibronectin domains (6 and 7). Forms of the protein found in muscle also contain the small muscle specific domain MSD domain (see text).

4 is inserted in a peptide loop on the surface of the protein. The MSD domain is inserted between the two fibronectin domains (Harpaz and Chothia, unpublished data).

CONCLUSION: PROTEIN STRUCTURES AND THE EVOLUTION OF DEVELOPMENT

The evolution of development has involved the development of new proteins. The evidence reviewed here points to most current proteins being the descendants of no more than 1000 ancestors. The process by which these descendants were produced involved gene duplications followed by mutations and, for large proteins, gene fusion. In metazoa, gene fusion has largely occurred through intronic recombination of the genes for whole protein domains.

The relatively small number of protein families in biology does not arise from any intrinsic physical or chemical limitation on number of protein folds but from history. In the earliest stages of evolution, there seems to have developed a set of proteins with a range of functional properties that was sufficiently wide for it to be much easier to evolve new proteins by the duplication, modification and recombination of old proteins than by the ab initio invention of new ones.

The view of protein evolution presented here suggests a programme for understanding the molecular basis of the evolution of development. This involves: (1) the complete description of the sequences that form organisms at different levels of developmental complexity; (2) the classification of the sequences into families, and (3) the determination of the evolutionary history of the appearance of new properties within protein families.

The dates proposed for the completion of the most advanced of the present genome sequencing projects mean that within a few years we will have a complete description of the sequences that form organisms at several different levels of developmental complexity:

Genome project	Number of genes	Estimated date of completion
Bacteria: *Bacillus subtilus*	4,000	1997-8
Yeast: *Saccharomyces cerevisae*	6,000	1996
Nematode: *Caenorhabditis elegans*	18,000	1998
Humans	65,000	2005 (?)

A detailed knowledge of protein structure will play a central role in the interpretation of this sequence data. The relationship of strongly diverged proteins can not be recognised from sequence similarities at present but it can be seen, in many cases, from their three dimensional structures. Structural descriptions also make it easier to understand how evolutionary changes have modified the properties of proteins and produced new ones.

At the moment more than 270 protein structures are being determined each year by X-ray crystallography and NMR, and the rate is increasing (Hendrickson and Wüthrich, 1993). A significant proportion of these structures are homologous to those known from previous work. The current rate at which the structures for new families are becoming known, however, means the combination of crystallography, NMR and molecular modelling will produce, at least in outline, structures for the large majority protein families within a few years.

Thus, the combined information from the genome projects and protein structures should allow the determination of the evolutionary history of protein families and hence a description of the molecular events that have produced the evolution of development.

I thank Samantha Barré for comments on the paper.

REFERENCES

Adams, M. D., Dubnick, M., Kerlavage, A. R., Moreno, R., Kelley, J. M., Utterback, T. R., Nagle, J. W., Fields, C. and Venter, J. C. (1992). Sequence identification of 2,375 human brain genes. *Nature* **355**, 632-634.

Adams, M. D., Kerlavage, A. R., Fields, C. and Venter, J. C. (1993). 3,400 new expressed sequence identify diversity of transcripts in human brain. *Nature Genetics* **4**, 256-267.

Blake, C. C. F. (1978). Do genes-in-pieces imply proteins in pieces? *Nature* **273,** 267.

Brew, K., Vanaman, T. C. and Hill, R. L. (1967). Comparison of the amino-acid sequence of bovine α-lactalbumin and hen's egg white lysozyme. *J. Biol. Chem.* **242,** 3747-3749.

Chothia, C. (1992). One thousand families for the molecular biologist. *Nature* **357,** 543-544.

Chothia, C. and Lesk, A. M. (1986). The relation between the divergence of sequence and structure in proteins. *EMBO J.* **5,** 823-826.

Chothia, C. and Lesk, A. M. (1987) The evolution of protein structures. *Cold Spring Harbor Quant. Biol.* Vol. **LII,** 399-405.

Dickson, G., Gower, H. J., Barton, C. H., Prentice, H. M., Elsom, V. L., Moore, S. E., Cox, R. D., Quinn, C., Putt, W. and Walsh, F. S. (1987). Human muscle neural cell adhesion molecule (N-CAM): identification of a muscle-specific sequence in the extracellular domain. *Cell* **50,** 1119-1130.

Doherty, P., Moolenaar, C. E. C. K., Ashton, S. V., Michalides, R. J. A. M. and Walsh F. S. (1992). The VASE exon down regulates the neurite growth promoting activity of NCAM 140. *Nature* **356,** 791-793.

Dujon, B. Alexandaki, D., Audré, B. and 103 others. (1994). Complete DNA sequence of yeast chromosome XI. *Nature* **369,** 371-378.

Fields, C., Adams, M. D., White, O. and Venter, J. C. (1994). How many genes in the human genome? *Nature Genetics* **7**, 345-346.

Flaherty, K. M., McKay, D. B., Kabsch, W. and Holmes K. C. (1991). Similarity of the three-dimensional structures of actin and the ATPase fragment of a 70-kDa heat shock cognate protein. *Proc. Natl. Acad. Sci. USA* **88,** 5041-5045.

Gilbert, W. (1978). Why genes in pieces? *Nature* **272,** 501.

Glaser, P., Kunst, F., Arnaud, M., Coudart, M.-P., Gonzales, W., Hullo, M.-F., Ionescu, M., Lubochinsky, B., Marcelino, L., Moszer, I., Presecan, E., Santana, M., Schneider, E., Schweizer, J., Vertes, A., Rapoport, G. and Danchin, A. (1993). Bacillus subtilis genome project: cloning and sequencing of the 97 kb region from 325° to 333°. *Mol. Microbiol.* **10**, 371-384.

Harpaz, Y. and Chothia, C. (1994). Many of the immunoglobulin superfamily domains in cell adhesion molecules and surface receptors belong to a new structural set which is close to that containing variable domains. *J. Mol. Biol.* **238,** 528-539.

Hartley, B. S., Brown, J. R., Kauffman, D. L. and Smillie, L. B. (1965). Evolutionary similarities between proteolytic enzymes. *Nature* **207,** 1157-1159.

Hendrickson, W. A. and Wüthrich, K. (1993). In *Macromolecular Structures 1993*. London: Current Biology Ltd.

Ingram, V. (1961). Gene evolution and the haemoglobins. *Nature* **189**, 704-708.

Jones, E. Y., Davis, S. J., Williams, A. F., Harlos, K. and Stuart, D. I. (1992). Crystal structure at 2.8 resolution of a soluable form of the cell adhesion molecule CD2. *Nature* **369**, 232-239.

Juy, M., Amit, A. G., Alzari, P. M., Poljak, R. J., Claeyssens, M., Beguin, P. and Aubert, J.-P. (1992). Three dimensional structure of a thermostable bacterial cellulase. *Nature* **357,** 89-91.

Klein, C. and Schulz, G. E. (1994). Structure of cyclodextrin glycosyltransferase refined at 2 Å resolution. *J. Mol. Biol.* **217,** 737-750.

Koonin, E. V., Bork, P. and Sander, C. (1994). Yeast chromosome III: new gene functions. *EMBO J.* **13,** 493-503.

Lesk, A. M. and Chothia, C. (1980). How different amino acid sequences determine similar protein structures: the structure and evolutionary dynamics of the globins. *J. Mol. Biol.* **136,** 225-270.

Maniatis, T. (1991). Mechanisms of alternative pre-mRNA splicing. *Science* **251,** 33-34.

McCombie, W. R., Adams, M. D., Kelley, J. M., FitzGerald, M. G., Utterback, T. R., Khan, M., Dubnick, M., Kerlavage, A. R., Venter, J. C. and Fields, C. (1992). Caenorhabditis elegans expressed sequence tags identify gene families and potential disease gene homologues. *Nature Genetics* **1,** 124-131.

Murzin, A and Chothia, C. (1992). Protein architecture: new superfamilies. *Curr. Opinion Struct. Biol.* **2,** 895-903.

Murzin, A. (1993). Can homologous proteins evolve different enzymatic activities? *Trends Biochem. Sci.* **18,** 403-405.

Oliver, S. G. van der Aart, Q. J. M., Agostoni-Carbone, M. L. and 144 others. (1992). The complete DNA sequence of yeast chromosome III. *Nature* **357,** 38-46.

Owens, G. S., Edelman, G. M. and Cunningham, B. A. (1987). Organisation of the neural cell adhesion molecule (N-CAM) gene: alternative exon usage as the basis for different membrane associated domains. *Proc Natl. Acad. Sci. USA* **84,** 294-298.

Pascarella, S. and Argo, P. (1992). A data bank merging related protein structures and sequences. *Protein. Eng.* **5,** 121-137.

Patthy, L. (1991a). Exons - original building blocks of proteins? *BioEssay* **13,** 187-192.

Patthy, L. (1991b). Modular exchange principles in proteins. *Curr. Opinion Struct. Biol.* **1,** 351-361.

Patthy, L. (1994). Introns and Exons. *Curr. Opinion Struct. Biol.* **4,** 383-392.

Prediger, E. A., Hoffman, S., Edelman, G. M. and Cunningham, B. A. (1988). Four exons encode a 93-base-pair insert in three neural cell adhesion molecule mRNAs specific for chicken heart and skeletal muscle. *Proc. Natl. Acad. Sci. USA* **85,** 9610-9620.

Sander, C. and Schneider, R. (1991). Database of homology-derived protein structures and structural meaning of sequence alignment. *Proteins* **9,** 56-68.

Smith, C. W. J., Patton, J. G. and Nadal-Ginard, B. (1989). Alternative splicing in the control of gene expression. *Annu. Rev, Genet.* **23,** 527-577.

Sulston, J., Du, Z., Thomas, K., Wilson, R., Hillier, L., Staden, R., Halloran, N., Green, P., Thierry-Mieg, J., Qiu, L., Dear, S., Coulson, A., Craxton, M., Durbin, R., Berks, M., Metzstein, M., Hawkins, T., Ainscough, R. and Waterson, R. (1992). The C. elegans genome sequencing project: a beginning. *Nature* **356,** 37-41.

Walsh, K. A. and Neurath, H. (1964). Trypsinogen and chymotrypsinogen as homologous proteins. *Proc. Natl. Acad. Sci. USA* **52,** 884-888.

Waterson, R., Martin, C., Craxton, M., Huynh, C., Coulson, A., Hillier, L., Durbin, R., Green, P., Showkeen, R., Halloran, N., Metzstein, M., Hawkins, T., Wilson, R., Berks, M., Du, Z., Thomas, K., Thierry-Mieg, J. and Sulston, J (1992). A survey of expressed genes in *Caenorhabditis elegans. Nature Genetics* **1,** 114-123.

Williams, A. F. and Barclay, A. N. (1988) The immunoglobulin superfamily - domains for surface recognition, *Ann. Rev. Immunol.* **6,** 381-405.

Wilson, R., Ainscough, R., Anderson, K. and 52 others. (1994). 2.2 Mb of contiguous nucleotide sequence from chromosome III of C. elegans. *Nature* **368,** 32-38.

Development 1994 Supplement, 35-42 (1994)
Printed in Great Britain © The Company of Biologists Limited 1994

Domain organizations of extracellular matrix proteins and their evolution

Jürgen Engel, Vladimir P. Efimov and Patrik Maurer

Department of Biophysical Chemistry, Biozentrum, University of Basel, Switzerland

SUMMARY

The astonishing diversity in structure and function of extracellular matrix (ECM) proteins originates from different combinations of domains. These are defined as autonomously folding units. Many domains are similar in sequence and structure indicating common ancestry. Evolutionarily homologous domains are, however, often functionally very different, which renders function prediction from sequence difficult. Related and different domains are frequently repeated in the same or in different polypeptide chains. Common assembly domains include α-helical coiled-coil domains and collagen triple helices. Other domains have been shown to be involved in assembly to other ECM proteins or in cell binding and cell signalling. The function of most of the domains, however, remains to be elucidated. ECM proteins are rather recent 'inventions', and most occur either in plants or mammals but not in both. Their creation by domain shuffling involved a number of different mechanisms at the DNA level in which introns played an important role.

Key words: extracellular matrix proteins, protein domains, intron

INTRODUCTION

The extracellular matrix (ECM) is not just the glue between cells as believed for a long time. It is instead a highly elaborate association of proteins, proteoglycans and glycosaminoglycans, each of which has a specialized function in fulfilling the manifold purposes that the ECM has. The main purpose is serving the cell as a substrate for growth and providing a stable structure around them. This is a fundamental precondition for the existence of multicellular organisms. The central systems in eukaryotes (neural, circulatory, digestive and fertilization systems) evolved within and along with the ECM.

The ECM has to serve two masters: it must be a pleasant living space for the cell and a suitable scaffold for functional elements of the organism. To fulfil these purposes, a huge set of proteins and proteoglycans of unusual size and shape have evolved, which furnish the tissue with its distinct features and anchor the cell in its surroundings. Electron microscopy gives insight into a strange microcosm of crosses, spiders, strings of pearls, brushes, dumb-bells, rods and other oddities. The astonishing diversity in structure and function of proteins in this bizarre arsenal, however, originates from a building set of a limited number of modules.

Here we review the present knowledge of the fundamentals of domain organization and scrutinize functional assignments derived from experimental data and from sequence homology. The complex multidomain organization of the multifunctional ECM proteins offers a fascinating view on mechanisms of evolution. The apparent redundancy of certain ECM proteins opens questions on selective forces in evolution.

In order to provide a concise overview, only the more recent primary publications and review articles could be cited. References to the original literature can be found in these.

EXTRACELLULAR MATRIX PROTEINS ARE MOSAIC MULTIDOMAIN PROTEINS BUILT OF MODULAR UNITS

Extracellular matrix proteins are typical multidomain proteins (Table 1). Most domains show identity with domains of the same protein or with domains found either in other ECM proteins or in multidomain proteins not normally classified as ECM proteins. These include cell adhesion molecules (CAMs and cadherins), many cellular receptors including integrins, and proteins of the immune and complement system and of the blood clotting cascade. Because of this wide and repeating distribution of domains, these types of proteins have been termed mosaic proteins built of modular units (Doolittle 1985, 1992; Doolittle et al., 1986).

Table 1 summarizes the domain organizations as revealed by sequence information for a large number of ECM proteins. Table 1 is not complete and additional compilations can be found in Bork (1991, 1992), Baron et al. (1991), Engel (1991), Patthy (1991a,b), Bork and Doolittle (1992) and Kreis and Vale (1993). Such comparisons demonstrate the widespread distribution of domains in different classes of proteins. Examples are the EGF domains in proteins of the ECM, in blood clotting and complement systems and in a number of cell-surface receptors. IgG-domains occur not only in the immunoglobulin family, but also in proteoglycans, in cell-adhesion molecules such as N-CAM, and in receptors recognizing growth factors and carbohydrates. One of the most widespread module is the fibronectin type 3 (F3) domain of which more than 300 variants in about 70 proteins (not counting species redundancies) have been detected so far. They are found in both extracellular and intracellular proteins; for example, in the muscle protein twitchin and in the cytosolic domain of the integrin subunit β4.

Table 1. Domain organization of extracellular matrix proteins

Protein	Domain order†	Reference
fibronectin	$F1_{6}$ $F2_{2}$ $F1_{3}$ $F3_{15\text{-}18}$ $F1_{3}$	1
tenascin	TL $\underline{CC}$ EG_{13} $F3_{11\text{-}15}$ FG	2
thrombospondin 1, 2	TA $\underline{CC}$ PN PR_{3} EG_{3} EF_{7} TC	3
thrombospondin 3, 4	TA $\underline{CC}$ EG_{4} EF_{7} TC	3
cartilage oligomeric matrix protein (COMP)	$\underline{CC}$ EG_{4} EF_{7} TC	
laminin α1 (Ae)‡	LA EG'_{4} EG″ EG'_{8} EG" EG'_{3} $\underline{CC}$ LG_{5}	5
laminin β1 (B1e)	LA EG'_{5} LB EG'_{8} $\underline{CC}$	5
laminin γ1 (B2e)	LA EG'_{4} EG" EG'_{6} $\underline{CC}$	5
kalinin γ2 (B2t)	EG'_{3} EG" EG'_{4} $\underline{CC}$	6
unc 6	LA EG'_{3} UA	7
perlecan	PA EG_{4} IG $(EG'' EG'_{3})_{3}$ $IG_{14\text{-}21}$ LG EG_{2} LG EG_{2} LG	8
agrin	KA_{8} EG'_{2} KA ST_{2} EG LG EG_{2} LG EG_{1} LG	9
nidogen/entactin	N1 EG N2 EG_{5} N3	10
fibulin/BM-90	AN_{3} EG $EG*_{8}$ FB	11
fibrillin	FA EG_{3} CR $EG*_{2}$ TB PR EG $EG*_{4}$ TB $EG*_{3}$ CR $EG*_{1}$ TB $EG*_{12}$ TB $EG*_{2}$ TB $EG*_{7}$ TB $EG*_{5}$ TB $EG*_{7}$ FB	12
osteonectin/SPARC/BM40	GR KA AL EF	13
procollagen I	PN $\underline{TH}$ PC	14
collagen IV	7S $\underline{TH}$ N4	14
collagen VI (α3 chain)	VA_{9} $\underline{TH}$ VA_{2} ST F3 PI	14
collagen XII	F3 VA F3 VA $F3_{6}$ VA $F3_{10}$ VA N9 $\underline{TH}$	15
aggrecan	LI LI $KS_{0\text{-}1}$ $C1_{0\text{-}1}$ C2 EG LE CO	16

†Designations of domains are explained in Table 2, linear triple helical and coiled coil domains are underlined.
‡New chain designations for laminins are used (Burgeson et al., 1994), old designations are shown in brackets.
1, Hynes, 1990; 2, Spring et al., 1989; 3, Lawler et al., 1993; 4, Oldberg et al., 1992; 5, Beck et al., 1990; 6, Kallunki et al., 1992; 7, Ishii et al., 1992; 8, Noonan and Hassel, 1993; Kallunki and Tryggvason, 1991; 9, McMahan et al., 1992; 10, Mann et al., 1989; 11, Pan et al., 1993; Agraves et al., 1990; 12, Corson et al., 1993; 13, Engel et al., 1987; 14, Bork, 1992; 15, Yamagata et al., 1991; 16, Mörgelin et al., 1994.

It has to be pointed out that all classifications contain a degree of ambiguity and uncertainty, in some cases because of very low sequence identities, which might not reflect common evolutionary descent. It has been argued that in some cases similar domains were produced by convergent evolution, for example as discussed for EF-hand domains (Kretsinger, 1987).

DOMAINS ARE AUTONOMOUS STRUCTURAL UNITS

Domains may be defined by the sequence blocks which are repeated in the same protein or reoccur in different proteins. Often, however, it is difficult to define the exact starts and ends of domains on this basis. Recognition of a linker sequence which is normally hydrophilic may help in some cases. Domains are often encoded by single exons, but this cannot be an absolute rule since introns can be secondarily introduced into exons during evolution. In the present work a domain is defined as an autonomous, independently-folded, structural unit. The most stringent proof for the structural independence of domains comes from three-dimensional structures. These have been derived by NMR and X-ray diffractions for a number of domains (Table 2; Baron et al., 1991). Earlier indications of a conformational independence of fibronectin and laminin domains were based on circular dichroism studies, which indicated additivity of the spectra of different fragments (Odermatt et al., 1982; Ott et al., 1982). A powerful method for distinguishing individual domains in regions with sequence repeats is based on the resistance of recombinantly prepared fragments against proteolytic susceptibility (Winograd et al., 1991). The structural integrity of separated domains has also been demonstrated for many ECM proteins by the preserved biological functions of domains and fragments. Recently, the three dimensional structures of a pair of fibronectin-type 1 (F1) domains in fibronectin (Williams et al., 1993), a pair of complement control (CO) domains in factor H (Barlow et al., 1993) and a lectin-EGF-module pair (Graves et al., 1994) were resolved. The secondary structure of each module within a pair conformed closely with the structure of the separated single domains, implying that modules fold entirely autonomous within intact proteins.

THE DOMAIN ORGANIZATION OF ECM PROTEINS

The example of the F1 pair in fibronectin demonstrated the potential of NMR for elucidation of the geometry of domain organizations. The NMR technique is limited, however, to structures of smaller molar mass than about 20 000. X-ray analysis is applicable to larger structures but in this case crystallisation of large ECM molecules is a severe problem. Electron microscopy, therefore, is one of the most powerful techniques for elucidation of larger domain organizations (Engel, 1994). Fibronectin (Hynes, 1990; Odermatt et al., 1982), laminin (Beck et al., 1990), thrombospondin (Lawler, 1986) and tenascin (Spring et al., 1989; Erickson, 1993) are examples in which this technique, in combination with hydrodynamic data and sequence analyses, yielded detailed information (Fig. 1). EGF domains are found in all four of these proteins, in linear arrangements with a repeat of 2-2.5 nm per domain. The F1, F2 and F3 domains in fibronectin are also in a linear array which is, however, strongly dependent on ionic strength (Markovic et al., 1983), indicating a solvent-dependent internal association of domains. Likewise, the arms of thrombospondin and cartilage oligomeric matrix protein

Table 2. Domains in extracellular matrix proteins

Code	Name	Structure	Function
AN	Anaphylatoxin	X-ray (Huber et al., 1980)	Occurs in complement components C3, C4, C5
AL	α-helical domain BM-40		Collagen binding (Pottgiesser et al.,1994)
CC	Heptad repeat region	Forms α-helical coiled coil structures with partner chains, X-ray of leucine zipper in GCN4 (Harbury et al., 1993)	Connects 3 laminin chains (Beck et al., 1990), 3 tenascin chains (Spring et al., 1989), 3 thrombospondin (Lawler, 1986), 3 fibrinogen and 5 COMP chains (Efimov et al., 1994)
C1 C2	Chondroitin sulfate binding domains 1, 2	Stretched polypeptide chains (Mörgelin et al., 1994)	Glycosaminoglycan attachment to gly-ser sequences
CO	Complement control domain	NMR (Baron et al., 1991)	Frequent in complement proteins
EF	EF hand-like domain	X-ray structures of cytosolic proteins	Ca^{2+} binding to Ca^{2+} modulated proteins, to osteonectin (Pottgiesser et al., 1994) and thrombospondin (Lawler, 1986)
EG	EGF-like domain	NMR of EGF (Baron et al.,1991), X-ray of EG domain in E-selectin (Graves et al., 1994)	Growth promotion in EGF, putative binding domain for proteins (Davis, 1990)
EG*	EGF domain, Ca2+ binding	NMR of first EG* domain in factor IX (Baron et al., 1991)	Ca^{2+} binding demonstrated for factor IX, for fibrillin (Corson et al., 1993) and other proteins
EG′	Laminin type EGF domain	X-ray and NMR work in progress†	EG′ domain in laminin γ1 chain is a specific binding site for nidogen (Mayer et al., 1993), growth promoting functions for unspecified EG′ domains in laminin were proposed (Panayotou et al., 1989)
EG″	Laminin type extended EGF domain		Cleavage during activation of RGD site in the short arms of murine laminin (Beck et al., 1990)
FA	N-terminal domain in fibrillin		
FB	C-terminal domain in fibrillin		
F1	Fibronectin type 1 domain	NMR of F1 and a pair of F1 domains in fibronectin (Baron et al., 1991; Williams et al., 1993)	N-terminal fragment of 5 F1 domains in fibronectin binds fibrinogen and heparin (Hynes, 1990)
F2	Fibronectin type 2 domain	NMR (Constantine et al.,1992)	The first and second F2 domain in fibronectin are involved in gelatine binding (Hynes, 1990)
F3	Fibronectin type 3 domain	NMR of 10th F3 domain in fibronectin (Baron et al., 1991); X-ray of F3 domain in tenasin (Leahy et al., 1992)	10th F3 domain in fibronectin is the major RGD dependent cell binding domain. Another more C-terminal F3 domain is also involved in cell binding (Hynes, 1990)
FG	Fibrinogen γ domain		Cell binding in tenascin (Joshi et al., 1993)
GR	Glutamic acid rich domain		
IG	IgG domain	X-ray of many immunoglobulins	Antigen binding and other recognition functions in IgG
KA	Kazal inhibitor-like domain	X-ray of Kazal inhibitor of Kazal family (Bode and Huber, 1992)	Homology with follistatin, ovomucoid. Kazal inhibitors inhibit proteases by binding to their active site but no such affinity was found for KA domains in ECM proteins
KS	Keratan sulfate binding domain	Attachment of keratan sulfate	
LE	Mammalian lectin domain	X-ray of LE in E-selectin (Graves et al., 1994), mannose binding protein (Weis et al., 1992)	LE in aggrecan binds saccharides with low activity but physiological target unknown
LG	C-terminal domains of laminin α-chain		Homology with steroid binding hormone (Beck et al., 1990), splice variants in agrin cluster acetylcholine receptor (McMahan et al., 1992)‡
LA	N-terminal laminin domain		Involved in self-association (Beck et al., 1990)
LI	Link protein domain		The first LI domain in aggrecan binds hyaluronan, the second not (Mörgelin et al., 1994)
N1	N-terminal nidogen domain		
N2	Central nidogen domain		Binds collagen IV and perlecan (Beck et al., 1990)
N3	C-terminal nidogen domain		Bind to a distinct EG′ domain of laminin γ1-chain (Mayer et al., 1993)
N4	Non-collagenous domain in collagen IV		
N9	Non-collagenous domain in collagen IX		
PA	N-terminal perlecan domain		
PC	C-terminal procollagen I domains		
PN	N-terminal procollagen I domains		
PI	BPTI-like domain	X-ray of of inhibitors of the class of the bovine pancreatic trypsin inhibitor family (Bode and Huber, 1992)	Small inhibitors inhibit specific proteases by binding to their active site but no such affinity was found for PI domains in ECM proteins
PR	Properdin domain		Binding to membranes?
ST	Serine/threonine rich domain		
TB	TGF-β binding protein domain		Proposed to be involved in TGF-β binding (Corson et al., 1993)
TA	N-terminal domain of thrombospondin		
TC	C-terminal domain of thrombospondin		Involved in cell binding (Lawler, 1986)
TH	Gly-X-Y repeat region	Three chains form a collagen triple helix	Connects 3 chains in various collagens
TL	N-terminal domain of tenascin		Probably involved in linking trimers to hexamers (Erickson, 1993)
UA	C-terminal unc-6 domain		
VA	von Willebrand factor A domain		VA domain in von Willebrand factor binds collagen (Colombatti and Bernaldo, 1991)
7S	N-terminal domain in collagen IV		Connects 4 collagen IV molecules in an antiparallel arrangement

For simplicity all domains are designated by a code of two letters or a letter and a number. Very obvious variants of domains within each homology class are indicated by dashes or asterisks. The subscripts indicate the number of repeated domains with internal homology. Ranges of numbers indicate the existence of splice variants.
†R.Huber and R. Timpl, personal communication. ‡Binding to α-dystroglycan (Sealock and Froehner, 1994).

(Mörgelin et al., 1992) are extended. The stretched arms of laminin and fibronectin exhibit a limited flexibility comparable with that of actin filaments (Engel et al., 1981); hence these domains are not loosely connected but interact with each other (as exemplified by the rigid end to end structure of the F1 pair). In contrast,only small constraints on the flexibility of domains were seen for the CO- and the LE-EGF-pairs.

The extended arrangements mentioned so far are formed by sequential arrangements of small globular domains in a single chain. In two types of domain, however, long linear structures are formed by several chains. These are the collagen triple helices formed by three chains with Gly-X-Y repeats, and the α-helical coiled-coil structures in which two to five chains with heptad repeats of nonpolar residues are connected (Cohen and Parry, 1990, 1994; Lupas et al., 1991). The length of these structures is highly variable. Coiled-coil structures in COMP and thrombospondin are not longer than 50 residues/chain (7.5 nm; Efimov et al., 1994). They are of similar size in tenascin (Spring et al., 1989) but longer, consisting of 600 residues (76 nm), in laminin (Beck et al., 1990). Collagen triple helices range from 45 residues/chain in the N-terminal small triple helix of collagen III (Bruckner et al., 1978) to 8 000 residues/chain (2400 nm) in some worm collagens (Gaill et al., 1991). Thus, an amazing diversity of forms of ECM proteins can be built up from the modular pool.

binding in native tenascin (Spring et al., 1989). As it was pointed out by Ruoslahti (1988) and Hynes (1990), but ignored by many others, attachment is usually highly conformation dependent (Deutzmann et al., 1990) and, as for fibronectin (Hynes, 1990), more than one binding site may be involved.

Another example of inaccurate prediction of functions based on sequence similarity relates to the EGF domains. It is an appealing concept that some of the EGF-like domains in ECM proteins may act as localized signals for growth and differentiation, which may act in a specific and vectorial way on adjacent cells. Indeed, growth-promoting functions have been experimentally shown for laminin, thrombospondin and tenascin (Engel, 1989). For laminin, which is amongst the first ECM molecules expressed in mammalian embryonic development, it was possible to localize this function to fragment P1 (Panayotou et al., 1989) which comprises short-arm regions of the α, β and γ chains with about 25 EGF-like repeats in total. Unambigous proof is missing, however, for an EGF domain being the active functional site in the very large fragment P1. Furthermore, it is clear that not all EGF domains in ECM and other proteins exhibit growth factor-like functions. The best demonstrated function of the laminin type EGF (EG′) domains (Table 2) is to provide a very specific binding site for the C-terminal nidogen domain N3 (Mayer et al., 1993). The three-dimensional structure of the nidogen binding EGF-domain will

FUNCTIONAL PREDICTIONS FOR MODULES BASED ON PRIMARY SEQUENCE HOMOLOGY ARE OFTEN WRONG

Elucidation of the functions of individual domains in ECM proteins is a challenging but very time consuming and difficult task. An outstanding success was the identification of the cell binding site containing Arg-Gly-Asp in fibronectin. In the pioneering work by Ruoslahti (1988), this site was identified in the 10th F3 domain of fibronectin. An exposed and flexible three dimensional structure has been recently demonstrated for the Arg-Gly-Asp region both in this domain (Baron et al., 1991) and in disintegrins (Blobel and White, 1992). When it was found that cell attachment by several other ECM proteins could be inhibited by Arg-Gly-Asp peptides, it was initially thought that a universally valid principle had been discovered. As a consequence, many putative cell attachment sites were predicted from sequence data. It is now realized that this does not hold true: many cell attachment processes are Arg-Gly-Asp independent and many major cell attachment sites do not contain this sequence. It was even found that an F3 domain in tenascin, which contains an Arg-Gly-Asp sequence at a similar location as the classic domain in fibronectin is not involved in attachment, although in the isolated recombinantly prepared domains the tripeptide sequence was active (Aukhil et al., 1993). Instead, another domain mediates Arg-Gly-Asp independent cell

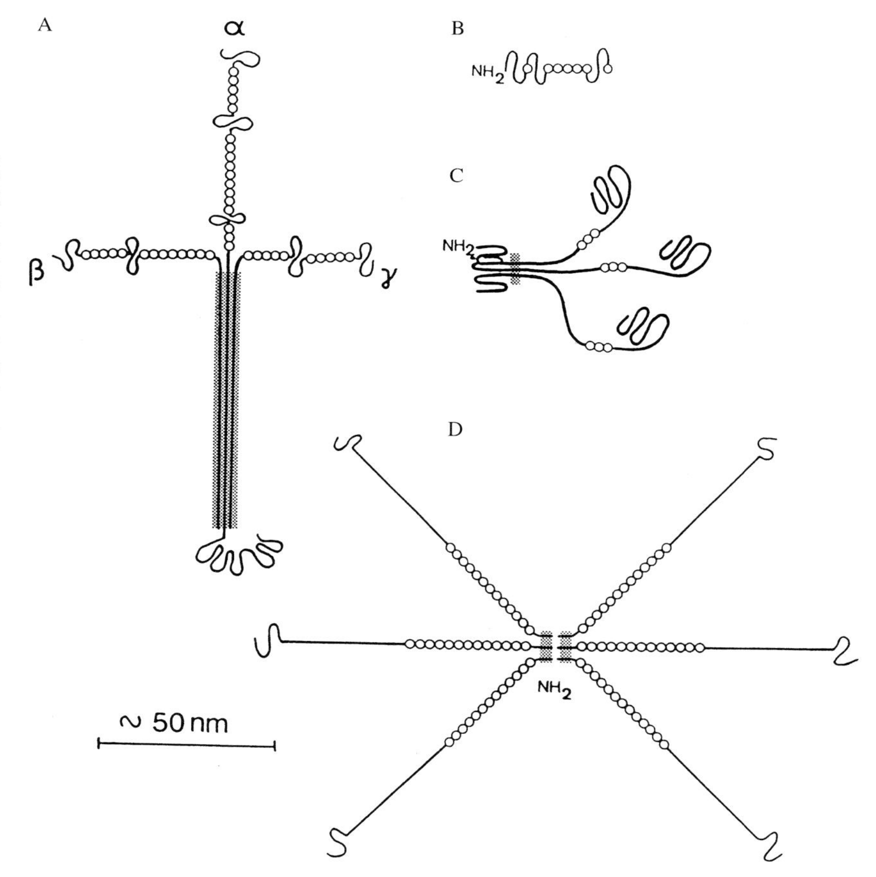

Fig. 1. Schematic representations of the shapes of (A) laminin, (B) nidogen/entactin, (C) thrombospondin and (D) tenascin, according to electron microscopic information. EGF-like domains are represented by open circles, hatched fields mark coiled-coil structures. The proteins are approximately drawn to scale (for references see text).

be known soon and it is hoped that details of its specific function will be explored. Other functions, like Ca^{2+} binding of specialized EGF domains (EG*), have been demonstrated (Table 2) and have been correlated with the three dimensional structure of the EG* domains (Baron et al., 1991). The functions of most other EGF-like domains remain unexplored.

Functional predictions that are entirely based on recognition of a general sequence motif are usually wrong. Very specific information like calcium-binding motifs in EGF- or EF-hand domains might be helpful but even in these cases there have been many disappointing experiences. We urge, therefore, that the frequently used term 'putative functional domains' should be avoided, since it can lead to confusion when 'putative' is inadvertantly omitted (eg. in the next review).

Another argument for the functional promiscuity of domains comes from estimates of the number of protein families. Recent genome sequencing efforts show that about one third of sequenced open reading frames belong to families that already have members in the databanks. From these data Chothia (1992) estimated that about 1500 different protein families exist. Even if there are ten times more families, because of biases in the databanks, the number of functions greatly excedes the number of basic protein structures. Thus the prototype of each family is modelled differently to fulfil specific functions. Extracellular modules provide examples of great functional variety being achieved from a few basic structures. Just as no one would claim to predict the antigen from the sequence of an antibody, we feel the elucidation of functions of protein modules should rely principally on experimental effort, not sequence comparisons.

DOMAINS MAY FUNCTION INDEPENDENTLY OR IN COMBINATION WITH OTHER DOMAINS

Perhaps the most important function of coiled-coil and collageneous domains is to connect subunits within a single molecule, in which they may exhibit a concerted function. This is clearly demonstrated by thrombospondins (Lawler 1986; Lawler et al., 1993) and COMP (Mörgelin et al., 1992) in which 3 or 5 identical chains are combined. These all point in the same direction and hence the C-terminal cell binding domains of these molecules and other domains are brought in close vicinity (Fig. 1). This alignment may be important for simultaneous recognition of multiple receptor sites at the cell surface. Although details of the binding mechanism have not yet been explored, the situation may be comparable to the binding of the hexameric 'flower bouquet'-shaped first component of complement C1q, that binds to clusters of IgG. In this example it was demonstrated that sufficient binding strength is only produced by multivalent binding (Tschopp et al., 1980). This affords a mechanism for discriminating between clustered and isolated IgG molecules at a cell surface.

In many collagens, several globular domains are combined by association of three chains in the collagen triple helix (for example, collagens IV, VI and XII; Table 1). Von Willebrand type A (VA) domains are involved in the self-assembly of some collagens and have frequently been designated as collagen-binding domains, although direct proof for this activity is missing in most cases (Colombatti and Bonaldo, 1991).

Laminin is comparable, in that three different chains α, β and γ are connected by a coiled-coil domain. Many genetically distinct variants of these chains have been found (Paulsson, 1993) and these are combined to give distinct laminin isoforms. Some isoforms are transiently expressed at restricted sites, suggesting specialized functions. The assembly of the three different chains is highly specific and correct assembly is crucial for cell binding of laminin by α6β1 integrin and for the promotion of neurite outgrowth (Hunter et al., 1992, Deutzmann et al., 1990; Sung et al., 1993).

THE TIME COURSE OF EVOLUTION OF ECM PROTEINS

Doolittle (1985, 1992) attempted to group proteins according to their time of invention. He classifies ECM proteins as very recent inventions, each of which is found in animals or plants but not in both, nor in prokaryotes. This suggests that ECM proteins arose around the time that plants and animals diverged, perhaps 1 billion years ago (Doolittle, 1985) It has been proposed that modern mosaic proteins are the result of efficient mechanisms of exon shuffling (Patthy, 1991b). For several ECM proteins for which sufficient sequences from phylogenetically distant organisms were available, phylogenetic trees were constructed. The construction of the dendrograms utilized the method of maximum parsimony, which determines the tree requiring the minimum number of base substitutions (alternative phylogenetic reconstruction methods are possible) As an example, the phylogenetic tree of the thrombospondin gene family is shown (Fig. 2; modified from Lawler et al., 1993, by addition and inclusion of COMP). The dendrogram is based on a comparison of the C-terminal six EF domains and the TC domain (Table 1). Lawler et al. (1993) were able to assign a very rough time scale to the dendrogram by calibration with two phylogenetic events. This is possible by

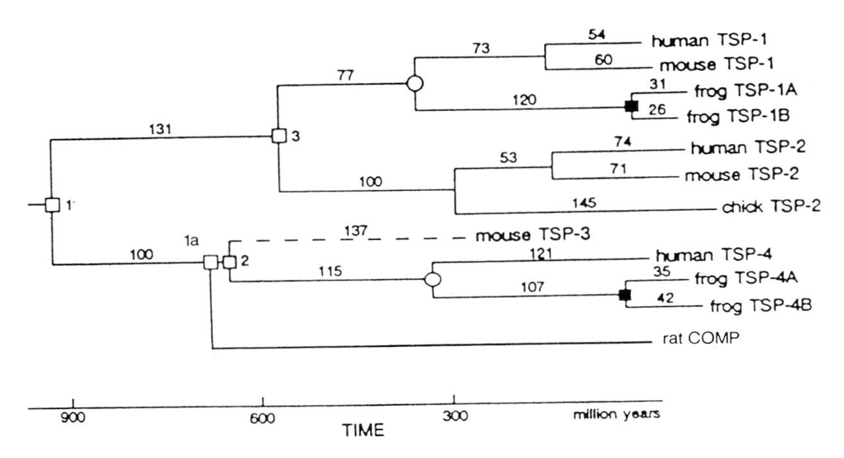

Fig. 2. Phylogenetic tree of the thrombospondin (TSP) gene family. Data for TSPs are according to Lawler et al. (1993) and the branch for cartilage oligomeric matrix protein (COMP) has been added on the basis of UPGMA distance matrix (Sneath and Sokal, 1973) and parsimony analyses (Felsenstein, 1993). The open squares mark the putative positions of gene duplications.

assuming a constant average rate of evolutionary divergence for the protein region under consideration. Different proteins change at very different rates but each rate is approximately constant (Doolittle, 1992); this rate constancy may apply to individual domains within mosaic proteins,but not for the entire protein. From Fig. 2, we can infer that the C-terminal portions of thrombospondins and COMP had a common ancestor earlier than 900 million years ago. At this time a gene duplication (open box 1) resulted in the two branches: one with a precursor for thrombospondins 1 and 2, and the other with a precursor for thrombospondin 3 and 4 plus COMP. One can also deduce from the phylogenetic tree, and the domain distribution in different branches, that either the N-terminal domains PR_3 EG of thrombospondins 1 and 2 were inserted into the thrombospondin 1/2 branch between 900 and 600 million years ago, or alternatively, they were already present in the common precursor and the thrombospondin type 3/4/COMP branch subsequently lost these domains. The resolution of such phylogenetic trees is insufficient to resolve whether COMP diverged from thrombospondins 3 and 4 before or after the branching off of thrombospondin type 3. However, an early branching of COMP would support a model that evolution proceeded in the direction of increasing domain complexity. It is important to note, however, that phylogenetic trees derived from certain domains of a multidomain protein are not able to predict the history of other domains in the same protein, hence it remains unclear whether the three EGF domains common to all proteins (Table 1) were present in the precursor. It will be interesting to scrutinize the phylogeny of different domains within one protein and hence gain more insight into the pattern and timing of domain acquisition within multidomain proteins.

MECHANISMS OF THE EVOLUTION OF ECM PROTEINS

New proteins come from old proteins as the result of gene duplications followed by base substitutions (Doolittle, 1992). This very general statement also applies to mosaic proteins. It is obvious from Table 1, however, that in their case individual domains can also be rearranged extensively, somewhat like mobile elements (Doolittle,1992).

Mechanisms of gene duplication by unequal crossing-over between sister chromosomes containing the genes are described in textbooks. Unequal cross-overs can also readily extend tandemly repeated genes into long series. Duplications, deletions, inversions, conversions, slippages and translocations of DNA segments can arise as the result of erratic rejoining of fragments. The genomic sequence of chromosome III of *C. elegans* (Wilson et al., 1994) and a comparative study of large DNA sequences of mouse and man (Koop and Hood, 1994) revealed an intriguing view on gene organization, with evidence for duplications, inversions and other gene rearrangements. Gene rearrangements are rare events catalyzed by the enzymes that mediate normal recombination processes; in the example of thrombospondins they resulted in gene duplications at time intervals in the range of 100 million years (see Fig. 2). For mosaic proteins it is generally believed that these processes are speeded up by the presence of introns. The most trivial reason for the higher speed of this process is the possibility of breaking and rejoining the DNA anywhere in the long introns on either side of an exon. Thus a large number of different possible breakages could lead to exon shuffling, in each case the exon is left intact, which in turn encodes a stable protein domain in the majority of extracellular modules.

Exon shuffling could involve transposable elements (which make up about 10% of the genome of higher eukaryotes); examples can be seen 'in flagranti' in several human genetic disorders. Duplication or deletion of exons are also the cause of several human genetic disorders (Bates and Lehrach, 1994, Makalowski et al., 1994). The possibility of an exon variation at the DNA level by reverse transcription of an alternative splice variant at the RNA level may also be considered. Reverse transcription may occur by mammalian enzymes with reverse transcriptase activity or by the help of virus systems (Fink, 1987).

Details of alternative possible mechanisms for exon shuffling have been discussed by Rogers (1990) and Patthy (1991b). Several examples strongly suggest that exons can be inserted into preexisting introns. However, domain shuffling is clearly not the result of just one mechanism, nor is it the only process operating in multidomain protein evolution. This becomes evident with the increasing number of observations in which exons do not correspond to protein domains, or in which domains consist of several exons (e.g. F3 domains are encoded by two exons, yet no half F3s has been found to date). Even harder to reconcile with a dominant role for exon shuffling is the observation that exon-intron boundaries, within the same domain organization, can differ from one species to another. Extensive rearrangements following the presumed duplication of a common primordial gene were shown for the genes for β and γ chains of laminin (Kallunki et al., 1991).

An example in which the relative contributions of exon shuffling and other processes were compared relates to the EF-hand calcium modulated proteins. Extensive analysis revealed a random distribution of introns over 'domain' and 'interdomain' space, and that some introns were acquired after a four domain precursor was formed. It was therefore concluded that in the evolution of the widely distributed EF-hand protein family, exon shuffling played little if any role (Kretsinger and Nakayama, 1993).

The evolution of genes for the modular ECM proteins may have been further complicated by horizontal gene transfer; for example, Bork and Doolittle (1992) suggested that bacteria may have acquired a F3 domain from animals.

WHAT WERE THE SELECTION PRESSURES AFFECTING THE EVOLUTION OF ECM PROTEINS?

Unequal crossing-over may lead to either an increase or a decrease in the number of repeated domains. This suggests that the large number of repeated domains present in ECM proteins may be the result of natural selection. One reason for the large number of domains in an ECM protein may be the need to prevent diffusion of domains with specific activities into the otherwise open extracellular space: this could be achieved simply by making the proteins very large. For example, this allows localization of domains with cell signalling activity at specific sites, and the variability of the extracellular environment by time-dependent and vectorial expressions of different proteins (Engel, 1989). Another common feature of the large and extended ECM proteins is their ability to bridge between distant sites, for example between cellular receptors and other parts of the matrix. Clearly the development of specialized assembly domains was a prerequisite to develop multifunctional large molecules with the potential of forming higher macromolecular organisations. The α-helical coiled-coil domains are also found in many cytoskeletal proteins but collagen triple helices are specific for extracellular proteins. In addition to other functions, collagen triple helices are essential for the formation of collagen fibres, cuticle structures and networks. They contribute essentially to the mechanical properties of tissues of larger organisms.

Of course, ECM proteins contain a large number of domains with much more specific functions than spacing, assembly or support; a few are listed in Table 2. In addition, it must be stressed that such specific functions have been elucidated for only a small percentage of the known domains.

This rather simplified interpretation of selection pressures is apparently contradictory to the complete lack of phenotype resultant from genetic elimination of certain ECM proteins (even those implicated in important functions). In contrast to fibronectin, which is absolutely required in early stages of embryonic development, no phenotype was detectable in transgenic mice after knock-outs of tenascin and S-type lectin (George et al., 1993, Poirier and Robertson, 1993, Saga et al., 1992). One explanation may be that hitherto unrecognized subtle functions of these proteins may cause a small increase of fitness, which is not obvious in the phenotype, and would only be apparent in the appropriate population size and natural environment (as seen for transgenic mice lacking metallothioneins; Michalska and Choo, 1993). Alternatively, there may be functional redundancies between some ECM proteins with similar domains, which can at least in part fulfil the function of the deficient protein. Even in this case, however, selective advantages may be necessary to maintain such a redundancy in a population. These could either be selection for a subtle divergent function, an increased fidelity for a certain process or an enhanced efficiency of a cumulative function.

REFERENCES

Agraves, W. S., Tran, H., Burgess, W. H. and Dickerson, K. (1990). Fibulin is an extracellular matrix and plasma glycoprotein with repeated domain structure. *J. Cell Biol.* **111**, 3155-3164.

Aukhil, I., Joshi, P., Yan, Y. and Erickson, H. P. (1993). Cell- and heparin-binding domains of the hexabrachion arm identified by tenascin expression proteins. *J. Biol. Chem.* **268**, 2542-253.

Barlow, P. N., Steinkasserer, A., Norman, D. G., Kieffer, B., Wiles, A. P., Sim, R. B. and Campbell, I. D. (1993). Solution structure of a pair of complement modules by nuclear magnetic resonance. *J. Mol. Biol.* **232**, 268-284.

Baron, M., Nordman, D. G. and Campbell, I. D. (1991). Protein modules. *Trends Biochem.* **16**, 13-17.

Bates, G. and Lehrach, H. (1994). Trinucleotide repeat expansions and human genetic disease. *Bioessays* **16**, 277-284

Beck, K., Hunter I. and Engel, J. (1990). Structure and function of laminin: Anatomy of a multidomain glycoprotein. *FASEB J.* **4**, 148-160.

Blobel, C. P. and White, J. M. (1992). Structure, function and evolutionary relationship of proteins containing a disintegrin domain. *Curr. Opin. Cell Biol.* **4**, 760-765.

Bode, W. and Huber, R. (1992). Natural protein proteinase inhibitors and their interaction with proteinases. *Eur. J. Biochem.* **204**, 433-451.

Bork, P. (1991). Shuffled domains in extracellular proteins. *FEBS Lett.* **286**, 47-54.

Bork, P. (1992). The modular architecture of vertebrate collagens. *FEBS Lett.* **307**, 49-54.

Bork, P. and Doolittle, R. F. (1992). Proposed acquisition of an animal protein domain by bacteria. *Proc. Natl. Acad. Sci. USA* **89**, 8990-8994.

Bruckner, P., Bächinger, H. P., Timpl, R. and Engel, J. (1978). Three conformationally distinct domains in the amino-terminal segment of type III procollagen and its rapid triple helix = coil transition. *Eur. J. Biochem.* **90**, 595-603.

Burgeson, R. E., Chiquet, M., Deutzmann, R., Ekblom, P., Engel, J., Kleinman, H., Martin, G. R., Meneguzzi, G., Paulsson, M., Sanes, J., Timpl, R. Tryggvason, K., Yamada, Y. and Yurchenco, P. D. (1994). A new nomenclature for the laminins. *Matrix Biol.*, in press.

Chothia, C. (1992). One thousand families for the molecular biologist. *Nature* **357**, 543-544

Cohen, C. and Parry, D. A. D. (1990). α-helical coiled- coils and bundles: How to design an α-helical protein. *Proteins* **7**, 1-15.

Cohen, C. and Parry, D. A. D. (1994). Alpha-helical coiled coils: more facts and better predictions. *Science* **263**, 488-489.

Colombatti, A. and Bonaldo, P. (1991). The superfamily of proteins with von Willebrand factor type A-like domains: one theme common to components of extracellular matrix. *Blood* **77**, 2305-2315.

Constantine, K. L., Madrid, M., Banyai, L., Trexler, M., Patthy, L. and Llinas, M. (1992). Refined solution structure and ligand-binding properties of PDC-109 domain b. A collagen-binding type II domain. *J. Mol. Biol.* **223**, 281-298.

Corson, G. M., Chalberg, S. C., Dietz, H. C., Charbonneau, N. L. and Sakai, L. Y. (1993). Fibrillin binds calcium and is coded by cDNAs that reveal a multidomain structure and alternatively spliced exons at the 5′ end. *Genomics* **17**, 476-484.

Davis, C. G. (1990). The many faces of epidermal growth factor repeats. *New Biologist* **2**, 410-419.

Deutzmann, R., Aumailley, M., Wiedemann, H., Pysny, W., Timpl, R. and Edgar, D. (1990). Cell adhesion spreading and neurite stimulation by laminin fragment E8 depend on maintenance of secondary and tertiary structure in its rod and globular domain. *Eur. J. Biochem.* **191**, 513-522.

Doolittle, R. F. (1985). The geneology of some recently evolved vertebrate proteins. *Trends Biochem.* **10**, 233-237.

Doolittle, R. F. (1992). Reconstructing history with amino acid sequences. *Protein Science* **1**, 192-200.

Doolittle, R. F., Feng, D. F., Johnson, M. S. and McClure, M. A. (1986). Relationship of human protein sequence to those of other organisms. *Cold Spring Harbor Symp Quant. Biol.* **51**, 447-455.

Efimov, V. P., Lustig, A. and Engel, J. (1994). The thrombospondin-like chains of cartilage oligomeric matrix protein are assembled by a five-stranded α-helical bundle between residues 20 and 83. *FEBS Lett.* **341**, 54-58.

Engel, J. (1989). EGF-like domains in extracellular matrix proteins: localized signals for growth and differentiation? *FEBS Lett.* **251**, 1-7.

Engel, J. (1991). Common structural motifs in proteins of the extracellular matrix. *Curr. Opin. Cell Biol.* **3**, 779-785.

Engel, J. (1994). Electron microscopy of extracellular matrix components. *Meth. Enzymol.* (in press).

Engel, J., Odermatt, E., Engel, A., Madri, J. A., Furthmayr, H., Rohde, H. and Timpl, R. (1981). Shapes, domain organizations and flexibility of laminin and extracellular matrix. *J. Mol. Biol.* **150**, 97-120.

Engel, J., Taylor, W., Paulsson, M., Sage, H. and Hogan B. (1987). Calcium binding domains and calcium induced conformational transition of SPARC (osteonectin, BM-40), an extracellular glycoprotein expressed in mineralized and non mineralized tissues. *Biochemistry* **26**, 6958-6965.

Erickson, H. P. (1993). Tenascin C, tenascin R and tenascin Y: a family of talented proteins in search of functions. *Curr. Opin. Cell Biol.* **5**, 869-876.

Felsenstein, J. (1993). PHYLIP (Phylogeny Interference package) Version 3. 5 (Computer software package distributed by the author, Dept. Genetics, University of Washington, Seattle.

Fink, G. R. (1987). Pseudogenes in yeast ? *Cell* **49**, 5-6.

Gaill, F., Wiedemann, H., Mann, K., Kühn, K., Timpl, R. and Engel, J. (1991). Molecular characterization of cuticle and interstitial collagens from worms collected at deep sea hydrothermal vents. *J. Mol. Biol.* **221**, 209-223.

George, E., Georges-Labouesse, E., Patel-King, R., Rayburn, H. and Hynes, R. (1993). Defects in mesoderm, neural tube and vascular development in mouse embryos lacking fibronectin. *Development* **119**, 1079-1091.

Graves, B. J., Crowther, R. L., Chandran, C., Rumberger, J. M., Li, S., Huang, K.-S., Pesky, D. H., Familletti, P. C., Wolitzky, B. A. and Burns,

D. K. (1994). Insight into E-selectin/ligand interaction from the crystal structure and mutagenesis of the lec/EGF domains. *Nature* **367**, 532-538.

Harbury, P. B., Zhang, T., Kim, P. S. and Alber, T. (1993). A switch between two-, three- and four-stranded coiled coils in GCN4 leucine zipper mutants. *Science* **262**, 1401-1407.

Huber, R., Scholze, H., Paques, E. P. and Deisenhofer, J. (1980). Crystal structure analysis and molecular model of human C3a anaphylatoxin. *Hoppe-Seylers Z. Physiol. Chem.* **361**, 1389-1399.

Hunter, I., Schulthess, T. and Engel, J. (1992). Lamininin chain assembly by triple and double stranded coiled-coil structures. *J. Biol. Chem.* **267**, 6006-6011.

Hynes, R. O. (1990). *Fibronectins* (ed. A. Rich), New York/Berlin: Springer Verlag.

Ishii, N., Wadsworth, W. G., Stern, B. D., Culotti, J. G. and Hedgecock, E. M. (1992). UNC-6, a laminin-related protein guides cell and pioneers axon migration in C. elegans. *Neuron* **9**, 873-881.

Joshi, P., Chung C. Y., Aukhil, I. and Erickson, H. P. (1993). Endothelial cells adhere to the RGD domain and the fibrinogen-like terminal knob of tenascin. *J. Cell. Sci.* **106**, 389-400.

Kallunki, P. and Tryggvason, K. (1991). Human basement membrane sulfate proteoglycan core protein: A 467 kD protein containing multiple domains resembling elements of the low density lipoprotein receptor, laminin, neural cell adhesion molecules and epidermal growth factor. *J. Cell Biol.* **116**, 559-571.

Kallunki, P., Sainio, K., Eddy, R., Byers, M., Kallunki, T., Sariola, H., Beck, K., Hirvonen, H., Shows, T. B. and Tryggvason, K. (1992). A truncated laminin chain homologous to the B2 chain: structure, spatial expression, and chromosomal assignment. *J. Cell Biol.* **119**, 679-693.

Kallunki, T., Ikonen, J., Chow, L. T., Kallunki, P. and Tryggvason, K. (1991). Structure of the human laminin B2 chain reveals extensive divergence from the laminin B1 chain gene. *J. Biol. Chem.* **266**, 221-228.

Koop, B. and Hood, L. (1994). Striking sequence similarity over almost 100 kilobases of human and mouse T-cell receptor DNA. *Nature Genetics* **7**, 48-53.

Kreis, T. and Vale, R. (1993). *Guidebook to the Extracellular Matrix and Adhesion Proteins.* Oxford: Oxford University Press.

Kretsinger, R. H. (1987). Calcium coordination and the calmodulin fold: divergent versus convergent evolution. *Cold Spring Harb. Symp. Quant. Biol.* **52**, 499-510.

Kretsinger, R. H. and Nakayama, S. (1993). Evolution of EF-hand calcium-modulated proteins. IV. Exon shuffling did not determine the domain composition of EF-hand proteins. *J. Mol. Evol.* **36**, 477-488.

Lawler, J. (1986). The structural and functional properties of thrombospondin. *Blood* **67**, 1197-1209.

Lawler, J., Duquette, M., Urry, L., McHenry, K. and Smith, T. F. (1993). The evolution of the thrombospondin gene family. *J. Mol. Evol.* **36**, 509-516.

Leahy, D. J., Hendrickson, W. A., Aukhil, I. and Erickson, H. P. (1992). Structure of a fibronectin type III domain from tenascin phased by MAD analysis of the selenomethionyl protein. *Science* **258**, 987-991.

Lupas, A., Van-Dyke, M. and Stock J. (1991). Predicting coiled coils from protein sequence. *Science* **252**, 1162.

Makalowski, W., Mitchell, G., and Labuda, D. (1994). Alu sequences in the coding regions of mRNA: a source of protein variability. *Trends Genet.* **10**, 188-193.

Mann, K., Deutzmann, R., Aumailley, M., Timpl, R., Raimondi, L., Yamada, Y., Pan, T., Conway, D. and Chu, M.-L. (1989). Amino acid sequence of mouse nidogen, a multidomain basement membrane protein with binding activity for laminin, collagen IV and cells. *EMBO J.* **8**, 65-72.

Markovic, Z., Lustig, A., Engel, J., Richter, H. and Hörmann, H. (1983). Shape and stability of fibronectin in solutions of different pH and ionic strength. *Hoppe-Seyler's Z. Physiol. Chem.* **364**, 1795-1804.

Mayer, U., Nischt, R., Poschl, E., Mann, K., Fukuda, K., Gerl, M., Yamada, Y. and Timpl, R. (1993). A single EGF-like motif of laminin is responsible for high affinity nidogen binding. *EMBO J.* **12**, 1879-1885.

McMahan, U. J., Horton, S. E., Werle, M. J., Honig, L. S., Kröger, S., Ruegg, M. A. and Escher, G. (1992). Agrin isoforms and their role in synaptogenesis. *Curr. Opin. Cell Biol.* **4**, 869-874.

Michalska, A. E. and Choo, K. H. A. (1993). Targeting and germ-line transmission of a null mutation at the metallothionein I and II loci in mouse. *Proc. Natl. Acad. Sci. USA* **90**, 8088-8092.

Moergelin, M., Heinegard, D., Engel, J. and Paulsson, M. (1992). Electron microscopy of native COMP (cartilage oligomeric matrix protein) purified from the swarm rat chondrosarcoma reveals a five-armed structure. *J. Biol. Chem.* **2367**, 6137-6141.

Moergelin, M., Heinegard, D., Engel, J. and Paulsson, M. (1994). The cartilage proteoglycan aggregate: Assembly through combined protein-carbohydrate and protein-protein interactions. *Biophys. Chem.* (in press).

Noonan, D. M. and Hassel, J. R. (1993). Proteoglycans of basement membranes. In *Molecular and Cellular Aspects of Basement Membranes* (ed. Rohrbach, D. H. and Timpl, R.) pp. 189-210, New York/London: Academic Press.

Odermatt, E., Engel, J., Richter, H. and Hörmann, H. (1982). Shape, conformation and stability of fibronectin fragments determined by electron microscopy, circular dichroism and ultracentrifugation. *J. Mol. Biol.* **159**, 109-123.

Oldberg, A, Antonsson, P., Lindblom, K. and Heinegard, D. (1992). COMP (cartilage oligomeric matrix protein) is structurally related to the thrombospondins. *J. Biol. Chem.* **267**, 22346-22350.

Ott, U., Odermatt, E., Engel, J., Furthmayr, H. and Timpl, R. (1982). Protease resistance and conformation of laminin. *Eur. J. Biochem.* **123**, 63-72.

Pan, T.-C., Kluge, M., Zhang, R.-Z., Mayer, U., Timpl, R. and Chu, M.-L. (1993). Sequence of extracellular mouse protein BM-90/fibulin and its calcium dependent binding to other basement-membrane ligands. *Eur. J. Biochem.* **215**, 733-740.

Panayotou, G., End, P., Aumailley, M. Timpl, R. and Engel, J. (1989). Domains of laminin with growth-factor activity. *Cell* **56**, 93-101.

Patthy, L. (1991a). Exons – original building blocks of proteins? *Bioessays* **13**, 187-192.

Patthy, L. (1991b). Modular exchange principles in proteins. *Curr. Opin. Struct. Biol.* **1**, 351-361.

Paulsson, M. (1993). Laminin and collagen IV variants and heterogeneity in basement membrane composition. In *Molecular and Cellular Aspects of Basement Membranes* (ed. Rohrbach, D. H. and Timpl, R.) pp. 177-185, New York/London: Academic Press.

Poirier, F. and Robertson, E. (1993). Normal development of mice carrying a null mutation in the gene encoding the L14 S-type lectin. *Development* **119**, 1229-1236.

Pottgiesser, J., Maurer, P., Mayer, U., Nischt, R., Mann, K., Timpl, R., Krieg, T. and Engel J. (1994). Changes in calcium and collagen IV binding caused by mutations in the EF hand and other domains of extracellular matrix protein BM-40 (SPARC, osteonectin). *J. Mol. Biol.* (in press).

Rogers, J. H. (1990). The role of introns in evolution. *FEBS Lett.* **268**, 339-343.

Ruoslahti, E. (1988). Fibronectin and its receptors. *Ann. Rev. Biochem.* **57**, 375-413.

Saga, Y., Yagi, T., Ikawa, Y., Sakakura, T., and Aizawa, S. (1992). Mice develop normally without tenascin. *Genes Dev.* **6**, 1821-1831.

Sealock, R. and Froehner, S. C. (1994). Dystrophin-associated proteins and synapse formation: is a-dystroglycan the agrin receptor? *Cell* **77**, 617-619.

Sneath, P. H. A. and Sokal, R. R. (1973). *Numerical Taxonomy.* San Francisco: Freemann.

Spring, J., Beck, K. and Chiquet-Ehrismann, R. (1989). Two contrary functions of tenascin: Dissection of the active sites by recombinant tenascin fragments. *Cell* **59**, 325-334.

Sung, U., O'Rear, J. J. and Yurchenco, P. D. (1993). Cell and heparin binding in the distal long arm of laminin: identification of active and cryptic sites with recombinant and hybrid glycoprotein. *J. Cell. Biol.* **123**, 1255-1268.

Tschopp, J., Schulthess, T., Engel, J. and Jaton, J.-C. (1980). Antigen-independent activation of the first component of complement C1 by chemically cross-linked rabbit IgG-oligomers. *FEBS Lett.* **112**, 152-154.

Weis, W. I., Drickamer, K. and Hendrickson, W. A. (1992). Structure of a C-type mannose-binding protein complexed with an oligosaccharide. *Nature* **360**, 127-134.

Williams, M. J., Phan, I., Baron, M., Driscoll, P. C. and Campbell, I. D. (1993). Secondary structure of a pair of fibronectin type 1 modules by two-dimensional nuclear magnetic resonance. *Biochemistry* **32**, 7388-7395.

Wilson, R. et al. (1994). 2. 2 MB of contiguous nucleotide sequence from chromosome III of *C. elegans*. *Nature* **368**, 32-38.

Winograd, E., Hume, D. and Branton, D. (1991). Phasing the conformational unit of spectrin. *Proc. Natl. Acad. Sci. USA* **88**, 10788-10791.

Yamagata, M., Yamada, K. M., Yamada, S. S., Shinomura, T., Tanaka, H., Nishida, Y., Obara, M. and Kimata, K. (1991). The complete primary structure of type XII collagen shows a chimeric molecule with reiterated fibronectin type III motifs, von Willebrand factor A motifs, a domain homologous to a non-collagenous region of type IX collagen and short collagenous domains with an Arg-Gly-Asp site. *J. Cell Biol.* **115**, 209-221.

Development 1994 Supplement, 43-51 (1994)
Printed in Great Britain © The Company of Biologists Limited 1994

The *hedgehog* gene family in *Drosophila* and vertebrate development

Michael J. Fietz[1], Jean-Paul Concordet[1], Robert Barbosa[1,*], Randy Johnson[2], Stefan Krauss[1,†], Andrew P. McMahon[3], Cliff Tabin[2] and Philip W. Ingham[1]

[1]Molecular Embryology Laboratory[‡], ICRF Developmental Biology Unit, Department of Zoology, South Parks Road, Oxford, UK
[2]Dept. of Genetics, Harvard University Medical School, 200, Longwood Avenue, Boston, Mass 02115, USA
[3]Dept. of Cellular and Developmental Biology, Harvard University, 16, Divinity Avenue, Cambridge, Mass 02138, USA
*Present address: Wellcome/CRC Institute, Tennis Court Road, Cambridge, CB2 1QK, UK
†Present address: Institute of Medical Biology, Dept. of Biochemistry, University of Tromsø, 9037 Tromsø, Norway.
‡New address: Imperial Cancer Research Fund, Lincoln's Inn Fields, London, WC2 3PX UK

SUMMARY

The segment polarity gene *hedgehog* plays a central role in cell patterning during embryonic and post-embryonic development of the dipteran, *Drosophila melanogaster*. Recent studies have identified a family of *hedgehog* related genes in vertebrates; one of these, *Sonic hedgehog* is implicated in positional signalling processes that show interesting similarities with those controlled by its *Drosophila* homologue.

Key words: *hedgehog*, cell-signalling, floor-plate induction, limb patterning, imaginal discs, segment polarity genes

INTRODUCTION

Although the role of signalling factors in organising cell populations in developing embryos has long been recognised, it is only fairly recently that the molecular nature of these signals has begun to be elucidated. Some of the most notable examples to date are the various proteins found to mimic the mesoderm-inducing capacity of cells of the vegetal hemisphere of early *Xenopus* embryos. These include members of the FGF (Slack et al., 1987) and TGFβ (Green et al., 1990; Kimmelman and Kirshner, 1987) growth factor families; in addition, members of the Wnt family of growth factor-like proteins have been implicated in this process (Christian et al., 1992; Smith and Harland, 1991). While the genes encoding these various protein families have been highly conserved at the structural level throughout evolution, few similarities in their deployment during the embryonic development of species from different phyla have been reported. One possible exception is provided by the *Wnt-1* gene and its *Drosophila* orthologue, the segment polarity gene *wingless*. Activity of *Wnt-1* in the mid-brain of vertebrate embryos appears to be required for the expression of the *Engrailed* genes (McMahon et al., 1992), a regulatory relationship that recalls the interaction between *wingless* and *engrailed*-expressing cells in the developing *Drosophila* embryo (discussed below).

The recent molecular characterisation of the segment polarity gene *hedgehog*, (Lee et al., 1992; Mohler and Vani, 1992; Tabata et al., 1992; Tashiro et al., 1993) has led to the discovery of a new family of putative signal-encoding genes in various vertebrate species that are highly homologous to the *Drosophila* gene (Echelard et al., 1993; Krauss et al., 1993; Riddle et al., 1993; Roelink et al., 1994). The deployment of one of these, *Sonic hedgehog* (*Shh*), in embryos of several different species presents some striking parallels with that of its invertebrate homologue. The *hedgehog* gene family thus provides the first clear example of a conserved signalling factor that regulates analogous processes in species of different phyla.

THE *hedgehog* FAMILY: A NEW CLASS OF SIGNALLING MOLECULES

The *Drosophila hedgehog* gene contains a 471 codon open reading frame (ORF) capable of encoding a polypeptide of M_r 52,147 (Lee et al., 1992). Hydropathy analysis identifies a highly hydrophobic region near the N terminus between residues 63 and 83. In vitro translation analysis suggests that this region may act either as a conventional signal sequence, leading to a secreted form of the protein, or as a membrane spanning domain, anchoring the protein in the cells in which it is expressed (Lee et al., 1992). The results of immunolocalisation analysis on fixed *Drosophila* tissues are consistent with both of these possibilities (Tabata and Kornberg, 1994; Taylor et al., 1993). Thus the properties of the Hh protein may implicate it in either short or long range signalling.

Using a combination of reduced stringency hybridisation and polymerase chain reaction, we have identified a number of *hh*-related genes in the genomes of several vertebrate species including mouse (Echelard et al., 1993), chick (Riddle et al., 1993), *Xenopus* (J-P. C. and P.W.I. unpublished results) and zebrafish (Krauss et al., 1993). The proteins encoded by these genes show a high degree of sequence identity both within and between species which is reflected at the functional level by the ability of the zebrafish *Shh* gene to activate the *Drosophila hh* signal transduction pathway (Krauss et al., 1993; M.J.F. and P.W.I., in preparation).

Alignment of the predicted amino acid sequences of the *Drosophila* Hh protein with those of the mouse Dhh, Ihh and Shh proteins and the chick and zebrafish Shh proteins reveals several interesting features of the *hh* family (see Fig. 1). Like the *Drosophila* protein, all the vertebrate proteins possess an amino-terminal hydrophobic region of approximately 20 residues; however, the initiation codon is located immediately upstream of this region, in contrast to *Drosophila* Hh which initiates some 60 residues upstream of this region. Thus it is likely that in vertebrates this sequence acts exclusively as a signal peptide sequence giving rise to secreted and not membrane spanning proteins. Interestingly, the *Drosophila* gene has a second ATG at a similar position, raising the possibility that it generates different forms of the protein via the control of translational initiation (Lee et al., 1992).

Sequence conservation between the proteins is highest in their amino-terminal ends; indeed, from position 85 (immediately after the predicted shared cleavage site) to position 249, 62% of the residues are completely invariant among the *Drosophila* and vertebrate proteins. Comparison of the different mouse proteins in this more conserved region, indicates that Ihh and Shh are more closely related to each other (90% amino acid identity) than to Dhh (80% identity). Comparison of Shh between species reveals a 99% identity between mouse and chick and 94% identity between mouse and fish in the same region. Conservation falls off rapidly after residue 266, apart from a short stretch at the C terminus.

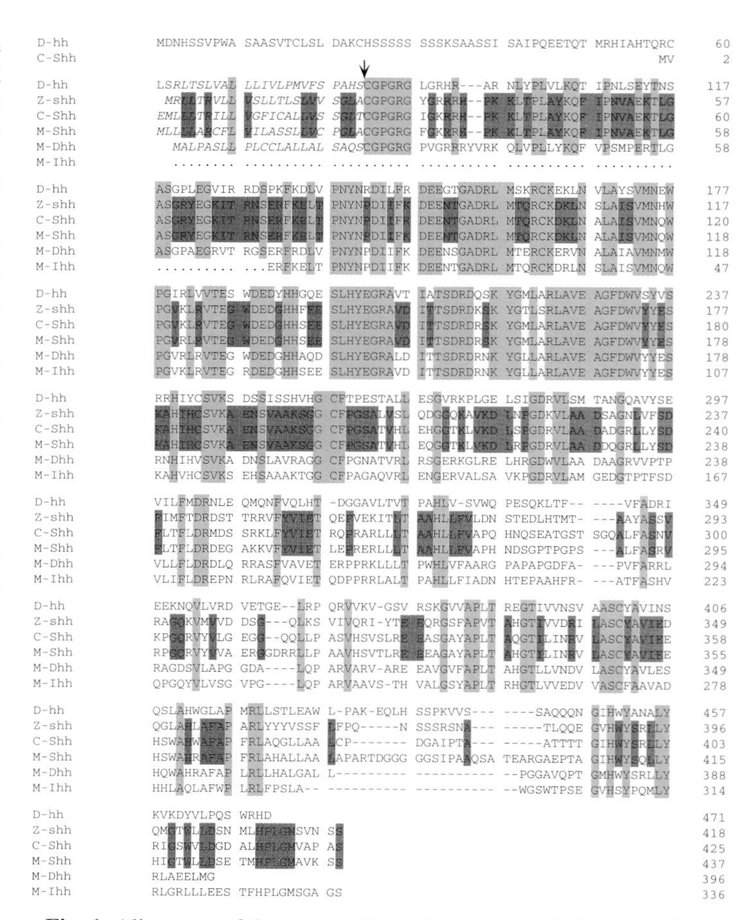

Fig. 1. Alignment of the *Drosophila* and vertebrate *hh*-family amino acid sequences. The predicted hydrophobic transmembrane/signal sequences are indicated in italics; the arrowhead indicates the predicted signal sequence processing site. Amino acids shared by all six proteins are shown in blue; identities between the mouse, chicken and zebrafish Shh proteins are shown in red. The amino acid sequence shown for the zebrafish Shh protein differs slightly from that previously published by Krauss et al., (1993); the corrected nucleic acid sequence from which this is derived is deposited in the EMBL data base under accession number Z35669.

SIGNALLING CENTRES AND *hh* FUNCTION IN THE *DROSOPHILA* EMBRYO

During the early stages of its development, the *Drosophila* embryo is subdivided into a series of repeating units, the parasegments. This subdivision is marked by the activation of the segment polarity genes *wingless* and *engrailed* in a series of discrete bands of cells along the anteroposterior axis of the embryo. Each *wg* domain abuts an adjacent *en* domain and these interfaces define the parasegment boundaries. Genetic studies have shown that parasegment boundaries have special properties, acting as sources of signals that organise the patterning and polarity of the cellular fields which they define (reviewed by Ingham and Martinez Arias, 1992). One of these signals is encoded by *wg* itself: in the absence of *wg* activity, *en* expression is lost from neighbouring cells (Di Nardo et al., 1988; Martinez Arias et al., 1988) and the positional specification of all the cells in each parasegment is disrupted, each cell now adopting a similar fate; this effect is clearly manifested at the end of embryogenesis in the cuticular pattern secreted by the epidermal cells.

Several lines of evidence indicate that the signal produced by *en*-expressing cells is encoded by *hedgehog*. Like *wg*, *hh* activity influences the development of the entire parasegment and embryos homozygous for loss of function *hh* alleles display a phenotype very similar to that seen in *wg* mutants. In the absence of *hh* activity, *wg* transcription is activated normally, but disappears rapidly after gastrulation (Ingham and Hidalgo, 1993). Thus one of the principal functions of *hh* is to maintain the transcription of *wg* in the cells of neighbouring parasegments. Notably, the maintenance of *wg* is restricted to a single row of cells immediately apposed to those expressing *hh*. This characteristic suggests that the range of *hh* activity is extremely limited, perhaps even contact dependent; alternatively, it could be that only these cells are able to respond to the *hh*-encoded signal. This latter possibility can however, be ruled out since in transgenic embryos in which *hh* is expressed ubiquitously, transcription of *wg* is activated ectopically (Ingham, 1993; Tabata and Kornberg, 1994; see Fig. 2). Significantly, this ectopic activation is limited to a subset of cells in each parasegment, immediately anterior to those that normally express *wg*. The capacity of cells to express *wg* in response to the *hh* signal depends upon the activity of the *sloppy paired* (*slp*) gene, a transcription factor belonging to the *forkhead* related family. Activity of *slp* is necessary but not sufficient for *wg* transcription, the *slp* expression domain defining an equivalence group of "*wg*-competent" cells (Cadigan et al., 1994). Thus in normal development, *hh* acts to trigger expression of *wg* in a subset of the cells of this equivalence group, thereby restricting its expression to the parasegment boundary.

The importance of the restricted range of *hh* activity is illus-

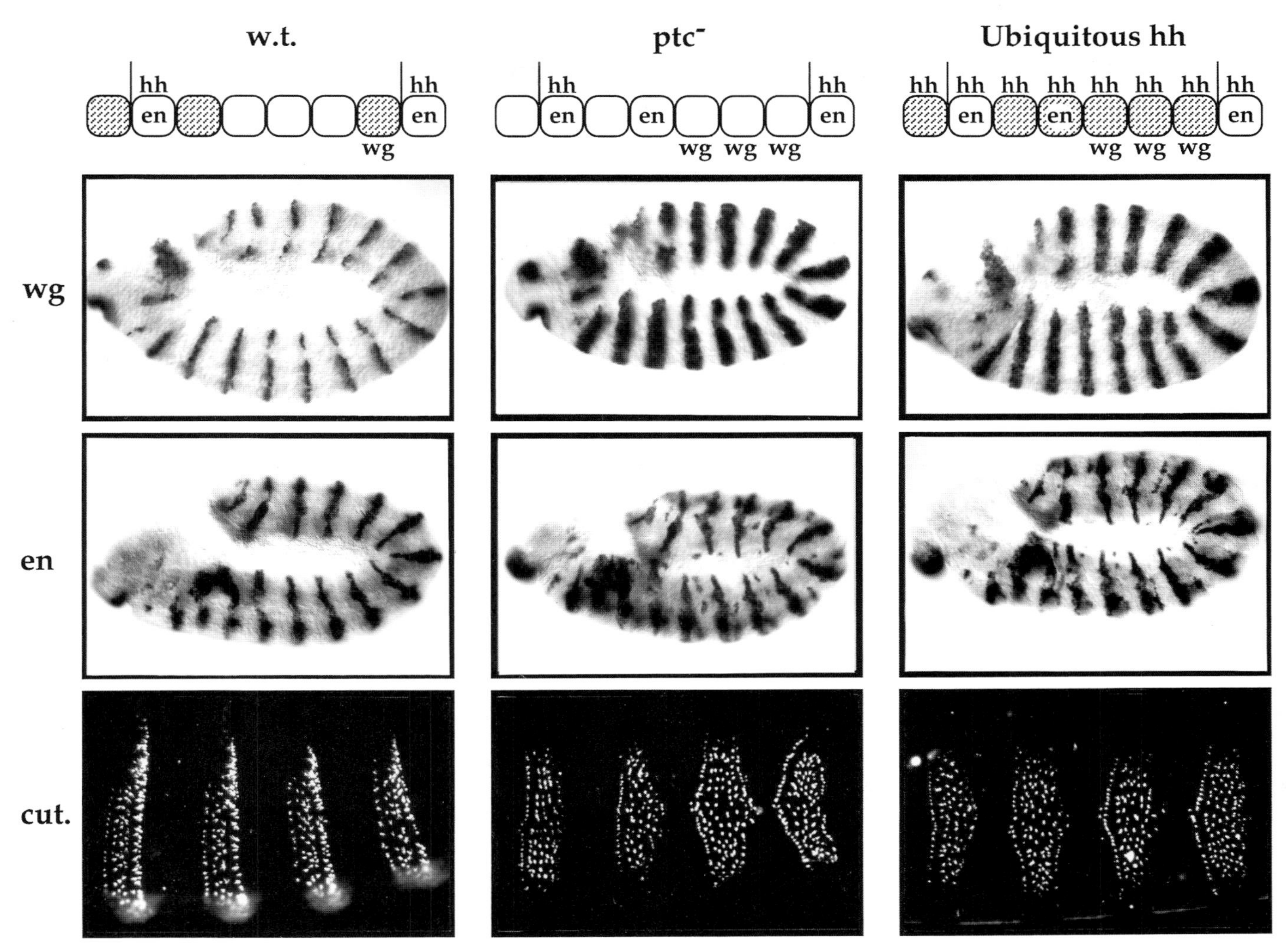

Fig. 2. Patterns of *wingless* (*wg*) and *engrailed* (*en*) expression and ventral cuticular (cut.) differentiation in wild type (w.t.) (left) and *patched* (*ptc*) mutant (centre) embryos and in embryos in which *hedgehog* (*hh*) is ubiquitously expressed (right). The expression domains of each gene in a single parasegment are represented schematically at the top of the figure. Ubiquitous expression of *hh* or absence of *ptc* activity leads to the expansion of the *wg* domain relative to wild-type and the ectopic induction of *en* expression in the centre of each parasegment. These changes in gene activity result in the duplication and deletion of specific pattern elements as manifested in the ventral cuticle.

trated by the pattern defects that ensue when it is overexpressed. Expansion of the *wg* domain results in the ectopic induction of *en* (Tabata and Kornberg, 1994) (Fig. 2). The interface between these ectopically located *en*-expressing cells and their anterior neighbours in turn induces the formation of an additional segment border in each parasegment and this is accompanied by the elimination of certain denticle types and their replacement by others with reversed polarity. These effects mimic precisely the phenotype of mutations of the another segment polarity gene named *patched* (*ptc*) (Martinez Arias et al., 1988; Fig. 2). This finding could suggest a role for *ptc* in restricting the range of the Hh protein and indeed, Hh is much more widely distributed in *ptc* mutant embryos than in wild type (Tabata and Kornberg, 1994; Taylor et al., 1993). Notably, however, activation of *wg* is rendered independent of *hh* activity in the absence of *ptc* function (Ingham and Hidalgo, 1993; Ingham et al., 1991) suggesting instead that the normal role of *ptc* is to suppress the *hh* signalling pathway, leaving it constitutively active in the absence of *ptc*. Since *ptc* encodes an integral membrane protein (Hooper and Scott, 1989; Nakano et al., 1989), one possibility is that the Ptc and Hh proteins interact at the cell surface, the latter inactivating the former and hence triggering the pathway that controls *wg* transcription. Despite this close functional relationship, no homologue of the *ptc* gene has yet been identified in any vertebrate species.

MIDLINE SIGNALLING AND *Sonic hedgehog* EXPRESSION IN VERTEBRATE EMBRYOS

One of the best characterised sources of signalling activity in developing vertebrate embryos is the notochord, the derivative of the axial mesoderm. Several processes have been associated with the inductive properties of this tissue including the induction of specialised ventral neural cells that form the floorplate (Placzek et al., 1990; van Straaten et al., 1989), the specification of neuronal differentiation (Placzek et al., 1991;

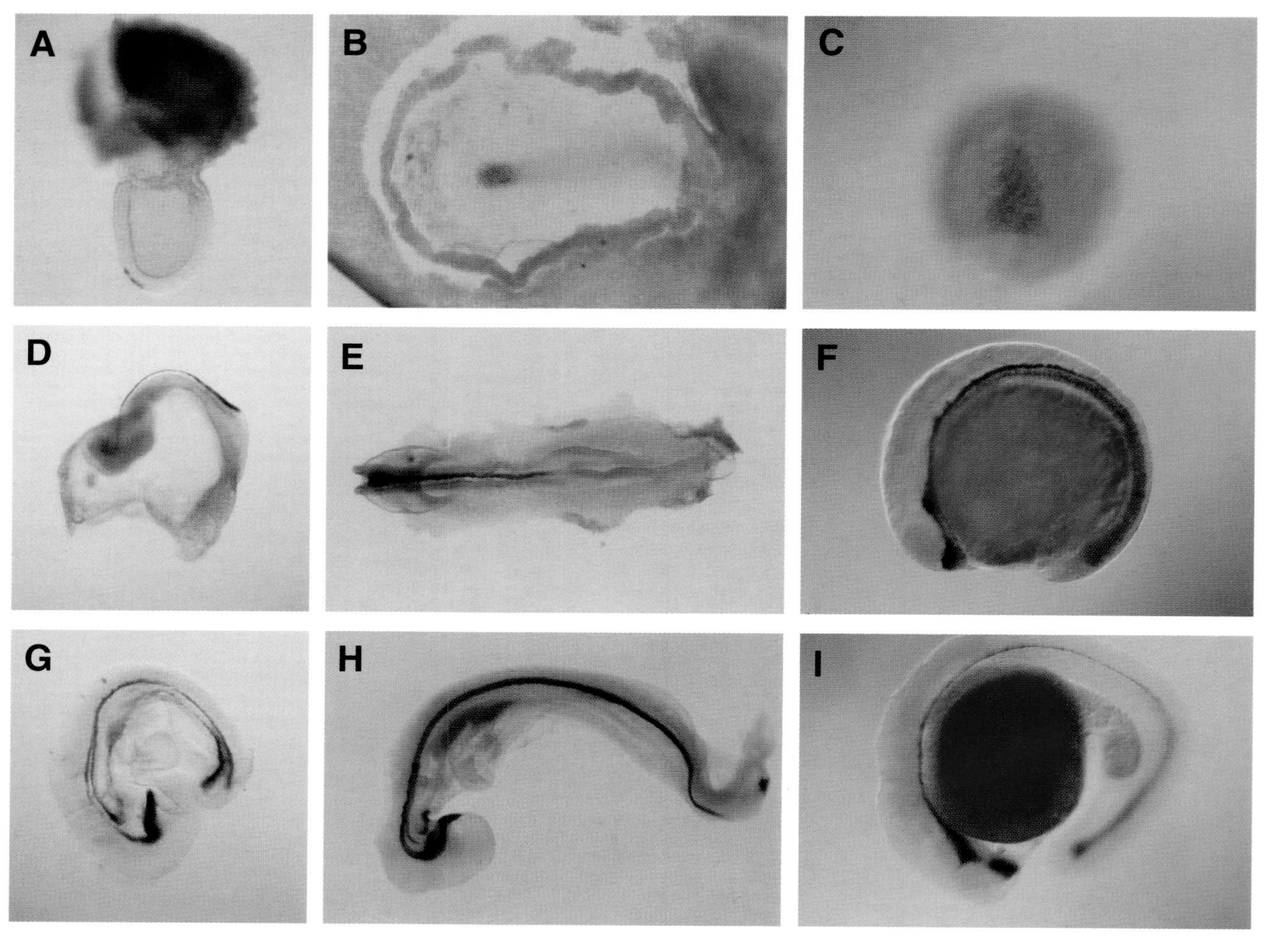

Fig. 3. A comparison of the expression of *Shh* in developing mouse (left), chicken (centre) and zebrafish (right) embryos. (A-C) Onset of *Shh* expression in gastrulation stages; (A) expression in the mouse is restricted to the head process; (B) in the chick, expression is limited to Hensen's node and in the fish (C) to the embryonic shield. (D-F) Early somitogenesis (~8-10 somites); expression is seen throughout the axial mesoderm (presumptive notochord) in all three species and is already detectable in the presumptive floorplate of the fish. (G-I) At later stages of somitogenesis, expression is detectable throughout the ventral floor of the central nervous system; note that in the fish embryo expression has already disappeared from most of the notochord.

Yamada et al., 1991) and the induction of paraxial mesoderm to form scleretome (Dietrich et al., 1993; Koseki et al., 1993; Pourquie et al., 1993).

Evidence for these interactions comes principally from experimental manipulations of developing mouse and chick embryos; in embryos of both species, ablation of the notochord results in a failure of floor plate and motor neuron differentiation, whereas grafting of notochord to ectopic locations in chick embryos results in the induction of ectopic floor plate and motor neurons in close proximity to the graft. Since notochord is closely apposed to floor plate cells both in normal development and in the experimentally manipulated embryos, it has been suggested that the inductive signal must be contact dependent (Placzek et al., 1990), a conclusion supported by the results of in vitro studies (Placzek et al., 1993). Motor neuron differentiation, by contrast, depends upon diffusible factors that act in a contact independent manner (Yamada et al., 1993) and which emanate both from the notochord and the floorplate cells induced by the notochord. Thus the patterning of the neural tube in amniotes can be seen in terms of a sequence of inductive interactions, in which one signalling centre, the notochord, induces another, the floorplate, the activity of which alone can pattern the ventral half of the neural tube. We have found that the putative signal encoding *hh* family gene, *Sonic hedgehog*, is expressed in both the axial mesoderm and the floorplate of mouse and chick embryos (Echelard et al., 1993; Riddle et al., 1993), thus implicating it in at least some of the signalling activities associated with these tissues. Moreover, the spatiotemporal expression pattern of *Shh* is remarkably similar in zebrafish embryos (Krauss et al., 1993) suggesting that the molecular basis of mid-line signalling may be conserved between fishes and amniotes.

Expression of *Shh* is first detectable during gastrulation stages of each species: in the fish embryo, transcripts are restricted to the inner cell layer of the embryonic shield, the equivalent of the amphibian organiser, while in chick, expression is detectable in the homologous structure, Hensen's node (Figure 3B,C). A slight difference is apparent in the mouse at this stage, where expression can first be detected in the midline mesoderm of the head process that arises from the node, though not in the node itself (Figure 3A); however, expression is detectable in the node soon thereafter.

Extension of the body axis of embryos of each species is accompanied by an extension of the *Shh* expression domain. In the zebrafish, by 9.5 hours of development, the *Shh* expression domain constitutes a continuous band of cells that extends from the tail into the head, the anterior boundary of expression being positioned in the centre of the animal pole anterior to the presumptive midbrain. In the mouse and chick, expression similarly extends rostrally from the node, although expression appears limited to the level of the midbrain. Whilst the early phase of *Shh* expression is restricted to the midline mesoderm a new phase of expression in the overlying neuroectoderm is initiated during early somitogenesis. In the mouse, neural expression is first seen at around the 8 somite stage when it is initiated at the ventral midline of the midbrain, above the rostral limit of the head process. Expression extends rapidly both rostrally, into the forebrain, and caudally into the hindbrain and spinal cord. In the chick, neural expression of *Shh* is initiated at the 7-8 somite stage and, in contrast to the mouse embryo, appears simultaneously along almost the entire length of the neural fold. In zebrafish, *Shh* expression is apparent in the embryonic CNS at the 5 somite stage extending from the tip of the forebrain caudally through the hindbrain and rapidly extends caudally along the length of the neural keel. Expression in each species is restricted in the hindbrain and spinal cord to the ventral midline, whilst in midbrain and forebrain , it extends more laterally. Up to the mid-brain forebrain boundary the expressing cells correspond to the morphologically identifiable floorplate; the rostral extension of the *Shh* domain suggests that the ventral forebrain may be functionally homologous to the floorplate in all vertebrates.

The spatiotemporal expression pattern of *Shh* together with the strong conservation of this pattern during vertebrate evolution provides good circumstantial evidence implicating *Shh* in the induction of floorplate and/or motor-neuron differentiation. In line with this possibility, overexpression of *Shh* in fish, frog or mouse embryos is sufficient to induce ectopic expression of the floorplate markers *axial/HNF3*β, *F-spondin* as well as *Shh* itself (Echelard et al., 1993; Krauss et al., 1993; Roelink et al., 1994; J.-P.C. and P.W.I., unpublished data). Furthermore, in vitro assays have shown that the rat *Shh* orthologue, *vhh1* is capable of inducing floorplate and motorneuron differentiation in neural tube explants (Roelink et al., 1994).

Despite the strong similarities between the initial phases of *Shh* expression an interesting difference arises after its induction in the ventral CNS. Whereas in chick and mouse, expression persists in the notochord at least until the end of somitogenesis, in fish, mesodermal expression begins to fade away soon after transcription is activated in the floor plate (Fig. 3G-I). This down-regulation proceeds, like the CNS induction, in a rostral to caudal sequence, coinciding with the changes in cell shape that accompany notochord differentiation. Thus by the 22 somite stage, while *Shh* expression is maintained at high levels throughout the ventral CNS, expression in the mesoderm is restricted to the caudal region of the notochord and to a bulge of undifferentiated cells in the tail bud. Although the significance of this difference is unclear it could reflect a divergence in the mechanisms of CNS patterning between fish and amniotes. One possibility is that floorplate induction represents the original function of *Shh* in vertebrates and that subsequently it has been recruited to an additional midline signalling role, including secondary motorneuron induction, in amniotes. Certainly, the presence of a floorplate in the nerve cord of cephalochordates (Lacalli et al., 1994) implies an ancient origin for floorplate induction, predating the vertebrate radiation. By contrast, whereas signals from both the floorplate and notochord have been implicated in motorneuron differentiation in chick and mouse embryos, the differentiation of primary and secondary motorneurons appears to be independent of any floorplate-derived signal in zebrafish. This conclusion is based upon studies of the *cyclops* mutation, in which floorplate differentiation is blocked but motorneuron differentiation is unaffected (Hatta, 1992); because of the rapid decay of *Shh* transcripts in the notochord, such embryos are devoid of all midline *Shh* expression at the time of motorneuron differentiation, a situation that contrasts with the persistent expression of *Shh* in both floorplate and notochord in amniotes at the equivalent developmental stage. Thus, whereas *Shh* is capable of inducing motorneurons and is expressed at the appropriate time and place in amniote embryos, it appears dispensable for their differentiation in the fish. The persistent expression of *Shh* in the floorplate of fish embryos may reflect some other function in this tissue or it may simply be redundant. Clearly, mutations of *Shh* in fish and mouse will be required to resolve these issues.

Shh AND LIMB PATTERNING IN VERTEBRATE EMBRYOS

In addition to its expression in axial midline structures, *Shh* is transcribed in a cluster of posterior mesenchymal cells in the limb buds of mouse and chick embryos (Echelard et al., 1993; Riddle et al., 1993; Fig. 4). The temporal and spatial pattern of *Shh* expression in these structures suggests a close association between the gene and the organising activity possessed by posterior mesenchymal cells that constitute the so-called zone of polarising activity or ZPA. Transplantation of cells from the *Shh*-expressing region of the limb bud to its anterior margin has long been known to result in the duplication of digits with reversed polarity. This phenomenon has been interpreted in terms of the ZPA acting as a source of a morphogen, a diffusible signal, different levels of whose activity would act to instruct cells to differentiate appropriate to their position within the developing limb field. The pattern duplicating activity of the ZPA can be reproduced by overexpression of *Shh* in cells at the anterior limb margin (Riddle et al., 1993; Fig. 5) strongly suggesting that *Shh* represents the molecular basis of the ZPA. Notably, *Shh* is similarly expressed in the posterior mesenchyme of the pectoral fin buds in fish embryos (Krauss et al., 1993; Fig. 4C), suggesting that the same patterning mechanism operates in these homologous structures.

Since the number and character of duplicated structures caused by ectopic *Shh* expression seems to vary as a function of the level of its activity, one possibility is that Shh protein itself acts as a morphogen. Alternatively, like its postulated floorplate inducing activity in the notochord, Shh may act at short range in the limb, inducing the expression of another sig-

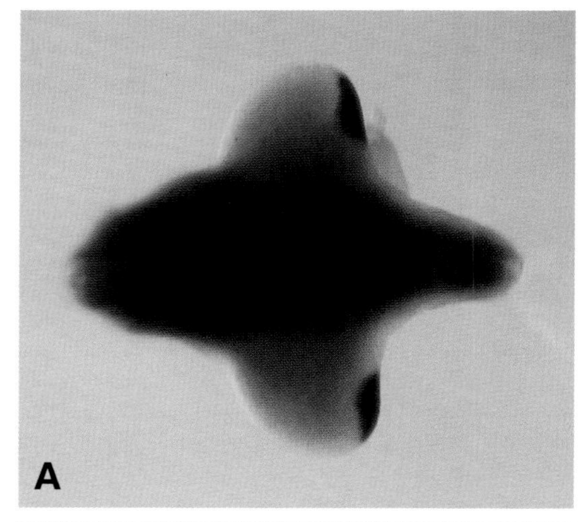

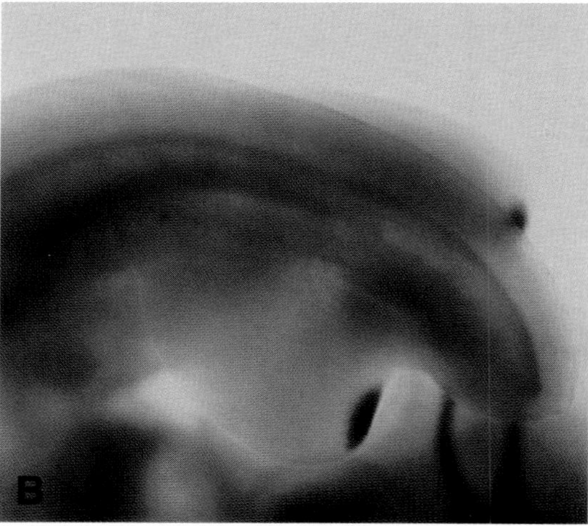

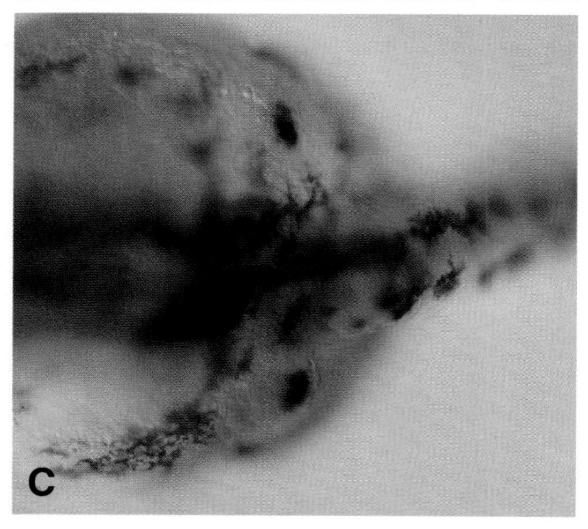

Fig. 4 *Shh* expression in mouse (A) and chick (B) limb buds and in the pectoral fin buds of the zebrafish (C). In all three species, expression is restricted to the posterior mesenchyme.

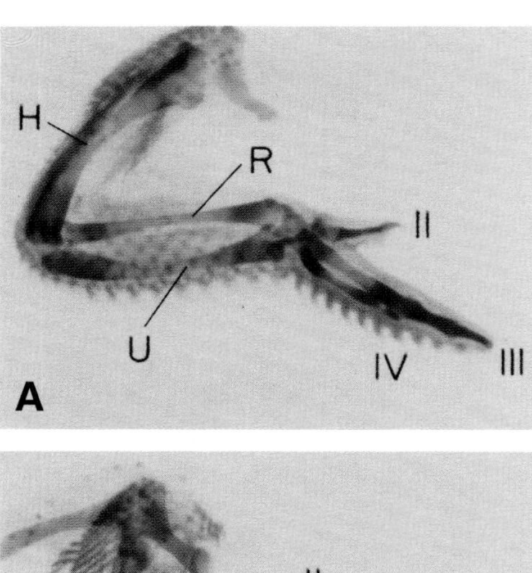

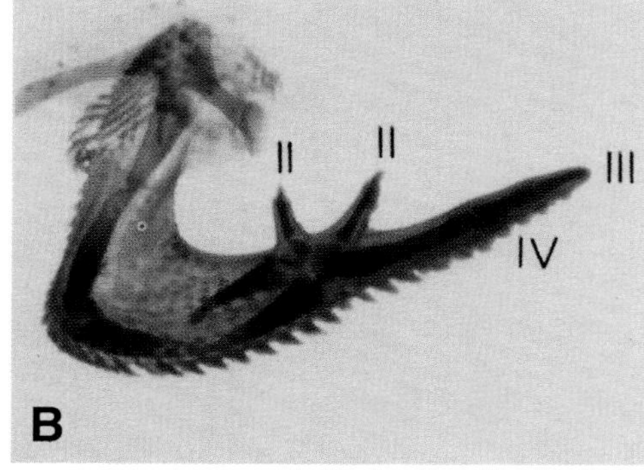

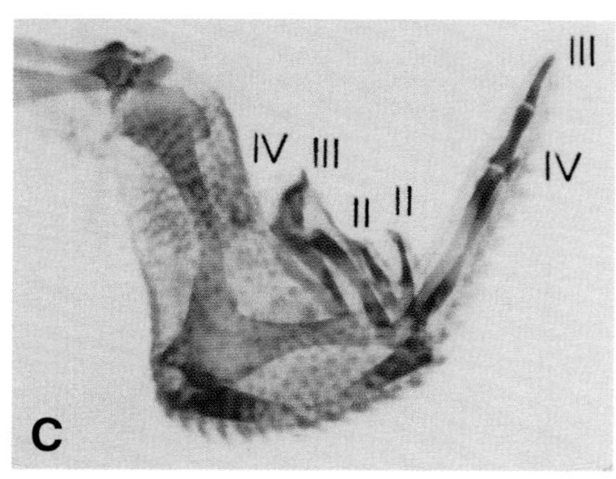

Fig. 5. Digit duplications induced by ectopic *Shh* expression in chick limb buds. (A) Normal limb. (B,C) Examples of the variable pattern duplication induced by grafting of *Shh*-expressing cells into the anterior margin of the limb bud.

nalling molecule or molecules in neighbouring cells. One possible candidate for such a molecule is the TGFβ family member BMP2; the gene encoding this protein is initially transcribed in a restricted domain in the posterior limb mesenchyme (Francis et al., 1994) that overlaps and surrounds the *Shh*-expressing cells (Fig. 6). Moreover *BMP2* transcription is first detectable just after the onset of *Shh* expression (R.J and C.T., unpublished results) and can be induced ectopically in the anterior half of the limb bud both by ZPA grafts (Francis et al., 1994) and by ectopic *Shh* expression (R.J., E. Laufer and C.T. unpublished results). While these observations are consistent with a role for *Shh* in inducing *BMP2* expression, presenting BMP2 as a possible effector of *Shh* activity in limb patterning, functional studies have so far failed to establish such a role for *BMP2* (Francis et al., 1994). Remarkably, however, a similar relationship between *hh* and the *Drosophila BMP2* homologue *decapentaplegic* (*dpp*) appears to underlie the patterning of imaginal discs, the fly equivalent of limb buds.

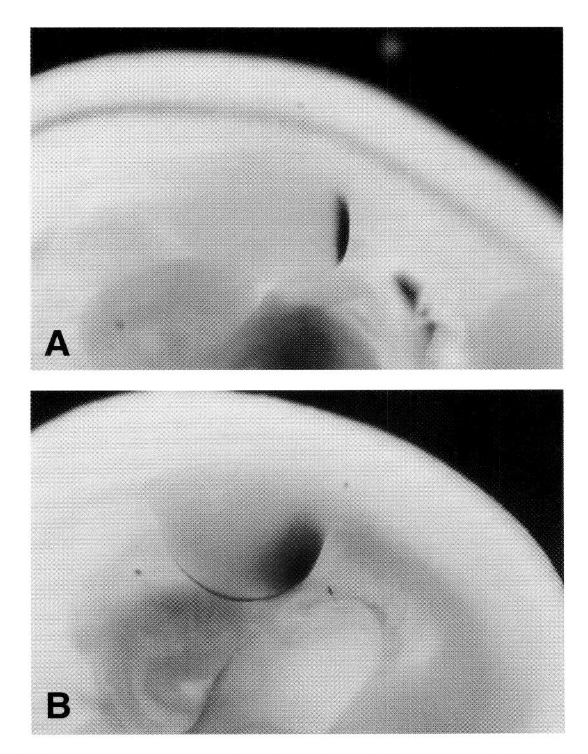

Fig. 6. Overlapping expression domains of *Shh* (A) and *BMP2* (B) in the forelimb of a stage 23 chicken embryo.

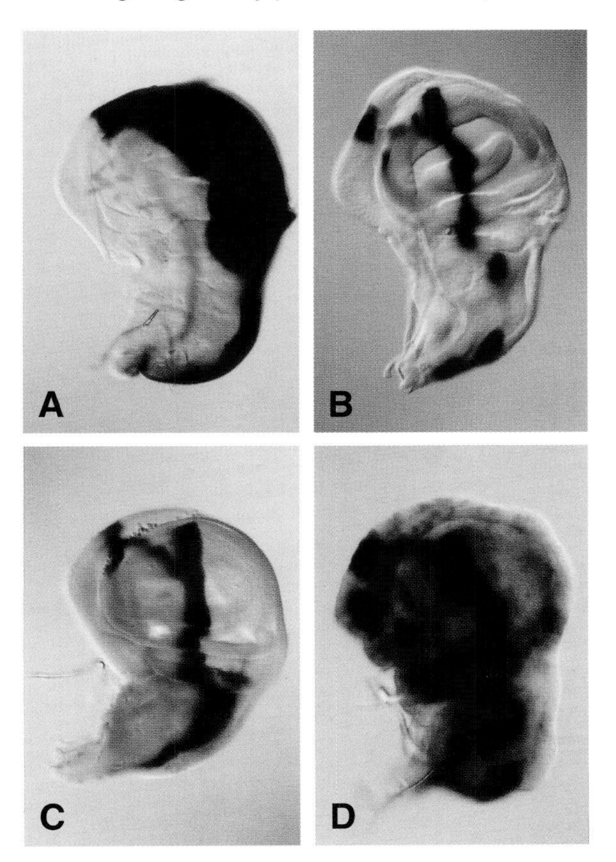

Fig. 7. Expression domains of *hh*, *decapentaplegic* (*dpp*) and *patched* (*ptc*) in wing imaginal discs of third instar *Drosophila* larvae. The expression of *hh* is restricted to the posterior compartment of the wing imaginal disc, revealed here (A) by β-galactosidase staining of an animal carrying an *en-lacZ* reporter gene. *dpp* (B) and *ptc* (C) by contrast are expressed in the anterior compartment, in a stripe of cells that runs along the compartment boundary. Transient ubiquitous expression of *hh* results in the ectopic expression of *dpp* throughout most of the anterior compartment (D).

hh AND THE PATTERNING OF *DROSOPHILA* LIMBS

The limbs or appendages of holometabolous insects develop from imaginal discs, simple epithelial cell sheets whose primordia arise at the parasegment borders of the developing embryo (Bate and Marinez Arias, 1991). This origin means that each disc incorporates and propagates portions of the cell populations that define the parasegmental borders, their progeny forming distinct polyclonal lineages that subdivide the appendages into developmental compartments. The posterior compartment of each disc is thus characterised by the expression of *hh* (Lee et al., 1992; Tabata et al., 1992), whereas *ptc* is expressed in cells of the anterior compartment (Phillips et al., 1990; see Fig. 7).

The function of *hh* in imaginal disc development was first analysed by Mohler (1988) using genetic mosaic techniques to remove the activity of *hh* from cells in different regions of the discs. These experiments demonstrated a requirement for *hh* activity in posterior compartment cells for the correct development of genetically wild-type cells in the neighbouring anterior compartment. We have investigated further this aspect of *hh* function using transgenic animals carrying an *HS-hh* construct to induce transient ectopic expression of *hh* in the anterior compartments of the wing discs. Such ectopic expression results in the duplication of anterior wing structures with mirror image symmetry (see Fig. 8) an effect that shows a striking analogy to the digit duplications induced by ZPA grafts or ectopic *Shh* expression in vertebrate limbs (compare with Fig. 5). The same kinds of duplications have also recently been reported by Basler and Struhl (1994), who used the "flip-out" technique to generate clones of cells expressing *hh* constitutively.

In some cases, ectopic *hh* activity results in duplication of only the most anterior structures, such as the wing margin and veins I and II, (Fig. 8B), whereas in other instances, differentiation of the anterior margin is almost completely suppressed, being replaced by veins II and III (Fig. 8C). As in the case of the chick limb, these variable effects could be indicative of a role for *hh* as a morphogen, different pattern elements being specified by different thresholds of *hh* activity. Several lines of evidence suggest, however, that in the imaginal disc, as in the embryo, *hh* acts in the wing to regulate the transcription of another signal-encoding gene.

Expression of *dpp*, which is absolutely required for normal wing morphogenesis (Posakony et al., 1991; Spencer et al., 1982), is restricted to a narrow band of cells that runs along the antero-posterior compartment boundary of the wing disc (Blackman et al., 1991; Masucci et al., 1990; see Fig. 7), closely apposed to the *hh*-expressing cells of the posterior compartment. In discs in which *hh* has been ectopically activated, *dpp* is similarly inappropriately expressed (Basler and Struhl, 1994; M.J.F. and P.W.I. in preparation; see Fig. 7), implying the latter to be a target of *hh* activity. Ectopic expression of

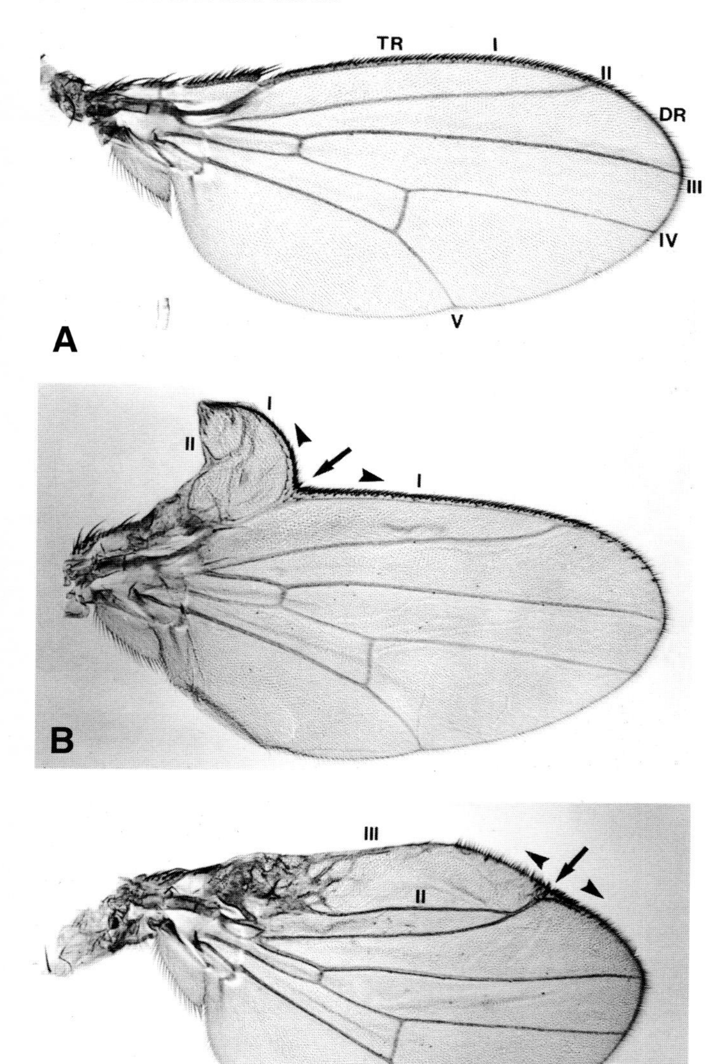

Fig. 8. Duplication and deletion of anterior compartment structures in the wing following transient ubiquitous expression of *hh*. (A) normal wild-type wing showing the characteristic venation pattern. The anterior margin is distinguished by the triple row (TR) and double row (DR) bristles. Veins I, II and III reside in the anterior compartment, veins IV and V in the posterior. (B,C) Examples of the variable mirror image duplications of anterior compartment structures induced by ectopic *hh* activity. The arrowheads indicate the proximodistal polarity of the normal and duplicated structures. the arrow indicates the boundary between normal and duplicated structures.

dpp is similarly induced in imaginal discs from animals with reduced activity of *ptc* (Capdevila et al., 1994; M.J.F and P.W.I. in preparation); thus as in the embryo, over-expression of *hh* has the same effect as the reduction or removal of *ptc* activity, suggesting that the same signalling mechanism acts to regulate *dpp* and *wg* at different stages of development. Thus in both cases, *hh* appears to act to regulate the source of other signalling molecules.

CONCLUSIONS

The parallels between the expression and function of *hh* family genes in *Drosophila* and vertebrate development are indeed striking. In the *Drosophila* embryo, *hh* acts as a localised signal that organises the patterning of each parasegment at least in part by regulating the expression of another signal-encoding gene *wg*. Ectopic expression of *hh* causes inappropriate activation of *wg* which in turn induces the expression of *en* in the middle of each parasegment; the result is a duplication of pattern elements and reversal of polarity that is reminiscent of the polarity reversals and ectopic differentiation induced by notochord grafts in chick embryos. Intriguingly, we have found that a close relative of *hh*, the *Shh* gene is expressed in the developing notochord, the activity of which is likely to be responsible for the inducing properties of this tissue. Thus molecules that have been highly conserved through evolution are deployed in different phyla to effect similar processes in the patterning of secondary fields.

The expression of *hh* family genes in the developing limbs of vertebrates and insects provides a yet more striking example of such functional similarity. In both cases, a member of the *hh* family is expressed in the posterior half of the limb primordium - and in each instance, its ectopic expression results in the duplication of pattern elements. Moreover, in both cases, activity of *Shh* and *hh* appears intimately associated with the expression of closely related members of the TGFβ family, namely *BMP2* and *dpp*. Whereas functional analysis of *dpp* has clearly implicated it in appendage morphogenesis, no such role for *BMP2* has yet been established. Nevertheless, it is difficult to escape the conclusion that, despite their apparently independent evolutionary origin, the limbs of vertebrate and invertebrates may be patterned by very similar mechanisms. Whether the remarkable similarities in the deployment of *hh* genes in the development of deuterostome and protostome embryos reflects a common origin for these various patterning processes or an example of evolutionary convergence remains to be seen. The isolation of *hh* family genes and analysis of their expression in organisms of other phyla should provide important new insights into the origin of the signalling mechanisms that underlie pattern formation in all metazoa.

We are grateful to Ron Blackman for making the *dpp-lacZ* reporter strain available to us. The authors' work was supported by a Human Frontiers Science Programme grant to A.P.M., P.W.I. and C.T. and by the Imperial Cancer Research Fund (P.W. I.) and the National Institutes of Health (C.T.). M.J.F. is a C.J. Martin Fellow of the Australian M.R.C.

REFERENCES

Basler, K. and Struhl, G. (1994). Compartment boundaries and the control of Drosophila limb pattern by hedgehog protein. *Nature* **368**, 208-214.

Bate, C. M. and Martinez. Arias, A. (1991). The embryonic origin of imaginal discs in *Drosophila*. *Development* **112**, 755-761.

Blackman, R. K., Sanicola, M., Raferty, L. A., Gillevet, T. and Gelbart, W. M. (1991). An extensive 3′ cis-regulatory region directs the imaginal disc expression of *decapentaplegic*, a member of the TGF-β family in *Drosophila*. *Development* **111**, 657-665.

Cadigan, K. M., U. Grossniklaus and Gehring, W. J. (1994). Localized expression of sloppy paired protein maintains the polarity of *Drosophila* parasegments. *Genes Dev.* **8**, 899-913.

Capdevila, J., Estrada, M. P., Sanchez-Herrero, E. and Guerrero, I. (1994). The *Drosophila* segment polarity gene patched interacts with decapentaplegic in wing development. *EMBO J.* **13**, 71-82.

Christian, J. L., Olson, D. J. and Moon, R. T. (1992). *XWnt-8* modifies the character of mesoderm induced by bFGF in isolated *Xenopus* ectoderm. *EMBO J.* **11**, 33-41.

Di Nardo, S., Sher, E., Heemskerk-Jongens, J., Kassis, J. and O'Farrell, P. H. (1988). Two tiered regulation of spatially patterned *engrailed* gene expression during *Drosophila* embryogenesis. *Nature* **332**, 604-609.

Dietrich, S., Schubert, F. R. and Gruss, P. (1993). Altered pax gene-expression in murine notochord mutants - the notochord is required to initiate and maintain ventral identity in the somite. *Mech. Dev.* **44**, 189-207.

Echelard, Y., Epstein, D. J., St.-Jacques, B., Shen, L., Mohler, J., McMahon, J. A. and McMahon, A. P. (1993). *Sonic hedgehog*, a member of a family of putative signaling molecules is implicated in the regulation of CNS and limb polarity. *Cell* **75**, 1417-1430.

Francis, P. H., Richardson, M. K., Brickell, P. M. and Tickle, C. (1994). Bone morphogenetic proteins and a signalling pathway that controls patterning in the developing chick limb. *Development* **120**, 209-218.

Green, J. B. A., Howes, G., Symes, K., Cooke, J. and Smith, J. C. (1990). The biological effects of XTC-MIF: quantative comparisons with Xenopus bFGF. *Development* **108**, 173-183.

Hatta, K. (1992). Role of the floor plate in axonal patterning in the zebrafish CNS. *Neuron.* **9**, 629-642.

Hooper, J. and Scott, M. P. (1989). The Drosophila *patched* gene encodes a putative membrane protein required for segmental patternine. *Cell* **59**, 751-765.

Ingham, P. W. (1993). Localised *hedgehog* activity controls spatially restricted transcription of *wingless* in the *Drosophila* embryo. *Nature* **366**, 560-562.

Ingham, P. W. and Hidalgo, A. (1993). Regulation of *wingless* transcription in the *Drosophila* embryo. *Development* **117**, 283-291.

Ingham, P. W. and Martinez Arias, A. (1992). Boundaries and fields in early embryos. *Cell* **68**, 221-235.

Ingham, P. W., Taylor, A. M. and Nakano, Y. (1991). Role of the *Drosophila patched* gene in positional signalling. *Nature* **353**, 184-187.

Kimmelman, D. and Kirshner, M. (1987). Synergistic induction of mesoderm by FGF and TGF-β and the identification of an mRNA coding for FGF in the early Xenopus embryo. *Cell* **51**, 869-877.

Koseki, H., Wallin, J., Wilting, J., Mizutani, Y., Kispert, A., Ebensperger, C., Herrmann, B. G., Christ, B. and R. Balling (1993). A role for pax-1 as a mediator of notochordal signals during the dorsoventral specification of vertebrae. *Development* **119**, 649-660.

Krauss, S., Concordet, J.-P. and Ingham, P. W. (1993). A functionally conserved homolog of the Drosophila segment polarity gene *hedgehog* is expressed in tissues with polarising activity in zebrafish embryos. *Cell* **75**, 1431-1444.

Lacalli, T. C., Holland, N. D. and West, J. E. (1994). Landmarks in the anterior central nervous system of amphioxus larvae. *Philos. Trans R. Soc. Lond. (Biol).* **344**, 165-185.

Lee, J. J., von Kessler, D. P., Parks, S. and Beachy, P. A. (1992). Secretion and localised transcription suggests a role in positional signalling for products of the segmentation gene *hedgehog. Cell* **70**, 777-789.

Martinez Arias, A., Baker, N. E. and Ingham, P. W. (1988). Role of segment polarity gene in the definition and maintenance of cell states in the *Drosophila* embryo. *Development* **103**, 157-170.

Masucci, J. D., Miltenberger, R. J. and Hoffmann, F. M. (1990). Pattern-specific expression of the *Drosophila decapentaplegic* gene in imaginal discs is regulated by 3′ cis regulatory elements. *Genes Dev.* **4**, 2011-2023.

McMahon, A., Joyner, A. L., Bradley, A. and McMahon, J. A. (1992). The midbrain-hindbrain phenotype of $Wnt\text{-}1^{-}/Wnt\text{-}1^{-}$ mice results from stepwise deletion of *engrailed* expressing cells by 9.5 days post-coitum. *Cell* **69**, 581-595.

Mohler, J. (1988). Requirements for *hedgehog*, a segment polarity gene, in patterning larval and adult cuticle of Drosophila. *Genetics* **120**, 1061-1072.

Mohler, J. and Vani, K. (1992). Molecular organisation and embryonic expression of the *hedgehog* gene involved in cell-cell communication in segmental patterning in *Drosophila. Development* **115**, 957-971.

Nakano, Y., Guerrero, I., Hidalgo, A., Taylor, A. M., Whittle, J. R. S. and Ingham, P. W. (1989). The *Drosophila* segment polarity gene *patched* encodes a protein with multiple potential membrane spanning domains. *Nature* **341**, 508-513.

Phillips, R., Roberts, I., Ingham, P. W. and Whittle, J. R. S. (1990). The *Drosophila* segment polarity gene *patched* is involved in a position-signalling mechanism in imaginal discs. *Development* **110**, 105-114

Placzek, M., Jessel, T. M. and Dodd, J. (1993). Induction of floor plate differentiation by contact dependent, homeogenetic signals. *Development* **117**, 205-218.

Placzek, M., Tessier-Lavigne, M., Yamada, T., Jessel, T. and Dodd, J. (1990). Mesodermal control of neural cell identity: floor plate induction by the notochord. *Science* **250**, 985-988.

Placzek, M., Yamada, T., Tessier-Lavigne, M., Jessel, T. and Dodd, J. (1991). Control of dorsoventral pattern in vertebrate neural develpment: induction and polarising properties of the floor plate. *Development* **Supplement**, 105-122.

Posakony, L. G., Raftery, L. A. and Gelbart, W. M. (1991). Wing formation in *Drosophila melanogaster* requires *decapentaplegic* gene function along the antero-posterior compartment boundary. *Mech. Dev.* **33**, 69-82.

Pourquie, O., Coltey, M., Teillet, M. A., Ordahl, C. and Ledouarin, N. M. (1993). Control of dorsoventral patterning of somitic derivatives by notochord and floor plate. *Proc. Nat. Acad. Sci. USA* **90** (11), 5242-5246.

Riddle, R., Johnson, R. L., Laufer, E. and Tabin, C. (1993). *Sonic Hedgehog* mediates the polarizing activity of the ZPA. *Cell.* **75**, 1401-1416.

Roelink, H., Augsburger, A., Heemskerk, J., Korzh, V., Norlin, S., Ruiz i Altaba, A., Tanabe, Y., Placzek, M., Edlund, T., Jessell, T. M. and Dodd, J. (1994). Floor plate and motor neuron induction by *vhh-1*, a vertebrate homolog of *hedgehog* expressed by the notochord. *Cell* **76**, 761-775

Slack, J. M. W., Darlington, B. G., Heath, J. K. and Godsave, S. F. (1987). Mesoderm induction in early *Xenopus* embryos by heparin binding growth factors. *Nature* **326**, 197-200.

Smith, W. and Harland, R. (1991). Injected Xwnt-8 acts early in Xenopus embryos to promote formation of a vegetal dorsalising center. *Cell* **67**, 753-765.

Spencer, F. A., Hoffman, F. M. and Gelbart, W. M. (1982). *Decapentaplegic*: a gene complex affecting morphogenesis in Drosophila melanogaster. *Cell* **28**, 451-461.

Tabata, T., S. Eaton and T. B. Kornberg. (1992). The *Drosophila hedgehog* gene is expressed specifically in posterior compartment cells and is a target of *engrailed* regulation. *Genes Dev.* **6**, 2635-2645.

Tabata, T. and Kornberg, T. B. (1994). Hedgehog is a signalling protein with a key role in patterning drosophila imaginal discs. *Cell* **76**, 89-102.

Tashiro, S., Michiue, T., Higashijima, S., Zenno, S., Ishimaru, S., Takahashi, F., Orihara, M., Kojima, T. and Saigo, K. (1993). Structure and expression of *hedgehog*, a *Drosophila* segment-polarity gene required for cell-cell communication. *Gene* **124** (2), 183-189.

Taylor, A. M., Nakano, Y., Mohler, J. and Ingham, P. W. (1993). Contrasting distributions of patched and hedgehog proteins in the *Drosophila* embryo. *Mech. Dev.* **43**, 89-96.

van Straaten, H. W. M., Hekking, J. W. M., Beursgens, J. P. W. M., Terwindt-Rouwenhorst, E. and Drukker, J. (1989). Effect of the notochord on proliferation and differentiation in the neural tube of the chick embryo. *Development* **107**, 793-803.

Yamada, T., Pfaff, S. L., Edlund, T. and Jessel, T. M. (1993). Control of cell pattern in the neural tube: motor neuron induction by diffusible factors from notochord and floor plate. *Cell* **73** (4), 673-686.

Yamada, T., Placzek, M., Tanaka, H., Dodd, J. and Jessell, T. M. (1991). Control of cell patterning in the developing nervous system: polarizing activity of the floor plate and motochord. *Cell* **64**, 635-647.

Development 1994 Supplement, 53-60 (1994)
Printed in Great Britain © The Company of Biologists Limited 1994

Growth factors in development: the role of TGF-β related polypeptide signalling molecules in embryogenesis

Brigid L. M. Hogan*, Manfred Blessing, Glenn E. Winnier, Noboru Suzuki and C. Michael Jones†

Howard Hughes Medical Institute and Department of Cell Biology, Vanderbilt Medical School, Nashville, TN 37232-2175, USA

*Author for correspondence
†Present address: Laboratory of Developmental Biology, National Institute for Medical Research, The Ridgeway, Mill Hill, London NW7 1AA, UK

SUMMARY

Embryonic induction, the process by which signals from one cell population influence the fate of another, plays an essential role in the development of all organisms so far studied. In many cases, the signalling molecules belong to large families of highly conserved proteins, originally identified as mammalian growth factors. The largest known family is related to Transforming Growth Factor-β (TGF-β) and currently consists of at least 24 different members. Genetic studies in *Drosophila* on the TGF-β related gene, *decapentaplegic* (*dpp*), reveal the existence of conserved mechanisms regulating both the expression of the protein during development and the way in which it interacts with other signalling molecules to generate pattern within embryonic tissues. Comparative studies on another TGF-β related gene, known as *Bone Morphogenetic Protein-4* (*BMP-4*), in *Xenopus* and mouse point to a conserved role in specifying posteroventral mesoderm during gastrulation. Analysis of other polypeptide signalling molecules during gastrulation suggests that their interaction in the generation of the overall body plan has also been conserved during vertebrate evolution.

Key words: TGF-β, embryonic induction, BMP, polypeptide signalling molecules

INTRODUCTION

An essential feature of embryogenesis is the process known as embryonic induction, by which signals produced by one cell population change the developmental fate of another. Inductive events may occur many times during the elaboration of an embryo, and involve several different levels of complexity. At one extreme, the inductive signal works over a relatively short distance and elicits a simple switch in the fate of the responding cells. At the other extreme, the signalling cells act as an organizing center, producing diffusible morphogens, which induce different responses in the target cells depending on their distance from the organizer and the concentration of signal to which they are exposed.

Over the past few years considerable progress has been made in identifying the signalling molecules mediating embryonic induction in both vertebrates and invertebrates. However, much less is known about their receptors and the downstream pathways eliciting cellular responses (for reviews see Smith, 1989; Slack, 1993, 1994). In many cases the signalling molecules belong to large, highly conserved families of proteins related to growth factors, such as fibroblast growth factor (FGF), epidermal growth factor (EGF), the Wnt gene products, and transforming growth factor-β (TGF-β). An attractive hypothesis is that the prototypic ancestral multicellular organism used a relatively small number of such signalling molecules, and associated receptors and signal transduction pathways, to co-ordinate embryogenesis. During evolution, these intercellular communication systems appear to have been conserved and elaborated upon to bring about increasingly complex morphogenetic processes.

THE TGF-β PROTEIN SUPERFAMILY: STRUCTURE, PROCESSING AND CLASSIFICATION INTO DIFFERENT SUBFAMILIES

All members of the TGF-β superfamily, of which there are currently at least 24, are synthesized as large prepro precursor molecules, which are cleaved at an RXXR site to release a C-terminal peptide of 110-140 amino acids (Fig. 1A). In most cases this region contains 7-9 cysteine residues, and studies on the crystal structure of the mature region of TGF-β2 have shown that one of these cysteines is involved in the intermolecular disulfide bonding associated with the formation of biologically active homo- or heterodimers (Fig. 1B; Daopin et al., 1992; Schlunegger and Grutter, 1992). Comparison of the crystal structure of TGF-β2 with that of three other growth factors/hormones, PDGF-BB (platelet-derived growth factor BB), NGF (nerve growth factor) and hCG (human chorionic gonadotrophin) has revealed a remarkable similarity in the three-dimensional structure of the proteins, with two pairs of antiparallel β strands and a conserved arrangement of intertwined disulfide bridges known as the 'cystine knot' (McDonald and Hendrickson, 1993; Lapthorn et al.,1994). Interestingly, all the proteins also form dimers, but the

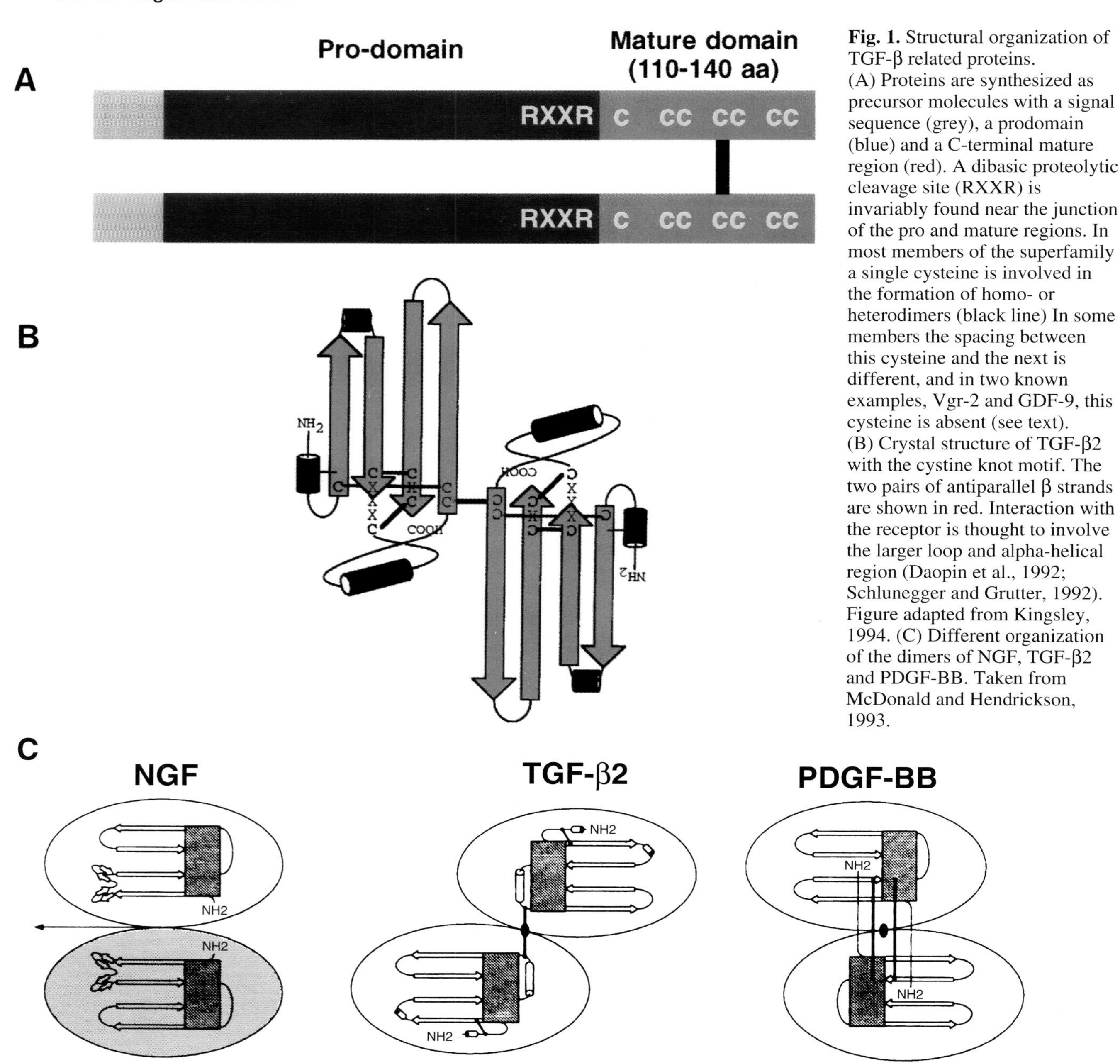

Fig. 1. Structural organization of TGF-β related proteins. (A) Proteins are synthesized as precursor molecules with a signal sequence (grey), a prodomain (blue) and a C-terminal mature region (red). A dibasic proteolytic cleavage site (RXXR) is invariably found near the junction of the pro and mature regions. In most members of the superfamily a single cysteine is involved in the formation of homo- or heterodimers (black line) In some members the spacing between this cysteine and the next is different, and in two known examples, Vgr-2 and GDF-9, this cysteine is absent (see text). (B) Crystal structure of TGF-β2 with the cystine knot motif. The two pairs of antiparallel β strands are shown in red. Interaction with the receptor is thought to involve the larger loop and alpha-helical region (Daopin et al., 1992; Schlunegger and Grutter, 1992). Figure adapted from Kingsley, 1994. (C) Different organization of the dimers of NGF, TGF-β2 and PDGF-BB. Taken from McDonald and Hendrickson, 1993.

protomers interact in very different ways. An intriguing possibility is that all cystine knot proteins have evolved from a common ancestral molecule that was monomeric (Fig. 1C; McDonald and Hendrickson, 1993) but convergent evolution of a very stable structure is also possible. Two TGF-β-related proteins have been identified that lack the cysteine involved in intermolecular disulfide bonding. These are Vgr-2/GDF-3 and GDF-9 (Jones et al., 1992b; McPherron and Lee, 1993). Recent preliminary studies have shown that Vgr-2 RNA injected into *Xenopus* embryos is able to elicit a strong biological response (C. M. J. unpublished results). This suggests that Vgr-2 protein can either act as a monomer or, more likely, form stable dimers through hydrophobic bonding of protomers in the absence of an intermolecular disulfide bridge.

It is assumed, without a great deal of supporting evidence, that all TGF-β precursor molecules are proteolytically cleaved in vivo to release biologically active C-terminal protein. This proteolysis is thought to involve both the RXXR sequence adjacent to the mature region and possibly the degradation of the propeptide that, at least in the case of TGF-β1-3, can form a latent complex with the C-terminal region. However, remarkably little is known about the precise in vivo mechanisms of dimer formation, cleavage of the C-terminal region and further processing, and how these different steps may be regulated particularly during embryonic development. Recent studies with *Xenopus* Vg-1 suggest that under some circumstances post-translational modification is a step at which important regulation can be exerted (Dale et al., 1993;

Thomsen and Melton,1993). Initial experiments in which Vg-1 RNA was injected into *Xenopus* eggs failed to show any biological response. Although synthesis of full length, monomeric protein could be detected with an antibody, no proteolytic cleavage and dimer formation occurs. By contrast, mesoderm induction is seen if the embryo is tricked into processing mature Vg-1 by injecting RNA encoding a chimeric protein, in which the pro region of BMP-2 or BMP-4 (including or not including the cleavage site) is joined to the mature region of Vg-1. This raises the possibility that in the unfertilized egg the Vg-1 protein is blocked from processing or dimer formation or that a Vg-1-specific processing protease is inactive. According to this model, after cortical rotation there is local activation of Vg-1 modification in the dorsal-vegetal blastomere, releasing a small amount of mature protein which then induces the Nieuwkoop center (Thomsen and Melton, 1994).

At present the factors involved in producing active Vg-1 have not been characterised. However, genetic studies in *Drosophila* have suggested that the product of the *tolloid* gene is involved in the activation of DPP, the TGF-β related protein encoded by the *decapentaplegic* gene. The sequence of the tolloid protein is closely related to that of mammalian BMP-1 (Bone Morphogenetic Protein-1), a protein that copurified with the TGF-β related BMPs from demineralized bone (Wozney et al., 1988; Shimell et al., 1991) Both tolloid and BMP-1 have an N-terminal domain related to the astacin family of metalloendopeptidases, as well as CUB and EGF repeats, and proteins related to tolloid have also been found in sea urchins (Hwang et al.,1994). Mutational analysis of *Drosophila* tolloid suggests that the protein does not directly activate DPP but may form a multiprotein complex with it and work indirectly by activating another accessory protein(s) in the processing pathway (Finelli et al., 1994; Childs and O'Connor, 1994). The story is likely to be complicated, however, since tolloid related proteins have recently been found in *Drosophila* (R. Padgett and M. O'Connor, personal communication) suggesting that a gene family may exist. Moreover, it is unclear whether tolloid interacts only with DPP or with other TGF-β related proteins in *Drosophila* such as 60A and screw (Childs and O'Connor, 1994).

Recently, both *Xenopus* and mouse genes closely related to human BMP-1 have been cloned and their expression patterns studied (Maeno et al., 1993; Fukagawa et al., 1994). The mouse protein differs from both human and *Xenopus* BMP-1 in having additional CUB and EGF repeats, making it more similar to tolloid. BMP-1 transcripts are found in the mouse embryo from 7.5 days p.c. on, but are distributed at low levels throughout the mesoderm rather than being localized to a few cell types. However, the gene is expressed at high levels in the floor plate of the spinal cord and midbrain/hindbrain from about 9.5 days where it is regulated directly or indirectly by the transcription factor, HNF-3β (Sasaki and Hogan, 1994). At present, the significance of the localized expression of BMP-1 in the floorplate is unclear. If BMP-1 is playing a role in the activation of a TGF-β related protein one might expect the genes encoding the protease and the substrate to be co-expressed, as seen for *tolloid* and *dpp* in *Drosophila*. However, although several TGF-β related genes are expressed locally in the developing spinal cord and brain (for example, dorsalin in the chick embryo roof plate, as shown by Basler et al., 1993), none has so far been found to be expressed specifically in the floor plate. Further elucidation of the role of putative processing enzymes like tolloid/BMP-1 must await gene knock-out studies in the mouse and the identification of more members of what appears to be a family of evolutionarily conserved proteins.

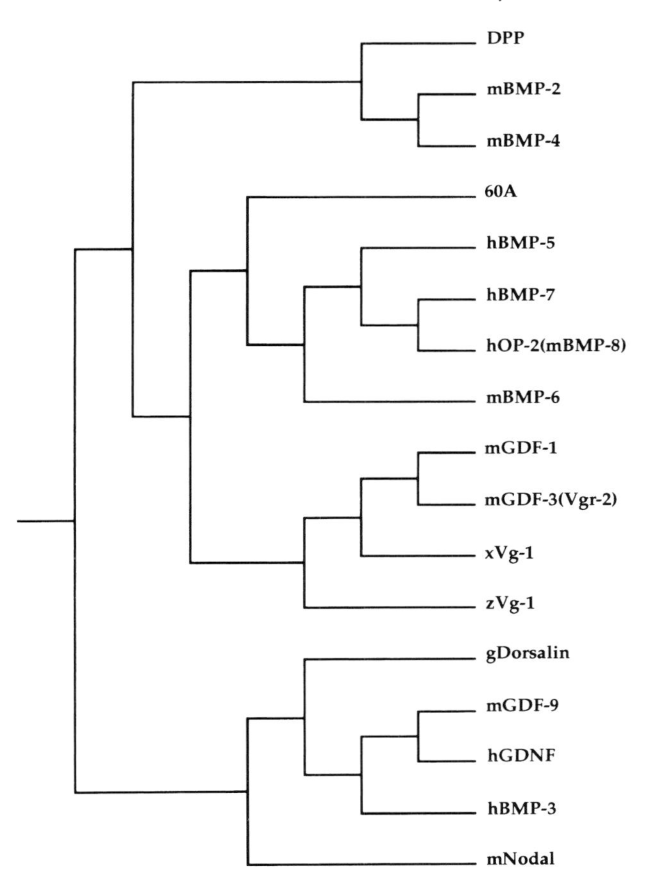

Fig. 2. Possible phylogenetic relationships between members of the DVR (decapentaplegic-Vg-related) subfamily of signalling molecules. This analysis is based on comparison of C-terminal amino acid sequences using PAUP 3.1 (Swofford and Berlocher, 1987) m, mouse; h, human; x, *Xenopus*; z, zebrafish; g, *Gallus* (chick); OP, osteogenic protein; GDF, growth and differentiation factor; Vgr, Vg related.

Comparison of the C-terminal mature regions of TGF-β superfamily proteins suggests that the members can be classified into a number of different subgroups (for recent review see Kingsley, 1994). By far the largest of these is the so-called DVR (Decapentaplegic-Vg-related) subfamily. The evolutionary relationship of members of this subfamily is shown in Fig. 2. Although preliminary, and almost certainly incomplete, this figure suggests that the family has evolved from a few ancestral proteins present before the divergence of vertebrates and invertebrates. In some cases the amino acid sequences of related proteins from insects and mammals are so similar that they can substitute for each other functionally. For example, both *Drosophila* DPP and 60A induce bone when injected subcutaneously into rat skin (Sampath et al., 1993), and Padgett et al. (1993) showed that the C-terminal region of human BMP-4 can substitute for that of DPP in the rescue of mutants in dorsoventral patterning of the embryo.

RECEPTORS

The nomenclature for the receptors of TGF-β-like ligands is based on that originally developed for the cell associated proteins which bind TGF-β itself (for review see Artisano et al., 1992; Massague, 1992). Thus, the type III receptor is a high molecular mass membrane-associated proteoglycan known as betaglycan (Wang et al., 1991), while the type I and II receptors are transmembrane proteins of much smaller size (53 and 70-85×10^3 M_r, respectively). The initial cloning of the *C. elegans daf-1* receptor (Georgi et al., 1990) and the type II receptors for TGF-β and activin (Mathews and Vale, 1991; Lin et al., 1992) finally opened up studies on the signal transduction pathways activated by TGF-β-like molecules. These studies revealed that the type II receptors belong to a novel family of transmembrane serine/threonine kinases. It was then shown that type I receptors also belong to a closely related family of serine/threonine kinases (Attisano et al., 1993; Estevez et al., 1993) and that type I and type II receptors form heteromeric complexes which bind different ligands and regulate different intracellular responses. Stated very simply, it appears at present that type I receptors are somewhat promiscuous and can bind different ligands depending on the more limited, but not totally inflexible, ligand specificity of the type II receptor with which they associate. However, while type II receptors are responsible for initiating certain intracellular signalling pathways, they are inactive in this process without associating with a type I receptor. Clearly, there is a great deal more to be learnt about the functional significance of combinatorial associations of type I and type II receptors and about the role they play in determining the different response of embryonic cells to TGF-β related ligands. Meanwhile, it is likely that the results of experiments in which a dominant-negative type II receptor is overexpressed in embryos (see, for example, Hemmati-Brivanlou and Melton, 1992) will have to be interpreted with caution since the truncated receptor protein may interact with more than one kind of type I receptor and so alter the response of cells to more than one ligand.

GENETIC ANALYSIS OF *decapentaplegic (dpp)* FUNCTION IN *DROSOPHILA* PROVIDES CLUES FOR MECHANISM OF ACTION OF RELATED GENES IN VERTEBRATE EMBRYOS

One of the best studied TGF-β related genes is *decapentaplegic* (*dpp*) in *Drosophila*. Genetic analysis has shown that it is required for cell-cell interaction during several different morphogenetic processes. These include dorsal-ventral patterning of the blastoderm embryo, proximal-distal patterning of the imaginal discs, and midgut morphogenesis. Each example provides an important paradigm for understanding the equally diverse roles of TGF-β related genes in vertebrate embryos. Indeed, as more is learnt about the role of TGF-β related genes in vertebrate embryogenesis, the factors that regulate their expression and the downstream genes that they control, a unifying concept may emerge based on the evolutionary conservation of not just a protein structure but of a whole signalling pathway.

During dorsal-ventral patterning of the *Drosophila* blastoderm embryo *dpp* is transcribed in the dorsal half of the embryo, but is repressed ventrally by the action of the *dorsal* gene product (Irish and Gelbart, 1987; Ferguson and Anderson, 1992; Wharton et al., 1993). Homozygous null mutants of *dpp* show complete ventralization of the embryo, while heterozygotes and hypomorphic mutants show graded deletions of dorsal structures and expansion of the ventral domain. By contrast, adding extra copies of the *dpp* gene, either to wild-type or to mutant embryos, increases the number of amnioserosal cells, which normally differentiate from the most dorsal ectoderm. These observations are all consistent with a model in which a gradient of DPP activity determines cell fate along the dorsal-ventral axis. In other words, in dorsal-ventral patterning, DPP acts as a morphogen. The crucial question is how this activity gradient is established. Among the many possibilities are the following: differential translation of a uniformly distributed mRNA, differential proteolytic processing, dimer formation or binding of accessory proteins that modify the activity of DPP, and graded synthesis of other TGF-β related protein(s) that can form heterodimers with DPP, which are more or less active than homodimers and/or elicit different responses in target cells. The recent identification of a putative type I DPP receptor should also throw new light on the way in which the protein regulates early development (Xie et al., 1994).

Midgut morphogenesis is a second process in the *Drosophila* embryo where *dpp* is required. In many ways it provides a model for understanding the role of TGF-β proteins in epithelial/mesenchymal interactions in vertebrates, since DPP is synthesized by the visceral mesodermal cells and then interacts locally with adjacent epithelial cells of the embryonic gut to influence their differentiation (Immergluck et al., 1990; Panganiban et al., 1990; Reuter et al., 1990). Recent studies have shown that in the mesoderm the *dpp* gene is activated directly by the product of the HOM gene *Ultrabithorax* (*Ubx*), which binds to multiple sites in a 5′ upstream regulatory region. By contrast, abdominal-A (abd-A) inhibits *dpp* expression in the mesoderm by interfering with this Ubx binding (Hursh et al., 1993; Masucci and Hoffmann, 1993; Capovilla et al., 1994). In the endoderm, binding of DPP protein leads to the activation of the HOM gene, *labial*, via a specific response element in the 5′ upstream region of the gene (Tremml and Bienz, 1992). These studies clearly show that *dpp* acts during development both upstream and downstream of homeotic genes, and, as we shall see, this has become an important paradigm for understanding the role of TGF-β related proteins in tissue patterning in higher organisms.

In the leg imaginal disc of *Drosophila*, *dpp* has been shown by genetic analysis to play a role in establishing the proximal-distal axis, and different mutations in *dpp* progressively reduce distal elements. Paradoxically, *dpp* is not expressed along the future P-D axis of the disc, nor in the distal tip, but in a stripe, immediately adjacent to, and just anterior of, the dorsal posterior compartment, which expresses engrailed (en) (Raftery et al., 1991). An important discovery is that *dpp* expression is maintained in this boundary by the short range signalling molecule, hedgehog (hh), produced by the posterior cells (Basler and Struhl, 1994). Another signalling molecule, Wingless (wg), a member of the Wnt family, is expressed in a similar stripe to dpp, but in the ventral half of the disc. Recent evidence suggests a model in which the expression domain of *dpp* can be reconciled with its patterning function in the leg disc. According to this model, local interaction of DPP and wg, at the intersection of the A-P and D-V axes, induces focal expression of the homeobox gene, *aristaless* (*al*) and this establishes an organizing center promoting P-D growth and patterning (Campbell et al., 1993; French and Daniels, 1994). This model, if correct, shows how the interaction of at least two polypeptide signalling molecules, wg and dpp, can set up pattern in an epithelial layer of cells. The recent discovery of a family of hh related proteins in vertebrates (Echelard et al., 1993; Krauss et al., 1993; Riddle et al., 1993; Roelink et al.,1994; Smith, 1994) raises the possibility that they regulate the expression patterns of *dpp*-related genes such as *BMP-2* and *BMP-4*, which are described below.

ESTABLISHING THE VERTEBRATE BODY PLAN: THE ROLE OF TGF-β RELATED SIGNALLING MOLECULES IN *XENOPUS* MESODERM INDUCTION AND SPECIFICATION

Mesoderm formation and patterning in the *Xenopus* embryo is perhaps the best understood example of embryonic induction in vertebrates. To date, several different TGF-β related proteins have been implicated in the overall process, namely Vg-1, activin (β_A and/or β_B), BMP-2 and BMP-4, and two nodal-related proteins, xNR-1 and xNR-2 (Slack, 1993, 1994; Smith,1989; C. M. J,. unpublished results). These are thought to work in conjunction with other polypeptide signalling molecules, including FGFs and Wnts. The complex sequence of interactions starts around the 32-64 cell stage when signals from the vegetal hemisphere induce two types of mesoderm - dorsoanterior and ventroposterior - in the overlying cells of the equatorial zone. Further regionalization of the mesoderm occurs during late blastula and gastrula stages, after which the basic body plan is established.

A detailed analysis of the experimental evidence in favor of the current model of *Xenopus* mesoderm induction and patterning is beyond the scope of this article (for most recent review, see Beddington and Smith, 1993; Slack, 1994). Very briefly, the most favored candidates for inducers of the ventroposterior mesoderm phenotype in the equatorial zone are Vg-1, Wnt-11, FGFs, BMPs and activin. In fact, it seems most likely that they act in a 'cascade', with Vg-1 and Wnt-11 acting as primary mesoderm inducers and FGF, BMPs and activin behaving as secondary factors. Vg-1 may also initiate the formation of the dorsovegetal Nieuwkoop centre. This, in turn, induces the Spemann organizer which releases factors responsible for dorsalizing the mesoderm. Current candidates for potent dorsalizing factors are noggin, nodal-related proteins and activin. Again, these factors may act in combination or in a cascade.

Most simple models for mesoderm patterning in *Xenopus* propose that ventroposterior mesoderm is a kind of 'default' phenotype, assumed by the marginal zone mesoderm furthest removed from the dorsalizing influence of the Spemann organizer. However, studies on the effect of misexpressing BMP-4 have suggested an alternative hypothesis in which BMP-4 (alone, or in combination with Wnt-8) is an active ventralizing factor counteracting or attenuating the dorsalizing factors produced by the organizer (Dale et al., 1992; Jones et al., 1992a). One problem with this model is that although maternal BMP-4 mRNA is present in the *Xenopus* blastula prior to mesoderm induction, it is not localized to any specific region. It is therefore necessary to invoke some differential activation of the mRNA or protein. However, studies have shown that BMP-4 autoinduces its own expression (Jones et al., 1992a), so that a small local production of active protein in the ventral vegetal hemisphere could subsequently be translated into activation of gene transcription in the posterior mesoderm of the marginal zone.

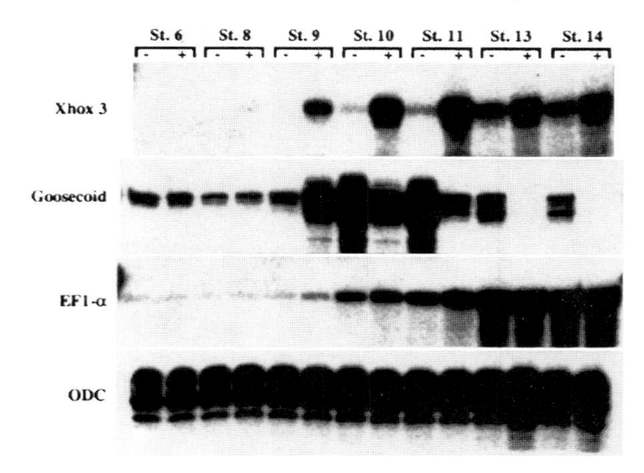

Fig. 3. Effect of BMP-4 on *goosecoid* and *Xhox3* expression in *Xenopus* embryos. BMP-4 RNA was injected into one-cell embryos (0.5-1.0 ng/embryo) as described by Jones et al. (1992a) and incubation continued through different stages. RNAse protection was carried out using approximately three embryo equivalents of RNA for each sample. *Goosecoid* expression is detected maternally (Stages 6 and 8) and through early gastrula (Stage 10). Subsequently, *goosecoid* transcripts rapidly decline in injected embryos (+) compared with controls (−) and become undetectable by the end of gastrulation. In contrast, *Xhox 3* transcipts accumulate to much higher levels in the same BMP-4-injected embryos compared with controls. The precocious expression of *Xhox 3* could mediate the ventralizing effect of BMP-4 during gastrula stages, resulting in the down regulation of goosecoid expression. Accumulation of EF-1α transcripts marks the beginning of zygotic transcription, and ODC serves as a loading control for all lanes.

Studies on the effect of injecting BMP-4 RNA into the fertilized *Xenopus* egg show that, while the injected embryos become ventralized, they apparently develop normally to the early gastrula stage, complete with the formation of a dorsal blastopore lip. This suggests that BMP-4 ventralizes mesoderm during gastrulation, in addition to any possible earlier effects on ventroposterior mesoderm induction. In support of this hypothesis it has recently been shown that *goosecoid*, a gene first expressed in the organizer region of the *Xenopus* embryo (Cho et al., 1991) is expressed at normal (or even slightly higher) levels in BMP-4-injected embryos at the initiation of gastrulation, but is then rapidly down-regulated during mid-gastrula stages (Fig. 3). Furthermore, in the same experiment, transcripts for *Xhox 3*, a homeobox containing gene previously shown to be preferentially expressed in posterior regions (Ruiz i Altaba and Melton, 1989a,b), accumulate to much higher levels in BMP-4-injected embryos than in control embryos of the same stage (Fig. 3). These data support the idea that the dorsal lip that forms in BMP-4-injected embryos initially has properties of a normal organizer, but that signals induced by BMP-4, possibly through regulation of the homeobox gene, *Xhox 3*, subsequently block differentiation of dorsal mesoderm. Further support for this hypothesis comes from experiments using a DNA construct in which the regulatory elements of the cytoskeletal actin gene drive expression of BMP-4 only after the initiation of zygotic transcription (C. M.

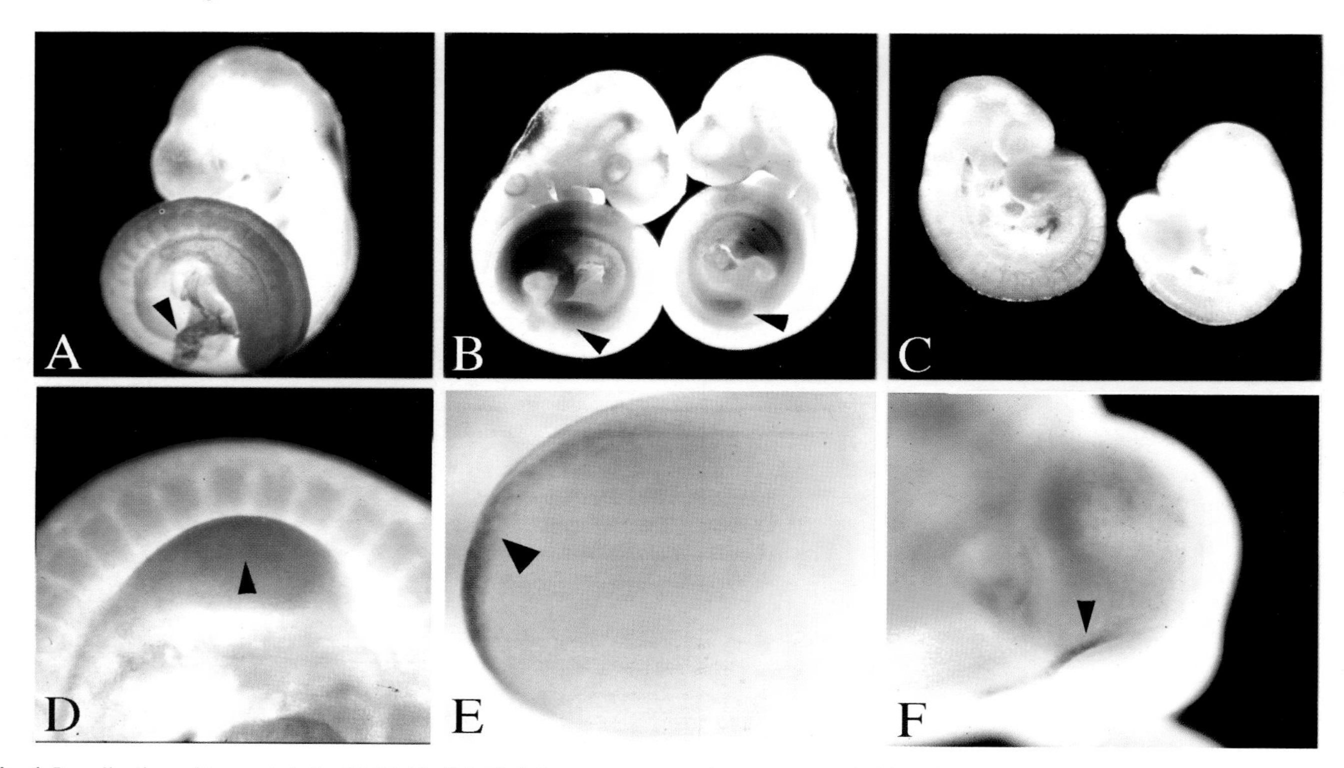

Fig. 4. Localization of transcripts for BMP-4 in 9.0-10.5 day p.c. mouse embryos, as revealed by whole-mount in situ hybridization. (A) 9.0-day embryo showing strong expression in the posterior mesoderm and around the umbilical vessels (arrowhead). (B) Two 9.5-day embryos, showing expression in the posterior mesoderm and in the mesenchyme of the forelimb bud (arrowheads). (C) Control embryos hybridized with sense strand probe. (D) High power magnification of the limb shown in B, with expression throughout the mesenchyme. (E) By 10.5 days strong BMP-4 expression is seen in the apical ectodermal ridge of the forelimb bud (arrowhead). (F) In the 10.5-day embryo BMP-4 transcripts are also seen in the ectoderm of the nasal pits derived from the first branchial arch (arrow). For probe construction, see Jones et al. (1991). The approximately 1 kb insert contains 360 bp of 5′ noncoding sequence, followed by coding sequence up to the beginning of the conserved region. It contains no sequences coding for the mature region. The whole-mount in situ hybridization was carried out essentially as described by Sasaki and Hogan, (1993).

J., unpublished observations). In this case, BMP-4 has the same ventralizing effect, and *Xhox 3* transcripts accumulate with the same kinetics, as when BMP-4 RNA is injected into the fertilized egg. The conclusion from these different studies is that BMP-4 elicits its ventralizing effect on mesoderm during gastrulation. As we shall see, this conclusion is compatible with the proposed role for BMP-4 during late gastrulation in the mouse embryo and suggests that this function of BMP-4 has been conserved during vertebrate evolution.

ESTABLISHING THE VERTEBRATE BODY PLAN: BMP-4 EXPRESSION IN THE EARLY MOUSE EMBRYO

BMP-4 is one of the first DVR genes to be expressed in the mouse embryo. cDNAs were isolated from a 6.5-day p.c. cDNA library (Jones et al., 1991) but whole-mount in situ hybridization does not detect transcripts until around 7.5 days, at low levels in the posterior primitive streak, allantois and amnion. By 8.5 days, higher levels are present in the posterior of the embryo, specifically in the mesoderm of the primitive streak and around the hindgut and in the ventral mesoderm caudal to the last somite (Jones et al., 1991 and data not shown). At approximately this time BMP-4 also begins to be

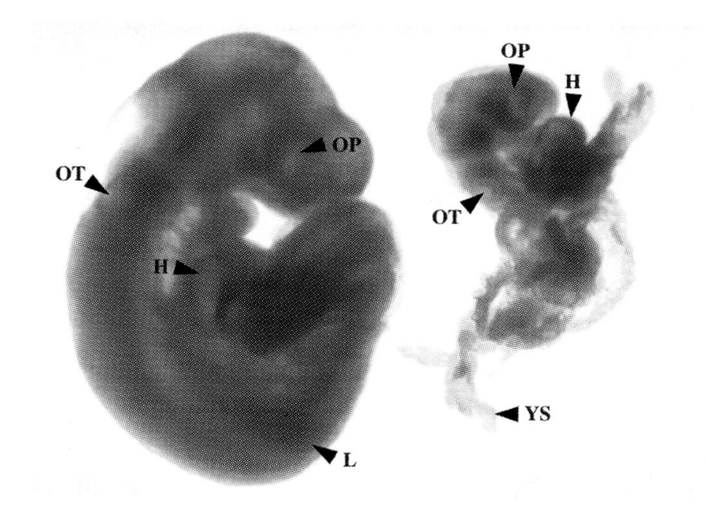

Fig. 5. Morphology of a homozygous null BMP-4 mouse embryo. Male and female (129 × C57BL/6) F_2 mice, heterozygous for a null mutation in the *BMP-4* gene, were mated and embryos collected at 9.5 days and genotyped by analysis of DNA from extraembryonic tissues. Shown here is a heterozygous embryo (left) and a homozygous mutant littermate (right). Note the disorganized posterior region of the homozygous embryo OT, otic vesicle; OP, optic pit; H, heart; YS, yolk sac.

expressed in the myocardium of the heart. At 9.0-9.5 days, whole-mount in situ hybridization clearly shows BMP-4 transcripts in the posterior and around the umbilical vessels (see Fig. 4). Sectioning these embryos reveals that posterior transcripts are localized in the somatopleure and splanchnopleure mesoderm. Expression is also detected in the mesenchyme of the limb bud and in the neuroepithelium of the diencephalon, in the ectoderm of the branchial arches, and in the myocardium of the heart (Fig. 4 and Jones et al., 1991). As the limbs develop, strong expression is seen in the apical ectodermal ridge, and transcripts in the mesenchyme become localized to the posterior, before shifting anteriorly. Some details of this limb bud expression are shown in Fig. 4 and are discussed fully elsewhere (Francis et al., 1994).

In summary, high levels of BMP-4 expression are first seen in the mouse embryo in the posterior primitive streak, and in ventral mesoderm around the posterior gut and umbilical blood vessels, and in the body wall (i.e. in the splachnopleure and somatopleure). This pattern of expression is consistent with BMP-4 playing a role in the specificiation of posterior and ventral mesoderm, as hypothesised from experiments described earlier with *Xenopus*. Later, expression of BMP-4 is seen in a variety of tissues undergoing mesenchymal/epithelial interactions, and models have been proposed by us (Jones et al., 1991; Lyons et al., 1991; Jones et al., 1992a) and by others (Francis et al., 1994; Vainio et al., 1993) in which BMP-4 and the closely related protein, BMP-2, play key roles in mediating the intercellular signalling involved.

In order to explore the role of BMP-4 in vivo, a null mutation has been introduced into the mouse gene by homologous recombination in embryonic stem cells (M. B., unpublished results). The targeting construct was designed to eliminate most of the first coding exon, and therefore most of the pro region of the protein, and to introduce a stop codon into all three reading frames of the second coding exon. Heterozygous mice appear normal, but homozygous embryos die between about 7.5 and 10.5 days p.c., with a rather variable phenotype, which probably depends upon genetic background. A 9.5-day p.c. homozygous embryo is shown in Fig. 5 along with a heterozygous littermate. It is retarded in overall growth, but anterior structures such as fore-, mid- and hind-brain, optic and otic vesicles, and heart are all present. However, the regions of the embryo posterior to the heart are disorganized. The abnormalities also include the extraembryonic mesoderm of the yolk sac, which shows fewer blood islands than normal and the mesoderm does not adhere closely to the endoderm. It is possible that some of the growth retardation of the homozygotes is due to anaemia resulting from a deficiency of blood cells and abnormalities of the blood vessels connecting the yolk sac with the embryo. In addition, there appears to be a general deficiency of posterior structures and splanchnopleure mesoderm. Further analysis is underway of both the abnormal phenotype of homozygous mutant embryos, and the effect of genetic background on the penetrance of the null mutation. However, while preliminary, these findings support the idea that BMP-4 is required for the normal differentiation of posterior and ventral mesoderm of the mouse embryo, and suggests that this function has been conserved during vertebrate evolution.

Work from the authors' laboratory was supported in part by NIH grants HD 28955 and CA48799. Brigid Hogan is an Investigator of the Howard Hughes Medical Institute.

REFERENCES

Artisano, L., Wrana, J. L., Cheifetz, S. and Massague, J. (1992). Novel activin receptors: distinct genes and alternative mRNA splicing generate a repertoire of serine/threonine kinase receptors. *Cell* **68,** 97-108.

Attisano, L., Carcamo, J., Ventura, F., Weis, F. M. B., Massague, J. and Wrana, J. L. (1993). Identification of human activin and TGFβ type I receptors that form heteromeric kinase complexes with Type II receptors. *Cell* **75,** 671-680.

Basler, K., Edlund, T., Jessell, T. M. and Yamada, T. (1993). Control of cell pattern in the neural tube: regulation of cell differentiation by dorsalin-1, a novel TGFβ family member. *Cell* **73,** 687-702.

Basler, K. and Struhl, G. (1994). Compartment boundaries and the control of Drosophila limb pattern by hedgehog protein. *Nature* **368,** 208-214.

Beddington, R. S. P. and Smith, J. C. (1993). The control of vertebrate gastrulation: inducing signals and responding genes. *Curr. Opin. Genet. Dev.* **3,** 655-661.

Campbell, G., Weaver, T. and Tomlinson, A. (1993). Axis specification in the developing Drosophila appendage: the role of wingless, decapentaplegic, and the homeobox gene aristaless. *Cell* **74,** 1113-1123.

Capovilla, M., Brandt, M. and Botas, J. (1994). Direct regulation of decapentaplegic *by Ultrabithorax* and its role in Drosophila midgut morphogenesis. *Cell* **76,** 461-475.

Childs, S. R. and O'Connor, M. B. (1994). Two domains of the *tolloid* protein contribute to its unusual genetic interaction with *decapentaplegic*. *Dev. Biol.* **162,** 209-220.

Cho, K. W. Y., Blumberg, B., Steinbeisser, H. and De Robertis, E. M. (1991). Molecular nature of Spemann's organizer: the role of the Xenopus homeobox gene goosecoid. *Cell* **67,** 1111-1120.

Dale, L., Howes, G., Price, B. M. J. and Smith, J. C. (1992). Bone morphogenetic protein 4: a ventralizing factor in early *Xenopus* development. *Development* **115,** 573-585.

Dale, L., Matthews, G. and Colman, A. (1993). Secretion and mesoderm-inducing activity of the TGF-β-related domain of Xenopus Vg1. *EMBO J.* **12,** 4471-4480.

Daopin, S., Piez, K. A., Ogawa, Y. and Davies, D. R. (1992). Crystal structure of transforming growth factor-β2: an unusual fold for the superfamily. *Science* **257,** 369-373.

Echelard, Y., Epstein, D. J., St-Jacques, B., Shen, L., Mohler, J., McMahon, J. A. and McMahon, A. P. (1993). Sonic Hedgehog, a member of a family of putative signaling molecules, is implicated in the regulation of CNS polarity. *Cell* **75,** 1417-1430.

Estevez, M., Attisano, L., Wrana, J. L., Albert, P. S., Massague, J. and Riddle, D. L. (1993). The *daf-4* gene encodes a bone morphogenetic protein receptor controlling *C. elegans* dauer larva development. *Nature* **365,** 644-649.

Ferguson, E. L. and Anderson, K. V. (1992). *decapentaplegic* acts as a morphogen to organize dorsal-ventral pattern in the *Drosophila* embryo. *Cell* **71,** 451-461.

Finelli, A. L., Bossie, C. A., Xie, T. and Padgett, R. W. (1994). Mutational analysis of the Drosophila tolloid gene, a human BMP-1 homolog. *Development* **120,** 861-870.

Francis, P. H., Richardson, M. K., Brickell, P. M. and Tickle, C. (1994). Bone morphogenetic proteins and a signalling pathway that controls patterning in the developing limb bud. *Development* **120,** 209-218.

French, V. and Daniels, G. (1994). The beginning and the end of insect limbs. *Current Biology* **4,** 34-37.

Fukagawa, M., Susuki, N., Hogan, B. L. M. and Jones, C. M. (1994). Embryonic expression of mouse bone morphogenetic protein-1 (BMP-1) which is related to the Drosophila dorsoventral gene tolloid and encodes a putative astacin metalloendopeptidase. *Dev. Biol.* **163,** 175-183

Ferguson, E. L. and Anderson, K. V. (1992). Localized enhancement and repression of the activity of the TGFβ family member, decapentaplegic, is necessary for dorsal-ventral pettern formation in the *Drosophila* embryo. *Development* **114,** 583-597.

Georgi, L. L., Albert, P. S. and Riddle, D. L. (1990). daf-1, a C. elegans gene controlling dauer larva development, encodes a novel receptor protein kinase. *Cell* **61,** 635-645.

Hemmati-Brivanlou, A. and Melton, D. A. (1992). A truncated activin receptor inhibits mesdoderm induction and formation of axial structures in Xenopus embryos. *Nature* **359,** 609-614.

Hursh, D., Padgett, R. W. and Gelbart, W. M. (1993). Cross regulation of *decapentaplegic* and *ultrabithorax* transcription in the embryonic visceral mesoderm of Drosophila. *Development* **117,** 1211-1222.

Hwang, S.-P. L., Partin, J.S. and Lennarz, W.J. (1994). Characterization of a homolog of human bone morphogenetic protein 1 in the embryo of the sea urchin, *Strongylocentrotus purpuratus*. *Development* **120,** 559-568.

Immergluck, K., Lawrence, P. A. and Bienz, M. (1990). Induction across germ layers in Drosophila mediated by a genetic cascade. *Cell* **62,** 261-268.

Irish, V. F. and Gelbart, W. M. (1987). The *decapentaplegic* gene is required for doral-ventral pattering of the *Drosophila* embryo. *Genes Dev.* **1,** 868-879.

Jones, C. M., Lyons, K. M. and Hogan, B. L. M. (1991). Involvement of Bone Morphogenetic protein-4(BMP-4) and Vgr-1 in morphogenesis and neurogenesis in the mouse. *Development* **111,** 531-542.

Jones, C. M., Lyons, K. M., Lapan, P. M., Wright, C. V. E. and Hogan, B. L. M. (1992a). DVR-4 (Bone Morphogenetic Protein-4) as a posterior-ventralizing factor in *Xenopus* mesoderm induction. *Development* **115,** 639-647.

Jones, C. M., Simon-Chazottes, D., Guenet, J.-L. and Hogan, B. L. M. (1992b). Isolation of Vgr-2, a novel member of the transforming growth factor-β-related gene family. *Molec. Endocrinol.* **6,** 1961-1968.

Kingsley, D. M. (1994). The TGF-β superfamily: new members, new receptors, and new genetic tests of function in different organisms. *Genes Dev.* **8,** 133-146.

Krauss, S., Concordet, J.-P. and Ingham, P. W. (1993). A functionally conserved homolog of the Drosophila segment polarity gene *hh* is expressed in tissues with polarizing activity in zebrafish embryos. *Cell* **75,** 1431-1444.

Lapthorn, A. J., Harris, D. C., Littlejohn, A., Lustbader, J. W., Canfield, R. E., machin, K. J., Morgan, F. J. and Isaacs, N. W. (1994). Crystal structure of human chorionic gonadotrophin. *Nature* **369,** 455-461.

Lin, H. Y., Wang, X.-F., Ng-Eaton, E., Weinberg, R. A. and Lodish, H. F. (1992). Expression cloning of the TGF-β type II receptor, a functional transmembrane serine/threonine kinase. *Cell* **68,** 775-785.

Lyons, K. M., Jones, C. M. and Hogan, B. L. M. (1991). The DVR gene family in embryonic development. *Trends Genet.* **1992,** 408-412.

Maeno, M., Xue, Y., Wood, T. I., Ong, R. C. and Kung, H. (1993). Cloning and expression of CDNA encoding Xenopus laevis bone morphogenetic protein-1 during early embryonic development. *Gene* **134,** 257-261.

Massague, J. (1992). Receptors for the TGF-β family. *Cell* **69,** 1067-1070.

Masucci, J. D. and Hoffmann, M. F. (1993). Identification of two regions from the *Drosophila decapentaplegic* gene required for embryonic midgut development and larval viability. *Dev. Biol.* **159,** 276-287.

Mathews, L. S. and Vale, W. W. (1991). Expression cloning of an activin receptor, a predicted transmembrane serine kinase. *Cell* **65,** 973-982.

McDonald, N. Q. and Hendrickson, W. A. (1993). A structural superfamily of growth factors containing a cystine knot motif. *Cell* **73,** 421-424.

McPherron, A. C. and Lee, S.-J. (1993). GDF-3 and GDF-9: Two new members of the transforming growth factor-β superfamily containing a novel pattern of cysteines. *J. Biol. Chem* **268,** 3444-3449.

Panganiban, G. E. F., Reuter, R., Scott, M. P. and Hoffmann, F. M. (1990). A *Drosophila* growth factor homolog, decapentaplegic, regulates homeotic gene expression within and across germ layers during midgut morphogenesis. *Development* **110,** 1041-1050.

Padgett, R.W., Wozney, J. and Gelbart, W.M. (1993). Human BMP sequences can confer normal dorsal-ventral patterning in the Drosophila embryo. *Proc. Natl. Acad.Sci. USA* **90,** 2905-2909.

Raftery, L. A., Sanicola, M., Blackman, R. K. and Gelbart, W. M. (1991). The relationship of decapentaplegic and engrailed expression in *Drosophila* imaginal disks: do these genes mark the anterior-posterior compartment boundary? *Development* **113,** 27-33.

Reuter, R., Panganiban, G. E. F., Hoffmann, F. M. and Scott, M. P. (1990). Homeotic genes regulate the spatial expression of putative growth factors in the visceral mesoderm of *Drosophila* embryos. *Development* **110,** 1031-1040.

Riddle, R. D., Johnson, R. L., Laufer, E. and Tabin, C. (1993). *Sonic hedgehog* mediates the polarizing activity of the ZPA. *Cell* **75,** 1401-1416.

Roelink, H., Augsburger, A., Heemskerk, J., Korzh, V., Norlin, S., Altaba, R. i., Tanabe, Y., Placzek, M., Edlund, T., Jessell, T. M. and Dodd, J. (1994). Floor plate and motor neuron induction by *vhh-1*, a vertebrate homolog of *hedgehog* expressed by the notochord. *Cell* **76,** 761-775.

Ruiz i Altaba, A. and Melton, D. A. (1989a). Interaction between peptide growth factors and homeobox genes in the establishment of antero-posterior polarity in frog embryos. *Nature* **341,** 33-38.

Ruiz i Altaba, A. and Melton, D. A. (1989b). Involvement of the Xenopus homeobox gene Xhox3 in pattern formation along the anterior-posterior axis. *Cell* **57,** 317-326.

Sampath, T. K., Rashka, K. E., Doctor, J. S., Tucker, R. F. and Hoffmann, F. M. (1993). Drosophila transforming growth factor beta superfamily proteins induce endochondral bone formation in mammals. *Proc. Natl. Acad. Sci. USA* **90,** 6004-6008.

Sasaki, H. and Hogan, B. L. M. (1993). Differential expression of multiple fork head related genes during gastrulation and axial pattern formation in the mouse embryo. *Development* **118,** 47-59.

Sasaki, H. and Hogan, B. L. M. (1994). HNF-3β as a regulator of floorplate development. *Cell* **76,** 103-115.

Schlunegger, M. P. and Grutter, M. G. (1992). An unusual feature revealed by the crystal structure at 2.2 A resolution of human transforming growth factor-β2. *Nature* **358,** 430-434.

Shimell, M. J., Ferguson, E. L., Childs, S. R. and O'Connor, M. B. (1991). The *Drosophila* dorsal-ventral patterning gene *tolloid* is related to human bone morphogenetic protein 1. *Cell* **67,** 469-481.

Slack, J. M. W. (1993). Embryonic induction. *Mech. Dev.* **41,** 91-107.

Slack, J. M. W. (1994). Inducing factors in Xenopus early embryos. *Current Biology* **4,** 116-126.

Smith, J. C. (1989). Induction and early amphibian development. *Curr. Opin. Cell Biol.* **1,** 1061-1070.

Smith, J. C. (1994). Hedgehog, the floor plate, and the zone of polarizing activity. *Cell* **76,** 193-196.

Swofford, D. L. and Berlocher, S. H. (1987). Inferring evolutionary trees from gene frequency data under the principle of maximum parsimony. *Systematic Zoology* **36,** 293-325.

Thomsen, G. H. and Melton, D. A. (1993). Processed Vg1 protein is an axial mesoderm inducer in Xenopus. *Cell* **74,** 433-441.

Tremml, G. and Bienz, M. (1992). Induction of labial expression in the *Drosophila* endoderm: response elements for dpp signalling and for autoregulation. *Development* **116,** 447-456.

Vainio, S., Karavanova, I., Jowett, A. and Thesleff, I. (1993). Identifcation of BMP-4 as a signal mediating secondary induction between epithelial and mesenchymal tissues during early tooth development. *Cell* **75,** 45-58.

Wang, X.-F., Lin, H. Y., Ng-Eaton, E., Downward, J., Lodish, H. F. and Weinberg, R. A. (1991). Expression cloning and characterization of the TGFβ type III receptor. *Cell* **67,** 797-805.

Wharton, K. A., Ray, R. P. and Gelbart, W. M. (1993). An activity gradient of decapentaplegic is necessary for the spcification of dorsal pattern elements in the *Drosophila* embryo. *Development* **117,** 807-822.

Wozney, J. M., Rosen, V., Celeste, A. J., Mitsock, L. M., Whitters, M. J., Kris, R. W., Hewick, R. M. and Wang, E. A. (1988). Novel regulators of bone formation: molecular clones and activities. *Science* **242,** 1528-1534.

Xie, T., Finelli, A. L. and Padgett, R. W. (1994). The Drosophila saxophone gene: a serine-threonine kinase receptor of the TGF-β superfamily. *Science* **263,** 1756-1759.

Development 1994 Supplement, 61-71 (1994)
Printed in Great Britain © The Company of Biologists Limited 1994

A class act: conservation of homeodomain protein functions

J. Robert Manak and Matthew P. Scott

Departments of Developmental Biology and Genetics, Howard Hughes Medical Institute, Stanford University School of Medicine, Stanford, California 94305-5427, USA

SUMMARY

Dramatic successes in identifying vertebrate homeobox genes closely related to their insect relatives have led to the recognition of classes within the homeodomain superfamily. To what extent are the homeodomain protein classes dedicated to specific functions during development? Although information on vertebrate gene functions is limited, existing evidence from mice and nematodes clearly supports conservation of function for the Hox genes. Less compelling, but still remarkable, is the conservation of other homeobox gene classes and of regulators of homeotic gene expression and function. It is too soon to say whether the cases of conservation are unique and exceptional, or the beginning of a profoundly unified view of gene regulation in animal development. In any case, new questions are raised by the data: how can the differences between mammals and insects be compatible with conservation of homeobox gene function? Did the evolution of animal form involve a proliferation of new homeodomain proteins, new modes of regulation of existing gene types, or new relationships with target genes, or is evolutionary change largely the province of other classes of genes? In this review, we summarize what is known about conservation of homeobox gene function.

Key words: homeobox, homeodomain, homeotic, Hox, conserved, evolution

INTRODUCTION

We celebrate this year two anniversaries. A century has passed since Bateson described and named homeotic mutations (Bateson, 1894). A decade has passed since the homeobox was discovered (McGinnis et al., 1984; Scott and Weiner, 1984). The protein fragment encoded by the homeobox, the DNA-binding homeodomain, is now viewed as a hallmark of transcription factors which control development in organisms as diverse as yeast, plants, insects and mammals.

The remarkable conservation of protein structures among developmental regulators is now so well established that apparent cases of lack of conservation are viewed with skepticism. Does conserved structure mean conserved function? The examples given below strongly suggest in several cases that homeodomain proteins of certain classes became dedicated to particular functions long ago and have maintained their dedication in ways that we do not fully understand. The homeodomain proteins, particularly the HOM/Hox group, remain the most dramatic example of retention of both protein structure and function during the evolution of developmental processes. Yet animals have enormous variety both in final form and how they develop. Attention therefore turns to learning how universal classes of developmental regulators give rise to diversity of form. In this review we focus on known or potentially conserved functions of homeodomain proteins and their regulators and targets.

HOMEODOMAINS ARE MEMBERS OF A STRUCTURAL SUPERFAMILY

Two crystal structures of homeodomains bound to DNA, and one NMR solution structure, have revealed a conserved structure for proteins sharing only 25% amino acid identity, or 15 amino acids of 60 (Qian et al., 1989; Kissinger et al., 1990; Wolberger et al., 1991). Because most homeodomains are more similar than this, the presently recognized homeodomains almost certainly have nearly identical backbone structures. The three alpha helical parts of the homeodomain create an internal hydrophobic core. One of these helices inserts into the major groove of the DNA and makes sequence-specific contacts. The N terminus of the homeodomain makes contact with the minor groove and stabilizes the association with DNA.

When we reviewed the extant 82 homeodomain sequences in 1989 (Scott et al., 1989), four amino acids were found to be diagnostic for homeodomains. The only exception was one of the yeast MAT proteins, which had three of the four relevant residues and a conservative change in the fourth. The definition has been useful, in that no protein clearly outside the homeodomain group has been found to have the critical four residues. However, the definition is arbitrary and reflects our limited ability to infer protein structure from primary sequences. Now there is evidence for an extended family of proteins which use an alpha helix to contact DNA in the major groove (Steitz, 1990; Pabo and Sauer, 1992; Schwabe and Travers, 1993).

Homeodomains are related in structure to helix-turn-helix proteins of bacteria (Laughon and Scott, 1984 ; Qian et al., 1989; Kissinger et al., 1990) and also to the POU-specific domain which is found in a family of proteins adjacent to a characteristic POU type of homeodomain. The POU-specific domain is astonishingly similar to the DNA-binding domain of the cI repressor protein of bacteriophage lambda (Assa-Munt et al., 1993). POU proteins therefore have two DNA-binding domains, each of which is similar to helix-turn-helix proteins. The structure of a member of a third group of proteins, the *forkhead/HNF3* group, reveals yet another set of relationships. The DNA-binding forkhead domain is most similar to eukaryotic histone H5 and to the catabolite activator protein (CAP) of *E. coli*. (Clark et al., 1993a) and is therefore yet another variation on helix-turn-helix. All of the proteins in the family use a single alpha helix to make the major groove contacts with DNA, but use somewhat different framework structures to form the rest of the domain. The three groups of proteins, homeodomain, POU and forkhead, are related at the structural level without the primary sequence being discernably similar. As more protein structures are determined, the family may expand. The structural relationships between the different DNA-binding domains make it difficult to rigorously define separate families.

Within the homeodomain group different classes can be defined based on primary sequence and these classes are remarkably distinctive in their functions; those functions are, in at least some cases, conserved across vast evolutionary distances. Representative sequences of a variety of classes have been described in Rubenstein and Puelles (1994) and in Duboule (1994).

FROM HEAD TO TAIL: HOX GENE ORGANIZATION

In all animals studied to date there is a cluster of homeobox genes known as HOM-C in insects and nematodes and HOX in mammals (McGinnis and Krumlauf, 1992). We will use the term HOX to refer to all such clusters (there are many homeobox genes; only those in these particular homologous clusters are called Hox genes). Key features of HOX clusters were first observed in studies of a *Drosophila* cluster, the bithorax complex (BX-C), by E. B. Lewis (Lewis, 1963). In *Drosophila* the primordial cluster appears to have split into two major parts, the second of which is the Antennapedia complex (ANT-C) whose organization and similarity to the BX-C was recognized and analyzed by T. C. Kaufman and colleagues (Kaufman et al., 1980). Consistent with the idea of a single primordial cluster, the flour beetle *Tribolium* contains a single complex of homeotic genes (Beeman et al., 1989) as does the chordate *Amphioxus* (P. W. H. Holland, personal communication). The nematode *Caenorhabditis elegans* has a cluster composed of at least five genes (Kenyon and Wang, 1991; Wilson et al., 1994)

The organization of Hox genes in mice (and humans), *Drosophila* and *C. elegans* is summarized in Fig. 1. At least four types of Hox gene appear to have existed prior to the divergence of the ancestors of these diverse animals: the *labial*, *Deformed*, *Antennapedia* and *AbdominalB* types, each of which is represented in these animals. In addition, genes related to the fly *empty spiracles* (*ems*) or *Distal-less* (*Dll*) genes are found in or near the worm and mammalian complexes (Boncinelli et al., 1993; Wang et al., 1993). The worm *ems*-like gene is about equally similar to the *ems* and *Distal-less* (*Dll*) genes of flies, and therefore it is striking that two of the mammalian relatives of the *Dll* gene, *Dlx1* and *Dlx2*, are found near the Hoxd complex (McGuinness et al., 1992; Ozcelik et al., 1992; Simeone et al., 1994). The fly *ems* and *Dll* genes are not found in either homeotic gene cluster (Cohen et al., 1989; Dalton et al., 1989), nor are two mammalian *ems*-related genes *Emx1* and *Emx2* (Kastury et al., 1994). The presence of two *Evx* genes in the mammalian complexes (D' Esposito et al., 1991; Faiella et al., 1991) suggests still more dispersal, as the most similar fly gene, *even skipped*, is also not in either fly cluster. We are therefore left with a tentative view of an ancestral cluster containing representatives of the *lab*, *Dfd*, *Antp*, *AbdB*, *eve* and *ems* (*Dll*) types. One candidate for a seventh member of the ancestral cluster would be the genes represented in flies by *orthodenticle* (*otd*), another homeobox gene expressed in a discrete region along the anterior-posterior axis (Finkelstein et al., 1990) and in mammals by the related *Otx* genes (Simeone et al., 1992). However, the mammalian *Otx1* and *Otx2* genes do not map near any of the HOX complexes (Kastury et al., 1994).

Starting, then, from the possible ancestral cluster, what happened during the evolution of each animal type? The fly complex has been split at least once (ANT-C and BX-C) and possibly three other times (dispersion of *ems*, *eve* and *Dll*). Ironically, the fly clusters that gave rise to the mystery of how the genes are bound together may be the most dispersed of any complex yet studied. In addition, three additional homeobox genes and two other types of genes, cuticle protein genes and one encoding an immunoglobulin superfamily protein, exist in the ANT-C and may have formed by duplication or invasion, respectively (Kaufman et al., 1990). The additional homeobox genes are a pair of closely related *zen* genes, one of which is required for dorsal-ventral differentiation and the anterior-posterior maternal morphogen gene *bicoid* (*bcd*) (Berleth et al., 1988). The vertebrate protein with some similarity to *bcd* protein, *goosecoid* (Cho et al., 1991), is not known to be located near the HOX complexes. Similarly, the nematode complex is interrupted by other genes (Salser and Kenyon, 1994).

The mammalian cluster duplicated twice or three times to form four copies (Kappen et al., 1989). There are now 14 identifiable types of gene in the Hox complexes (apart from the nearby *Dlx* genes), called paralogs 1-13 and *Evx*. Paralog grouping is based on homeodomain sequence similarities as well as position within the cluster. Sequence alignments suggest the loss of paralogs from each of the clusters as shown (Fig. 1), plus the proliferation of members of the *AbdB* (paralog 9-13) group. The *Otx* genes may also have been lost, and the Hoxb and c clusters lack *Evx* representatives. The worm cluster is interrupted by other genes, like the fly cluster, and lacks any known representatives of the *eve* type.

The intriguing relationship between expression of Hox and HOM genes along the anteroposterior axis and their order along the chromosomes to which they map has been termed 'colinearity' and suggests a connection between structure of the HOM/Hox complexes and function of the genes within them. Perhaps this relationship reflects a requirement to position homeotic regulatory information of several genes in a well-defined order so that such information could influence more than one gene of the complex. If this were so, in addition to conservation of DNA encoding homeotic proteins, one would expect to find conservation of regulatory DNA within the HOM/Hox clusters. This has been shown in the case of the fly and mouse *Dfd* homologs, since regulatory sequences from either are able to respond to appropriate spatial cues when introduced into the other species (Awgulewitsch and Jacobs, 1992; Malicki et al., 1992; see below).

HOX GENE FUNCTIONS

Several rules governing homeotic gene function have been fairly well conserved. (1) Genes are ordered along the chromosome in the same order as their expression and function along the anterior-posterior axis of the animal. (2) More genes are usually expressed in more posterior regions. (3) Loss of gene function leads to loss of structures or to development of anterior structures where more posterior structures should have formed. (4) Activation of genes where they should be off, i.e. gain-of-function mutations, leads to posterior structures developing where more anterior structures would normally be found. To these generalizations we may add some molecular data. (5) Each homeotic gene contains a single homeobox and encodes a sequence-specific DNA-binding protein which acts as a transcription factor. Some encode a family of proteins with alternatively spliced mRNAs. (6) Most of the homeotic genes are transcribed in the same direction, with the 5′ ends of transcription units oriented toward the posterior end of the HOX cluster.

In flies, the clustered homeotic genes of the Antennapedia and bithorax complexes determine segment identity by promoting the morphogenesis of appropriate anatomical structures within particular segmental or parasegmental domains of the body. These domains, which are reiterated units along the anterior-posterior axis, are established before the homeotic genes are active. The fly homeotic genes are not required for establishment of the segmental body plan, but only to govern segmental form. Mutations in fly homeotic genes lead to alterations in cell fate decisions, not changes in segment number. The epidermal expression patterns and sites of action are summarized in Fig. 2. A single fly homeotic protein, rather than a combination, is in some cases sufficient to activate a morphogenetic pathway. For example, ubiquitous expression during early embryogenesis of the *Ultrabithorax* homeotic gene, which normally promotes anterior abdomen identity and is transcribed primarily there, leads to specification of head and thoracic segments as anterior abdomen-like segments (Mann and Hogness, 1990). *Ubx* does not activate any other homeotic genes and represses some more anteriorly acting ones, so the *Ubx* protein is sufficient to organize pattern without other Hox proteins.

The *C. elegans* HOM-C or Hox genes (Kenyon and Wang, 1991) provide fascinating information about the effects of Hox genes on individual identified cells (reviewed in Salser and Kenyon, 1994). For example *mab-5* and *lin-39* govern migration of certain neuroblasts (Clark et al., 1993b; Salser et al., 1993); heat shock promoter activation of *mab-5* causes cells to change direction during their migration (Salser and Kenyon, 1992). The nematode genes are expressed in the order along the anterior-posterior axis expected from mouse and fly studies, even though the order of two of the nematode genes on the chromosome is inverted compared to other Hox genes (Fig. 1) (Cowing and Kenyon, 1992). Cross-regulatory interactions among the homeotic genes limit their domains of expression (Salser et al., 1993). Cell fates are controlled largely autonomously by the Hox genes (Salser and Kenyon, 1992; Clark et al., 1993b). There are clear examples of combinatorial actions of the genes, as when cells fuse in response to *mab-5* and *lin-39* but not to either alone (Salser et al., 1993). In other cases the presence of one homeotic gene activity precludes evident action of another.

The vertebrate Hox genes also instruct cells to undergo appropriate developmental decisions. Hox genes are expressed in nearly every cell type, but have been most extensively studied for their roles in the developing central nervous system and axial skeleton. They are transcribed in limited regions along the anterior-posterior axis, like the fly genes, although in more substantially overlapping patterns. Fewer Hox genes are expressed in the anterior than posterior, suggesting that progressively more caudal structures may depend upon concerted actions of multiple homeotic proteins. However, the posterior prevalence model argues against this possibility (see below). The scarcity of antibody studies makes it difficult to ascertain whether multiple Hox proteins are found in the same cell, but this seems likely to be true given that this is the case in flies (Carroll et al., 1988). The picture is further complicated by the four sets of Hox genes (Fig. 1). Because it is often the case that corresponding paralogs in different Hox clusters are expressed in similar patterns, particularly for the Hoxa and Hoxd clusters, redundant or partially redundant function of the genes may make interpretation of mutations in only one gene difficult. This problem will soon be addressed by the engineering of doubly mutant transgenic mice, but in the meantime a considerable amount can be learned from single mutations. The mutations have been engineered in embryonic stem cells in culture and then introduced into the mouse germline. These mutants provide convincing evidence of homeosis and leave little doubt of the power of Hox genes to control cell fates during development.

In the central nervous system, the anterior border of expression of many of the Hox genes coincides with rhombomeric boundaries, the rhombomeres being transient, reiterated bulges of the hindbrain thought to be indicative of segmental organization of the brain. Loss-of-function mutations of *Hoxa-1* and *Hoxa-3* cause defects in hindbrain and branchial regions of the mouse, but do not appear to cause homeotic transformations of the affected regions (Chisaka and Capecchi, 1991; Lufkin et al., 1991; Carpenter et al., 1993). The cells affected by both mutations are derived from the cranial neural crest. A reduction in the number of rhombomeres was observed in *Hoxa-1* mutants (Lufkin et al., 1991; Carpenter et al., 1993). Mutations in the two fly homeobox genes *ems* and *otd*, not located in the HOX clusters but possibly there ancestrally, cause embryos to develop with altered fates in the head and with reduced numbers of segments. It has been suggested that deletion of body parts, in contrast to homeotic transformation, occurs when the absence of one homeotic gene function does not result in the expression of another in it's place (McGinnis and Krumlauf, 1992). This idea may explain why both types of phenomena are observed in flies and in mice.

Targeted gene disruption of *Hoxc-8*, *Hoxb-4*, or *Hoxa-2*, as well as constitutive overexpression of *Hoxa-7*, *Hoxc-6*, *Hoxc-8* or *Hoxd-4*, causes homeotic transformations in mouse embryos (Kessel et al., 1990; Jegalian and De Robertis, 1992; Le Mouellic et al., 1992; Lufkin et al., 1992; Pollock et al., 1992; Gendron-Maguire et al., 1993; Ramirez-Solis et al., 1993; Rijli et al., 1993). In the *Hoxc-8* mutant, the first lumbar

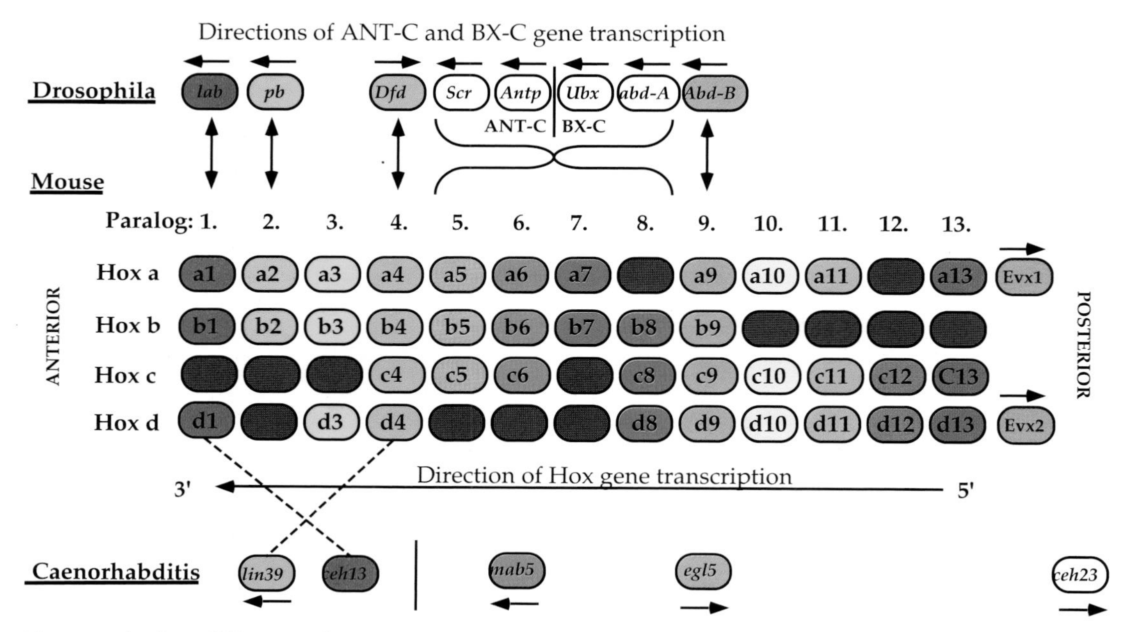

Fig. 1. Comparitive organization of Hox complexes. The alignments of the *Drosophila*, nematode and mouse complexes are shown. Colors indicate similar homeodomain sequences. The order of the genes along the chromosome is as shown, with the order roughly corresponding to where along the anterior-posterior body axis a gene is expressed or, in the case of mammals, the anterior-most limit of expression. Mammalian nomenclature is as described in Scott (1992). The four mammalian clusters are thought to have arisen by two duplication events. Categories 1-13 are called 'paralog' groups and are indicative of homeodomain sequence similarity. Gray ovals indicate the lack of a paralog from a cluster. The Evx genes encode homeodomains most closely related to the *even skipped* segmentation gene of *Drosophila*. The assignment of *Scr*, *Antp*, *Ubx* and *abd-A* to any particular paralog group is uncertain. The *ceh-23* gene, located 30 kb from *egl-5* (Wang et al., 1993) is related to the *empty spiracles* or *Distal-less* genes of flies and mammals, and its presence near the nematode Hox cluster may be indicative of the ancestral linkage of this type of gene. Two related mammalian genes have been mapped near the Hox complexes with cytological studies but the molecular distance is unknown and could be substantial (McGuinness et al., 1992; Ozcelik et al., 1992; Simeone et al., 1994). The direction of *ceh-13* transcription is unknown. See text for additional references.

vertebra is converted to a thoracic vertebra, thus producing a 14th pair of ribs (Le Mouellic et al., 1992). Mice lacking *Hoxa-2* function have anterior transformations of skeletal elements derived from the second branchial arch (Rijli et al., 1993; Gendron-Maguire et al., 1993). Two *Hoxb-4* mutations have been introduced into mice, both of which cause transformations of the second cervical vertebra from axis to atlas (Ramirez-Solis et al., 1993). All of these phenotypes are consistent with loss-of-function homeotic transformations observed in flies: posterior structures are converted to more anterior ones.

Overexpression of either *Hoxa-7* or *Hoxd-4*, analogous to gain-of-function homeotic mutations in flies, leads to transformations of anterior structures into posterior structures (Kessel et al., 1990; Lufkin et al., 1992). *Hoxd-4* overexpression transforms occipital bones into structures that resemble cervical vertebrae whereas *Hoxa-7* overexpression transforms the basioccipital bone into a proatlas structure. Overexpression of *Hoxa-4* causes a condition similar to congenital megacolon, probably due to abnormalities in the enteric nervous system (Tennyson et al., 1993). Although this gene is expressed in a variety of tissues including spinal cord, ganglia and gut mesoderm, abnormalities were only observed in the terminal colon.

Mice that overexpress *Hoxc-6* (Jegalian and De Robertis, 1992), or *Hoxc-8* (Pollock et al., 1992), develop rib-bearing vertebrae in lieu of one (or more) of the lumbar vertebrae, a transformation of posterior to anterior similar to the *Hoxc-8* loss-of-function phenotype (Le Mouellic et al., 1992). Because *Hoxc-8* is normally expressed in that region, the *Hoxc-8* overexpression phenotype is presumably due to heightened or temporally incorrect expression. More work is needed to understand the origins of these effects, but the observed phenotypes could be partially explained if overexpression of *Hoxc-6* or *Hoxc-8* blocks function of *Hoxc-8* in its normal domain of expression.

A hierarchy of homeotic protein function exists amongst homeotic genes. Most fly HOM genes are unable to transform the identity of segments posterior to their normal domains of expression when activated ubiquitously, even though they are capable of transforming anterior segments (Gonzales-Reyes and Morata, 1990; Gonzales-Reyes et al., 1990; Mann and Hogness, 1990). There are exceptions: either *Antp* or *Scr* can prevail in the thorax (Gibson et al., 1990). Since the inability to transform posterior regions is dependent on posterior homeotic protein function, more posterior-acting homeotic proteins may be 'dominant' with respect to function over more anterior-acting proteins. This phenomenon has been termed

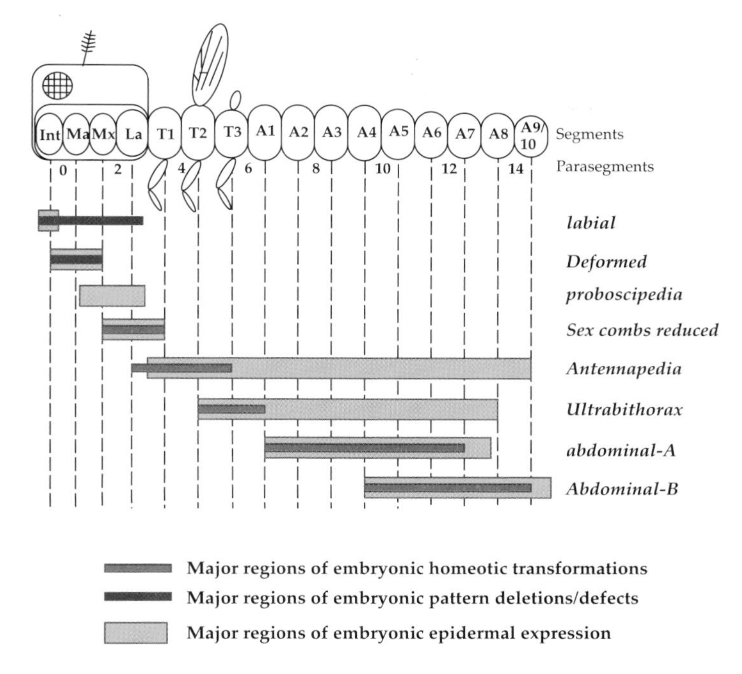

Fig. 2. Expression and function of the *Drosophila* HOM genes. Fly body segments include the intercalary (Int), mandibular (Ma) and labial (La) segments of the presumptive head, T1 to T3 segments of the presumptive thorax, and A1 to A10 segments of the presumptive abdomen. Normal expression patterns of *labial* (*lab*; Diederich et al., 1989), *Deformed* (*Dfd*; Jack et al., 1988; Mahaffey et al., 1989), *proboscipedia* (*pb*; Pultz et al., 1988; Mahaffey et al., 1989), *Sex combs reduced* (*Scr*; (Riley et al., 1987; Pattatucci and Kaufman, 1991), *Antennapedia* (*Antp*; Carroll et al., 1986; Wirz et al., 1986), *Ultrabithorax* (*Ubx*; Beachy et al., 1985; White and Wilcox, 1985; Carroll et al., 1988), *abdominal-A* (*abd-A*; Karch et al., 1990) and *Abdominal-B* (*Abd-B*; Celniker et al., 1989; DeLorenzi and Bienz, 1990) are indicated. Mutations in *Scr*, *Antp*, *Ubx*, *abd-A* and *Abd-B* produce homeotic transformations in the embryo whereas mutations in *lab* and *Dfd* produce pattern deletions. Alterations in mandibular, maxillary and labial segments in *lab* mutants may be due to the secondary effect of failure of head involution. *pb* is expressed in the embryo but no functions for it have been detected there. *pb* is located between *lab* and *Dfd* in the ANT-C complex.

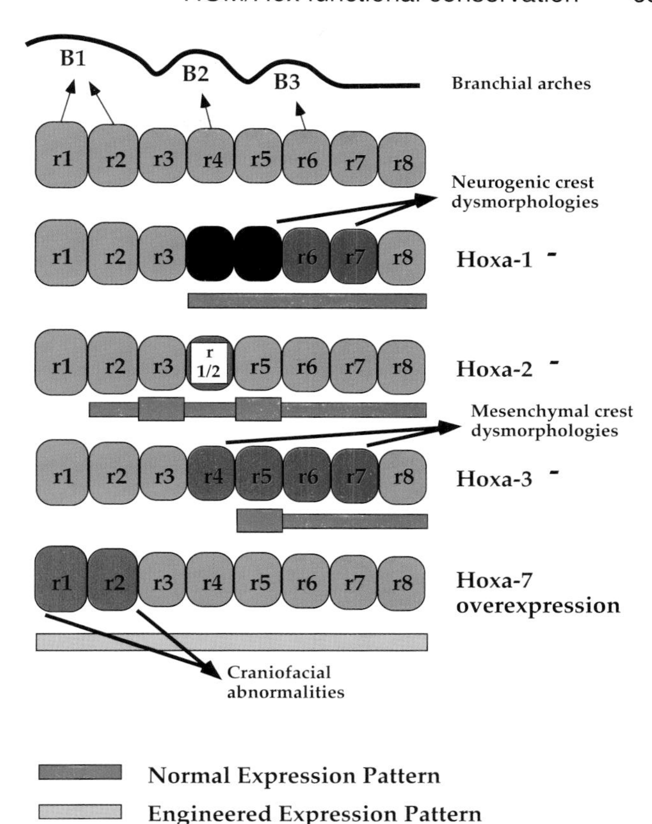

Fig. 3. Mouse Hox mutations affect cranial development. Rhombomeres 1 to 8 (r1-r8) are thought to represent segmental organization of the hindbrain. The first three branchial arches (b1-b3) are derived from neural crest cells which originate from specific rhombomeres (arrows) and produce head mesodermal structures such as bone and connective tissue. *Hoxa-1* mutations affect neurogenic crest-derived structures such as sensory and motor ganglia from the region encompassing rhombomeres 4 to 7. *Hoxa-3* mutations also affect this region of the hindbrain; however, dysmorphologies are specifically observed for mesenchymal crest-derived structures. For example, both the thymus and parathyroid glands fail to develop. Mice deficient in *Hoxa-2* show homeotic transformations of 2nd branchial arch derivatives to 1st branchial arch derivatives. Reichert's cartilage, which forms the stapes bone of the middle ear as well as other structures, is replaced with Meckel's cartilage, which forms the malleus and incus. *Hoxa-7* overexpression throughout the hindbrain produces craniofacial abnormalities such as cleft palate, open eyes and non-fused pinnae. These structures are derived from 1st branchial arch neural crest cells. *Hoxa-7* neural expression is normally confined to the spinal cord. For references, see text. Normal expression patterns are indicated, when appropriate, for *Hoxa-1* (Murphy and Hill, 1991), *Hoxa-2* (Hunt et al., 1991; Tan et al., 1992), *Hoxa-3* (Gaunt, 1988; Gaunt et al., 1988) and *Hoxa-7* (Mahon et al., 1988). See Krumlauf et al. (1993) for a review. Rhombomeres indicated by red color are affected by the indicated mutation. Rhombomeres indicated in black are deleted in the specified mutant. The anterior homeotic transformation of skeletal elements derived from the second branchial arch to structures normally derived from the first branchial arch (thus adopting fates of neural crest cells emanating from rhombomeres 1 and 2) in *Hoxa-2* mutants is indicated by the white box.

'phenotypic suppression' (Gonzales-Reyes and Morata, 1990). A strikingly similar phenomenon is observed in mice. In general, mice mutant for Hox genes show homeotic transformations in the anteriormost region where that Hox gene is normally expressed and not within regions where a more posterior Hox gene is expressed (see Figs 4 and 5). Therefore, the more posterior Hox genes are able to promote the appropriate morphological responses without the assistance of more anterior genes (termed 'posterior prevalence' by Duboule, 1991). Consistent with these results, overexpression of two Hox genes in the mouse leads to posteriorization of anterior structures (Kessel et al., 1990; Lufkin et al., 1992); however, this is not the case for two other Hox genes, where overexpression leads to anteriorization in posterior regions (Fig. 5; Jegalian and De Robertis, 1992; Pollock et al., 1992). In the latter two cases levels or timing of expression may lead to the different outcome.

Elegant experiments by the McGinnis and Morata groups have demonstrated a functional relationship between the fly HOM and vertebrate Hox genes (Malicki et al., 1990; McGinnis et al., 1990; Bachiller et al., 1994). Ubiquitous expression of mouse *Hoxb-6*, most closely related to the fly *Antp* gene, causes homeotic transformations in embryos that are similar in nature to those produced by ectopic expression of *Antp* (Malicki et al., 1990). In addition, antenna-to-leg transformations, the classic *Antp* gain-of-function phenotype, are observed in the adult head when *Hoxb-6* is ubiquitously expressed in larvae. Along the same lines, ubiquitous expression of the *Dfd* human homolog *Hoxd-4* can provide some functions attributed to *Dfd* such as autoactivation of the *Dfd* gene in embryos and larvae, producing phenotypic alterations of adult head structures similar to those observed with a dominant allele of *Dfd* (McGinnis et al., 1990).

Ubiquitous expression of several mouse Hoxd genes in flies (*Hoxd-8* to *Hoxd-11*) suppresses fly homeotic gene function, but in a most intriguing way (Bachiller et al., 1994). *Hoxd-8* is most closely related to the *Antp/Ubx/abd-A* class of homeobox genes whereas *Hoxd-9* to *Hoxd-11* are most closely related to the *Abd-B* gene. *Hoxd-8* to *Hoxd-11* are sequentially arranged along the chromosome 3′ to 5′ and act in progressively more posterior regions in the mouse. When expressed in flies, the more posterior acting Hox genes are better able to overcome the effects of the clustered fly homeotic genes than the anterior-acting genes. The result of such experiments is the transformation of affected segments to a thoracic ground-state character, even though endogenous fly homeotic genes in these regions are expressed at slightly reduced to normal levels (Bachiller et al., 1994). Thus, *Hoxd-8* or *Hoxd-9* prevails over head-specific homeotic genes, *Hoxd-10* prevails over head- and thorax-specific homeotic genes, and *Hoxd-11* prevails over head-, thorax- and abdominal-specific homeotic genes. In addition, *Hoxd-11* can activate the endogenous *Abd-B* gene as well as an *Abd-B* target *(empty spiracles)* even in the absence of endogenous *Abd-B*. Filzkorper, morphological readouts of *Abd-B* function, are ectopically induced by *Hoxd-11* in these experiments. Even though *Hoxd-9* to *Hoxd-11* all show an equal degree of similarity with respect to *Abd-B*, they still differ in terms of their ability to suppress the fly homeotics. Thus, the posterior prevalence rule holds: the farther back a mammalian gene is expressed, the better it is at overriding more anterior homeotic genes.

Most of the genes discussed so far act in the trunk of the animal. Additional Hox genes, and homeobox genes not presently found in all Hox complexes, act primarily in head development. At least some of these head homeobox genes may have once been located in a primordial homeotic complex. As mentioned previously, the vertebrate and nematode complexes provide evidence for an original Hox complex with more types of genes than the present fly complex. The head genes may have functions analogous to those of the Hox genes acting in the trunk, but the complexities of anterior structures make the regulatory roles of these genes less clear.

HOMEOBOX GENE FUNCTIONS IN THE ANTERIOR EMBRYO: EVIDENCE FOR MORE DIVERSE ORIGINAL HOX CLUSTERS

The Hox cluster genes *lab*, *Dfd* and *Scr* contribute to fly head patterning (Fig. 2). However none of the Hox genes 'cover' the most anterior parts of the embryo. The most anterior cells employ other genes, including at least three types of fly homeobox genes (*Dll, otd* and *ems*) not present in the Hox cluster (Cohen and Jürgens, 1991; Finkelstein and Perrimon, 1991). *Dll* is expressed in the primordia of the limbs, in the brain and in head ectoderm, in particular the anlage of the facial sensory appendages within the labral, antennal, maxillary and labial segments (Cohen et al., 1989). *otd* and *ems* are expressed in the cephalic region of the head (Dalton et al., 1989; Finkelstein et al., 1990; Wieschaus et al., 1992), *otd* expression overlapping with *ems* but extending anterior to it. *otd* mutants have partial and complete deletions in pre-antennal and antennal segments, respectively, whereas mutants of *ems* lack antennal and intercalary segments. A smaller region of the pre-antennal segment is also deleted in *ems* mutants. Both *otd* and *ems* expression is dependent on bicoid and torso expression to define their posterior and anterior boundaries, respectively, but does not require gap or pair-rule segmentation gene input (Finkelstein and Perrimon, 1990). Based on expression patterns as well as genetic data, Cohen and Jurgens (1991) have proposed that both segmentation and segmental identity are controlled by *ems* and *otd* as well as *buttonhead* (*btd*), whose expression reaches more posteriorly than the other two. The model proposes that a segment's identity is determined by the particular combination of such genes expressed in a segment. For example, the antennal segment expresses *otd*, *ems* and *btd*, whereas the intercalary segment expresses only *ems* and *btd*.

The vertebrate homeobox genes related to the fly head genes are expressed in patterns similar to those of the fly genes, suggesting evolutionary conservation of function underlying dramatically different final morphologies. Four vertebrate genes, two homologous to *otd* (*Otx-1*, *Otx-2*) and two similar to *empty spiracles* (*Emx-1*, *Emx-2*), are primarily expressed in the anterior developing brain (Simeone et al., 1992). There are distinct overlaps in the patterns of expression of the four genes. *Otx-2* is most broadly expressed, from telencephalon to mesencephalon, inclusive. *Otx-1* expression is contained within that of *Otx-2*. Similarly,the *Emx* genes are expressed in a portion of the *Otx-1* domain and *Emx-1* is expressed within the *Emx-2* domain. In mice *Otx-2* expression appears earliest, at 7.5 days d.p.c., followed by *Otx-1* and *Emx-2* and then *Emx-1*. Thus different genes may be used for different stages of fate determination.

Six mouse *Distal-less* homologs have been isolated (Price et al., 1991). The expression patterns of four of them, *Dlx-1*, *Dlx-2*, *Dlx-5* and *Dlx-6*, have been reported (Dollé et al., 1992; Simeone et al., 1994). All four genes are expressed in the forebrain, the primordia of the face and neck (branchial arches), and the ectoderm of the limb buds, a pattern strikingly conserved from fly to mouse. In addition, *Dlx5* and *Dlx-6* are expressed in the developing skeleton after early cartilage formation (Simeone et al., 1994). The first branchial arch, which has high levels of *Dlx* expression, gives rise to the mouth/jaw region of the mouse. In flies, the maxillary, labral and labial segments, domains of *Dll* expression, also give rise to mouth structures. There is no HOX expression in forebrain regions, so perhaps the *Dlx*, *Otx* and *Emx* gene activities determine cellular identity there. The idea of a commonality of gene functions in face and limbs between insects and vertebrates is nearly beyond belief, given the utterly different terminal morphologies. However, it was not long ago that homologies now generally accepted, such as Hox gene relationships, were rightly viewed with enormous skepticism! We will simply have to wait and see what is truly conserved.

The Hox clusters remain the most remarkable example of conservation of regulatory genes involved in development. While the parallels described here are striking, the differences in Hox gene expression and function among animal types are notable as well. The major problem that we face is in trying to understand what it means for a gene to define a certain region of the body. What is in common between the thorax of a fly and a human?

POTENTIALLY CONSERVED FUNCTIONS OF OTHER CLASSES OF HOMEOBOX GENES

In addition to the clustered Hox genes, other distinctive types of homeodomain proteins appear to have evolutionarily conserved functions in development. The evidence is often weak but suggestive enough to warrant summarizing. We emphasize that many homeodomain proteins are used at different times and in different tissues, and only some of their functions may be conserved.

forkhead and *caudal* in the gut

The *fork head* domain has a structure related to that of histone H5 and the bacterial CAP protein, and is more distantly related to homeodomains (Clark et al., 1993a). Like the homeodomain proteins, proteins containing forkhead/HNF3 domains may have evolutionarily conserved functions (Clevidence et al., 1993; Hromas et al., 1993). *forkhead* (*fkh*) is required for formation of the gut endoderm in flies (Jürgens and Weigel, 1988; Weigel et al., 1989). The HNF3 proteins were discovered as factors needed for transcription of genes in mammalian liver (Lai et al., 1991). Because the forkhead class of proteins has many other functions, however (Lai et al., 1993), the apparent similarity in some of those functions may or may not be meaningful.

A second homeobox gene likely involved in gut development is represented by *caudal* in flies (Mlodzik et al., 1985; Macdonald and Struhl, 1986) and by the *Cdx* genes in vertebrates (Duprey et al., 1988; Joly et al., 1992; Frumkin et al., 1993, 1994). In flies, *cad* is required for development of posterior cuticle structures. *cad* is also expressed in the developing gut. *Cdx* genes are expressed in the intestinal epithelium and mesenchyme and may be involved in gut closure.

prospero and *cut* in the nervous system

The *prospero* 'family' of proteins is as yet represented by only two genes, but because they exist in both mammals and insects the family is probably genuine. These proteins have among the most divergent homeodomains known in higher animals. The mouse and fly proteins share both homeodomain and C-terminal protein sequences (Oliver et al., 1993). The *Drosophila pros* gene was identified as a regulator of neuron differentiation required after the completion of mitotic divisions (Doe et al., 1991; Vaessin et al., 1991). The gene is transcribed in embryonic neuroblasts and the mRNA is translated in descendant ganglion mother cells and young neurons. It is also expressed in the cone cells which secrete lens in the developing eye and in the midgut. The mammalian gene *Prox1*, which may or may not be the only mammalian version of *pros*, is also active during neural development. The similarity extends to the types of cells in which both genes are active: undifferentiated neurons, eye lens, heart, pancreas, and liver. The strongest argument for genuine conservation is rather weak: expression in neurons after cell division but prior to differentiation.

The fly gene *cut* encodes a distinctive type of homeodomain protein (Blochlinger et al., 1988). *cut* is expressed in many tissues, governing neural differentiation as well as other developmental events (Bodmer et al., 1987; Blochlinger et al., 1990; Jack et al., 1991; Liu et al., 1991). A related mammalian gene called *Clox* is expressed in diverse tissues in mice (Andres et al., 1992) including cartilage, liver, skeletal muscle, brain, lung and heart. Thus both the vertebrate and invertebrate genes act in a diverse set of tissues; what links the different tissues together in terms of *cut* function is not yet clear. Each of the two proteins contains three 73 aa repeats outside the homeodomain, the *cut* repeats.

MADS boxes and homeoboxes: the heart of the matter

The earliest known marker of vertebrate heart development is expression of the cardiac-specific homeobox gene *Csx* (Komuro and Izumo, 1993). *Csx* is expressed in the myocardium in 7.5 day mouse embryos, the late primitive streak stage when the future heart is just a small flattened plate. Thus the gene is active long before overt heart differentiation. Expression of *Csx* persists through adulthood. Although no functional studies are yet available, the great specificity of *Csx* expression strongly suggests a role in controlling heart muscle development.

A second early marker of heart development, though one that is also expressed in many other tissues, is the transcription factor MEF-2C (E. Olson, personal communication). MEF-2C is one of a family of transcription factors active in a broad range of mesodermal tissues. The protein contains a MADS box (Pollock et al., 1991), a sequence found in a larger family of proteins including serum response factor and some plant homeotic genes. The MEF group contains a MEF-specific domain adjacent to the MADS box. The best studied MADS box protein is a yeast transcription factor called MCM1. MCM1 associates with the repressor MATα2 to repress **a** mating type-specific genes (Johnson, 1992). The two proteins together therefore contribute to the control of yeast differentiation. It is the 70 amino acid MADS box region that binds to MATα2 and cooperates with it in binding DNA (Johnson, 1992).

A fly heart, or 'dorsal vessel', is completely different in appearance from a vertebrate heart. The embryonic heart is composed of several parallel rows of cells which later form loosely knit valves to move hemolymph through the larva without the use of ducts. Despite the drastic difference in morphology, the fly and vertebrate hearts have common genetic regulators. A MEF2-like gene, called DMEF2, has been found in flies and is expressed in the developing heart as well as in visceral mesoderm (Lilly et al., 1994). Not to be outdone, a homolog of *Csx* called *tinman* is expressed in the developing fly heart. *tinman* is expressed in both visceral mesoderm and heart earlier in embryogenesis, but comes to be restricted to the heart. Mutant animals that lack *tinman* function develop neither the visceral mesoderm nor the heart (Azpiazu and Frasch, 1993; Bodmer, 1993).

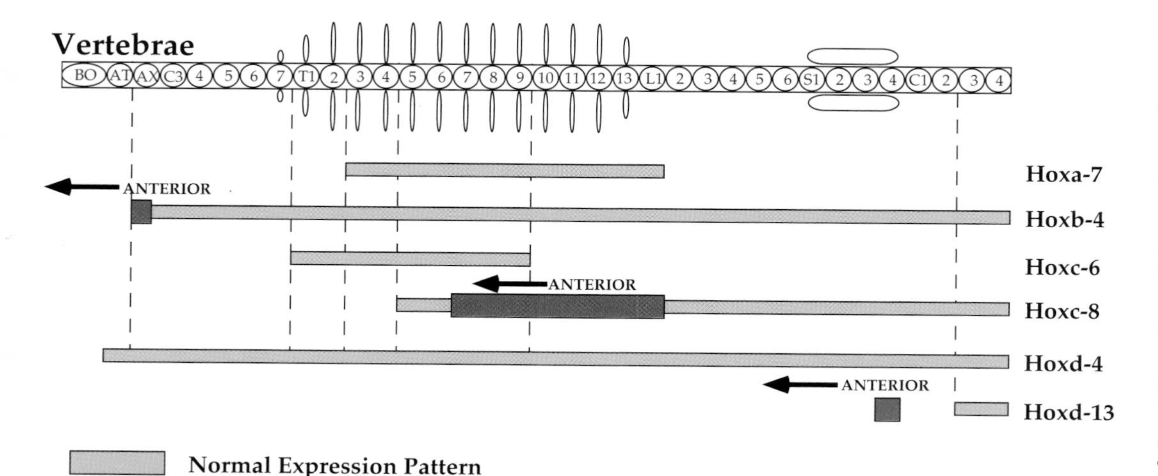

Fig. 4. Mouse Hox mutations result in anteriorly directed homeotic transformations in the paraxial mesoderm. A schematic representation of the vertebrae of the mouse is shown along with the expression patterns of several hox genes prior to vertebral specification. The basioccipital bone (BO) is connected to the cervical vertebrae (C3 through 7), which includes the atlas (AT) and axis (AX). Thirteen rib-bearing thoracic vertebrae (T1 to 13) are followed by six lumbar (L1 to 6), four sacral (S1 to 4) and four caudal (C1 to 4) vertebrae. Normal expression patterns of *Hoxa-7* (Mahon et al., 1988), *Hoxb-4* (Gaunt et al., 1989), *Hoxc-6* (Sharpe et al., 1988), *Hoxc-8* (Gaunt, 1988; Gaunt et al., 1988), *Hoxd-4* (Gaunt et al., 1989) and *Hoxd-13* (Dolle et al., 1991) are indicated, although the precise location of boundaries of expression, especially posteriorly, is unclear at this time. In general, knockouts of *Hoxb-4*, *Hoxc-8* and *Hoxd-13* affect the most anterior region in which the gene is normally expressed. The homeotic transformations which result are in the anterior direction, as observed for loss-of-function homeotic mutations in flies. Although the phenotype observed for *Hoxd-13* mutations is consistent with a homeotic transformation, the authors argue that the phenotype could be due to altered growth properties of cells instead. It is also important to note that the reported expression pattern of *Hoxd-13* does not include the region of the vertebrae that is affected in the mutation. Expression may extend further anteriorly. Alternatively, cell non-autonomous events may dictate the phenotype. See text for additional references.

The presence of both types of regulatory molecule in corresponding tissues in animals with a common ancestor some 600 million years ago suggests the dedication of the two molecules to creation of a heart-like organ in the ancestors of insects and vertebrates. The relationship is remarkable given the utter disparity in structure. The results also suggest a possible intimate relationship between the MADS and *tinman* proteins, as in the yeast homeodomain-MADS box association.

The *engrailed* group and neural development

The *engrailed* (*en*) gene in *Drosophila* is one of the few segment polarity genes that encodes a homeodomain-containing transcription factor. *En* is expressed in the posterior half of each segment in the ectoderm as well as in the developing nervous system (DiNardo et al., 1985; Kornberg et al., 1985). Flies homozygous for a special allele that allows adult survival have posterior wings similar in appearance to the anterior wing (Eker, 1929; Morata and Lawrence, 1975). Similarly, clones of cells in various adult tissues that lack *en* function exhibit posterior-to-anterior transformations of fate (Morata et al., 1983). Embryos homozygous for a null *en* allele have disruptions of pattern in each segment (Nüsslein-Volhard and Wieschaus, 1980; Kornberg, 1981). The closely linked *invected* gene is highly homologous to *en* and is expressed in a very similar or identical pattern.

En expression is initially under the control of pair-rule segmentation genes, but later comes under the influence of segment polarity genes such as *wingless*, the closest fly relative to the mammalian *Wnt-1* gene (DiNardo and O'Farrell, 1987; Ingham et al., 1988; Heemskerk et al., 1991). The gene also autoregulates. The successive waves of regulation may be typical for genes that remain active during a broad period of development and which must be responsive to changing arrangements of cells.

In addition to its role in metamere specification, *en* is expressed in elaborate and precise segmentally reiterated patterns in the developing nervous system (Doe, 1992). Like the striped expression pattern of *en* in the epidermis, the neural pattern is exquisitely conserved in arthropods from flies to crayfish (Patel et al., 1989). Neural expression seems to be a more general phenomenon than the metameric expression since no other higher organisms included in this study express *en* in developing metameres.

Two mouse genes with significant homologies to the fly *en* genes have been identified and named EN-1 and EN-2 (Joyner and Martin, 1987). The genes are expressed in a similar pattern at the midbrain-hindbrain border. *Wnt-1*, which is activated earlier in development than EN-1 or 2 and is expressed at the midbrain-hindbrain border, is required for activation of both EN genes (Bally-Cuif et al., 1992). This relationship between these vertebrate genes is reminiscent of the *en-wg* interaction in flies. The result of the *Wnt-1* mutation in mice is the absence of the entire midbrain and part of the hindbrain (McMahon et al., 1992). Thus, *Wnt-1* is required in regions where it does not appear to be expressed. Several other vertebrate *en* genes have been isolated, including 3 from zebrafish, and their expression patterns are conserved to a large extent (see (Rubenstein and Puelles, 1994) for review). Loss-of-function mutations in EN-2 result in subtle disruptions of the cerebellum (Joyner et al., 1991; Millen et al., 1994), suggesting that EN-1 may be able to partially substitute for EN-2 in the mouse to mask the full phenotype. Nonetheless, in EN-2 mutants, several features of the phenotype are worth noting: (1) all major morphological defects occur in the posterior cerebellum, and (2) cerebellar

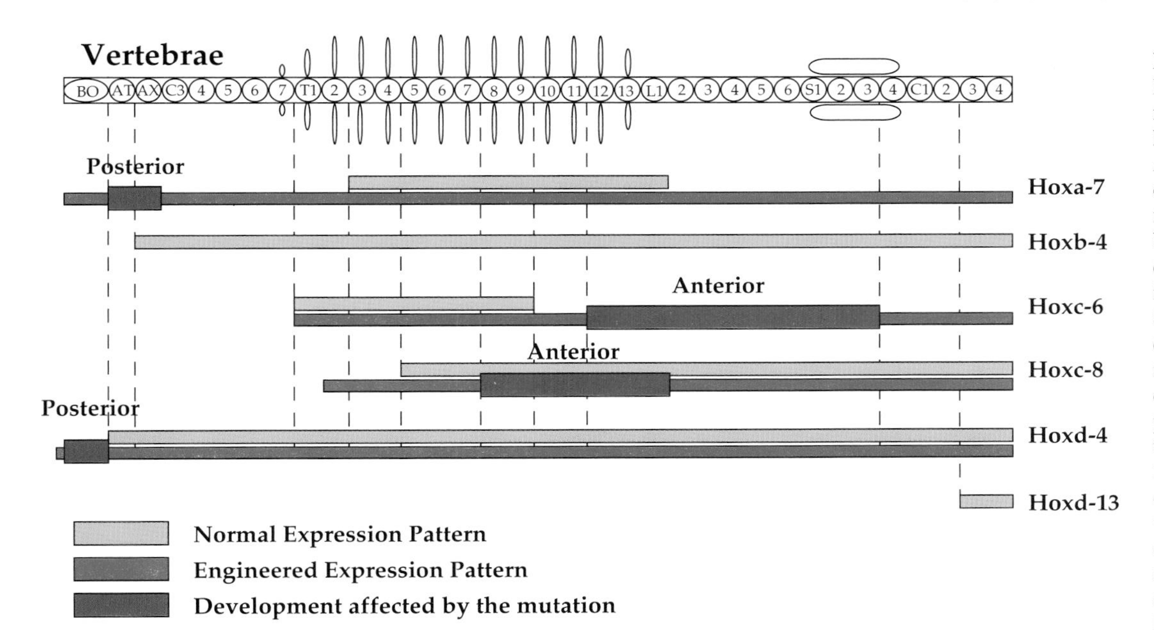

Fig. 5. Overexpression of Hox genes in the paraxial mesoderm causes homeotic transformations. A schematic representation of the vertebrae of the mouse is shown along with both normal and engineered hox expression in the paraxial mesoderm. Some Hox genes expressed anterior to their normal domains of expression cause posteriorized homeotic transformations (eg, *Hoxa-7* and *Hoxd-4* overexpressing mice) whereas other Hox genes expressed posterior to their normal domains of expression or at elevated levels cause anteriorized homeotic transformations (eg, *Hoxc-6* and *Hoxc-8*). Results obtained with *Hoxa-7* and *Hoxd-4* overexpressing mice are consistent with the posterior prevalence model whereas results from *Hoxc-6* and *Hoxc-8* overexpressing mice are not (see text).

lobe identities appear to be changed. Two of the lobes are transformed from posterior to anterior identities whereas one lobe is transformed from an anterior to posterior identity. Thus, EN-2 subdivides the brain in vertebrates, as *en* subdivides *Drosophila* segments, and plays a role in cell fate determination. In fact, the fly *en* has been described both as a segmentation gene and as a homeotic gene.

Analyses of flies mutant for either of two segment polarity genes revealed that head expression of *en* is controlled differently than trunk expression (DiNardo et al., 1988; Heemskerk et al., 1991). It is possible that vertebrates have maintained anterior functions of *en* and have not adapted it for a role in segmentation. Moreover, consistent with its role in vertebrate brain development, flies mosaic for *engrailed*-lethal cells show abnormalities in the brain (Lawrence and Johnston, 1984). It will be interesting to learn whether the genes that regulate anterior and neural *en* expression in flies are similar in nature to those that regulate vertebrate en genes.

The POU group

Comparison of sequences from several homeobox genes from mammals (*Pit-1*, *Oct-1*, *Oct-2*) and worms (*unc-86*) led to the discovery that in addition to highly conserved and distinctive homeoboxes, these genes contain another stretch of homology 5′ to the homeobox (reviewed in Rosenfeld, 1991). These two regions of homology and the linker region between them encode what is now referred to as the POU domain. Subsequently, many other genes encoding POU-domain proteins were cloned, including several from *Drosophila* (Johnson and Hirsh, 1990; Dick et al., 1991; Lloyd and Sakonju, 1991; Treacy et al., 1991; Treacy et al., 1992; Affolter et al., 1993). The mammalian POU genes have been grouped into 6 classes based on sequence similarities (Wegner et al., 1993).

In addition to the homeodomain, these proteins share an approximately 70 amino acid domain amino terminal to the homeodomain, the POU-specific domain, whose structure is discussed above. The POU-specific domain, located amino terminal to the homeodomain, is capable of binding DNA in the absence of the homeodomain, albeit poorly (Verrijzer et al., 1992). Thus, POU proteins contain two autonomous DNA-binding domains which work together to bind DNA avidly, with greater discriminatory capabilities. No gene identified to date contains a POU-specific domain in the absence of a homeodomain. The POU-specific domain in conjunction with the homeodomain is also required for homodimerization and heterodimerization of POU proteins (Ingraham et al., 1990; Voss et al., 1991; Verrijzer et al., 1992). The biological relevance of such interactions is unclear at this point, although in one case, heterodimerization of a *Drosophila* transactivator POU protein (*Cf1a*) with another POU factor (*I-POU*) renders the complex unable to bind DNA in vitro (Treacy et al., 1991). POU domain proteins contain a variety of transactivation domains and have been shown to activate transcription in co-transfection assays (Theill et al., 1989; Ingraham et al., 1990; Muller et al., 1990; Tanaka and Herr, 1990).

In mammals, many genes encoding POU domain proteins are expressed during early embryogenesis. In addition to their expression in the developing nerve cord, several POU domain proteins are expressed in many regions of the developing brain, including forebrain (for review, see Rubenstein and Puelles, 1994). None of the HOX genes are expressed in the forebrain, so POU domain proteins may play a crucial role in patterning events in this region of the CNS. To date, the only vertebrate POU gene mutant analyzed, *Pit-1*, has its late expression in the anterior pituitary (Li et al., 1990; Radovick et al., 1992; Tatsumi et al., 1992; Pfaffle et al., 1992). In these mutants, several cell types of the anterior pituitary fail to develop, suggesting that *Pit-1* is important for their specification and maintenance. *Pit-1* target genes include those encoding growth hormone and prolactin.

In *Drosophila*, several genes encoding POU domain proteins have been cloned (for review, see Wegner et al., 1993). The fly proteins fall into similar classes as the mammalian proteins and may share at least part of their function. For example, the fly genes similar to *Oct-1* and *Oct-2* (*dOct-1*, *dOct-2*) are expressed in the brain and neural tube of the fly, as are their mammalian counterparts. Three other fly POU genes (*Cf1a, I-POU, tI-POU*) and their mammalian relatives (*Brn-1*, *Brn-2*, *Brn-3.0*) are all expressed in nervous tissue.

The Pax group

Vertebrate *Pax* genes are distinguished by the presence of a paired box, encoding a 128 amino acid DNA-binding domain, that is found in *Drosophila* pair-rule and segment polarity genes including *paired*, *gooseberry* and *gooseberry distal* (Bopp et al., 1986). In addition, two fly genes have been isolated that contain a paired box but no homeobox. One of these (*Pox meso*) is expressed in mesoderm whereas the other (*Pox neuro*) is expressed in the nervous system (Bopp et al., 1989). This second class of paired genes, whose products are localized to the nucleus and may act as transcriptional regulators, appears to be under the control of *paired* and other segmentation genes.

So far, ten *Pax* genes have been identified in mammals (for reviews, see Chalepakis et al., 1993; Rubenstein and Puelles, 1994). Out of eight mouse *Pax* genes, half (*Pax-1, Pax-2, Pax-5* and *Pax-8*) do not contain homeoboxes. The ones that contain both homeoboxes and paired boxes (*Pax-3, Pax-6, Pax-7*) are expressed earlier than the others, similar to what is observed in flies. The *Pax-1* gene, which is expressed in the mesoderm, shows a very high degree of sequence homology (90%) with *Pox meso* in the paired domain (Bopp et al., 1989). All *Pax* genes except *Pax-1* are expressed primarily in the nervous system, including neural tube, neural crest cells and brain, in distinct anteroposterior as well as dorsoventral patterns. *Pax* expression is seen in other tissues, including excretory system, muscle, pituitary, pancreas, thyroid, B-cells, ear, eye, limb bud and testes (reviewed in Chalepakis et al., 1993). Whether *Pax* genes are involved in establishing polarity of the embryo, or control specific differentiation events, is unclear at this time.

To date, three mouse mutant phenotypes have been shown to be caused by mutations in *Pax* genes. Mice that bear the *undulated* mutation, which consists of a missense mutation in the Paired box of *Pax-1*, show defects in the axial skeleton (Balling et al., 1988). *Splotch* alleles, mutations in the *Pax-3* gene, cause central nervous system disorders such as excencephalus and spina bifida in addition to abnormalities associated with structures derived from the neural crest (Epstein et al., 1991). Mutation of the *Pax-6* locus (*small eye*) results in mice that lack eyes and nasal structures (Hill et al., 1991). Two human conditions resembling *Splotch* and *small eye*, Waardenburg Syndrome I and Aniridia, are caused by mutations in the human *PAX3* and *PAX6* genes, respectively (Baldwin et al., 1992; Tassabehji et al., 1992). *PAX3* has also been implicated in human rhabdomyosarcomas (Shapiro et al., 1993). In an astonishing example of conservation, the gene *eyeless* of *Drosophila* has been found to encode a protein similar to *PAX6* (Quiring et al., 1994). Thus, *Pax-6* appears to have become dedicated to the visual part of the brain prior to the separation of vertebrate and invertebrate lineages. The strikingly distinct eye structures in mammals and insects apparently conceal a startling common ancestry, much as in the case of the insect and vertebrate hearts.

CONSERVATION UPSTREAM: ZINC FINGER PROTEINS AND LIGAND DEPENDENCE

In both vertebrates and insects, some mechanism must couple maternal or other positional information to the localized activation of the Hox genes. It is still not clear what if any components of this mechanism are similar in these two groups of animals. The apparently vast differences between the syncytial beginning of *Drosophila* embryogenesis and the formation of the inner cell mass in mammals suggests distinct mechanisms must exist, yet the outcome in both cases is differential anterior-posterior Hox expression. We face two mysteries: how Hox genes are regulated and how their different forms of regulation evolved.

There are few mammalian regulators of Hox genes known. The most intensively studied ones are retinoic acid (controlling transcription through its zinc finger protein receptor), the *Krox20* zinc finger protein and the *Sonic hedgehog* (*Shh*) signalling protein. Retinoic acid (RA) regulates Hox genes in cultured cells: the Hox genes are sequentially induced with time or with increased RA concentration (Mavilio et al., 1988). Elegant studies of *Krox20* regulation of *Hoxb2* in the hindbrain rhombomeres has established the high probability of a direct interaction with defined control elements (Sham et al., 1993). *Shh* activates Hox genes by acting as an apparent morphogen in limb development (Riddle et al., 1993). The *Shh* protein is a signalling protein and its receptor is unknown.

The regulation of vertebrate Hox genes by RA receptors and *Krox20* is reminiscent of some of the regulators of Hox gene expression in flies. The 'gap' segmentation genes have been found to regulate initiation of homeotic gene transcription (Harding and Levine, 1988; Irish et al., 1989). Gap genes encode zinc finger proteins, at least two of which, *knirps* and *tailless* proteins, are similar to steroid hormone receptors (Jäckle et al., 1985; Pignoni et al., 1990). There are no known ligands for the gap gene proteins. However, it is intriguing to suppose that ligands may be necessary in animals which, unlike *Drosophila*, do not pass through a syncytial stage of development. If the ligands are able to cross membranes they may function as RA may in mammals. Signalling molecules may govern transcription of Hox genes in both cases.

MAINTENANCE OF REPRESSION

Many of the regulators of homeotic genes are expressed early in *Drosophila* development and then disappear. Yet the genes continue to be spatially regulated. Some progress has been made on understanding genes involved in the maintenance of active or repressed gene states, the genome expression memory systems. Very little is yet known of the extent to which similar systems exist in other animals. A process as fundamental as gene expression maintenance seems likely to be conserved (although animals differ greatly in the degree to which they methylate DNA).

Regulators of homeotic gene transcription have been grouped according to their negative or positive effects.

Members of the *Polycomb* group (Pc-G) are negative regulators required to keep homeotic genes inactive where they should be inactive (reviewed in Paro, 1993). Conversely, members of the trithorax group (Tr-G) have activating functions necessary to maintain homeotic transcription after it has been initiated (reviewed in Kennison, 1993). While these generalities are a good starting guide, the individual properties of many members of both groups have not been fully explored and it is likely that the grouping into positive and negative regulators masks unique attributes of the loci.

The view that Pc-G genes are involved in maintenance of homeotic gene expression is based upon experiments in which homeotic gene transcription was monitored in Pc-G mutants. Initial patterns were normal, but later transcripts were observed where they are not normally found (Struhl and Akam, 1985; Wedeen et al., 1986).

Protein products of three of the Pc-G genes have been localized to about 100 sites on the salivary gland polytene chromosomes (Zink and Paro, 1989; Zink et al., 1991; DeCamillis et al., 1992; Martin and Adler, 1993). The colocalization of the three negative regulators *Polycomb*, *Posterior sex combs* and *polyhomeotic* suggests a very close functional relationship, and indeed evidence has been obtained for a biochemically detectable protein complex containing *Pc* and *ph* proteins (Franke et al., 1992). The *Polycomb* protein contains a short sequence similar to a part of the *Drosophila* HP-1 heterochromatin protein, termed the chromodomain (Paro and Hogness, 1991). Proteins containing this domain have also been observed in vertebrates (Singh et al., 1991; Pearce et al., 1992). Beyond strengthening the hypothesis that Pc-G proteins repress through a mechanism related to heterochromatin-mediated inactivation, the chromodomain has been shown to be sufficient for chromosome localization at the proper 100 sites. Nothing is known about the degree to which *Polycomb* is functionally conserved in vertebrates.

A second case of conservation has also been found, involving the two Pc-G members *Su(z)2* and *Posterior sex combs* (*Psc*). Proteins coded by these two genes share similar regions with the *bmi-1* mammalian proto-oncogene (Brunk et al., 1991; van Lohuizen et al., 1991). Part of the similar region may be a novel type of zinc finger. The regions of similarity are dispersed through much of the proteins, suggesting conservation of a large domain or multiple domains.

HRX, *TRX* AND LEUKEMIA

The *trithorax* (*trx*) gene in *Drosophila*, originally called *Regulator of bithorax*, is responsible for appropriate activation of Antennapedia and bithorax complex homeotic genes (Ingham and Whittle, 1980; Capdevila and Garcia-Bellido, 1981). Thus, *trx* mutant flies show homeotic transformations due to insufficient production of Hox proteins. The maintenance of expression of certain homeotic genes, eg *Ubx*, is more sensitive to the loss of *trx* than others such as *Antp* (Breen and Harte, 1993). Mutations in Polycomb group genes are suppressed by *trx* alleles (Ingham, 1983). Thus, reduced repressor function is balanced by reduced activator function. The *trx* gene has been cloned and shown to encode a very large protein (3759 amino acids) containing many cysteine-rich zinc finger-like domains which are found in proteins that bind DNA (Mazo et al., 1990).

The cloning of a gene involved in translocations associated with acute leukemias led to the discovery that it encodes a human *trx* homolog (*Hrx*; Djabali et al., 1992; Gu et al., 1992; Tkachuk et al., 1992). The encoded protein is 3968 amino acids in length and contains several zinc finger domains. The human protein contains several domains similar to DNA-binding AT hook motifs originally identified in HMG proteins, which are associated with active chromatin structures (Tkachuk et al., 1992). Three regions show homology to the fly *trx* protein, including a carboxy terminal domain that is 82% similar with 61% identity. *Hrx* is expressed during fetal development (Gu et al., 1992).

UPSTREAM ACTIVATORS? THE SWI/SNF COMPLEX AND *BRAHMA*

brahma is another activator of homeotic gene transcription (Kennison and Tamkun, 1988; Tamkun et al., 1992). The encoded protein is similar to a yeast protein called SWI2 or SNF2, whose roles in regulating transcription are reviewed in Winston and Carlson (1992). SWI2, which encodes an ATPase (Laurent et al., 1993), is a member of a 2×10^6 M_r complex containing about ten proteins (Cairns et al., 1994; Peterson et al., 1994). Flies and mammals may contain a SWI/SNF complex of similar size (Khavari et al., 1993; J. W. Tamkun, personal communication; G. R. Crabtree, personal communication). The yeast complex is thought to activate a subset of genes, possibly by opposing the repressive effects of chromatin (Peterson and Herskowitz, 1992). Proteins related to *brahma* have been identified in mammals, though their functions in vivo are unknown (Okabe et al., 1992; Soininen et al., 1992; Khavari et al., 1993; Muchardt and Yaniv, 1993).

COLLABORATORS: THE *EXD/PBX* GROUP

The fly gene *extradenticle* (*exd*) may encode a cofactor that can work together with Hox transcription factors. *exd* mutations result in homeotic transformations of body segments, even though the homeotic genes are expressed at the appropriate times and places (Peifer and Wieschaus, 1990). The cloning of *exd* led to the exciting discovery that the *exd* gene encodes a homeodomain-containing protein with significant similarity to the products of two different genes, the yeast transcriptional repressor MAT**a**1 (which itself is a cofactor for the homeodomain-containing MATα2 protein) and the human PBX homeobox gene family consisting of *PBX-1*, *PBX-2* and *PBX-3*. In yeast, the MAT**a**1/MATα2 heterodimer, which is responsible for repressing haploid-specific genes, binds a different set of target sites than the MATα2 homodimer, which represses **a**1-specific genes. Similarly, the combined action of *exd* and the homeotic gene products could determine binding site selectivity. This may explain the lack of binding site discrimination observed for many homeodomain proteins in in vitro DNA binding assays.

Recently, it has been shown that the proper expression patterns of three target genes regulated by fly homeotic genes (*dpp, wg, tsh;* see below) require *exd* function (Rauskolb and Wieschaus, 1994). The normalcy of Hox gene expression and the altered expression of the target genes regulated by Hox proteins is consistent with a model of *exd* protein as a cofactor of some sort.

PBX-1, one of three *exd*-like genes known in mammals (Monica et al., 1991), was cloned based on its involvement in chromosomal translocations associated with acute leukemias (Kamps et al., 1990; Nourse et al., 1990). One translocation results in the fusion of a portion of PBX-1 including its homeobox with a portion of a helix-loop-helix gene (E2A) with homologies to the *Drosophila daughterless* gene. This gene fusion was shown to be under the control of the E2A regulatory sequences. Further analysis has shown that the E2A/PBX fusion is necessary and sufficient to produce tumors when expressed in blood cells in the mouse(Dedera et al., 1993). The E2A region of the fusion contains a transcriptional activation domain and may be imparting on an otherwise 'inactive' PBX protein a transactivation capability. Since PBX does not activate transcription in co-transfection assays, it is intriguing to speculate that PBX must interact with other factors, perhaps Hox products, to regulate transcription of its target genes. Fusion of an activation domain to PBX might result in inappropriate regulation of these target genes.

INFORMATION FLOWING DOWNSTREAM: ARE TARGETS CONSERVED?

The *Dfd* autoregulatory element is a well-characterized homeotic regulatory element (HRE). *Dfd* protein can activate the element and requires the homeodomain-binding sites within it to do so (Kuziora and McGinnis, 1988; Bergson and McGinnis, 1990; Regulski et al., 1991). Remarkably, the autoregulatory element may be conserved in function to mice, as an element from a mouse *Dfd*-like gene gives appropriate head-specific expression in response to *Dfd* in flies (Malicki et al., 1992). Conversely, the fly element drives localized expression in mouse brain (Awgulewitsch and Jacobs, 1992).

Few direct (or probably direct) target genes of Hox proteins have been identified (reviewed in White et al., 1992; Botas, 1993). The cross-regulatory interactions observed between fly Hox genes have not as yet been seen in mammals. The list of other target genes in flies is short and includes: *decapentaplegic* (*dpp*) and *wingless* in the midgut (Immerglück et al., 1990; Reuter et al., 1990; Capovilla et al., 1994), *teashirt* in the epidermis (Röder et al., 1992) and midgut (Mathies et al., 1994), *spalt* in imaginal discs (Wagner-Bernholz et al., 1991), *connectin* in neuromuscular tissue (Gould and White, 1992), *ems* in the epidermis (Jones and McGinnis, 1993) and *Distal-less* in leg primordia (Vachon et al., 1992). The last two cases mentioned may represent a type of cross-regulation if these genes were originally members of an ancestral Hox cluster.

Virtually nothing is known about target genes in mammals. An intriguing parallel is seen, however, in midgut development of flies and mice (Roberts et al., unpublished data). In flies the transcription of *dpp*, which encodes a TGFβ-class secreted protein, is activated by the *Ubx* homeotic protein in the visceral mesoderm. The two vertebrate proteins most similar to the *dpp* protein are bone morphogenetic proteins 2 and 4 (BMP2, 4; Jones et al., 1991). BMP2 is expressed in the visceral mesoderm of the chick in a region of the gut where homeotic genes are also expressed. The distinct, non-overlapping domains of homeotic gene expression in the *Drosophila* midgut are reminiscent of the discrete domains of Hox expression in the chick midgut. The boundaries of Hox gene expression correspond to the boundaries between different tissue types in the gut, suggesting regulation of gut differentiation by Hox genes may be common to insects and mammals, and may even involve some of the same target genes. The most direct evidence for a role of Hox genes in gut differentiation comes from the phenotype of ectopically expressed *Hoxc-8*, which causes defects in stomach differentiation (Pollock et al., 1992).

CONCLUSIONS

We confront a remarkable array of conserved regulators of development. In some cases, but probably not all, the conservation goes beyond protein structure to conservation of the relationships between types of molecules and the parts of an animal that they control. In many cases the proteins are needed in a variety of tissues and cannot be viewed as dedicated to one organ or tissue. Functions common to many organisms may identify the original sites of gene action. Two questions of outstanding importance are: to what extent did proteins become dedicated to particular developmental processes more than half a billion years ago, and why? What special features of regulators allow them to play their roles? The tools to approach these questions are in hand and we can look forward to new views of developmental regulation when the next decade of homeobox research has passed.

We are grateful to Michael Akam, Laura Mathies and an anonymous reviewer for their helpful comments about an earlier draft. We thank Drs Edoardo Boncinelli, Gerald Crabtree, Walter Gehring, Peter Holland, Eric Olson, Cliff Tabin and John Tamkun for communication of results prior to publication. Research in our laboratory is supported by N.I.H. grant #18163 and by the Howard Hughes Medical Institute.

REFERENCES

Affolter, M., U. Walldorf, U. Kloter, A. F. Schier and W. J. Gehring (1993) Regional repression of a *Drosophila* POU box gene in the endoderm involves inductive interactions between germ layers. *Development* **117**, 1199-1210.

Andres, V., G. B. Nadal and V. Mahdavi (1992) *Clox*, a mammalian homeobox gene related to *Drosophila cut*, encodes DNA-binding regulatory proteins differentially expressed during development. *Development* **116**, 321-334.

Assa-Munt, N., R. J. Mortishire-Smith, R. Aurora, W. Herr and P. E. Wright (1993) The solution structure of the Oct-1 POU-specific domain reveals a striking similarity to the bacteriophage lambda repressor DNA-binding domain. *Cell* **73**, 193-205.

Awgulewitsch, A. and D. Jacobs (1992) *Deformed* autoregulatory element from *Drosophila* functions in a conserved manner in transgenic mice. *Nature* **358**, 341-344.

Azpiazu, N. and M. Frasch (1993) tinman and bagpipe: Two homeo box genes that determine cell fates in the dorsal mesoderm of Drosophila. *Genes Dev* **7**, 1325-1340.

Bachiller, D., A. Macias, D. Duboule and G. Morata (1994) Conservation of a functional hierarchy between mammalian and insect Hox/HOM genes. *EMBO J* **13**, 1930-1941.

Baldwin, C. T., C. F. Hoth, J. A. Amos, E. O. Da-Silva and A. Milunsky (1992) An exonic mutation in the *HuP2* paired domain gene causes Waardenburg's syndrome. *Nature* **355**, 637-638.

Balling, R., U. Deutsch and P. Gruss (1988) *Undulated*, a mutation affecting the development of the mouse skeleton, has a point mutation in the paired box of *Pax-1*. *Cell* **55**, 531-535.

Bally-Cuif, L., R. M. Alvarado-Mallart, D. K. Darnell and M. Wassef (1992) Relationship between *Wnt-1* and *En-2* expression domains during early development of normal and ectopic met-mesencephalon. *Development* **115**, 999-1009.

Bateson, W. (1894) *Materials for the Study of Variation.* MacMillan & Co.

Beachy, P. A., S. L. Helfand and D. S. Hogness (1985) Segmental distribution of bithorax complex proteins during *Drosophila* development. *Nature* **313**, 545-551.

Beeman, R. W., J. J. Stuart, M. S. Haas and R. E. Denell (1989) Genetic analysis of the homeotic gene complex (HOM-C) in the beetle Tribolium castaneum. *Dev. Biol.* **133**, 196-209.

Bergson, C. and W. McGinnis (1990) An autoregulatory enhancer element of the Drosophila homeotic gene Deformed. *EMBO J* **9**, 4287-4297.

Berleth, T., M. Burri, G. Thoma, D. Bopp, S. Richstein, G. Frigerio, M. Noll and V. C. Nusslein (1988) The role of localization of bicoid RNA in organizing the anterior pattern of the Drosophila embryo. *EMBO J* **7**, 1749-1756.

Blochlinger, K., R. Bodmer, J. Jack, L. Y. Jan and Y. N. Jan (1988) Primary structure and expression of a product from *cut*, a locus involved in specifying sensory organ identity in *Drosophila. Nature* **333**, 629-635.

Blochlinger, K., R. Bodmer, L. Y. Jan and Y. N. Jan (1990) Patterns of expression of cut, a protein required for external sensory organ development in wild-type and cut mutant Drosophila embryos. *Genes Dev* **4**, 1322-1331.

Bodmer, R. (1993) The gene *tinman* is required for specification of the heart and visceral muscles in *Drosophila. Development* **118**, 719-729.

Bodmer, R., S. Barbel, S. Sheperd, J. W. Jack, L. Y. Jan and Y. N. Jan (1987) Transformation of sensory organs by mutations of the cut locus of D. melanogaster. *Cell* **51**, 293-307.

Boncinelli, E., M. Gulisano and V. Broccoli (1993) Emx and Otx homeobox genes in the developing mouse brain. *J Neurobiol* **24**, 1356-1366.

Bopp, D., M. Burri, S. Baumgartner, G. Frigerio and M. Noll (1986) Conservation of a large protein domain in the segmentation gene *paired* and in functionally related genes of *Drosophila. Cell* **47**, 1033-1040.

Bopp, D., E. Jamet, S. Baumgartner, M. Burri and M. Noll (1989) Isolation of two tissue-specific *Drosophila* paired box genes, *pox meso* and *pox neuro. EMBO J.* **8**, 1183-1190.

Botas, J. (1993) Control of morphogenesis and differentiation by HOM/Hox genes. *Curr. Opinion in Cell Biol.* **5**, 1015-1022.

Breen, T. R. and P. J. Harte (1993) trithorax Regulates multiple homeotic genes in the bithorax and Antennapedia complexes and exerts different tissue-specific, parasegment-specific and promoter-specific effects on each. *Development* **117**, 119-134.

Brunk, B. P., E. C. Martin and P. N. Adler (1991) Drosophila genes *Posterior Sex Combs* and *Suppressor two of zeste* encode proteins with homology to the murine *bmi-1* oncogene. *Nature* **353**, 351-353.

Cairns, B. R., Y. J. Kim, M. H. Sayre, B. C. Laurent and R. D. Kornberg (1994) A multisubunit complex containing the SWI1/ADR6, SWI2/SNF2, SWI3, SNF5, and SNF6 gene products isolated from yeast. *Proc. Natl Acad. Sci. USA* **91**, 1950-1954.

Capdevila, M. P. and A. Garcia-Bellido (1981) Genes involved in the activation of the bithorax complex of *Drosophila. Wilhelm Roux Arch. Dev. Biol.* **190**, 339-350.

Capovilla, M., M. Brandt and J. Botas (1994) Direct regulation of *decapentaplegic* by *Ultrabithorax* and its role in midugt morphogenesis. *Cell* **76**, 461-475.

Carpenter, E. M., J. M. Goddard, O. Chisaka, N. R. Manley and M. R. Capecchi (1993) Loss of *Hox-A1* (*Hox-1.6*) function results in the reorganization of the murine hindbrain. *Development* **118**, 1063-1075.

Carroll, S. B., S. DiNardo, P. H. OFarrell, R. A. White and M. P. Scott (1988) Temporal and spatial relationships between segmentation and homeotic gene expression in Drosophila embryos: distributions of the *fushi tarazu, engrailed, Sex combs reduced, Antennapedia*, and *Ultrabithorax* proteins. *Genes Dev.* **2**, 350-360.

Carroll, S. B., R. A. Laymon, M. A. McCutcheon, P. D. Riley and M. P. Scott (1986) The localization and regulation of *Antennapedia* protein expression in *Drosophila* embryos. *Cell* **47**, 113-122.

Celniker, S. E., D. J. Keelan and E. B. Lewis (1989) The molecular genetics of the bithorax complex of *Drosophila*:: characterization of the products of the *Abdominal-B* domain. *Genes Dev.* **3**, 1424-1436.

Chalepakis, G., A. Stoykova, J. Wijnholds, P. Tremblay and P. Gruss (1993) Pax: Gene regulators in the developing nervous system. *J. Neurobiol.* **24**, 1367-1384.

Chisaka, O. and M. R. Capecchi (1991) Regionally restricted developmental defects resulting from targeted disruption of the mouse homeobox gene hox-1.5. *Nature* **350**, 473-479.

Cho, K. W., B. Blumberg, H. Steinbeisser and E. M. De Robertis (1991) Molecular nature of Spemann's organizer: the role of the Xenopus homeobox gene goosecoid. *Cell* **67**, 1111-1120.

Clark, K. L., E. D. Halay, E. Lai and S. K. Burley (1993a) Co-crystal structure of the HNF-3/fork head DNA-recognition motif resembles histone H5. *Nature* **364**, 412-420.

Clark, S. G., A. D. Chisholm and H. R. Horvitz (1993b) Control of cell fates in the central body region of C. elegans by the homeobox gene lin-39. *Cell* **74**, 43-55.

Clevidence, D. E., D. G. Overdier, W. Tao, X. Qian, L. Pani, E. Lai and R. H. Costa (1993) Identification of nine tissue-specific transcription factors of the hepatocyte nuclear factor 3/forkhead DNA-binding-domain family. *Proc. Natl Acad. Sci. USA* **90**, 3948-3952.

Cohen, S. and G. Jürgens (1991) Drosophila headlines. *Trends Genet* **7**, 267-272.

Cohen, S. M., G. Bronner, F. Kuttner, G. Jürgens and H. Jäckle (1989) *Distal-less* encodes a homoeodomain protein required for limb development in *Drosophila. Nature* **338**, 432-434.

Cowing, D. W. and C. Kenyon (1992) Expression of the homeotic gene *mab-5* during *Caenorhabditis elegans* embryogenesis. *Development* **116**, 481-490.

D' Esposito, M., F. Morelli, D. Acampora, E. Migliaccio, A. Simeone and E. Boncinelli (1991) EVX2, a human homeobox gene homologous to the *even-skipped* segmentation gene, is localized at the 5′ end of HOX4 locus on chromosome 2. *Genomics* **10**, 43-50.

Dalton, D., R. Chadwick and W. McGinnis (1989) Expression and embryonic function of *empty spiracles*: A *Drosophila* homeo box gene with two patterning functions on the anterior-posterior axis of the embryo. *Genes Dev.* **3**, 1940-1956.

DeCamillis, M., N. Cheng, D. Pierre and H. W. Brock (1992) The polyhomeotic gene of Drosophila encodes a chromatin protein that shares polytene chromosome-binding sites with Polycomb. *Genes Dev.* **6**, 223-232.

Dedera, D. A., E. K. Waller, D. P. LeBrun, A. Sen-Majumdar, M. A. Stevens, G. S. Barsh and M. L. Cleary (1993) Chimeric homeobox gene E2A-PBX1 induces proliferation, apoptosis, and malignant lymphomas in transgenic mice. *Cell* **74**, 833-843.

DeLorenzi, M. and M. Bienz (1990) Expression of Abdominal-B homeoproteins in *Drosophila* embryos. *Development* **108**, 323-329.

Dick, T., X. H. Yang, S. L. Yeo and W. Chia (1991) Two closely linked Drosophila POU domain genes are expressed in neuroblasts and sensory elements. *Proc. Natl Acad. Sci. USA* **88**, 7645-7649.

Diederich, R. J., V. K. L. Merrill, M. A. Pultz and T. C. Kaufman (1989) Isolation, structure, and expression of *labial*, a homeotic gene of the Antennapedia Complex involved in Drosophila head development. *Genes Dev.* **3**, 399-414.

DiNardo, S., J. M. Kuner, J. Theis and P. H. O'Farrell (1985) Development of embryonic pattern in *Drosophila melanogaster* as revealed by accumulation of the nuclear *engrailed* protein. *Cell* **43**, 59-69.

DiNardo, S. and P. H. O'Farrell (1987) Establishment and refinement of segmental pattern in the *Drosophila* embryo: spatial control of *engrailed* expression by pair-rule genes. *Genes Dev.* **1**, 1212-1225.

DiNardo, S., E. Sher, J. Heemskerk-Jorgens, J. Kassis and P. O'Farrell (1988) Two-tiered regulation of spatially patterned *engrailed* gene expression during *Drosophila* embryogenesis. *Nature* **332**, 604-609.

Djabali, M., L. Selleri, P. Parry, M. Bower, B. D. Young and G. A. Evans (1992) A trithorax-like gene is interrupted by chromosome 11q23 translocations in acute leukaemias. *Nature Genetics* **2**, 113-118.

Doe, C. Q. (1992) Molecular markers for identified neuroblasts and ganglion mother cells in the *Drosophila* central nervous system. *Development* **116**, 855-863.

Doe, C. Q., L. Q. Chu, D. M. Wright and M. P. Scott (1991) The *prospero* gene specifies cell fates in the Drosophila central nervous system. *Cell* **65**, 451-464.

Dolle, P., B. J. C. Izpisua, E. Boncinelli and D. Duboule (1991) The Hox-4.8 gene is localized at the 5′ extremity of the Hox-4 complex and is expressed in the most posterior parts of the body during development. *Mech. Dev.* **36**, 3-13.

Dollé, P., M. Price and D. Duboule (1992) Expression of the murine Dlx-1 homeobox gene during facial, ocular and limb development. *Differentiation* **49**, 93-99.

Duboule, D. (1991) Pattern formation in the vertebrate limb. *Curr. Opin. Genet. Dev.* **1**, 211-216.

Duboule, D. (1994) *Guidebook to the Homeotic Genes*. Oxford University Press, Cambridge.

Duprey, P., K. Chowdhury, G. R. Dressler, R. Balling, D. Simon, J. L. Guenet and P. Gruss (1988) A mouse gene homologous to the Drosophila gene caudal is expressed in epithelial cells from the embryonic intestine. *Genes Dev* **2**, 1647-1654.

Eker, R. (1929) The recessive mutant engrailed in Drosophila melanogaster. *Hereditas* **12**, 217-222.

Epstein, D. J., M. Vekemans and P. Gros (1991) splotch (Sp2H), a mutation affecting development of the mouse neural tube, shows a deletion within the paired homeodomain of Pax-3. *Cell* **67**, 767-774.

Faiella, A., M. D'Esposito, M. Rambaldi, D. Acampora, S. Balsofiore, A. Stornaiuolo, A. Mallamaci, E. Migliaccio, M. Gulisano, A. Simeone and E. Boncinelli (1991) Isolation and mapping of EVX1, a human homeobox gene homologous to *even-skipped*, localized at the 5′ end of HOX1 locus on chromosome 7. *Nucleic Acids Res* **19**, 6541-6545.

Finkelstein, R. and N. Perrimon (1990) The *orthodenticle* gene is regulated by *bicoid* and *torso* and specifies *Drosophila* head development. *Nature* **346**, 485-488.

Finkelstein, R. and N. Perrimon (1991) The molecular genetics of head development in *Drosophila melanogaster*. *Development* **112**, 899-912.

Finkelstein, R., D. Smouse, T. M. Capaci, A. C. Spradling and N. Perrimon (1990) The *orthodenticle* gene encodes a novel homeo domain protein involved in the development of the *Drosophila* nervous system and ocellar visual structures. *Genes Dev.* **4**, 1516-1527.

Franke, A., M. DeCamillis, D. Zink, N. Cheng, H. W. Brock and R. Paro (1992) Polycomb and polyhomeotic are constituents of a multimeric protein complex in chromatin of Drosophila melanogaster. *EMBO J* **11**, 2941-2950.

Frumkin, A., R. Haffner, E. Shapira, N. Tarcic, Y. Gruenbaum and A. Fainsod (1993) The chicken *CdxA* homeobox gene and axial positioning during gastrulation. *Development* **118**, 553-562.

Frumkin, A., G. Pillemer, R. Haffner, N. Tarcic, Y. Gruenbaum and A. Fainsod (1994) A role for *CdxA* in gut closure and intestinal epithelia differentiation. *Development* **120**, 253-263.

Gaunt, S. J. (1988) Mouse homeobox gene transcripts occupy different but overlapping domains in embryonic germ layers and organs: a comparison of *Hox-3.1* and *Hox-1.5*. *Development* **103**, 135-144.

Gaunt, S. J., R. Krumlauf and D. Duboule (1989) Mouse homeo-genes within a subfamily, *Hox-1.4*, *-2.6* and *-5.1*, display similar anteroposterior domains of expression in the embryo, but show stage- and tissue-dependent differences in their regulation. *Development* **107**, 131-141.

Gaunt, S. J., P. T. Sharpe and D. Duboule (1988) Spatially restricted domains of homeo-gene transcripts in mouse embryos: relation to a segmented body plan. *Development* **104 Supplement**, 169-179.

Gendron-Maguire, M., M. Mallo, M. Zhang and T. Gridley (1993) Hoxa-2 mutant mice exhibit homeotic transformation of skeletal elements derived from cranial neural crest. *Cell* **75**, 1317-1331.

Gibson, G., A. Schier, P. LeMotte and W.J. Gehring (1990) The specificities of Sex combs reduced and Antennapedia are defined by a distinct portion of each protein that includes the homeodomain. *Cell* **62**, 1087-1103.

Gonzalez-Reyes, A. and G. Morata (1990) The developmental effect of overexpressing a *Ubx* product in *Drosophila* embryos is dependent on its interactions with other homeotic products. *Cell* **61**, 515-522.

Gonzalez-Reyes, A., N. Urquia, W. Gehring, G. Struhl and G. Morata (1990) Are cross-regulatory interactions between homeotic genes functionally significant? *Nature* **344**, 78-80.

Gould, A. P. and R. White (1992) Connectin, a target of homeotic gene control in *Drosophila*. *Development* **116**, 1163-1174.

Gu, Y., T. Nakamura, H. Alder, R. Prasad, O. Canaani, G. Cimino, C. M. Croce and E. Canaani (1992) The t(4;11) chromosome translocation of human acute leukemias fuses the ALL-1 gene, related to Drosophila trithorax, to the AF-4 gene. *Cell* **71**, 701-708.

Harding, K. and M. Levine (1988) Gap genes define the limits of *Antennapedia* and bithorax gene expression during early development in Drosophila. *EMBO J.* **7**, 205-214.

Heemskerk, J., S. DiNardo, R. Kostriken and P. H. O'Farrell (1991) Multiple modes of engrailed regulation in the progression towards cell fate determination. *Nature* **352**, 404-410.

Hill, R. E., J. Favor, B. Hogan, C. Ton, G. F. Saunders, I. M. Hanson, J. Prosser, T. Jordan, N. D. Hastie and V. Van Heyningen (1991) Mouse Small eye results from mutations in a paired-like homeobox-containing gene. *Nature* **354**, 522-525.

Hromas, R., J. Moore, T. Johnston, C. Socha and M. Klemsz (1993) Drosophila forkhead homologues are expressed in a lineage-restricted manner in human hematopoietic cells. *Blood* **81**, 2854-2859.

Hunt, P., M. Gulisano, M. Cook, M. H. Sham, A. Faiella, D. Wilkinson, E. Boncinelli and R. Krumlauf (1991) A distinct Hox code for the branchial region of the vertebrate head. *Nature* **353**, 861-864.

Immerglück, K., P. A. Lawrence and M. Bienz (1990) Induction across germ layers in Drosophila mediated by a genetic cascade. *Cell* **62**, 261-268.

Ingham, P. W. (1983) Differential expression of bithorax complex genes in the absence of the *extra sex combs* and *trithorax* genes. *Nature* **306**, 591-593.

Ingham, P. W., N. E. Baker and A. A. Martinez (1988) Regulation of segment polarity genes in the Drosophila blastoderm by *fushi tarazu* and *even skipped*. *Nature* **331**, 73-75.

Ingham, P. W. and R. Whittle (1980) Trithorax: A new homeotic mutation of *Drosophila melanogaster* causing transformations of abdominal and thoracic segments. *Molec. Gen. Genet.* **179**, 607-614.

Ingraham, H. A., S. E. Flynn, J. W. Voss, V. R. Albert, M. S. Kapiloff, L. Wilson and M. G. Rosenfeld (1990) The POU-specific domain of Pit-1 is essential for sequence-specific, high affinity DNA binding and DNA-dependent Pit-1-Pit-1 interactions. *Cell* **61**, 1021-1033.

Irish, V. F., A. Martinez-Arias and M. Akam (1989) Spatial regulation of the *Antennapedia* and *Ultrabithorax* genes during *Drosophila* early development. *EMBO J.* **8**, 1527-1537.

Jack, J., D. Dorsett, Y. Delotto and S. Liu (1991) Expression of the *cut* locus in the *Drosophila* wing margin is required for cell type specification and is regulated by a distant enhancer. *Development* **113**, 735-747.

Jack, T., M. Regulski and W. McGinnis (1988) Pair-rule segmentation genes regulate the expression of the homeotic selector gene, *Deformed*. *Genes Dev.* **2**, 635-651.

Jäckle, H., U. B. Rosenberg, A. Preiss, E. Seifert, D. C. Knipple, A. Kienlin and R. Lehmann (1985) Molecular analysis of Kruppel, a segmentation gene of Drosophila melanogaster. *Cold Spring Har.b Symp. Quant. Biol.* **50**, 465-473.

Jegalian, B. G. and E. M. De Robertis (1992) Homeotic transformations in the mouse induced by overexpression of a human Hox3.3 transgene. *Cell* **71**, 901-910.

Johnson, A. (1992) A combinatorial regulatory circuit in budding yeast. In *Transcriptional Regulation*, **2**, (Ed. McKnight, S. L. and K. R. Yamamoto). pp. 975-1006. Cold Spring Harbor Laboratory Press, Cold Spring Harbor, NY.

Johnson, W. A. and J. Hirsh (1990) Binding of a Drosophila POU-domain protein to a sequence element regulating gene expression in specific dopaminergic neurons. *Nature* **343**, 467-470.

Joly, J. S., M. Maury, C. Joly, P. Duprey, H. Boulekbache and H. Condamine (1992) Expression of a zebrafish caudal homeobox gene correlates with the establishment of posterior cell lineages at gastrulation. *Differentiation* **50**, 75-87.

Jones, C. M., K. M. Lyons and B. L. M. Hogan (1991) Involvement of *Bone Morphogenetic Protein-4* (BMP-4) and *Vgr-1* in morphogenesis and neurogenesis in the mouse. *Development* **111**, 531-542.

Jones, B. and W. McGinnis (1993) The regulation of *empty spiracles* by *Abdominal-B* mediates an abdominal segment identity function. *Genes Dev* **7**, 229-240.

Joyner, A. L., K. Herrup, B. A. Auerbach, C. A. Davis and J. Rossant (1991) Subtle cerebellar phenotype in mice homozygous for a targeted deletion of the En-2 homeobox. *Science* **251**, 1239-1243.

Joyner, A. L. and G. R. Martin (1987) En-1 and En-2, two mouse genes with sequence homology to the Drosophila engrailed gene: expression during embryogenesis [published erratum appears in Genes Dev 1987 Jul;1(5):521]. *Genes Dev* **1**, 29-38.

Jürgens, G. and D. Weigel (1988) Terminal vs. segmental development in the *Drosophila* embryo: the role of the homeotic gene *fork head*. *Roux's Arch. Dev. Biol.* **197**, 345-354.

Kamps, M. P., C. Murre, X.-h. Sun and D. Baltimore (1990) A new homeobox gene contributes the DNA binding domain of the t(1;19) translocation protein in pre-B ALL. *Cell* **60**, 547-555.

Kappen, C., K. Schughart and F. H. Ruddle (1989) Two steps in the evolution of Antennapedia-class vertebrate homeobox genes. *Proc. Natl. Acad. Sci. USA* **86**, 5459-5463.

Karch, F., W. Bender and B. Weiffenbach (1990) *abdA* expression in *Drosophila* embryos. *Genes Dev* **4**, 1573-1587.

Kastury, K., T. Druck, K. Huebner, C. Barlett, D. Acampora, A. Simeone, A. Faiella and E. Boncinelli (1994) Chromosome locations of human *EMX* and *OTX* genes. *Genomics*, **in press**.

Kaufman, T. C., R. Lewis and B. Wakimoto (1980) Cytogenetic analysis of chromosome 3 in *Drosophila melanogaster*: the homeotic gene complex in polytene chromosomal interval 84A, B. *Genetics* **94**, 115-133.

Kaufman, T. C., M. A. Seeger and G. Olsen (1990) Molecular and genetic organization of the Antennapedia gene complex of *Drosophila melanogaster*. *Adv. Genet.* **27**, 309-362.

Kennison, J. A. (1993) Transcriptional activation of Drosophila homeotic genes from distant regulatory elements. *Trends Genet* **9**, 75-79.

Kennison, J. A. and J. W. Tamkun (1988) Dosage-dependent modifiers of *Polycomb* and *Antennapedia* mutations in Drosophila. *Proc. Natl. Acad. Sci. USA* **85**, 8136-8140.

Kenyon, C. and B. Wang (1991) A cluster of Antennapedia-class homeobox genes in a nonsegmented animal. *Science* **253**, 516-517.

Kessel, M., R. Balling and P. Gruss (1990) Variations of cervical vertebrae after expression of a Hox-1.1 transgene in mice. *Cell* **61**, 301-308.

Khavari, P. A., C. L. Peterson, J. W. Tamkun, D. B. Mendel and G. R. Crabtree (1993) BRG1 contains a conserved domain of the SWI2/SNF2 family necessary for normal mitotic growth and transcription. *Nature* **366**, 170-174.

Kissinger, C. R., B. Liu, B. E. Martin, T. B. Kornberg and C. O. Pabo (1990) Crystal structure of an engrailed homeodomain-DNA complex at 2.8 Å resolution: A framework for understanding homeodomain-DNA interactions. *Cell* **63**, 579-590.

Komuro, I. and S. Izumo (1993) Csx: a murine homeobox-containing gene specifically expressed in the developing heart. *Proc. Nat. Acad. Sci. USA* **90**, 8145-8149.

Kornberg, T. (1981) *engrailed*: a gene controlling compartment and segment formation in *Drosophila*. *Proc. Natl. Acad. Sci. USA* **78**, 1095-1099.

Kornberg, T., I. Siden, P. O'Farrell and M. Simon (1985) The *engrailed* locus of Drosophila: In situ localization of transcripts reveals compartment-specific expression. *Cell* **40**, 45-53.

Krumlauf, R., H. Marshall, M. Studer, S. Nonchev, M. H. Sham and A. Lumsden (1993) *Hox* homeobox genes and regionalisation of the nervous system. *J. Neurobiol.* **24**, 1328-1340.

Kuziora, M. A. and W. McGinnis (1988) Autoregulation of a Drosophila homeotic selector gene. *Cell* **55**, 477-485.

Lai, E., K. L. Clark, S. K. Burley and J. Darnell Jr. (1993) Hepatocyte nuclear factor 3/fork head or 'winged helix' proteins:. *Proc. Nat. Acad. Sci. USA* **90**, 10421-10423.

Lai, E., V. R. Prezioso, W. Tao, W. S. Chen and J. J. Darnell (1991) Hepatocyte nuclear factor 3alpha belongs to a gene family in mammals that is homologous to the Drosophila homeotic gene fork head. *Genes Dev* **5**, 416-427.

Laughon, A. and M. P. Scott (1984) Sequence of a *Drosophila* segmentation gene: protein structure homology with DNA-binding proteins. *Nature* **310**, 25-31.

Laurent, B. C., I. Treich and M. Carlson (1993) The yeast SNF2/SWI2 protein has DNA-stimulated ATPase activity required for transcriptional activation. *Genes Dev* **7**, 583-591.

Lawrence, P. A. and P. Johnston (1984) On the role of the *engrailed+* gene in the internal organs of *Drosophila*. *EMBO J* **3**, 2839-2844.

Le Mouellic, H., Y. Lallemand and P. Brûlet (1992) Homeosis in the mouse induced by a null mutation in the *Hox3.1* gene. *Cell* **69**, 251-264.

Lewis, E. B. (1963) Genes and developmental pathways. *Am. Zool.* **3**, 33-56.

Li, S., E. I. Crenshaw, E. J. Rawson, D. M. Simmons, L. W. Swanson and M. G. Rosenfeld (1990) Dwarf locus mutants lacking three pituitary cell types result from mutations in the POU-domain gene pit-1. *Nature* **347**, 528-533.

Lilly, B., S. Galewsky, A. B. Firulli, R. A. Schulz and E. N. Olson (1994) D-MEF2: A MADS box transcription factor expressed in differentiating mesoderm and muscle cell lineages during *Drosophila* embryogenesis. *Proc. Nat. Acad. Sci. USA*, **in press**,

Liu, S., E. McLeod and J. Jack (1991) Four distinct regulatory regions of the cut locus and their effect on cell type specification in Drosophila. *Genetics* **127**, 151-159.

Lloyd, A. and S. Sakonju (1991) Characterization of two Drosophila POU domain genes, related to oct-1 and oct-2, and the regulation of their expression patterns. *Mech. Dev.* **36**, 87-102.

Lufkin, T., A. Dierich, M. LeMeur, M. Mark and P. Chambon (1991) Disruption of the Hox-1.6 homeobox gene results in defects in a region corresponding to its rostral domain of expression. *Cell* **66**, 1105-1119.

Lufkin, T., M. Mark, C. P. Hart, P. Dollé, M. LeMeur and P. Chambon (1992) Homeotic transformation of the occipital bones of the skull by ectopic expression of a homeobox gene. *Nature* **359**, 835-841.

Macdonald, P. M. and G. Struhl (1986) A molecular gradient in early *Drosophila* embryos and its role in specifying the body pattern. *Nature* **324**, 537-545.

Mahaffey, J. W., R. J. Diederich and T. C. Kaufman (1989) Novel patterns of homeotic protein accumulation in the head of the *Drosophila* embryo. *Development* **105**, 167-174.

Mahon, K. A., H. Westphal and P. Gruss (1988) Expression of homeobox gene *Hox 1.1* during mouse embryogenesis. *Development* **104**, 187-195.

Malicki, J., K. Schughart and W. McGinnis (1990) Mouse *Hox-2.2* specifies thoracic segmental identity in Drosophila embryos and larvae. *Cell* **63**, 961-967.

Malicki, J., L. C. Cianetti, C. Peschle and W. McGinnis (1992) A human HOX4B regulatory element provides head-specific expression in *Drosophila* embryos. *Nature* **358**, 345-347.

Mann, R. S. and D. S. Hogness (1990) Functional dissection of Ultrabithorax proteins in D. melanogaster. *Cell* **60**, 597-610.

Martin, E. C. and P. N. Adler (1993) The Polycomb group gene *Posterior Sex Combs* encodes a chromosomal protein. *Development* **117**, 641-655.

Mathies, L. D., S. Kerridge and M. P. Scott (1994) Role of the *teashirt* gene in *Drosophila* midgut morphogenesis: secreted proteins mediate the action of homeotic genes. *Development* **120 (in press)**.

Mavilio, F., A. Simeone, E. Boncinelli and P. W. Andrews (1988) Activation of four homeobox gene clusters in human embryonal carcinoma cells induced to differentiate by retinoic acid. *Differentiation* **37**, 73-79.

Mazo, A. M., D. H. Huang, B. A. Mozer and I. B. David (1990) The trithorax gene, a trans-acting regulator of the bithorax complex in *Drosophila*, encodes a protein with zinc-binding domains. *Proc. Natl Acad. Sci. USA* **87**, 2112-2116.

McGinnis, W., R. L. Garber, J. Wirz, A. Kuroiwa and W. J. Gehring (1984) A homologous protein-coding sequence in *Drosophila* homeotic genes and its conservation in other metazoans. *Cell* **37**, 403-408.

McGinnis, N., M. A. Kuziora and W. McGinnis (1990) Human *Hox-4.2* and Drosophila *Deformed* encode similar regulatory specificities in Drosophila embryos and larvae. *Cell* **63**, 969-976.

McGinnis, W. and R. Krumlauf (1992) Homeobox genes and axial patterning. *Cell* **68**, 283-302.

McGuinness, T. L., G. P. MacDonald, T. K. Koch and J. L. R. Rubenstein (1992) Evidence for linkage of *Tes-1* and *Dlx-1*, two homeobox genes expressed in the developing mammalian forebrain. *Soc. Neurosci. Abstract* #404.5.

McMahon, A. P., A. L. Joyner, A. Bradley and J. A. McMahon (1992) The midbrain-hindbrain phenotype of Wnt-1−/Wnt-1− mice results from stepwise deletion of engrailed-expressing cells by 9.5 days postcoitum. *Cell* **69**, 581-595.

Millen, K. J., W. Wurst, K. Herrup and A. L. Joyner (1994) Abnormal embryonic cerebellar development and patterning of postnatal foliation in two mouse *Engrailed-2* mutants. *Development* **120**, 695-706.

Mlodzik, M., A. Fjose and W. J. Gehring (1985) Isolation of caudal, a Drosophila homeobox-containing gene with maternal expressions whose transcripts form a concentration gradient at the pre-blastoderm stage. *EMBO J* **4**, 2961-2969.

Monica, K., N. Galili, J. Nourse, D. Saltman and M. L. Cleary (1991) PBX2 and PBX3, new homeobox genes with extensive homology to the human proto-oncogene PBX1. *Mol. Cell Biol.* **11**, 6149-6157.

Morata, G., T. Kornberg and P. A. Lawrence (1983) The phenotype of *engrailed* mutations in the antenna of Drosophila. *Dev. Biol.* **99**, 27-33.

Morata, G. and P. A. Lawrence (1975) Control of compartment development by the *engrailed* gene in Drosophila. *Nature* **255**, 614-617.

Muchardt, C. and M. Yaniv (1993) A human homolog of *Saccharomyces cervisiae SNF2/SWI2* and *Drosophila brm* genes potentiates transcriptional activation by the glucocorticoid receptor. *EMBO J.* **12**, 4279-4290.

Muller, I. M. M., W. Schaffner and P. Matthias (1990) Transcription factor Oct-2A contains functionally redundant activating domains and works selectively from a promoter but not from a remote enhancer position in non-lymphoid (HeLa) cells. *EMBOJ* **9**, 1625-1634.

Murphy, P. and R. E. Hill (1991) Expression of the mouse labial-like homeobox-containing genes, *Hox 2.9* and *Hox 1.6*, during segmentation of the hindbrain. *Development* **111**, 61-74.

Nourse, J., J. D. Mellentin, N. Galili, J. Wilkinson, E. Stanbridge, S. D. Smith and M. L. Cleary (1990) Chromosomal translocation t(1;19) results in the synthesis of a homeobox fusion mRNA that codes for a potential chimeric transcription factor. *Cell* **60**, 535-545.

Nüsslein-Volhard, C. and E. Wieschaus (1980) Mutations affecting segment number and polarity in *Drosophila*. *Nature* **287**, 795-801.

Okabe, I., L. C. Bailey, O. Attree, S. Srinivasan, J. M. Perkel, B. C.

Laurent, M. Carlson, D. L. Nelson and R. L. Nussbaum (1992) Cloning of human and bovine homologs of SNF2/SWI2: A global activator of transcription in yeast *S. cerevisiae*. *Nuc. Acids. Res.* **20**, 4649-4655.

Oliver, G., B. Sosa-Pineda, S. Geisendorf, E. P. Spana, C. Q. Doe and P. Gruss (1993) Prox1, a *prospero*-related homeobox gene expressed during mouse development. *Mech. Dev.* **44**, 3-16.

Ozcelik, T., M. H. Porteus, J. L. R. Rubenstein and U. Francke (1992) *DLX2 (TES1)*, a homeobox gene of the *Distal-less* family, assigned to conserved regions on human and mouse chromosomes 2. *Genomics* **13**, 1157-1161.

Pabo, C. O. and R. T. Sauer (1992) Transcription factors: Structural families and principles of DNA recognition. *Ann. Rev. Biochem.* **61**, 1053-1095.

Paro, R. (1993) Mechanisms of heritable gene repression during development of *Drosophila. Curr. Opinion in Cell Biol.* **5**, 999-1005.

Paro, R. and D. S. Hogness (1991) The Polycomb protein shares a homologous domain with a heterochromatin-associated protein of *Drosophila. Proc. Nat. Acad. Sci. USA* **88**, 263-267.

Patel, N. H., E. Martin-Blanco, K. G. Coleman, S. J. Poole, M. C. Ellis, T. B. Kornberg and C. S. Goodman (1989) Expression of *engrailed* proteins in arthropods, annelids, and chordates. *Cell* **58**, 955-968.

Pattatucci, A. M. and T. C. Kaufman (1991) The homeotic gene *Sex combs reduced* of *Drosophila melanogaster* is differentially regulated in the embryonic and imaginal stages of development. *Genetics* **129**, 443-461.

Pearce, J., P. B. Singh and S. J. Gaunt (1992) The mouse has a Polycomb-like chromobox gene. *Development* **114**, 921-929.

Peifer, M. and E. Wieschaus (1990) Mutations in the Drosophila gene *extradenticle* affect the way specific homeo domain proteins regulate segmental identity. *Genes Dev* **4**, 1209-1223.

Peterson, C. L., A. Dingwall and M. P. Scott (1994) Five *SWI/SNF* gene products are components of a large multisubunit complex required for transcriptional enhancement. *Proc. Nat. Acad. Sci. USA* **91**, 2905-2908.

Peterson, C. L. and I. Herskowitz (1992) Characterization of the yeast SWI1, SWI2, and SWI3 genes, which encode a global activator of transcription. *Cell* **68**, 573-583.

Pfaffle, R. W., G. E. DiMattia, J. S. Parks, M. R. Brown, J. M. Wit, M. Jansen, H. Van Der Nat, J. L. Van Den Brande, M. G. Rosenfeld and H. A. Ingraham (1992) Mutation of the POU-specific domain of *Pit-1* and hypopituitarism without pituitary hypoplasia. *Science* **257**, 1118-1121.

Pignoni, F., R. M. Baldarelli, E. Steingrimsson, R. J. Diaz, A. Patapoutian, J. R. Merriam and J. A. Lengyel (1990) The Drosophila gene *tailless* is expressed at the embryonic termini and is a member of the steroid receptor superfamily. *Cell* **62**, 151-163.

Pollock, R. A., G. Jay and C. J. Bieberich (1992) Altering the boundaries of *Hox3.1* expression: evidence for antipodal gene regulation. *Cell* **71**, 911-923.

Pollock, R. and R. Treisman (1991) Human SRF-related proteins: DNA-binding properties and potential regulatory targets. *Genes Dev.* **5**, 2327-2341.

Price, M., M. Lemaistre, M. Pischetola, R. Di Lauro and D. DuBoule (1991) A mouse gene related to *Distal-less* shows a restricted expression in the developing forebrain. *Nature* **351**, 748-751.

Pultz, M. A., R. J. Diederich, D. L. Cribbs and T. C. Kaufman (1988) The proboscipedia locus of the Antennapedia complex: a molecular and genetic analysis. *Genes Dev.* **2**, 901-920.

Qian, Y. Q., M. Billeter, G. Otting, M. Müller, W. J. Gehring and K. Wüthrich (1989) The structure of the Antennapedia homeodomain determined by NMR spectroscopy in solution: Comparison with procaryotic repressors. *Cell* **59**, 573-580.

Quiring, R., U. Walldorf, U. Kloter and W. J. Gehring (1994) Homology of the *eyeless* gene of *Drosophila* to the *Small eye* gene in mice and *Aniridia* in humans. *Science*, **in press**.

Radovick, S., M. Nations, Y. Du, L. A. Berg, B. D. Weintraub and F. E. Wondisford (1992) A mutation in the POU-homeodomain of Pit-1 responsible for combined pituitary hormone deficiency. *Science* **257**, 1115-1118.

Ramirez-Solis, R., H. Zheng, J. Whiting, R. Krumlauf and A. Bradley (1993) *Hoxb-4 (Hox-2.6)* mutant mice show homeotic transformation of a cervical vertebra and defects in the closure of the sternal rudiments. *Cell* **73**, 279-294.

Rauskolb, C. and E. Wieschaus (1994) Coordinate regulation of downstream genes by *extradenticle* and the homeotic selector proteins. EMBO J, **in press**.

Regulski, M., S. Dessain, N. McGinnis and W. McGinnis (1991) High-affinity binding sites for the Deformed protein are required for the function of an autoregulatory enhancer of the Deformed gene. *Genes Dev* **5**, 278-286.

Reuter, R., G. E. F. Panganiban, F. M. Hoffmann and M. P. Scott (1990) Homeotic genes regulate the spatial expression of putative growth factors in the visceral mesoderm of *Drosophila* embryos. *Development* **110**, 1031-1040.

Riddle, R. D., R. L. Johnson, E. Laufer and C. Tabin (1993) *Sonic hedgehog* mediates the polarizing activity of the ZPA. *Cell* **75**, 1401-1416.

Rijli, F. M., M. Mark, S. Lakkaraju, A. Dierich, P. Dollé and P. Chambon (1993) A homeotic transformation is generated in the rostral branchial region of the head by disruption of Hoxa-2, which acts as a selector gene. *Cell* **75**, 1333-1349.

Riley, P. D., S. B. Carroll and M. P. Scott (1987) The expression and regulation of *Sex combs reduced* protein in Drosophila embryos. *Genes Dev.* **1**, 716-730.

Röder, L., C. Vola and S. Kerridge (1992) The role of the *teashirt* gene in trunk segmental identity in *Drosophila*. *Development* **115**, 1017-1033.

Rosenfeld, M. G. (1991) POU-domain transcription factors: Pou-er-ful developmental regulators. *Genes Dev* **5**, 897-907.

Rubenstein, J. L. R. and L. Puelles (1994) Homeobox gene expression during development of the vertebrate brain. In *Current Topics in Developmental Biology*, (Ed. Pederson, R.) New York: Academic Press.

Salser, S. J. and C. Kenyon (1992) Activation of a C. elegans Antennapedia homologue in migrating cells controls their direction of migration. *Nature* **355**, 255-258.

Salser, S. J. and C. Kenyon (1994) Patterning *C. elegans*: homeotic cluster genes, cell fates and cell migrations. *Trends Genet.* **10**, 159-164.

Salser, S. J., C. M. Loer and C. Kenyon (1993) Multiple HOM-C gene interactions specify cell fates in the nematode central nervous system. *Genes Dev.* **7**, 1714-1724.

Schwabe, J. W. R. and A. A. Travers (1993) What is evolution playing at? *Current Biology* **3**, 628-630.

Scott, M. P. (1992) Vertebrate homeobox gene nomenclature. *Cell* **71**, 551-553.

Scott, M. P., J. W. Tamkun and G. W. Hartzell III (1989) The structure and function of the homeodomain. *BBA Rev. Cancer* **989**, 25-48.

Scott, M. P. and A. J. Weiner (1984) Structural relationships among genes that control development: sequence homology between the *Antennapedia, Ultrabithorax,* and *fushi tarazu* loci of Drosophila. *Proc. Natl. Acad. Sci. USA* **81**, 4115-4119.

Sham, M. H., C. Vesque, S. Nonchev, H. Marshall, M. Frain, R. Das Gupta, J. Whiting, D. Wilkinson, P. Charnay and R. Krumlauf (1993) The zinc finger gene Krox20 regulates HoxB2 (Hox2.8) during hindbrain segmentation. *Cell* **72**, 183-196.

Shapiro, D. N., J. E. Sublett, B. Li, J. R. Downing and C. W. Naeve (1993) Fusion of PAX3 to a member of the forkhead family of transcription factors in human alveolar rhabdomyosarcoma. *Cancer Res.* **53**, 5108-5112.

Sharpe, P. T., J. R. Miller, E. P. Evans, M. D. Burtenshaw and S. J. Gaunt (1988) Isolation and expression of a new mouse homeobox gene. *Development* **102**, 397-407.

Simeone, A., D. Acampora, M. Gulisano, A. Stornaiuolo and E. Boncinelli (1992) Nested expression domains of four homeobox genes in developing rostral brain. *Nature* **358**, 687-690.

Simeone, A., D. Acampora, M. Pannese, M. D'Esposito, A. Stornaiuolo, M. Gulisano, A. Mallamaci, K. Kastury, T. Druck, K. Huebner and E. Boncinelli (1994) Cloning and characterization of two members of the vertebrate Dlx gene family. *Proc. Natl Acad. Sci. USA* **91**, 2250-2254.

Singh, P. B., J. R. Miller, J. Pearce, R. Kothary, R. D. Burton, R. Paro, T. C. James and S. J. Gaunt (1991) A sequence motif found in a Drosophila heterochromatin protein is conserved in animals and plants. *Nucleic Acids Res* **19**, 789-794.

Soininen, R., M. Schoor, U. Henseling, C. Tepe, B. Kisters-Woike, J. Rossant and A. Gossler (1992) The mouse *Enhancer trap locus (Etl-1)*: A novel mammalian gene related to *Drosophila* and yeast transcriptional regulator genes. *Mech. Dev.* **39**, 111-123.

Steitz, T. A. (1990) Structural studies of protein-nucleic acid interaction: the sources of sequence-specific binding. *Quart. Rev. Biophysics* **23**, 205-280.

Struhl, G. and M. Akam (1985) Altered distributions of *Ultrabithorax* transcripts in *extra sex combs* mutant embryos of Drosophila. *EMBO J.* **4**, 3259-3264.

Tamkun, J. W., R. Deuring, M. P. Scott, M. Kissinger, A. M. Pattatucci, T. C. Kaufman and J. A. Kennison (1992) brahma: A regulator of Drosophila homeotic genes structurally related to the yeast transcriptional activator SNF2/SWI2. *Cell* **68**, 561-572.

Tan, D. P., J. Ferrante, A. Nazarali, X. Shao, C. A. Kozak, V. Guo and M. Nirenberg (1992) Murine Hox-1.11 homeobox gene structure and expression. *Proc. Natl Acad. Sci. USA* **89**, 6280-6284.

Tanaka, M. and W. Herr (1990) Differential transcriptional activation by Oct-

1 and Oct-2: interdependent activation domains induce Oct-2 phosphorylation. *Cell* **60**, 375-386.

Tassabehji, M., A. P. Read, V. E. Newton, R. Harris, R. Balling, P. Gruss and T. Strachan (1992) Waardenburg's syndrome patients have mutations in the human homologue of the *Pax-3* paired box gene. *Nature* **355**, 635-638.

Tatsumi, K.-I., K. Miyai, T. Notomi, K. Kaibe, N. Amino, Y. Mizuno and H. Kohno (1992) Cretinism with combined hormone deficiency caused by a mutation in the *PIT1* gene. *Nature Genet.* **1**, 56-58.

Tennyson, V. M., M. D. Gershon, D. L. Sherman, R. R. Behringer, R. Raz, D. A. Crotty and D. J. Wolgemuth (1993) Structural abnormalities associated with congenital megacolon in transgenic mice that overexpress the *Hoxa-4* gene. *Devel. Dynamics* **198**, 28-53.

Theill, L. E., J. L. Castrillo, D. Wu and M. Karin (1989) Dissection of functional domains of the pituitary-specific transcription factor GHF-1. *Nature* **342**, 945-948.

Tkachuk, D. C., S. Kohler and M. L. Cleary (1992) Involvement of a homolog of Drosophila *trithorax* by 11q23 chromosomal translocations in acute leukemias. *Cell* **71**, 691-700.

Treacy, M. N., X. He and M. G. Rosenfeld (1991) I-POU: A POU-domain protein that inhibits neuron-specific gene activation. *Nature* **350**, 577-584.

Treacy, M. N., L. I. Neilson, E. E. Turner, X. He and M. G. Rosenfeld (1992) Twin of I-POU: a two amino acid difference in the I-POU homeodomain distinguishes an activator from an inhibitor of transcription. *Cell* **68**, 491-505.

Vachon, G., B. Cohen, C. Pfeifle, M. E. McGuffin, J. Botas and S. M. Cohen (1992) Homeotic genes of the bithorax complex repress limb development in the abdomen of the Drosophila embryo through the target gene *Distal-less*. *Cell* **71**, 437-450.

Vaessin, H., E. Grell, E. Wolff, E. Bier, L. Y. Jan and Y. N. Jan (1991) prospero Is expressed in neuronal precursors and encodes a nuclear protein that is involved in the control of axonal outgrowth in Drosophila. *Cell* **67**, 941-953.

van Lohuizen, M., M. Frasch, E. Wientjens and A. Berns (1991) Sequence similarity between the mammalian *bmi-1* proto-oncogene and the *Drosophila* regulatory genes *Psc* and *Su(z)2*. *Nature* **353**, 353-355.

Verrijzer, C. P., O. J. Van, Van, der, Vliet and Pc (1992) The Oct-1 POU domain mediates interactions between Oct-1 and other POU proteins. *Mol. Cell Biol.* **12**, 542-551.

Voss, J. W., L. Wilson and M. G. Rosenfeld (1991) POU-domain proteins Pit-1 and Oct-1 interact to form a heteromeric complex and can cooperate to induce expression of the prolactin promoter. *Genes Dev* **5**, 1309-1320.

Wagner-Bernholz, J. T., C. Wilson, G. Gibson, R. Schuh and W. J. Gehring (1991) Identification of target genes of the homeotic gene *Antennapedia* by enhancer detection. *Genes Dev.* **5**, 2467-2480.

Wang, B. B., M. M. Müller-Immergluck, J. Austin, N. T. Robinson, A. Chisholm and C. Kenyon (1993) A homeotic gene cluster patterns the anteroposterior body axis of C. elegans. *Cell* **74**, 29-42.

Wedeen, C., K. Harding and M. Levine (1986) Spatial regulation of *Antennapedia* and *bithorax* gene expression by the *Polycomb* locus in Drosophila. *Cell* **44**, 739-748.

Wegner, M., D. W. Drolet and M. G. Rosenfeld (1993) POU-domain proteins: structure and function of developmental regulators. *Curr. Op. Cell Biol.* **5**, 488-498.

Weigel, D., G. Jürgens, F. Küttner, E. Seifert and H. Jäckle (1989) The homeotic gene *fork head* encodes a nuclear protein and is expressed in the terminal reigons of the *Drosophila* embryo. *Cell* **57**, 645-658.

White, R., J. J. Brookman, A. P. Gould, L. A. Meadows, L. S. Shashidhara, D. I. Strutt and T. A. Weaver (1992) Targets of homeotic gene regulation in Drosophila. *J Cell Sci* **16**, 53-60.

White, R. A. H. and M. Wilcox (1985) Regulation of the distribution of *Ultrabithorax* proteins in Drosophila. *Nature* **318**, 563-567.

Wieschaus, E., N. Perrimon and R. Finkelstein (1992) Orthodenticle activity is required for the development of medial structures in the larval and adult epidermis of *Drosophila*. *Development* **115**, 801-811.

Wilson, R. and others (1994) 2.2 Mb of contiguous nucleotide sequence from chromosome III of *C. elegans*. *Nature* **368**, 32-38.

Winston, F. and M. Carlson (1992) Yeast SNF/SWI transcriptional activators and the SPT/SIN chromatin connection. *Trends Genet.* **8**, 387-391.

Wirz, J., L. Fessler and W. J. Gehring (1986) Localization of the *Antennapedia* protein in the Drosophila embryo and imaginal discs. *EMBO J.* **5**, 3327-3334.

Wolberger, C., A. K. Vershon, B. Liu, A. D. Johnson and C. O. Pabo (1991) Crystal structure of a MATα2 homeodomain-operator complex suggests a general model for homeodomain-DNA interactions. *Cell* **67**, 517-528.

Zink, B., Y. Engström, W. J. Gehring and R. Paro (1991) Direct interaction of the *Polycomb* protein with *Antennapedia* regulatory sequences in polytene chromosomes of Drosophila melanogaster. *EMBO J* **10**, 153-162.

Zink, B. and R. Paro (1989) In vivo binding pattern of a trans-regulator of homoeotic genes in *Drosophila melanogaster*. *Nature* **337**, 468-471.

Development 1994 Supplement, 79-84 (1994)
Printed in Great Britain © The Company of Biologists Limited 1994

The evolutionary origin of development: cycles, patterning, privilege and continuity

L.Wolpert

Department of Anatomy and Developmental Biology, University College, London, London, UK

SUMMARY

A scenario for the evolution of a simple spherical multicellular organism from a single eukaryotic cell is proposed. Its evolution is based on environmentally induced alterations in the cell cycle, which then, by the Baldwin effect, become autonomous. Further patterning of this primitive organism - a Blastaea, could again involve environmentally induced signals like contact with the substratum, which could then become autonomous, by, perhaps, cytoplasmic localization and asymmetric cell division. Generating differences between cells based on positional information is probably very primitive, and is well conserved; its relation to asymmetric cell division is still unclear. Differentiation of new cell types can arise from non equivalence and gene duplication. Periodicity also evolved very early on. The origin of gastrulation may be related to mechanisms of feeding.

The embryo may be evolutionarily privileged and this may facilitate the evolution of novel forms.

Larvae are secondarily derived and direct development is the primitive condition as required by the continuity principle.

Key words: evolution, embryo, cell patterning

INTRODUCTION

Now this is the next tale, and it tells how the camel got his big hump

Just So Stories, Rudyard Kipling

The evolution of the cell can be regarded as the 'big bang' of biological evolution even though it took a very long time. The origin of embryonic development from cells can be regarded as the 'little bang' since the cell was already there. So the general question is, what was required, given the eukaryotic cell, for development to occur (Wolpert, 1990)? How did the egg, patterning and change in form originate? Since embryonic development requires the formation of a multicellular organism from a single cell, the origin of the egg is a central and sadly neglected problem.

The evolution of development is intimately linked to the origin of multicellular forms. The earliest eukaryotic organisms are thought to have been present some 1400 million years ago, while the earliest evidence for metazoans is some 800 million years ago. Metazoan origins are generally thought to be monophyletic (Willmer, 1990; Slack et al., 1993). It may thus seem that the transition from single celled organisms to multicellularity was a difficult one, requiring hundreds of millions of years. However, the fossil record for such delicate organisms is undoubtedly fragmentary and incomplete but, as I have suggested, given the eukaryotic cell with its ability to replicate and move, all the basic elements required for development were already present, and the transition to multicellularity relatively easy. Nevertheless it took another few hundred million years, until the Cambrian, before fossils of recognizable animals can be found.

I will try to present a scenario whereby the eukaryotic cell could have evolved multicellular embryonic development. In doing so a central requirement is that each stage is required to have a selective advantage and that there is continuity between stages (Horder, 1983). Big jumps – hopeful monsters – are not allowed. Even so, I recognize that my scenario is only slightly better, perhaps, than one of Kipling's Just So Stories, like how the leopard got its spots, or the camel its hump.

There are two main theories as to the origin of the Metazoa (Salvini-Plawen, 1978; Willmer, 1990). The first proposes that there was a coming together of two or more cells to form a colony while the second proposes growth of a cell with nuclear division followed by the later establishment of cell boundaries. These theories are extended in various ways to account for the evolution of simple invertebrates and so invoke various sequences leading to Cnidaria, Porifera and Spiralia. In most of these scenarios a Gastrea-like organism appears either earlier or later. However, none of the theories give any attention to the evolution of developmental mechanisms.

ORIGIN OF THE EMBRYO

Of the two theories I shall pursue a modification of the one based on repeated division of a single cell. Plausible though the others may be – the cellular slime mould is an excellent example of aggregation and the early *Drosophila* embryo one for nuclear division, followed by cellularization - they do not easily give rise to what is characteristic of all animal development: the generation of a group of cells by cleavage of a largish cell. This process had to originate at some stage or another so I prefer to take it as my starting point. It is also the only way of reliably obtaining a set of genetically identical cells.

I suggest a mechanism based on a process that is somewhat similar to the Baldwin effect, which was proposed in the early years of this century and extended by Waddington into what he called 'genetic assimilation' (Waddington, 1957). (For a recent review in relation to the evolution of animal behaviour see Bateson, 1988). In essence it involves an environmentally produced effect becoming part of the developmental programme. An environmental signal is replaced by a developmental one.

Several changes are required in order for a single cell to give rise to a multicellular group by cell division (Nurse, personal communication). Firstly, the cell has to grow larger than its normal size and this requires a transient block to mitosis. Secondly, in the enlarged cell, the block over mitosis has to be released and the cell must then divide several times. And thirdly, the cells have to remain together.

The first two requirements involve modification of the cell cycle and it is relatively easy to propose mechanisms based on knowledge of controls in modern cells; but then our basic assumption is that the cell cycle was present in our original eukaryotic cell. In fission yeast, cell size at mitosis is enlarged in a good medium and reduced in a poor medium. This works by modulating when the p34 cd2 kinase is activated, activation in poor media being at a lower cell mass. Thus it is very possible that environmental cues could affect when the transition occurs. For example a single cell growing in a good medium could become enlarged and on a shift to a poor medium undergo two divisions without further growth. This actually occurs with *Chlamydomonas*: during fast rates of growth, cells divide into 4 or even 8 at each cell cycle.

We can thus imagine a cell increasing its diameter 2.5-fold and then dividing 4 times to give 16 cells (Fig. 1). Let us further assume that the third requirement is satisfied and the cells remain together, and, moreover, the cells form a hollow sphere. The latter may require oriented cell divisions or the maintenance of junctions between cells on the outer surface but that may not be too difficult.

The net result of such changes would be an increased cell size in good medium leading to a multicellular sphere in poor medium. It would be a selective advantage if the cells were ciliated so a sphere might swim faster and so find an environment with good food. Now there would be, for the first time, a positive selection for the multicellular state in a poor medium and the multicellular state would be environmentally induced. By the Baldwin effect we can imagine mutations such that no matter what the medium the cells grow large and divide many times. An environmental signal would have been taken over by the genes: the signal has in a sense become constitutive. The selective advantage for multicellularity could now be speed of swimming, sharing of metabolites, protection from predators.

Just one more step is required for the evolution of the embryo. The individual cells need to separate and start the programme again. This could occur when the cells grow and became too large to remain in contact. Each individual cell could now go through the same programme. Embryonic development had evolved. It could well have taken a long time to reach this simple, but crucial stage.

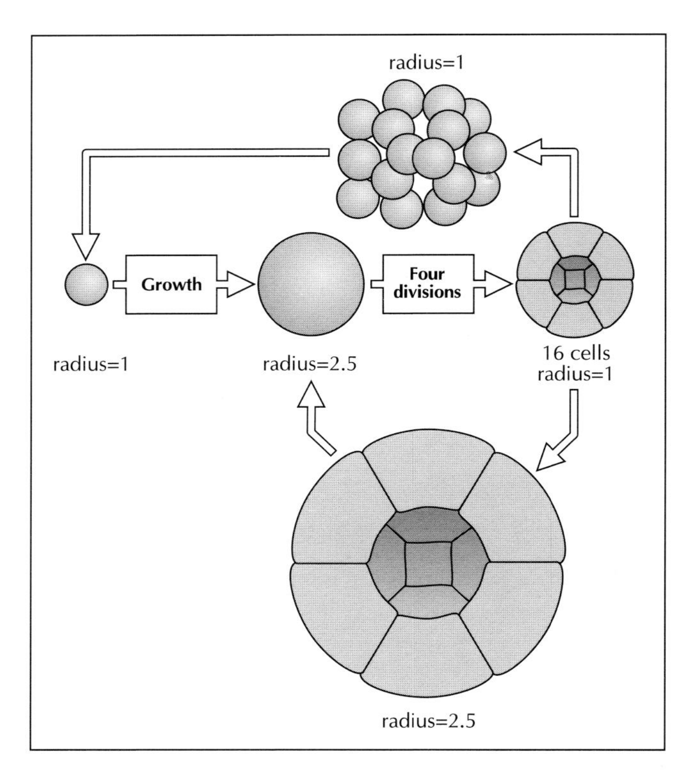

Fig. 1. The origin of the embryo. A single cell is assumed to have increased in size without dividing and then divided four times to form a Blastaea. This may have reflected external signals such as the nature of the medium in which it is growing. If this process became constitutive then the process could be repeated either by the cells separating and repeating the process, or growing in contact with the other cells and then separating.

In this primitive embryo, which, following Haeckel, we call a Blastaea, each cell becomes an egg and there is no spatial organization. Patterning is the next step.

THE ORIGIN OF PATTERNING

Given that the mechanism of altered cell cycles could have generated a hollow sphere formed by a single sheet of cells, we can now consider the origins of patterning, the origins of spatial organization.

The key to all development is the generation of differences between the cells, that is, making them non-equivalent (Lewis and Wolpert, 1976). Only if the cells are different can the organism be patterned so that there are organized changes in shape, and cells at specific sites differentiate into different cell types. How could this have evolved?

Consider that the primitive Blastaea came to rest on the ocean bottom (Fig. 2). At the site where it made contact with the substratum it is not unreasonable to assume that these cells will be differentially affected. It could alter metabolism or affect surface receptors. In time, this environmental signal could result in a cascade of activities starting at the point of contact, affecting both the cells at the contact site and signals that influence more distant cells. For example, there could be

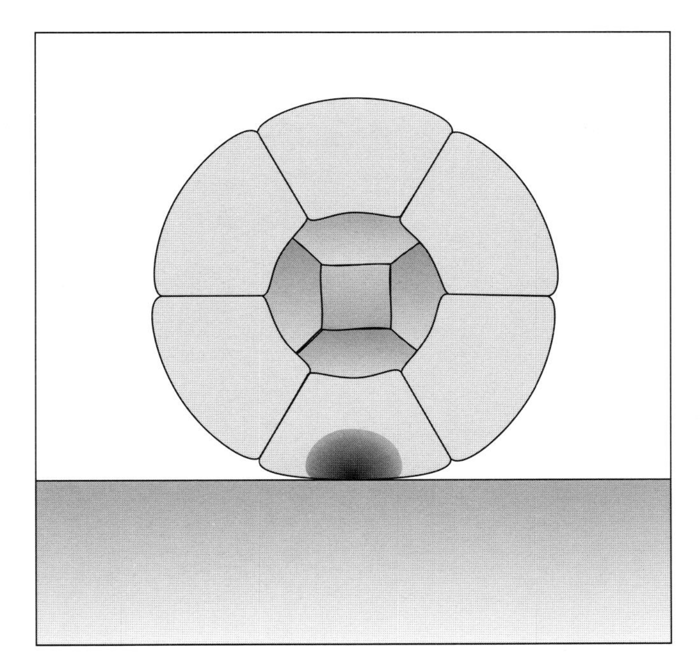

Fig. 2. The origin of an axis. If the Blastaea came to rest on a substratum the site of contact could induce a change in cell state which could specify an axis.

a selective advantage to the organism becoming attached to the point of contact. Or it may be an advantage for the cells to invaginate at the site of contact. Whatever the advantage, an environmental signal brings about a localized change in the organism which becomes elaborated with time. It could even lead to suppression of growth of adjacent cells and so the restriction to reproduction of cells at the opposite end of the embryo. An embryonic axis could be specified. Evolution of patterning has occurred.

There is of course the problem of the nature of the signal. Even now the identification of signals involved in development is restricted to a very few cases. However, one can imagine cells secreting a variety of molecules and by chance some of the cells responding to such signals. There would be little evolutionary pressure against such developmental explorations (see below). It is worth noting that most patterning occurs in sheets, that is in a two dimensional array of cells (Wolpert, 1992).

This initial patterning was entirely environmentally induced by an asymmetric external signal. Even today there are many such examples: light polarizing the *Fucus* egg; sperm entry determining the dorsoventral axis in amphibians; unknown environmental signals determining the anteroposterior axis in Hydrozoa and mammals.

Invoking again, the Baldwin effect, we can see how this process might be constitutive, that is the cells at the contact site could be genetically specified. This illustrates the advantages and economy for evolution of development that the Baldwin effect provides. In its absence it would be necessary to first genetically specify one group of cells as being different but without there being any selective advantage. Only then could there be changes in these cells similar to those described above. This sequence of events is unlikely in the extreme. The reason for invoking the Baldwin effect is that an environmental signal provides an initial basis for a developmental alteration, which could have a selective advantage.

CYTOPLASMIC LOCALIZATION AND ASYMMETRIC CELL DIVISION

Using the Baldwin effect we can imagine how the specification of an environmentally induced axis could be incorporated into the developmental programme. All the machinery for specifying the axis was now present and all that was required was to replace the environmental signal by generating special cells in which the signal would be constitutive. One way of doing this would be to have cytoplasmic localization in the egg so that only one cell or a few cells acquired this special cytoplasm following cleavage (Fig.3). This cell could then specify the axis. Thus asymmetric cell cleavages could be involved in defining a single or small group of cells originally specified by an environmental signal. Moreover, this cell could then become the only cell capable of development and a distinction between soma and germ cells would be established.

It remains unclear why today some organisms like nematodes rely quite heavily on highly determinate patterns of cell cleavage which are often coupled to asymmetric cell divisions, while others, like vertebrate and insects, rely almost exclusively on cell interactions. Why should these two different strategies have evolved? One possibility is that asymmetric cell divisions provide a more reliable means of specifying cell identity on a cell by cell basis, whereas cell signalling is preferable where groups of cells are being specified.

But the puzzle is compounded by the fact that organisms that are specified by determinate cell lineages, which probably involve asymmetric divisions, can also reproduce asexually and have considerable powers of regeneration. Such animals include Cnidaria, Platyhelminthes, and polychaete Annelida (Barnes et al., 1988). Since regeneration and asexual reproduction both absolutely require cell interactions - there is a problem of the relation between the mechanisms of early development and interactions in the adult. It seems that cell interactions represent the primitive condition and that the requirement for asymmetric cell divisions for specification was secondary. Both processes would set up a body plan in which interactions were dominant, and it is very difficult to see how cell interactions could evolve from asymmetric cell divisions. A general principle is that early development is very variable and the phylotypic stage is the more constant. It is worth noting that regeneration is essentially a meta phenomenon related to either asexual reproduction or the fortuitous maintenance of the embryonic state in which cell interactions dominate. Morphallactic regeneration is quite a common mechanism and this, as in *Hydra*, involves the specification of new boundary regions, and the respecification of intermediate values (Wolpert, 1989).

The specification of one axis imposes radial symmetry on a spherical larva (Fig. 4). In principle this only requires the patterning of one small group of cells which could then act as a boundary signal. It seems that bilateral symmetry is a very primitive character among invertebrates. Unfortunately we understand little about how bilateral symmetry is established in present day organisms though it can be related to the first cell cleavage and the specification of the dorsoventral axis. In principle it only requires the specification of another single

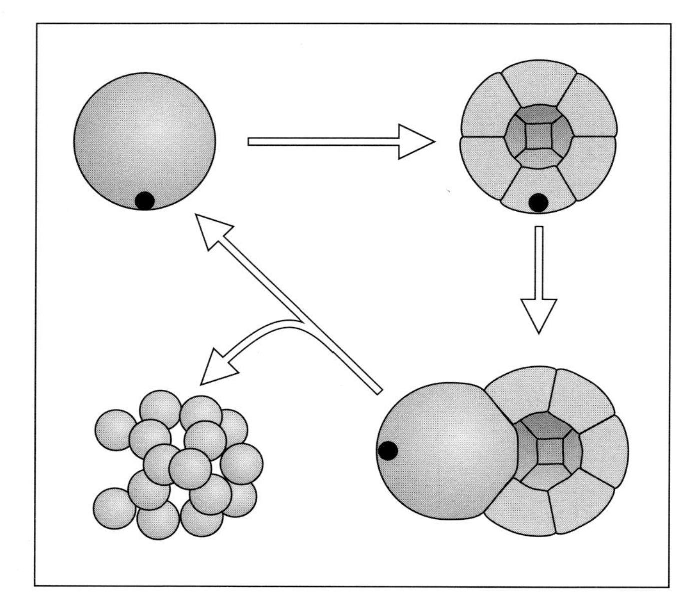

Fig. 3. Cytoplasmic localization and asymmetric cell division could specify a germ cell. Cytoplasmic localization of some factor could specify that only one cell could grow without dividing and so establish a difference between somatic and germ cells.

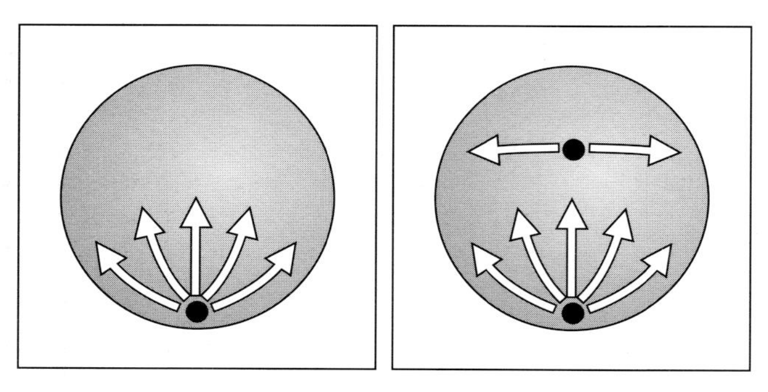

Fig. 4. Specification of axes. Specification of one group of cells could define the anteroposterior axis with radial symmetry; specification of another could define the dorsoventral axis and so produce bilateral symmetry.

location on the surface of the spherical embryo and antero-posterior, dorsoventral, and bilateral symmetry are defined

POSITIONAL INFORMATION AND CELL DIFFERENTIATION

A central feature of positional information is the dissociation of the specification of differences from the ultimate fate of a cell; that is, how it differentiates (Wolpert, 1989). For example, if cells acquire positional identities along the anteroposterior axis then how they differentiate further is almost unconstrained. The concept of the zootype (Slack et al., 1993) illustrates this nicely since apparently all animals have a set of positional identities along the antero posterior axis specified by similar homeobox containing genes. This very strongly suggests that this mechanism for patterning is very primitive.

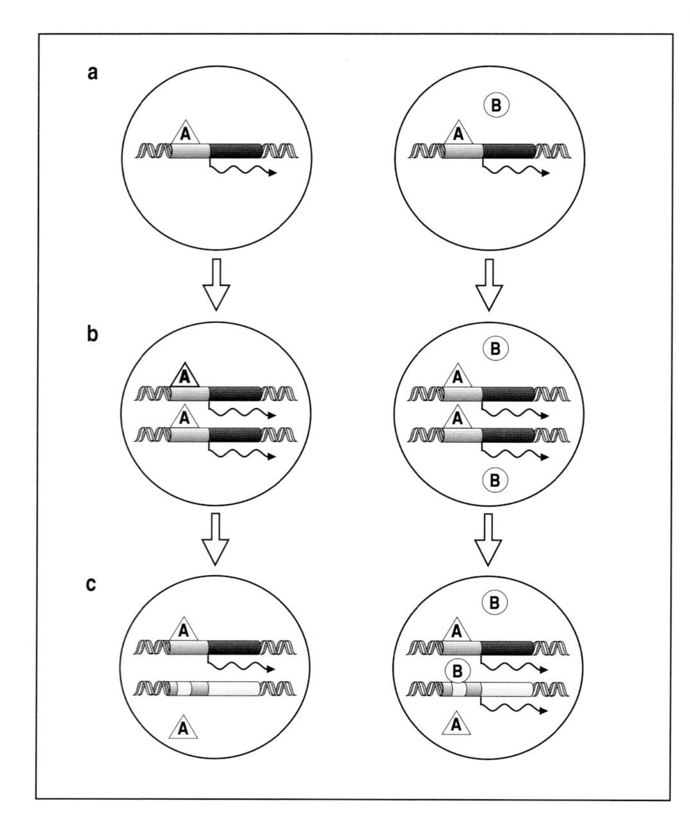

Fig. 5. (A) Two cells express a protein since both contain factor A, which activates the gene. Factor B is only present in the right hand cell. (B) The gene is duplicated and both cells express double amounts of the protein. (C) Changes in both the promoter and structural region result in the expression of the altered duplicated gene only in the right hand cell, because of the presence of B.

Presumably it was established when the first primary axis evolved, as suggested above.

Sternberg and Horwitz (1984) have suggested that cell lineages may evolve by a process of cell duplication followed by a modification of one of the duplicated cells, the analogy being with gene duplication. There are mutations in nematodes that give an extra cell division. Lineally equivalent cells with multiple potentials may thus be an intermediate step in the evolution of cell diversity. They may express different potentials as a result of extracellular signals and then later the potential for diversity might be lost. This requires either, that the two cells are exposed to different signals, or that they are made non equivalent at cell division, that is, the extra cell division is asymmetric.

Another way that new cell types can arise is by the cells originally being non equivalent and then diverging in development because of gene duplication (Fig. 5). Consider two adjacent cells containing transcription factors A and A, B respectively, both expressing a protein, since A activates the gene. If now the gene and its promoter are duplicated then initially both cells will produce double the amount of the protein. However, because of the non equivalence of the cells, and due to the presence of B in only one, small changes in the promoter region of the duplicated gene could result in it only being expressed in the cell with B, which offers the possibility of divergent differentiation.

PERIODICITY AND SEGMENTATION

As Francis Crick once observed, embryos seem to be very fond of stripes. By this he presumably meant that it is very common to find periodicity in embryos and animals, whether it be segmental arrangement of somites in vertebrates, segments in arthropods and worms, or even tentacles in Cnidaria. Development of periodic structures was undoubtedly primitive. Again we do not in general have a good understanding of the mechanisms whereby periodic structures are generated so it makes it difficult to envision an ancestral mechanism. The periodicity of parasegments in *Drosophila* comes from the positional information provided by the gap genes specifying each parasegment separately; in the leech, segmentation seems to be based on a temporal mechanism; and in the spacing of ommatidia in *Drosophila*, or feather buds in chick, it seems to be based on lateral inhibition.

The combination of a mechanism for making periodic structures like segments, together with each acquiring a positional identity opened the way to generating a very wide variety of structures. Using the same basic plan individual segments could diversify, rather like gene duplication.

MORPHOGENESIS AND THE GERM LAYERS

Morphogenesis involves cellular forces bringing about change in form of the embryo. Mechanisms for generating these forces were already present in the primitive eukaryotic cells: the contractile filaments in the furrow bring about cleavage, and the microtubules in the mitotic apparatus could be used for cell elongation.

A very important evolutionary step was gastrulation, which transformed the embryo from a single layered structure into a two-layered and later a three-layered system. I have considered elsewhere the possible origin of the gastrula from the Gastraea as originally suggested by Haeckel (Wolpert, 1992). It could have involved a simple invagination to assist with feeding. As already mentioned, patterning most often occurs in two dimensions, that is within sheets of cells, and gastrulation provides a mechanism for establishing the third dimension; this may be why the basic mechanism is so highly conserved.

It is conventionally thought that there are three basic germ layers, two of which, the mesoderm and endoderm, move internally during gastrulation. However, as Lawrence (1992) points out one must distinguish between the observed movement of layers during gastrulation and their actual fate. While the concept is useful and one can easily envision its evolutionary origin with the endoderm forming the gut, and mesoderm migrating in as single cells, their fate need not necessarily be fixed. For example, neural crest can give rise to 'mesodermal' cells like cartilage. Thus, while early patterning of the embryo usually divides it up into the three germ layers, their fates are not fully restricted.

Another important evolutionary step was the development of a further opening that could fuse with the primitive blind ending gut and so provide a genuine gut tube with mouth and anus that is characteristic of almost all animals apart from Cnidaria and flatworms. The formation of this second opening could originally have resulted from the invaginating gut making contact with the ectoderm and thus providing a signal for making these cells different – possibly the first inductive event in development. This description side steps the problem of whether the invagination occurs in the region of the future mouth as in Protostomes or the future anus as in Deuterstomes (Willmer, 1990).

THE EVOLUTIONARILY PRIVILEGED EMBRYO

What selection pressures will act on the developing embryo? Unlike the adult organism or larva, the embryo seems to be rather privileged. It need not, for example, seek food or mate, and so is not in direct competition for ecological niches. Its primary function is to develop reliably and this reliability is the main feature on which selection will act. This in no way excludes selection for aspects of development that relate more to reproduction and life cycles than structure. These modifications include yolkiness, rate of development, and the evolution of larval forms. Moreover, these can have considerable impact on developmental mechanism, for example, yolkiness can affect the mechanics of gastrulation (Wolpert, 1992). Again the mechanism of segmentation in short germ and long germ insects may be different and related to the rate of development. *Drosophila* with its long germ band develops very rapidly.

If the main selection pressure is for reliability, there is the possibility that the embryo is evolutionarily privileged in that invariants in development that do not affect reliability will not be subject to negative selection. For example, the expression of a gene in cells other than where it is required, the secretion of a variety of molecules, or a transient invagination will escape negative selection provided they do not interfere with development. There is, for example, no evidence that such processes are energetically costly and so not subject to negative selection. It is striking how costly it is just to be alive, the sodium and calcium pump alone using about 25% of total ATP and protein synthesis and breakdown another 50% (Wolpert, 1990). This lack of negative selection may offer the embryo the possibility to explore developmental pathways.

Consider for example the primitive Blastaea or Gastraea before the mesoderm had evolved. Imagine a small number of cells moving into the interior by a chance mutation. The cell could continue to be present for many generations and so could themselves differentiate in a variety of ways. Some will be selected against, while others may, for example, generate internal muscles. The selection is on the adult. Again the widespread secretion of signals and variety of gene expression offer greater opportunities for useful combinations to emerge.

An important consequence of the privileged position of the embryo is that different pathways of development, leading to a similar end result, are to be expected. No one way of gastrulation is 'better' than another, only the end result matters. And there are, indeed, many variations, for example, in the pathways leading to the simple two-layered planula of Cnidaria or to the patterning of the blastula in invertebrates (Wolpert 1992).

LARVAL EVOLUTION

A most remarkable feature of many lower invertebrates including annelids, molluscs and flatworms is the widespread presence of a larval form known in various modifications as the trochophore. The trochophore is rather simple with an outer ectodermal layer with ciliated bands, a gut with mouth and anus, and mesodermal tube like muscles. Unlike embryos, larvae are subject to quite different selections as they have to both feed and disperse before metamorphosis. At metamorphosis there is a profound morphological change – one only has to compare a mollusc with an annelid.

Could the larva be the ancestral form, which gave rise to a variety of invertebrates? This has been suggested by several authors (see Willmer, 1990) who regard the larva as primitive and direct development as a secondary phenomenon. A similar view is taken by Wray and Raff (1990) in relation to echinoderms and their larvae. However, from a developmental point of view this seems improbable, if not impossible.

The difficulty arises because of the continuity principle. How could a larva evolve the ability to metamorphose into the complex morphological forms of the adult? The intermediate steps required are completely improbable. By contrast, if the larval form is a secondary adaptation of development then it is far easier to see how it could have evolved. The similarity of larvae would then reflect convergence rather than a primitive condition.

How larvae could evolve is most easily seen with amphibians and insects. The tadpole is clearly a modification of the early embryo for swimming and feeding. It is not possible for a tadpole to evolve legs. Primitive development was direct and leg development delayed in the larva. Similar consideration applies to insects. Insect larvae could never evolve wings. Thus direct development is always the primitive condition and larvae evolve by modification of embryonic stages.

CONCLUSIONS

Given the eukaryotic cell it is not too difficult to imagine scenarios in which multicellular organisms develop by cleavage of a single cell. It is also possible to imagine how eggs became specific as distinct from body cells and how axes and patterning evolved. The evolution of these processes may have initially involved environmental signals which then became autonomous via the Baldwin effect. Gene duplication and segmentation involving duplication of structures, together with nonequivalence, based on positional identities and asymmetrical cell divisions, opened up pathways for the development of divergent cell patterns and structures. Larval evolution is an adaptation of embryonic forms. It is possible that the evolutionary privilege of the embryo made new developmental pathways easier to evolve.

The future of such evolutionary studies lies in greater understanding of the extent to which changes in single genes can alter cellular behaviour in development.

I am indebted to Matthew McClements for drawing the figures.

REFERENCES

Barnes, R. S. K., Calow, P. and Clive, P. J. W. (1988). *The Invertebrates.* Oxford: Blackwell.

Bateson, P. (1988). The active role of behaviour in evolution. In *Evolutionary Processes and Metaphors* (ed. M. W. Ho and S. W. Fox), pp. 191-207. London: John Wiley.

Horder, T. J. (1983). Embryological basis of evolution. In *Development and Evolution* (ed. B. C. Goodwin et al.,), pp. 315-352. Cambridge, UK: Cambridge University Press.

Lawrence (1992). *The Making of a Fly.* London: Blackwell.

Lewis, J. H. and Wolpert, L. (1976). The principle of non-equivalence in development. *J. Theor. Biol.* **62**, 479-490.

Salvini-Plawen, L. V. (1978). On the origin and evolution of the lower metazoa. *Z. zool. Syst. Evolut-forsch.* **16**, 40-88.

Slack, J. M. W., Holland, P. W. H. and Graham, C. F. (1993). The zootype and the phylotypic stage. *Nature* **361**, 490-492

Sternberg, P. W. and Horwitz, H. R. (1984). The genetic control cell lineage during nematode development. *Ann. Rev. Genet.***18**, 489-524.

Waddington, C. H. (1957). *The strategy of the Genes.* London: Allen and Unwin.

Willmer, P. (1990). *Invertebrate relationships* Cambridge, UK: Cambridge University Press.

Wolpert, L. (1989). Positional information revisited. *Development* **Supplement** 3-12.

Wolpert, L. (1990). The evolution of development. *Biol. J. Linn. Soc.* **39**, 109-124.

Wolpert, L. (1992). Gastrulation and the evolution of development. *Development* **Supplement** 7-13.

Wray, G. A. and Raff, R. A. (1990). Pattern and process heterochronies in the early development of sea urchins. *Semin. Dev. Biol.* **1**, 246-251.

Development 1994 Supplement, 85-95 (1994)
Printed in Great Britain © The Company of Biologists Limited 1994

The evolution of cell lineage in nematodes

Ralf J. Sommer, Lynn K. Carta and Paul W. Sternberg

Howard Hughes Medical Institute and Division of Biology, California Institute of Technology, Pasadena, California, CA 91125, USA

SUMMARY

The invariant development of free-living nematodes combined with the extensive knowledge of *Caenorhabditis elegans* developmental biology provides an experimental system for an analysis of the evolution of developmental mechanisms. We have collected a number of new nematode species from soil samples. Most are easily cultured and their development can be analyzed at the level of individual cells using techniques standard to *Caenorhabditis*. So far, we have focused on differences in the development of the vulva among species of the families Rhabditidae and Panagrolaimidae. Preceding vulval development, twelve Pn cells migrate into the ventral cord and divide to produce posterior daughters [Pn.p cells] whose fates vary in a position specific manner [from P1.p anterior to P12.p posterior]. In *C. elegans* hermaphrodites, P(3-8).p are tripotent and form an equivalence group. These cells can express either of two vulval fates (1° or 2°) in response to a signal from the anchor cell of the somatic gonad, or a non-vulval fate (3°), resulting in a 3°-3°-2°-1°-2°-3° pattern of cell fates. Evolutionary differences in vulval development include the number of cells in the vulval equivalence group, the number of 1° cells, the number of progeny generated by each vulval precursor cell, and the position of VPCs before morphogenesis. Examples of three Rhabditidae genera have a posterior vulva in the position of P9-P11 ectoblasts. In *Cruznema tripartitum*, P(5-7).p form the vulva as in *Caenorhabditis*, but they migrate posteriorly before dividing. Induction occurs after the gonad grows posteriorly to the position of P(5-7).p cells. In two other species, *Mesorhabditis* sp. PS 1179 and *Teratorhabditis palmarum*, we have found changes in induction and competence with respect to their presumably more *C. elegans*-like ancestor. In *Mesorhabditis*, P(5-7).p form the vulva after migrating to a posterior position. However, the gonad is not required to specify the pattern of cell fates 3°-2°-1°-2°-3°. Moreover, the Pn.p cells are not equivalent in their potentials to form the vulva. A regulatory constraint in this family thus forces the same set of precursors to generate the vulva, rather than more appropriately positioned Pn.p cells.

Key words: nematodes, evolution, cell lineage, induction, cell migration, cell death

INTRODUCTION

Most nematodes display invariant cell lineages, that is, a similar pattern of cell divisions in all individuals of a species. One hundred years ago, Boveri and zur Strassen used the invariance of *Ascaris* embryology in their morphological and developmental studies (Boveri 1899; zur Strassen, 1896). They described the early germline - soma differentiation in the embryo of *Ascaris*, visualized by the elimination of chromatin material only in the somatic cells. Together with centrifugation and polyspermy experiments of early *Ascaris* embryos these studies indicated for the first time the necessity of interactions between the cytoplasm and the nucleus in the generation of different cell fates during ontogeny (Boveri, 1910). As an extrapolation of the early germline - soma differentiation, pioneer developmental biologists considered nematodes as a very extreme example of preformistic development with autonomous cell specification.

Over the last twenty years a resurgence of interest in nematode development has been led by the establishment of the free-living nematode *Caenorhabditis elegans* as a model system for genetics, developmental biology, neurobiology and genome analysis (e.g., Brenner, 1974; Wilson et al. 1994). The invariant development of this species, combined with the small cell number, allowed description of the complete cell lineage (Sulston et al., 1983; Sulston and Horvitz, 1977; Kimble and Hirsh, 1979). The combination of genetics, cell lineage and experimental analysis by cell ablation gave insight into a variety of developmental processes. We now know that the invariant development in *Caenorhabditis* results from both autonomous and conditional cell specification, with highly reproducible cell-cell interactions occurring because homologous cells in different individuals have homologous neighbors (reviewed by Lambie and Kimble, 1991; Wood and Edgar, 1994).

The striking invariance of nematode development also made it possible to use this group of organisms for evolutionary developmental analysis. Comparative studies have been initiated in both embryonic and postembryonic development (Sternberg and Horvitz, 1981, 1982; Sulston et al., 1983; Ambros and Fixsen, 1987; Skiba and Schierenberg, 1992). Here we first describe the types of results obtained from genetic analysis in one species, and from comparison of the complete postembryonic cell lineages of two species representing different nematode families. We then focus on the results of our recent studies of vulval development in one family of nematodes (Sommer and Sternberg, 1994; R. Sommer and P. Sternberg, unpublished data).

POSSIBLE EVOLUTIONARY CHANGES INFERRED FROM GENETIC STUDIES

Variation in ontogeny ultimately arises from mutation. What types of changes in development can be caused by single or a few mutations? Cell lineage mutants in *Caenorhabditis* have defined some of the types of changes that can affect a cell lineage in one step. The best example is provided by the "heterochronic" genes, which control the relative timing of specific developmental events in several tissues (Ambros and Horvitz, 1984; reviewed by Ambros and Moss, 1994). Mutations in the genes *lin-4, lin-14, lin-28* and *lin-29* alter the timing of particular events in relation to events in other tissues. These temporal transformations lead to "heterochrony," considered as a possible major source of evolutionary novelty (De Beer, 1958; Gould, 1977). Heterochronic mutations can cause precocious or retarded development (Ambros and Moss, 1994). For example, recessive mutations in *lin-14* cause particular developmental events to occur earlier than normal. Vulva formation normally begins in the third larval stage (L3), whereas precocious *lin-14* alleles it cause to start during the second larval stage (L2). However, semidominant alleles of *lin-14* cause retarded development, resulting in supernumerary larval molts beyond the four wild-type molts.

Genetic and phenotypic analyses of the heterochronic genes revealed a phenotypic hierarchy of the genes *lin-4, lin-14, lin-28* and *lin-29* (Ambros and Moss, 1994). *lin-28* and *lin-29* affect only a subset of the developmental events controlled by *lin-4* and *lin-14*. Consistent with the phenotype described above, molecular analysis has shown that *lin-14* is highly active in the L1 stage and regulates a L1-specific program (Ruvkun and Giusto, 1989). *lin-4* encodes a small RNA that post-transcriptionally regulates *lin-14* by an antisense mechanism (Lee et al., 1993; Wightman et al., 1993). Thus molecular analysis of heterochronic genes in *Caenorhabditis* gives plausible mechanisms for how temporal aspects of developmental events are controlled and how they could be altered during evolution.

SPECIES COLLECTION AND THE COMPARATIVE APPROACH

One approach for the investigation of the evolution of development is comparison between species. Ideally, this approach requires a species collection that spans several different level of taxa. The nematode phylum fulfills this requirement. Together with insects, nematodes show the highest level of radiation in the animal kingdom. Average estimates calculate one million nematode species, many of which are free-living. We have collected nematodes species from soil samples, by placing soil samples on standard *C. elegans* Petri plates and picking the nematodes that crawl out. This approach allows for easy laboratory culture, but biases the worms we extract. So far, we have collected 33 new isolates of Rhabditidae, defining 18 species. Thirteen of these species are new (L. Carta, K. Thomas and P.W.S., unpublished data; L. Carta and P.W.S., unpublished observations). Some nematodes can survive years in dry soil after collection (e.g., Aroian et al., 1993). Most of these nematodes are easily cultured, and moreover, most strains can be frozen at −70°C for long term storage (L. Carta, Y. Hajdu and P.W.S., unpublished observations; M. Edgley, D. Riddle, T. Stiernagle and R. Herman, personal communication). Most of these species belong to the family Rhabditidae, like *Caenorhabditis elegans*, but species of the families Panagrolaimidae and Neodiplogasteridae were also found (see Table 1 for classification of species described in this review). These three families belong to the Order *Rhabditida*, one of approximately 20 nematode orders. Within the Rhabditidae, the collected species span several distinct branchpoints, according to the phylogeny of Sudhaus (1976).

The development of these free-living species can be analyzed at the level of individual cells using techniques standard to *Caenorhabditis* (cell lineage, cell ablation, genetic analysis, and potentially transgenic technology), because most nematodes have a similar bauplan.

POSTEMBRYONIC CELL LINEAGE COMPARISON

Comparison of the cell lineages of two related but morphologically distinct species reveals ways in which cell lineages can change. The complete postembryonic lineage of the free-living nematode *Panagrellus redivivus* of the familiy Panagrolaimidae was compared to *Caenorhabditis elegans* of the family Rhabditidae (Sternberg and Horvitz, 1981, 1982). The changes in the cell lineage between these two species define five classes of modification (Fig. 1). Each of these classes could involve either changes in cell-cell interactions or intrinsic programming.

(1) Switches in the fate of a cell to a fate associated with another cell (Fig. 1 part 1)

One example of a fate switch is in theV5.ppp lineage: in *C. elegans* males the cell V5.ppp generates a sensory ray as well as seam and hyp7 epidermal cells; in *P. redivivus* males the V5.ppp cell generates only seam and hyp7 epidermal cells. A striking example of a fate switch is that of the gonadal cell

Table 1. A very simplified classification of some genera of the phylum Nematoda, class Secernentea

Classification	Vulva position* (%)	Gonad type†	AC-dependence‡
Order Rhabditida			
Rhabditidae			
Mesorhabditis	80	M	–
Cruznema	80	M	+
Teratorhabditis	95	M	–
Pelodera	50	D	+
Caenorhabditis	50	D	+
Oscheius	50	D	+
Panagrolaimidae			
Panagrellus	60	M	+
Order Ascarida			
Ascaris§	35	D	?

*Position of vulva from anterior end (0%) to rectum (100%).
†M, monodelphic; D, didelphic.
‡+, dependent on anchor cell signal; –, independent of anchor cell signal.
§According to Schmidt et al. (1985).

Z4.pp, that becomes a distal tip cell (DTC) in *Caenorhabditis* hermaphrodites, but undergoes programmed cell death in *Panagrellus* females. This cell death would in principle be sufficient to cause the anatomical transformation of a two-armed gonad into a one-armed gonad (see below; Sternberg and Horvitz, 1981); ablation of the posterior distal tip cell in *Caenorhabditis* causes an essentially one-armed gonad to form (Kimble and White, 1981). There is a surprisingly wide range of interspecific differences in lineages in which programmed cell death occur (Sulston and Horvitz, 1977; Sulston et al., 1983; Sternberg and Horvitz, 1981, 1982; R. J. S. and P. W. S., in preparation). Raff's hypothesis for cell death as a default (Raff, 1992) provides an explanation for these promiscuous changes in programmed cell death: it is in principle relatively easy to program cell death merely by blocking response to a survival factor. Ellis and Horvitz (1991) have identified single gene mutations (defining the genes *ces-1* and *ces-2*) that prevent the programmed death of specific cells, as opposed to all programmed cell death; changes in such genes might underlie behavioral and morphological evolution.

(2) Reversal in the polarity of a sublineage (Fig. 1 part 2)

The fates of two sisters are interchanged; the consequence is a change in the spatial distribution of cells without change in cell number or type. Lineage reversals have been described in the male tail lineages between *Caenorhabditis* and *Panagrellus*, but also exist between the sexes of *Panagrellus* in parts of their gonadal lineage (Sternberg and Horvitz, 1981, 1982), and the early embryonic founder lineage, in the division of P2 between *Caenorhabditis* and *Cephalobus* (Skiba and Schierenberg, 1992). Genes have been identified that alter polarity of a lineage, for example, *lin-18* and *lin-44* (Ferguson et al., 1987; Herman and Horvitz, 1994; W. Katz and P. W. S., unpublished data).

(3) Alterations in the number of rounds of cell divisions (Fig. 1 part 3)

Addition (Fig. 1-3A) or suppression (Fig. 1-3B) of a cell division is the most common type of cell lineage alteration seen between *Caenorhabditis* and *Panagrellus*. Additional cell divisions can either be symmetric or asymmetric. Symmetric additional divisions form a duplication of sublineages (e.g. Z1.ppp lineage in *Panagrellus* female), whereas asymmetric additional divisions can produce a pattern reiteration (Fig. 1-3C; e.g., AR.pp gonadal lineage in *Panagrellus*). *unc-86* mutations lead to reiterations in particular lineages (Chalfie et al., 1981).

The generation of additional cells is likely to be of evolutionary importance. After duplication, the additional cell will be functionally equivalent to its sister. But as with the duplication of genes or gene-clusters, this cell might be able to adopt novel characters and functions because of different extrinsic (different neighbors) or different intrinsic (different gene expression) properties. Thus, cell duplication might create cell diversity over evolutionary time scales, as proposed by Goodman (1977) and Chalfie et al. (1981).

(4) Changes in the relative timing of cell divisions (Fig. 1 part 4)

Interesting examples include the early embryonic founder lineage, in which there is no known consequence (Skiba and Schierenberg, 1992) and Z1 in *P. redivivus* females, in which the delayed division is the first difference in the Z1 and Z4 lineages, ultimately involving the death of Z4.pp but not the homologous Z1.aa (Sternberg and Horvitz, 1981). This is a subtle example of heterochrony.

(5) Altered segregation of lineage potential (Fig. 1 part 5)

The developmental potential to generate particular cell types might be transferred from one cell to its sister. One examples is in the male gonadal lineage between *Caenorhabditis* and *Panagrellus* (Sternberg and Horvitz, 1981). Since fates might be specified by cell-cell interactions between sisters and cousins (e.g., Posakony, 1994), this type of change might involve a reversal in the polarity of division potential, with the same cell interactions occurring, such that the posterior daughter rather than the anterior daughter divides, but the middle cell is nonetheless of type B. At a molecular level, *mex-1* mutants are defective in localization of the SKN-1 determinant (Bowerman et al., 1993).

The finding that similar types of lineage transformations have also been observed in *Caenorhabditis* after cell ablation and in mutant animals suggests that relatively minor genetic changes might cause the observed changes in species characters. Thus, alterations at the level of cells or genes can cause the evolution of cell lineages.

COMPARISON OF VULVA DEVELOPMENT

The induction of the hermaphrodite vulva is one of the best

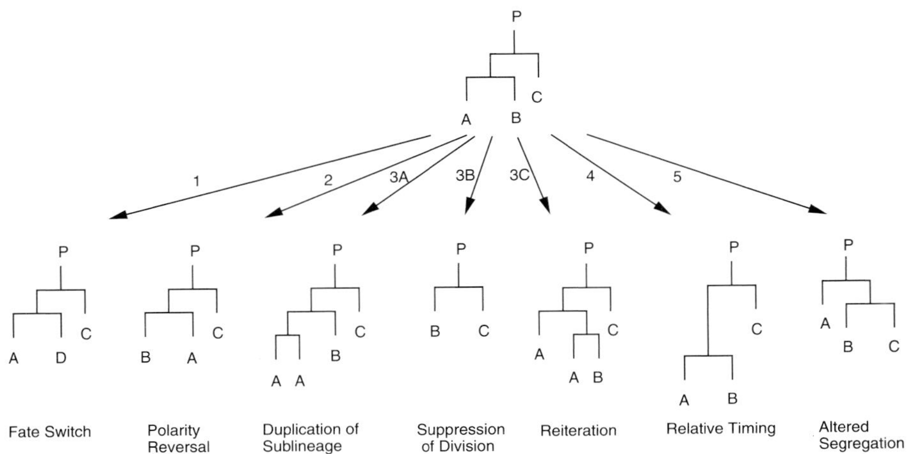

Fig. 1. Different types of cell lineage transformations observed between nematode species. P, a precursor cell. A, B and C, different cell types. 1. Fate Switch, cell with fate B instead has fate D. 2. Polarity Reversal, A and B are produced in different positions. 3. Changes in number of cell divisions: 3A. Duplication of A cell. 3B. Suppression of division; might change the fate of parent of A and B to A, B (as shown) or other fate. 3C. Reiteration, B has fate of its parent, i. e., and extra asymmetric cell division. 4. Change in the relative timing of divisions; such a change can affect subsequenct cell interactions. 5. Altered segregation (see text).

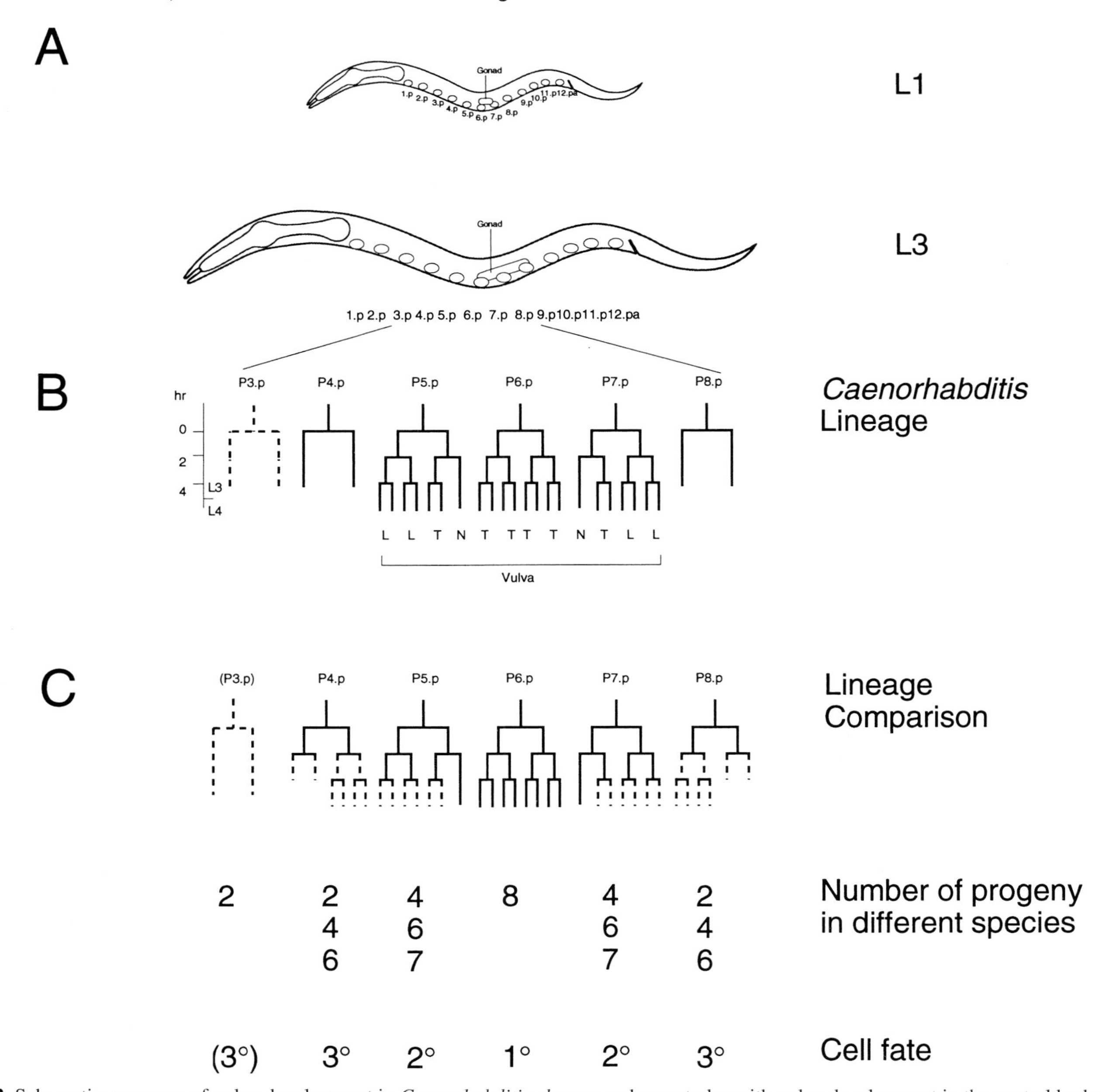

Fig. 2. Schematic summary of vulva development in *Caenorhabditis elegans* and nematodes with vulva development in the central body region. (A) Position of the Pn.p ectoblasts in the L1 stage and the early L3 stage. Cells are homogenously distributed between pharynx and anus. (B) Vulva cell lineage in *Caenorhabditis elegans*. L, longitudinal division; T, transverse division; N, non-dividing cell; according to Sternberg and Horvitz (1986). (C) Schematic comparison of the vulva cell lineage in other species. The plain lines refer to cell divisions occurring in all species. Dashed lines represent the cell divisions that occur only in one or some species. P6.p generates 8 progeny in all analyzed species, whereas P(5,7).p generate 4, 6 or 7 progeny and P(4,8).p generate 2, 4 or 6 progeny. Examples of the different number of progeny are: 2°, 4 in *Oscheius*, 6 in *Pelodera*, 7 in *Caenorhabditis*; 3°, 2 in *Caenorhabditis*; 4 in *Oscheius* and 6 in *Pelodera*. P3.p is 3° in 50% of *Caenorhabditis* hermaphrodites, but is not a VPC in the other species.

understood examples of postembryonic development in *Caenorhabditis* (reviewed by Hill and Sternberg, 1993). Twelve Pn ectoblast cells migrate into the ventral cord and divide to produce posterior daughters (Pn.p cells) whose fate varies in a position-specific manner (Sulston and Horvitz, 1977; Fig. 2A). Some Pn.p cells are competent to generate vulval cells; others are non-specialized epidermal cells. In *Caenorhabditis*, the Pn.p cells located in the central body region (P3.p-P8.p) are tripotent vulva precursor cells (VPCs) and form an equivalence group (Fig.2B). In intact animals, P(5-7).p respond to a signal from the gonadal anchor cell (AC) (Kimble, 1981). P6.p has the 1° cell fate generating 8 progeny; P(5,7).p have the 2° cell fate and generate 7 progeny; together these progeny form the vulva. The remaining cells, P(3,4,8).p have a non-vulval fate, called 3°, resulting in a 3°-3°-2°-1°-2°-3° pattern of cell fates. After ablation of individual Pn.p cells, more AC-proximal cells, within the equivalence group have the potential to replace more AC-distal cells and thereby regenerate the vulval pattern (Sulston and Horvitz, 1977; Sternberg and Horvitz, 1986).

Over the last few years molecular analysis has shown that one of the *Hom-C* genes, *lin-39*, contributes positional information to establish the vulva equivalence group within the linear array of the 12 Pn ectoblast cells (Wang et al., 1993; Clark et al., 1993) (Fig. 3A). Further patterning within the equivalence group is initiated by an EGF-like growth factor inductive signal (LIN-3) that stimulates a receptor tyrosine-kinase mediated signal transduction pathway (Hill and Sternberg, 1992, 1993; R. Hill, W. Katz, T. Clandinin and P. Sternberg, unpublished observations; Fig. 3B). Additional lateral signaling between the VPCs acts via a distinct receptor and signal transduction pathway (Greenwald et al., 1983; Sternberg, 1988; Yochem et al., 1988; Sternberg and Horvitz, 1989). A negative signal, presumably from the hyp7 epidermis, and requiring the action of *lin-15* as well as other genes, may block basal activity of the LET-23 receptor (Herman and Hedgecock, 1990; Huang et al., 1994). While *lin-15*-controlled signaling appears to convey no patterning information in *Caenorhabditis*, it is conceivable that it might do so in another taxa.

Caenorhabditis and many other species form the vulva in the central body region. Analysis of vulva development in the species *Oscheius sp.* PS1131, *Oscheius tipulae*, *Rhabditella axei*, *Rhabditoides regina*, *Pelodera strongyloides* and *Protorhabditis* sp. PS1010 by cell lineage observation and cell ablation experiments revealed that in all these species the vulva is formed by the progeny of P(5-7).p as in *Caenorhabditis* (R.J.S. and P.W.S., unpublished data). Nonetheless, evolutionary alterations are present at several levels.

(1) Number of VPCs constituting the equivalence group (Fig. 2C)

So far, only in *Protorhabditis* does P(3-8).p form the equivalence group as in *Caenorhabditis*. In all other analyzed species P(4-8).p are VPCs; the final pattern within the equivalence group is 3°-2°-1°-2°-3°. This is an example of a fate change with respect to P3.p (Fig. 1-1).

(2) The cell lineage generated by the 1°, 2° and 3° VPCs (Fig. 2C)

P6.p, which has the 1° cell fate, generates eight progeny in all analyzed Rhabditidae species with a central vulva. The cells with the 2° cell fate, P(5,7).p, generate between four and seven progeny in a species-specific manner. The 3° cell lineage of P(4,8).p varies between two and six progeny in different species. In the *Rhabditoides* and *Pelodera* 3° lineage the AC-proximal two cells of the four-cell stage undergo a third round of cell division and produce an asymmetric lineage. These are examples of changes in the number of rounds of cell division (Fig. 1, part 3).

How is this mirror image difference between P4.p and P8.p regulated? After ablation of the gonad, all VPCs express a 3° lineage with the asymmetry normally found in P4.p, suggesting this is the ground state. Thus, vulva induction by the gonad is directly, or indirectly involved in reversing the polarity of the asymmetric lineage of P8.p. A similar polarity reversal occurs in the 2° cell lineage in *Caenorhabditis*; genetic and cell ablation experiments indicate that a signal from the gonad reverses the polarity of the P7.p lineage (W. Katz and P.W.S., unpublished data).

(3) Variability of cell lineages

Pelodera strongyloides displays striking variability in its 3° vulva lineages. P3.p and P9.p divide in nearly 50% of the females, expressing a partial or complete 3° lineage, generating two, four or six progeny. Increasing the temperature changes the frequency of this variability, a phenomenon also known to occur in, for example, the occasional division of P5.ppp and P7.paa in the *Caenorhabditis* hermaphrodite at 25°C (Sternberg and Horvitz, 1986).

An important question is how invariant cell lineages change during evolution into other cell lineages. Variability might play a role in the evolutionary transition process from one invariant cell lineage to another. More detailed comparative studies based on a solid phylogeny might help test this hypothesis. A taxon that includes members with two distinct invariant cell lineages, and other members with variable lineages would support this hypothesis and provide the experimental material to analyze the genetic and molecular basis for the transition. Another hypothesis is that variability results from a failure to select precision in the underlying genetic program, for example in a lineage undergoing extensive proliferation (e.g., Sternberg and Horvitz, 1981, 1982).

FORMATION OF A POSTERIOR VULVA

One major difference between species exists in the body position of the developing vulva (central vs. posterior vulva formation) (Figs 4, 5). This difference coincides with different gonad morphology. Central vulva species form two ovaries symmetrically about the vulva (didelphic); posterior vulva

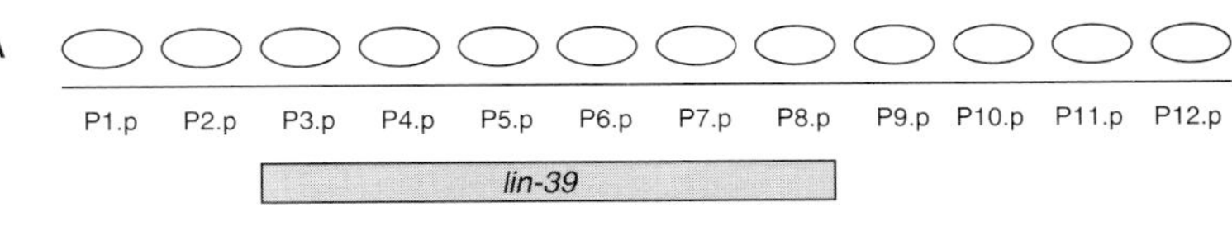

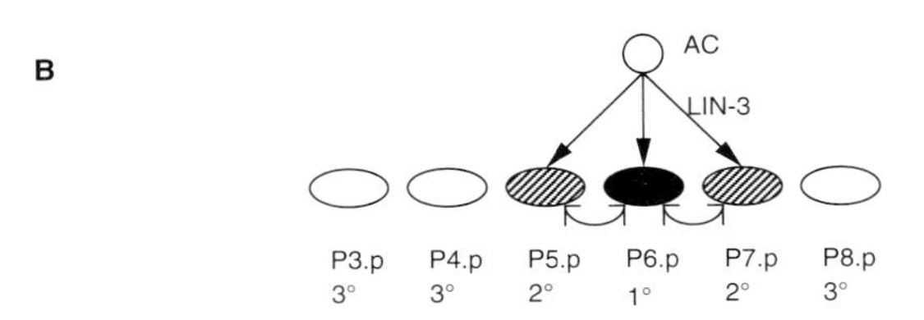

Fig. 3. Schematic summary of the early steps in vulva pattern formation in *Caenorhabditis elegans*. (A) The twelve Pn.p cells have different developmental potentials because of positional information generated by the Hom-C genes. Only the two Hom-C genes influencing the central body region are shown. *lin-39* is involved in the establishment of the vulva equivalence group. (B) After P(3-8).p have been specified as the vulval equivalence group, further pattern formation is initiated by an inductive signal from the gonadal anchor cell (AC) (indicated by bold arrows). Lateral signaling is one of the additional interactions necessary for proper vulva development (curved arrows).

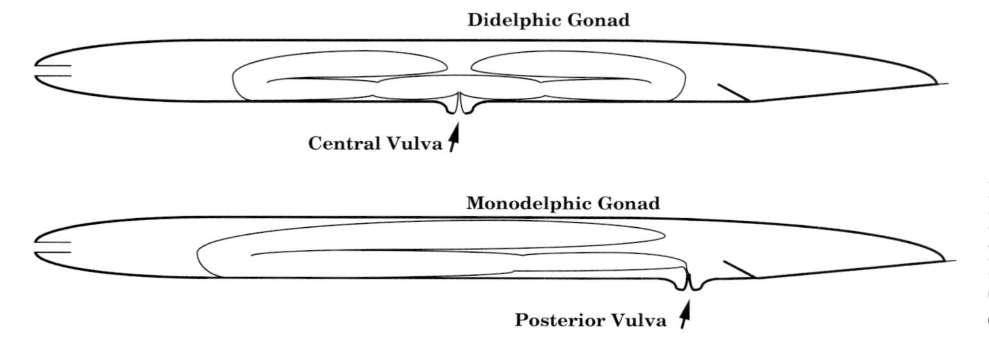

Fig. 4. Correlation of gonad type and vulval position. Central vulva species have two-armed (didelphic) gonads with two ovaries. Posterior vulva species have one-armed (mondelphic) gonads with a single ovary directd anterior from the vulva.

species form a single ovary, directed anterior from the vulva (monodelphic). A didelphic gonad with a vulva in the central body region is considered to be the ancestral character in the taxa we have considered so far (Chitwood and Chitwood, 1950).

Which set of precursor cells make the vulva in species with posterior vulva formation? In *Panagrellus redivivus*, which has a vulva at 60% body length (head = 0%; tail = 100%), the equivalence group and the gonad primordium are shifted posteriorly. In principle, a more extreme shift in the equivalence group, for example to P9.p-P11.p, and gonad primordium would allow posterior vulva formation. In three genera, *Mesorhabditis, Teratorhabditis* and *Cruznema*, cell lineage analysis revealed that the central P-ectoblasts still form the vulva equivalence group (Sommer and Sternberg, 1994). In *Cruznema tripartitum*, P(3-8).p, and in *Mesorhabditis* sp. PS 1179 and *Teratorhabditis palmarum* P(4-8).p migrate posteriorly during the L2 stage. As in *Caenorhabditis*, P(5-7).p form the vulva in intact animals of these species. An example of P-cell location before and after migration and of the vulva cell lineage is shown for *Mesorhabditis* (Fig. 6).

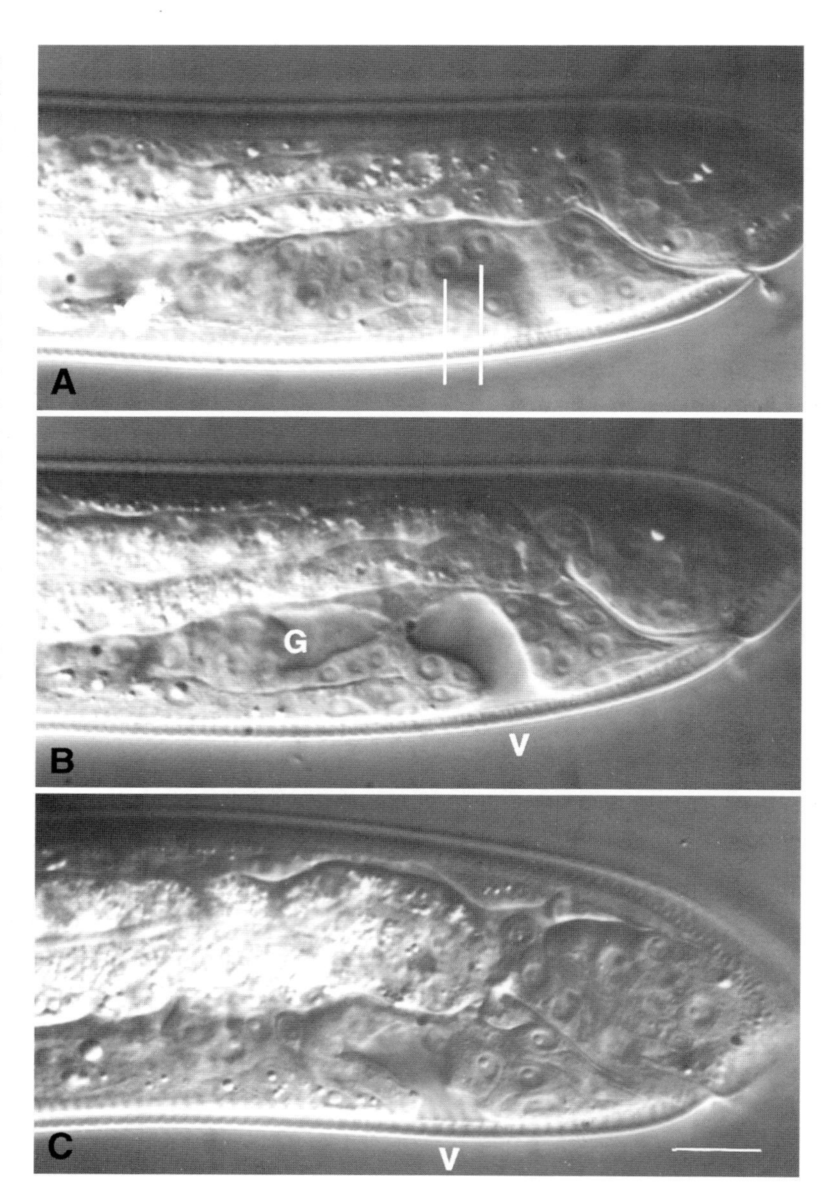

Fig. 5. Nomarski photomicrographs of lateral views of *Teratorhabditis* females at different stages of vulva development in intact (A,B) and cell-ablated (C) animals. (A) Late L3 stage of vulva development in an intact animal with two progeny of P6.p visible in this plane of focus. (B) Mid L4 stage showing a medial focal plane with the gonad (G), vulva (V) and the rectum. (C) Mid L4 stage of a gonad-ablated worm. No uterus is formed, the vulva is still present. Scale bar, 20 μm.

Cruznema

In *Cruznema,* induction of the vulva by the AC occurs after the gonad grows posteriorly to the position of the VPCs (Sommer and Sternberg, 1994). In the intact animal the AC forms a specific contact with P6.p, the cell adopting the 1° cell fate. After ablation of P(5,6).p, their neighbors P4.p and P7.p can assume the 1° cell fate. In these experiments, the AC stops migration in the region of P4.p or P7.p respectively, indicating that cell-cell interactions occur between the AC and the VPCs prior or simultaneously to induction. In central vulva species, the AC is born close to P6.p and thus it is more difficult to assess the role of such interactions. However, in *Caenorhabditis*, the anchor cell will extend a process towards the cluster of cells generated by a misplaced 1° VPC, presumably in response to a signal from the VPC grandprogeny (K. Tietze and P. S., unpublished data). Posterior vulva formation in *Cruznema* is thus gonad-dependent and occurs within a set of VPCs that seems to form an equivalence group as in *Caenorhabditis*.

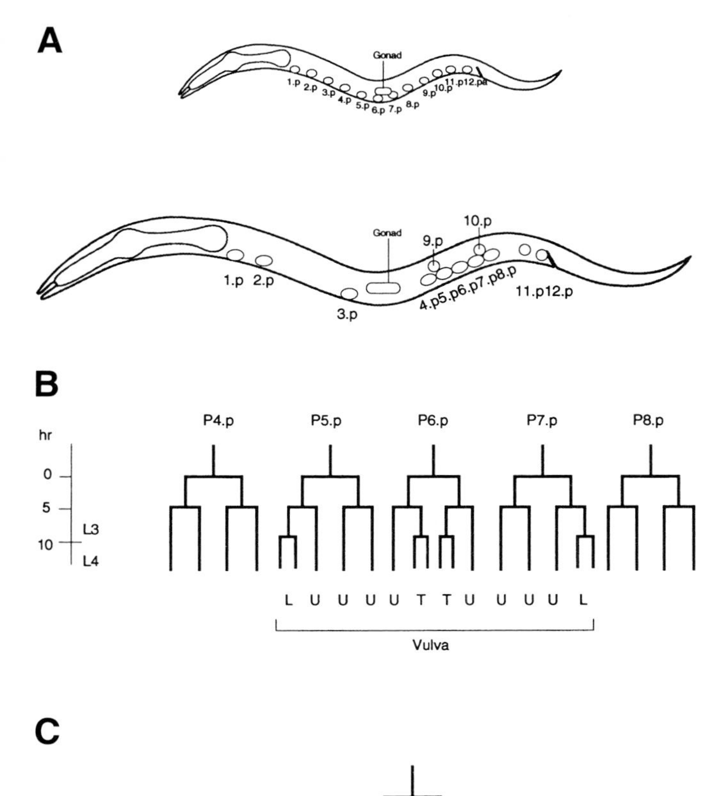

Fig. 6. Schematic summary of vulva development in *Mesorhabditis*. (A) Position of the Pn.p-ectoblasts in the L1 stage and after cell migration in the early L3 stage. P(4-8).p migrate to their final posterior position in the region of P(9,10).p. (B) Vulva cell lineage in *Mesorhabditis* . U, undivided cell; L, longitudinal division, T, transverse division; according to Sternberg and Horvitz (1986). (C) Intermediate cell lineages observed only after cell ablation. The third round of cell division was variable, in the way that 5, 6 or 7 progeny were observed.

Mesorhabditis/Teratorhabditis

In *Mesorhabditis* and *Teratorhabditis* vulva development is not induced by the AC (Sommer and Sternberg, 1994). The VPCs undergo two rounds of cell division before the AC contacts the forming vulva and ablation of the AC precursor does not affect vulva formation. Is vulva development still induced by other cells or do the VPCs self organize (Fig. 7)? We were unable to find other cells that induce the vulva (Sommer and Sternberg, 1994). Therefore, a simple hypothesis is that the VPCs form a correct pattern solely through interactions among themselves. The VPCs might secrete a diffusible inductive signal. The concentration of signal could be highest in the center of the 5-cell region, allowing P6.p to assume the 1° cell fate and promoting P5.p and P7.p towards a 2° cell fate. This hypothesis is testable by ablating all but two VPCs (Fig. 7). According to the model, both remaining cells should secrete and receive equal amounts of signal, resulting in random patterning (1°-2° or 2°-1°). However, ablation experiments revealed that only one specific cell in a pair adopts

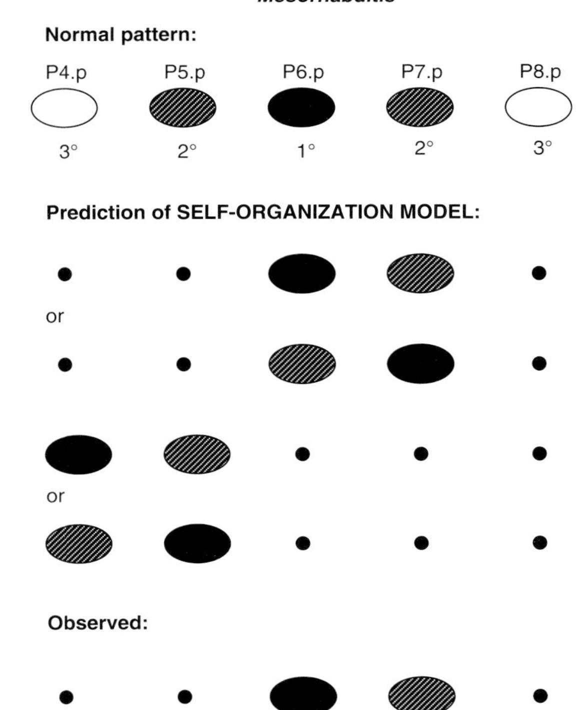

Fig. 7. Self organization of the VPCs: model, prediction and results. (See text for further details). If self organization is responsible for pattern formation within this group of cells, one would expect random specification after ablation of three of the five VPCs. The observed pattern after corresponding ablation experiments in *Mesorhabditis* show that only one cell, P6.p of P(6,7).p, and P5.p of P(4,5).p have the 1° cell fate, arguing against a simple self organization.

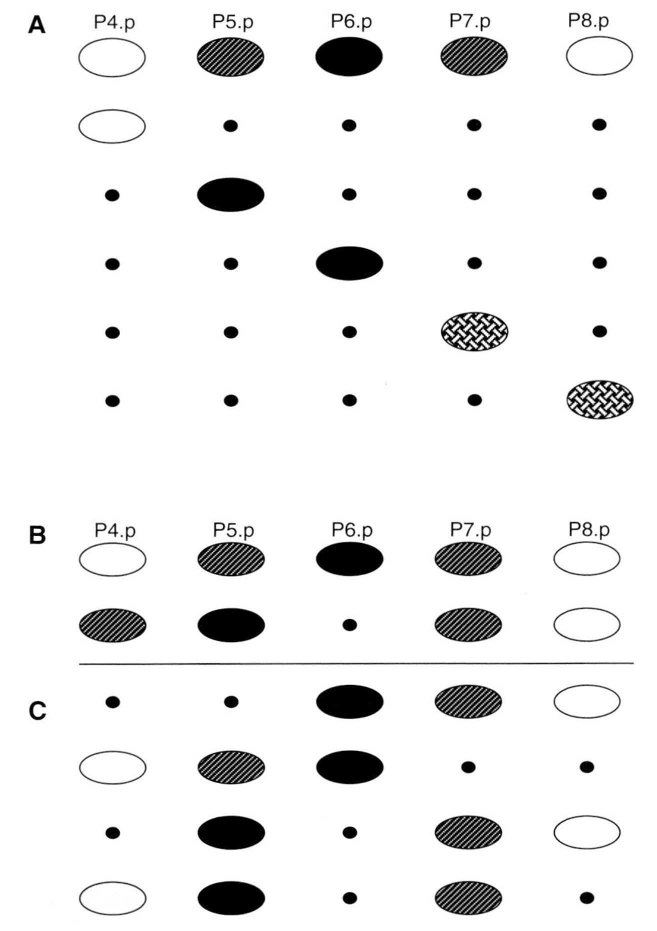

Fig. 8. Schematic summary of cell ablation experiments in *Mesorhabditis*. (A) After ablation of four of the five cells, the remaining cell has a different developmental potential depending on which cell is remaining. P4.p has only the 3° cell fate as an isolated cell. P(5,6).p can adopt the 1° cell fate as an isolated cell, whereas P(7,8).p have a fate, intermediate between that of a 1° and a 2° cell fate. (B) After ablation of P6.p, always P5.p has the 1° cell fate. In the corresponding experiment in *Caenorhabditis* P5.p or P7.p have the 1° cell fate. (C) Additional experiments where two or three VPCs were ablated. Only P(5,6).p can adopt the 1° cell fate. 1°, black oval; 2°, hatched oval; 3°, white oval; intermediate lineage, cross-hatched oval.

the 1° cell fate (Fig. 7). Therefore it is unlikely that vulva formation is due to simple self organization of equivalent VPCs.

Further analysis of *Mesorhabditis* revealed that the VPCs are not equivalent (Fig. 8). After ablation of VPCs only P5.p or P6.p can assume a 1° cell fate . In contrast, P7.p and P8.p assume 2°, 3° or an intermediate cell fate with characteristics of both 1° and 2° lineages. P4.p has the 3° cell fate as an isolated VPC. Thus, changes in induction and competence occur during the evolution of posterior vulva development in *Mesorhabditis* and *Teratorhabditis*. Vulval pattern formation does not require an inductive signal from the gonad, and the VPCs are not equivalent in their competence to form vulval tissue.

Two models can explain how the pattern of VPC fates is established in *Mesorhabditis*. After an initial specification process making some VPCs distinct from one another, final vulva patterning might result from either an inductive signal (most likely from the posterior body region) or solely by lateral interactions among the already different VPCs (Fig. 9). For example, P6.p, once specified, could prevent P5.p from becoming 1° (lateral inhibition) and signal P7.p to be 2° (induction).

A regulatory constraint

All the three posterior vulva genera, *Cruznema, Mesorhabditis* and *Teratorhabditis*, form the vulva with the progeny of P(5-7).p rather than the more appropriately positioned posterior Pn.p cells (Fig. 5). Early morphogenetic processes that regulate anterior-posterior pattern formation by the Hom-C genes (Wang et al., 1993; Clark et al., 1993) might create a regulatory constraint at the cellular level in the Rhabditidae that forces the same set of precursors to generate the vulva. In other families, such as the Panagrolaimidae, there is posterior shifting of the set of precursors, as discussed above.

The newly acquired character of the VPCs is the migration to a more posterior position. While it is unknown how this migration is regulated, in *Caenorhabditis*, there are several examples of genetic control over specific cell migrations. The migratory behaviour of the neuroblast QL is controlled by the *Antennapedia*-like gene *mab-5* acting in the neuroblasts (Salser and Kenyon, 1992). The circumferential migration of a variety of cell types is controlled by the UNC-6 system, with expression of the UNC-5 integrin determining dorsalward response to a global guidance cue (Hamelin et al., 1993).

The difference in induction and competence between *Cruznema* and *Mesorhabditis/Teratorhabditis* might give important evolutionary cues to their phylogenetic relationship. Central vulva development is considered ancestral in this family (Chitwood and Chitwood, 1950), but it is not known if posterior vulva development within the Rhabditidae evolved once or several times independently. This question is of evolutionary importance because it might imply different relationships and origins of the derived characters “posterior vulva” and “gonad independent vulva development.” According to the phylogram in Fig. 10B (based on Sudhaus, 1976), posterior vulva development in *Cruznema* and *Mesorhabditis* evolved independently. If so, we cannot determine whether posterior vulva formation and gonad independence evolved simultaneously or subsequently in *Mesorhabditis*.

Based on our results, the most parsimonious phylogram is as depicted in Fig. 10A, in which posterior vulva development evolved just once within the *Rhabditidae*, generating a *Cruznema*-like intermediate ancestor. This phylogram implies subsequent acquisition of gonad-independent vulva development. If this is true, the evolution of a posterior vulva in a *Cruznema*-like intermediate ancestor might have helped reset developmental conditions. In this context one can consider posterior vulva development as a heterotopic change that allows further changes of the developmental process. Changing boundary conditions by destroying the proximity of VPCs and AC, might thus be an important requirement for the acquistion of subsequent evolutionary novelty. Similar theoretical suggestions have been made concerning the importance of heterochronic changes during development (Buss, 1987).

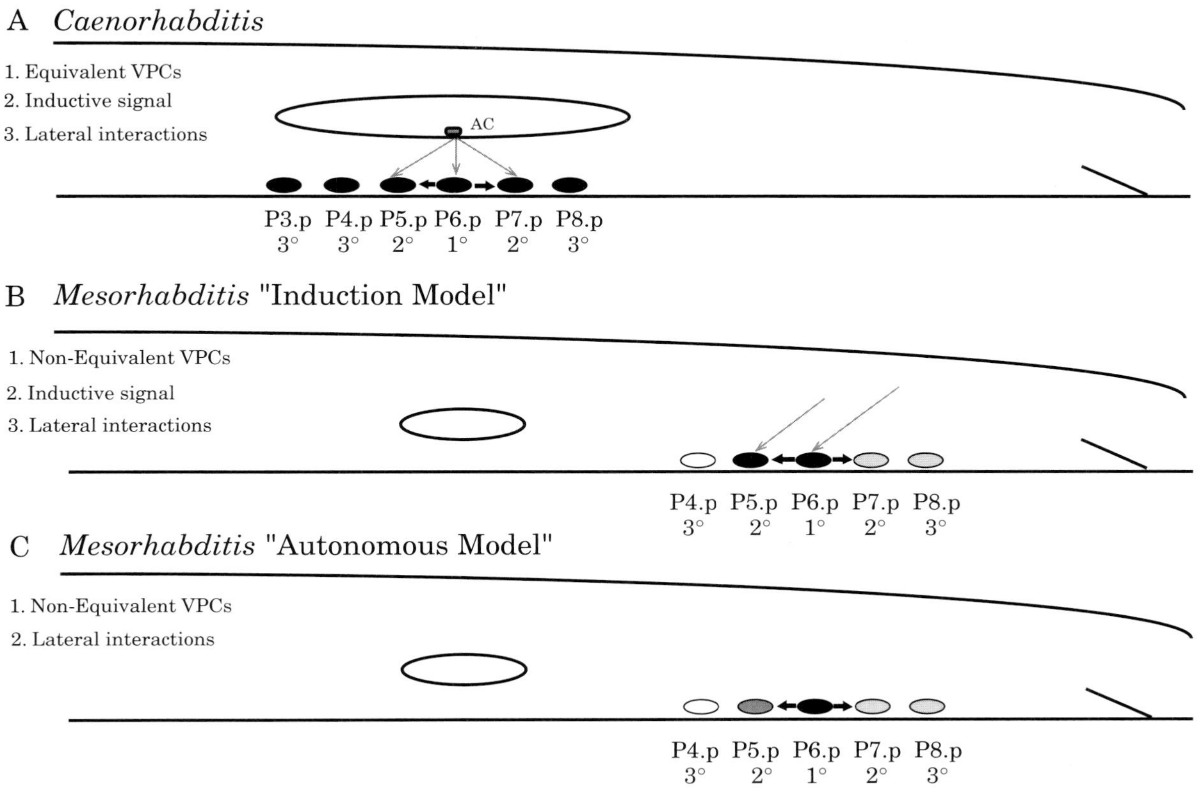

A *Caenorhabditis*

1. Equivalent VPCs
2. Inductive signal
3. Lateral interactions

AC

P3.p P4.p P5.p P6.p P7.p P8.p
3° 3° 2° 1° 2° 3°

B *Mesorhabditis* "Induction Model"

1. Non-Equivalent VPCs
2. Inductive signal
3. Lateral interactions

P4.p P5.p P6.p P7.p P8.p
3° 2° 1° 2° 3°

C *Mesorhabditis* "Autonomous Model"

1. Non-Equivalent VPCs
2. Lateral interactions

P4.p P5.p P6.p P7.p P8.p
3° 2° 1° 2° 3°

Fig. 9. Two models for the initiation of vulva pattern formation in *Mesorhabditis.* Two alternative models for the early steps of vulva pattern formation in *Mesorhabditis* (B, C) compared to the existing model in *Caenorhabditis* (A). According to the "Induction Model" a pre-bias between P4.p-P(5,6).p-P(7,8).p exists. An inductive signal, perhaps from the posterior body region initiates further cell specification. Lateral signaling ensures final cell fate of the VPCs. According to the "Autonomous Model" a pre-bias establishes four different types of VPCs; P4.p-P5.p-P6.p-P(7,8).p. P6.p is proposed to be intrinsically different from P5.p allowing it to adopt the 1° cell fate. After ablation of P6.p only P5.p has the ability to replace P6.p. Lateral signaling is involved in the determination of the final cell fates (plain arrows). Positional information is indicated by the different shading of the VPCs. The inductive signal is indicated by the grey arrows. The arrows between the VPCs (black) refer to the lateral inhibitory signal which exists in *Caenorhabditis*, and which we suggest exists in *Mesorhabditis.*

EQUIVALENCE GROUPS AND INVARIANT CELL LINEAGES

Cell fate specification in nematodes, as in most animals, proceeds by progressive restrictions in potential. Equivalence groups represent a transition state, and might be an intermediate between specification of cell type using oligopotent cells and using cells whose fates are tightly constrained by lineage. From comparative studies, there are two examples in which loss of multipotentiality of cells within an equivalence group appears to be a derived character: AC specification in *Panagrellus* and VPC fate specification in *Mesorhabditis.*

The anchor cell versus ventral uterine precursor equivalence group in *Caenorhabditis* hermaphrodites arises from the generation of two cells (Z1.ppp and Z4.aaa) with the same bipotentiality by two homologous lineages. The difference between the cells results solely from interactions among them via a lateral signaling mechanism (Kimble and Hirsh, 1979; Seydoux and Greenwald, 1989). In *Panagrellus* females, only the posterior homolog, Z4.aaa becomes the anchor cell; Z1.ppp is a ventral uterine precursor (Sternberg and Horvitz, 1981). Since *Panagrellus* monodelphic gonad development is likely an evolutionarily derived character, it is probable that the fixed lineage in *Panagrellus* evolved from an equivalence group. By contrast, the analogous equivalence group in the male, the linker cell/vas deferens precursor cell group, exists in both species (Kimble and Hirsh, 1979; Sternberg and Horvitz, 1981), further supporting the derived character of apparently autonomous anchor cell specification.

Vulva development in *Caenorhabditis*, *Panagrellus* and other nematode species starts with a set of equipotent precursors, the vulval equivalence group. Although the cell fate and cell lineage of each individual precursor is invariant in intact animals, these cells have the potential to replace each other after cell ablation. By contrast, in *Mesorhabditis*, vulval precursor cells are not equivalent. It might be possible during the course of evolution to transform initially equivalent blast cells into non-equivalent blast cells. The possibility that a *Cruznema*-like species was indeed the ancestor to *Mesorhabditis* would support this view. One could also imagine that in other derived evolutionary lines, all vulval precursors might already be committed, creating an autonomous mode of specification as is thought to exist in many cell lineages in the animal kingdom (see Davidson, 1991).

PROSPECTS

Many different developmental mechanisms are used to

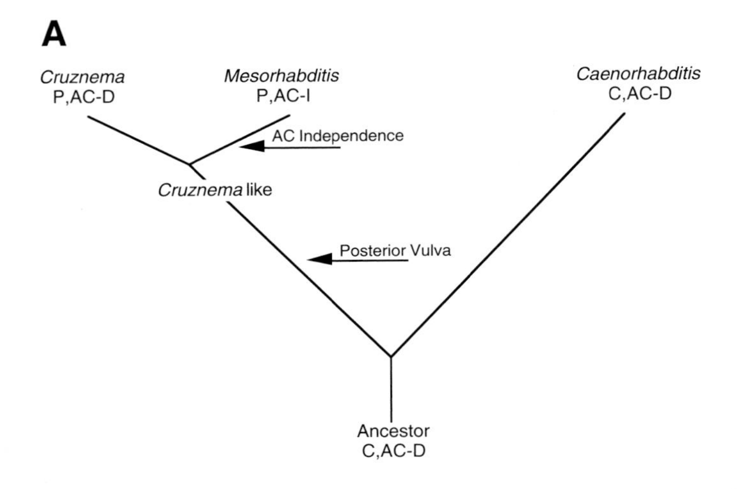

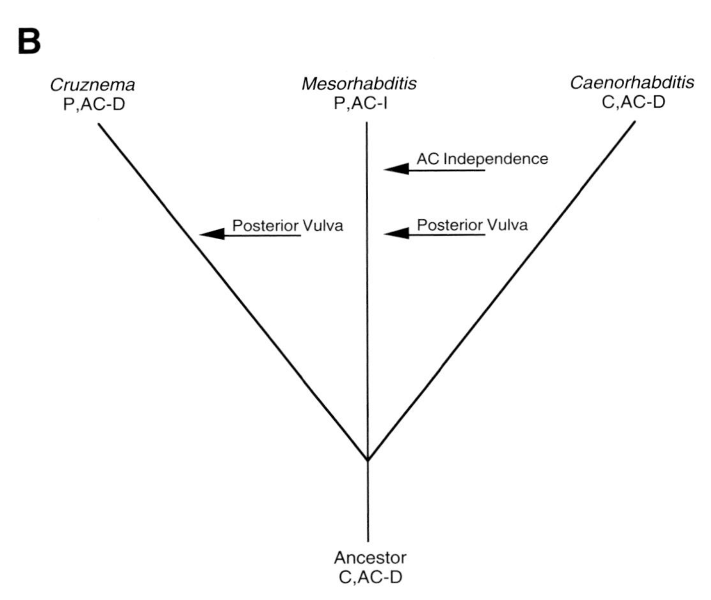

Fig. 10. Two phylogenetic trees indicating the different possible relationships between the posterior vulva species. (A) Posterior vulva development evolved just once, generating a *Cruznema*-like ancestor. AC-independence evolved later in the sublineage guiding to *Mesorhabditis*-like forms. (B) Posterior vulva development evolved several times independently in different lineage. According to this tree *Cruznema* and *Mesorhabditis* are no closer relatives to each other. Developmental evolutionary implications are discussed in the text.

generate cell identity during nematode development. Autonomous cell specification, conditional cell specification by induction and lateral signaling as well as the generation of positional information by Hom-C genes are present in *Caenorhabditis*. The analysis of these mechanisms and of the genes involved is becoming increasingly more detailed. Comparative developmental studies involving cell lineage and cell ablation experiments, are beginning to reveal the types of changes in development that have occurred during evolution of nematodes. By comparing what is known of the genetic control of development in *Caenorhabditis* to the inferred changes in phylogeny, we can formulate hypotheses as to the mechanistic differences in the development of the various nematodes. These hypotheses will be testable by genetics, molecular biology and transgenic nematode technology. Genetic analysis of other free-living nematodes will define the types of cell lineage changes that can occur in species other than *Caenorhabditis*. Molecular cloning, examination of the expression and gene transfer experiments with homologous genes from related species can test specific hypotheses concerning genes that might have mutated to cause the observed changes in development.

ACKNOWLEDGMENTS

We thank Kelly Thomas for discussions of Rhabditid molecular phylogeny, our many colleagues for contributing nematodes and soil samples, and Linda Huang, Tom Clandinin, Wendy Katz and Giovanni Lesa for comments on the manuscript. This research was supported by an NSF Presidential Young Investigator Award to P. W. S., an investigator of the Howard Hughes Medical Institute. R. J. S. is an EMBO long-term postdoctoral Fellow.

REFERENCES

Ambros, V. and Horvitz, H. R. (1984). Heterochronic mutants of the nematode *Caenorhabditis elegans*. *Science*, **226,** 409-416.

Ambros, V. and Fixsen, W. (1987). Cell lineage variation among nematodes. In *Development as an evolutionary process* (eds. R. A. Raff and E.C. Raff). New York, Liss.

Ambros, V. and Moss, E. (1994) Heterochronic genes and the temporal control of *C. elegans* development. *Trends Genet.* **10,** 123.

Aroian, R. V., Carta, L., Kaloshian, I. and Sternberg, P. W. (1993). A free-living *Panagrolaimus* sp. from Armenia can survive in anhydrobiosis for 8.7 years. *J. of Nematology* **25,** 500-502.

Boveri, T. (1899). Die Entwicklung von *Ascaris megalocephala* mit besonderer Rücksicht auf die Kernverhältnisse. In *Festschrift C. von Kuppfer*, Gustav Fischer, Jena, Germany.

Boveri, T. (1910). Die Potenzen der *Ascaris*-Blastomeren bei abgeänderter Furchung. Zugleich ein Beitrag zur Frage qualitativ-ungleicher Chromosomen-Teilung. In *Festschrift für R. Hertwig* Vol.III, Fischer Verlag, Jena, Germany.

Bowerman, B., Draper, B. W., Mello, C. C. and Priess, J. R. (1993). The maternal gene *skn-1* encodes a protein that is distributed unequally in early *C. elegans* embryos. Cell **74,** 443-452.

Brenner, S. (1974). The genetics of *Caenorhabditis elegans*. *Genetics* **77,** 71-94.

Buss, L. (1987). *Evolution of Individuality*. Princeton: Princeton Univ. Press.

Chalfie, M., Horvitz, H. R. and Sulston, J. E. (1981). Mutations that lead to reiterations in the cell lineages of *Caenorhabditis elegans*. *Cell* **24,** 59-69.

Chitwood, B. G. and Chitwood, M. B. (1950). *Introduction to Nematology*. Baltimore, USA: University Park Press.

Clark, S. G. Chisholm, A. D. and Horvitz, H. R. (1993). Control of cell fates in the central body region of *C.elegans* by the homeobox gene *lin-39*. *Cell* **74,** 43-55

Davidson, E.H. (1991). Spatial mechanisms of gene regulation in metazoan embryos. *Development* **113,** 1-26

DeBeer, G. R. (1958). *Embryos and ancestors*. Clarendon Press, Oxford, UK.

Ellis, R. E. and Horvitz, H. R. (1991). Two *C. elegans* genes control the programmed deaths of specific cells in the pharynx. *Development* **112,** 591-603

Ferguson, E. L., Sternberg, P. W. and Horvitz, H. R. (1987). A genetic pathway for the specification of the vulval cell lineages of *Caenorhabditis elegans*. *Nature* **326,** 259-267.

Goodman, C. S. (1977). Neuron duplications and deletions in locust clones and clutches. *Science* **197,** 1384-1386

Gould, S. J. (1977). *Ontogeny and Phylogeny*. The Belknap Press of Harvard University Press, Cambridge Mass.

Greenwald, I. S., Sternberg, P. W. and Horvitz, H. R. (1983). The *lin-12* locus specifies cell fates in *Caenorhabditis elegans*. *Cell* **34,** 435-444

Hamelin, M., Zhou, Y., Su, M.- W., Scott, I. M. and Culotti, J. G. (1993).

Expression of *unc-5* guidance receptor in the touch neurons of *C. elegans* steers their axons dorsally. *Nature* **364,** 327-330.

Herman, M. A. and Horvitz, H. R. (1994). The *Caenorhabditis elegans* gene *lin-44* controls the polarity of asymmetric cell divisions. *Development* **120,** 1035-1047.

Herman, R. K. and Hedgecock, E. M. (1990). The size of the *C. elegans* vulval primordium is limited by *lin-15* expression in surrounding hypodermis. *Nature* **348,** 169-171.

Hill, R. J. and Sternberg, P. W. (1992). The *lin-3* gene encodes an inductive signal for vulval development in *C.elegans*. *Nature* **358,** 470-476.

Hill, R. J. and Sternberg, P. W. (1993). Cell fate patterning during *C. elegans* vulval development. *Development 1993* **(Suppl.),** 9-18.

Huang, L.S ., Tzou, P. and Sternberg, P. W. (1994). The *lin-15* locus encodes two negative regulators of *C. elegans* vulval development. *Molec. Biol. Cell* **5,** 395-412.

Kimble, J. (1981). Lineage alterations after ablation of cells in the somatic gonad of *Caenorhabditis elegans. Dev. Biol.* **87,** 286-300.

Kimble, J. and Hirsh, D. (1979). Post-embryonic cell lineages of the hermaphrodite and male gonads in *Caenorhabditis elegans. Dev. Biol.* **70,** 396-417.

Kimble, J. and White, J. G. (1981). On the control of germ cell development in *Caenorhabditis elegans. Dev. Biol.* **81,** 208-219

Lambie, E. and Kimble, J. (1991). Genetic control of cell interactions in nematode development. *Ann. Rev. Genet.* **25,** 411-436.

Lee, R. C. Feinbaum, R. L. and Ambros, V. (1993). The *C.elegans* heterochronic gene *lin-4* encodes small RNAs with antisense complementarity to *lin-14*. *Cell* **75,** 843-854.

Posakony, J. W. (1994). Nature versus Nurture: Asymmetric cell divisions in *Drosophila* bristle development. *Cell* **76,** 415-418

Raff, M. C. (1992). Social controls on cell survival and cell death. *Nature* **356,** 397-400

Ruvkun, G. and Giusto, J. (1989). The *Caenorhabditis elegans* heterochronic gene *lin-14* encodes a nuclear protein that forms a temporal developmental switch. *Nature* **338,** 313-319

Salser, S. J. and Kenyon, C. (1992). Activation of a *C. elegans Antennapedia* homolog in migrating cells controls their direction of migration. *Nature* **355,** 255-258.

Schmidt, G. P. and Roberts, L. S. (1985). *Foundations of Parasitology*. Times Mirror Publishing, St. Louis, p. 488.

Seydoux, G. and Greenwald, I. (1989). Cell autonomy of *lin-12* function in a cell fate decision in *C. elegans. Cell* **57,** 1237-1245

Skiba, F. and Schierenberg, E. (1992). Cell lineages, developmental timing and spatial pattern formation in embryos of free-living soil nematodes. *Dev. Biol.* **151,** 597-610.

Sommer, R. J. and Sternberg, P. W. (1994). Changes of induction and competence during the evolution of vulva development in nematodes, *Science* **265**, 114-118.

Sternberg, P. W. (1988). Lateral inhibition during vulval induction in *Caenorhabditis elegans. Nature* **335,** 551-554.

Sternberg, P. W. and Horvitz, H. R. (1981). Gonadal cell lineages of the nematode *Panagrellus redivivus* and implications for evolution by the modification of cell lineage. *Dev. Biol.* **88,** 147-166.

Sternberg, P. W. and Horvitz, H. R. (1982). Postembryonic nongonadal cell lineages of the nematode *Panagrellus redivivus:* Description and comparison with those of *Caenorhabditis elegans. Dev. Biol.* **93,** 181-205.

Sternberg, P. W. and Horvitz, H. R. (1986). Pattern formation during vulval development in *C. elegans.Cell* **44,** 761-772.

Sternberg P. W. and Horvitz, H. R. (1989). The combined action of two intercellular signaling pathways specifies three cell fates during vulval induction in *C. elegans. Cell* **58,** 679-693.

Sudhaus, W. (1976). Vergleichende Untersuchungen zur Phylogenetik, Systematik, Ökologie, Biologie und Ethologie der *Rhabditidae* (*Nematoda*). *Zoologica*, Schweizerbart`sche Verlagsbuchhandlung Stuttgart, 43.Band, Heft 125.

Sulston, J. E. and Horvitz, H. R. (1977). Postembryonic cell lineage of the nematode *Caenorhabditis elegans. Dev. Biol.* **56,** 110-156.

Sulston, J. E., Schierenberg, E., White, J. G. and Thomson, J. N. (1983). The embryonic cell lineage of the nematode *Caenorhabditis elegans. Dev. Biol.* **100,** 64-119.

Wang, B. B., Mueller-Immergluck, M. M., Austin, J., Robinson, N. T., Chisholm, A. and Kenyon, C. (1993). A homeotic gene cluster patterns the anteroposterior body axis of *C. elegans. Cell* **74,** 29-42.

Wightman, B., Ha, I. and Ruvkun, G. (1993). Posttranscriptional regulation of the heterochronic gene *lin-14* by *lin-4* mediates pattern formation in *C. elegans. Cell* **75,** 855-862.

Wilson, R., et al. (1994). 2.2 Mb of contiguous nucleotide sequence from chromosome III of *C. elegans. Nature* **368,** 32-38.

Wood, W. B. and Edgar, L. G. (1994). Patterning in the *C. elegans* embryo. *Trends Genet.* **10,** 49-54.

Yochem, J., Weston, K. and Greenwald, I. (1988). *C.elegans lin-12* encodes a transmembrane protein similar to *Drosophila notch* and yeast cell cycle gene poducts. *Nature* **335,** 547-550.

zur Strassen, O. (1896). Embryonalentwicklung der *Ascaris megalocephala. Arch. für Entwicklungsmechanik,* **3**, 27-105.

Development 1994 Supplement, 97-106 (1994)
Printed in Great Britain © The Company of Biologists Limited 1994

The evolution of echinoderm development is driven by several distinct factors

Gregory A. Wray and Alexandra E. Bely

Department of Ecology and Evolution, State University of New York at Stony Brook, Stony Brook, NY 11794, USA

SUMMARY

We analyzed a comparative data base of gene expression, cell fate specification, and morphogenetic movements from several echinoderms to determine why developmental processes do and do not evolve. Mapping this comparative data onto explicit phylogenetic frameworks revealed three distinct evolutionary patterns. First, some evolutionary differences in development correlate well with larval ecology but not with adult morphology. These associations are probably not coincidental because similar developmental changes accompany similar ecological transformations on separate occasions. This suggests that larval ecology has been a potent influence on the evolution of early development in echinoderms. Second, a few changes in early development correlate with transformations in adult morphology. Because most such changes have occurred only once, however, it is difficult to distinguish chance associations from causal relationships. And third, some changes in development have no apparent phenotypic consequences and do not correlate with obvious features of either life history or morphology. This suggests that some evolutionary changes in development may evolve in a neutral or nearly neutral mode. Importantly, these hypotheses make specific predictions that can be tested with further comparative data and by experimental manipulations. Together, our phylogenetic analyses of comparative data suggest that at least three distinct evolutionary mechanisms have shaped early development in echinoderms.

Key words: evolution of development, regulation of gene expression, cell lineage, gastrulation, sea urchin, comparative method

INTRODUCTION

The processes of early development are often conserved among animals with similar adult body plans. Comparative embryology provides many examples of conserved patterns of cell lineage, gene expression and morphogenetic movement within phyla and classes (e.g., Kumé and Dan, 1968; Anderson, 1973; Patel et al., 1989). These empirical data support a widespread expectation: that any modification to embryogenesis will have grave phenotypic consequences, thereby constraining the evolution of early development (Gould, 1977; Arthur, 1988; Thomson, 1988). Together, the empirical data of comparative embryology and the assumption of constraint have produced a traditional view of how developmental processes evolve. This traditional view posits strong and direct links between early development and adult morphology.

But developmental processes often evolve in ways that violate this traditional view. A large body of empirical evidence demonstrates that the link between embryos and adult body plans is evolutionarily rather labile. First, early development can be conserved in species whose adults are very different. The embryos of clams and earthworms, for example, share distinctive features of axis formation and cell lineage, as well as cell movements during gastrulation and coelom formation (Kumé and Dan, 1968; Freeman and Lundelius, 1992). Similarities in the expression and function of homeotic genes in arthropods and mammals provide another example (McGinnis and Krumlauf, 1992). If processes of early development were required only to build features of adult morphology, they would not be conserved among phyla with very different body plans. Second, and more telling, early development can differ profoundly among species that are closely related and morphologically very similar as adults. This phenomenon is known from many phyla and encompasses a broad range of developmental processes (e.g., Lillie, 1895; Garstang, 1929; Berrill, 1931; Coe, 1949; Elinson, 1990; Byrne, 1991; Wray and Raff, 1991a; Jeffery and Swalla, 1992; Janies and McEdward, 1993; Henry and Martindale, 1994). If the processes of early development only evolved in order to produce a modified adult morphology, they would not differ among closely related and morphologically similar species.

Together, these two phenomena demonstrate that the relationship between early development and adult morphology is likely to be complex. Certainly, there is no simple one-to-one mapping of embryogenesis onto adult phenotype. But if adult body plans are not sufficient to explain evolutionary similarities and differences in early development, what are the evolutionary mechanisms that conserve and alter developmental processes? In this paper, our objective is to ask *why* developmental processes evolve. We examine the evolution of early development in echinoderms, a clade particularly well suited to addressing this question. We consider several distinct kinds of developmental processes, and posit testable hypotheses to explain why they have or have not evolved.

REPLICATE TRANSFORMATIONS AND HYPOTHESIS TESTING

In trying to understand evolutionary history, it is rarely feasible to carry out actual experiments. Fortunately, evolution has itself provided replicate "experiments" in the form of parallel phenotypic transformations. Examples include transformations between short and long germ band development in insects; between panoistic and meroistic oogenesis in insects; between egg laying and livebearing in vertebrates; between mono- and polyembryony in several phyla; between free-spawning and internal fertilization in most phyla; between parasitic and free-living life cycles in many invertebrates; and between feeding and non-feeding larvae in most phyla.

Besides their parallel origins, these replicate transformations share two important features. First, they are all associated with significant changes in developmental mechanisms. Thus, patterns of gene expression, cell division, and morphogenesis correlate with type of germ band development in insects (Sander, 1983; Patel et al., 1989, 1994; Brown et al., 1994) and with type of larvae (feeding versus non-feeding) in echinoderms (Wray, 1994). And second, these replicate transformations all reflect ecologically significant differences in life history strategies. Whether a vertebrate lays eggs or is live bearing, or whether a nematode is parasitic or free-living has profound ecological consequences. Together, these two features provide a way of relating changes in developmental mechanisms to altered selection regimes.

Parallel phenotypic transformations provide exceptional test cases for teasing apart the various factors that drive evolutionary change and stability in development. Replicate events allow one to determine whether the correlation between developmental change and some external factor, such as an altered life history strategy, is due to chance or reflects a causal relationship. By extending comparisons of development to several replicates, it is possible to rule out chance associations statistically. As illustrated in Fig. 1, comparative data for the same number species may be statistically nonsignificant if only one transformation has occurred, but highly significant when more than one replicate transformation is considered. Thus, studying replicate phenotypic transformations allows one to identify the probable external cause of the evolution of a developmental feature. Functional explanations can then be sought for such cases. We illustrate this approach by examining the evolution of several functionally distinct developmental processes in echinoderms.

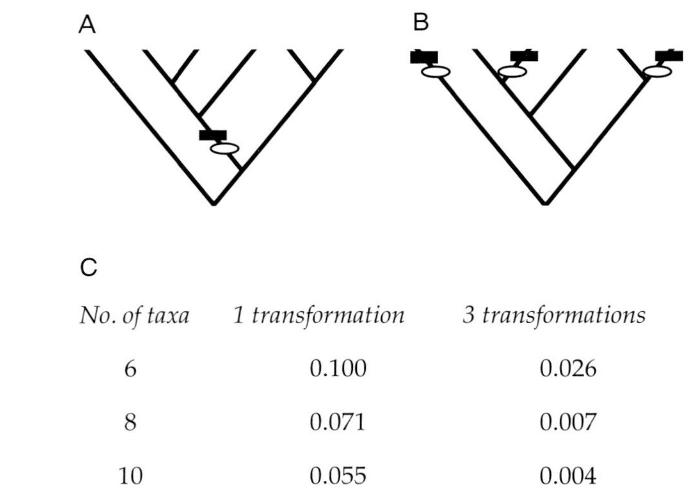

No. of taxa	*1 transformation*	*3 transformations*
6	0.100	0.026
8	0.071	0.007
10	0.055	0.004

Fig. 1. The importance of replicate transformations in testing for the association between an evolutionary change in two characters. (A,B) Two characters that change in concert are mapped onto a hypothetical phylogeny in two ways. The associated changes have occurred once in phylogeny A and three times independently in phylogeny B. Note that both phylogenies have the same number of taxa representing each character state. The concentrated changes test (Maddison, 1990) indicates that the association between the changes is significant in phylogeny B (P=0.02) but not phylogeny A (P=0.1). (C) The strength of association between two characters is affected by the number of replicate transformations and the number of taxa included in the analysis. P-values from the concentrated changes test are given for a series of hypothetical phylogenies. The association becomes more robust when replicate transformations have occurred and when more taxa are included in the analysis. These values emphasize the importance of examining developmental processes in several species and across replicate evolutionary transformations.

ECHINODERMS AS A MODEL CLADE FOR STUDYING DEVELOPMENTAL EVOLUTION

Features that make a single species suitable for studying development are not the same as those that make a group of species suitable for studying the *evolution* of development. Of particular importance in analyzing the evolution of developmental processes are the following: a detailed understanding of phylogenetic relationships, a fossil record, the existence of replicate transformations, and an extensive comparative data base of developmental studies. Echinoderms meet all of these criteria (Fig. 2) (Smith, 1984, 1992; Wray, 1992, 1994).

In particular, echinoderms provide a variety of replicate transformations in life history strategies (Strathmann, 1978; Raff, 1987; McEdward and Janies, 1993). Of these, the switch from feeding to non-feeding larvae has been studied in the most detail (Byrne, 1991; Wray and Raff, 1991a; Raff, 1992). Feeding and non-feeding larvae differ substantially in their morphology, behaviour and ecology (Strathmann, 1985; Wray, 1992; McEdward and Janies, 1993). In particular, non-feeding larvae nearly always lack a mouth and functional digestive tract, the arms used for feeding are lost or reduced, the supporting endoskeleton is lost or reduced, and the time to metamorphosis is greatly accelerated. Differences in the embryonic development of species with feeding and non-feeding larvae are extensive, and include functionally profound changes in oogenesis, axis formation, cell fate specification, cleavage geometry, gene expression, and morphogenetic movements (Wray and Raff, 1989, 1990, 1991b; Henry and Raff, 1990; Scott et al., 1990; Raff, 1992).

Mapping feeding versus non-feeding life history strategies onto a cladogram for echinoderms tells us three important things. First, the polarity of the transformation is invariably from feeding to non-feeding larvae (Strathmann, 1978; Raff, 1987). Second, lecithotrophy has evolved literally dozens of times from planktotrophy (Emlet, 1990; our unpublished tallies). Among sea urchins, the group we will focus on, this transformation has evolved at least 20 times. And third, this

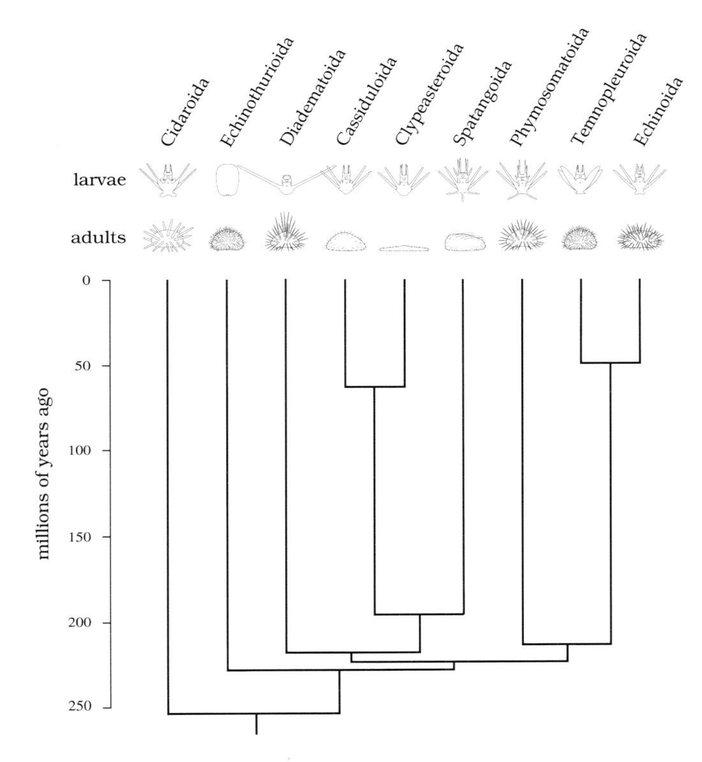

Fig. 2. Phylogenetic relationships, divergence times, and characteristic morphologies of major sea urchin orders. The estimated phylogenetic relationships of sea urchin orders is shown, following recent cladistic analyses based on adult morphology (Smith, 1984, 1988; Smith et al., 1993). Estimated divergence times are based on Smith (1984).

transformation is not associated with any specific changes in adult morphology (Wray and Lowe, unpublished). In particular, very closely related and morphologically similar species can differ in larval type.

WHY ECHINODERM DEVELOPMENT EVOLVES

In this section, we examine the various evolutionary mechanisms that influence the evolution of developmental mechanisms in echinoderms (Table 1). The examples cover three fundamental and functionally distinct developmental processes: regulation of embryonic gene expression, specification of cell fates, and morphogenetic movements.

Gene expression

The first example concerns evolutionary changes in the regulation of early zygotic gene expression. The expression of the gene *msp130* has been studied by antibody localization in several species (Parks et al., 1988, 1989; Wray and McClay, 1988, 1989; Amemiya and Emlet, 1992). msp130 protein is part of the mesodermally derived envelope that surrounds the endoskeleton (Decker et al., 1988; Parr et al., 1990). Although the endoskeleton is biochemically similar throughout the life cycle (Benson et al., 1987; Drager et al., 1989), it plays functionally distinct roles in larvae and adults: it supports arms that are used to capture food in larvae, while it provides protection and rigidity to the adult. Importantly, the endoskeleton is lacking or reduced in non-feeding larvae (Wray and Raff, 1991a). This makes *msp130* interesting from an evolutionary perspective: it plays a direct role in producing phenotypic features, these features have clearly defined functions, and they vary in functionally significant ways among species. These characteristics are important in understanding why evolutionary changes have occurred in gene expression.

The timing of *msp130* expression varies widely among the dozen species of sea urchin that have been examined. These differences in timing are mapped onto a cladogram of the study species in Fig. 3. The changes fall into two distinct categories. The first class of changes (white bars) have the following two features in common: they are relatively small timing shifts (1-2 hours), and they all involve an earlier onset of expression. The second class of changes in *msp130* expression (black bars) are quite different: they are relatively large timing shifts (10-20 hours), and they are all delays in the onset of expression.

Why have these changes in early zygotic gene expression evolved? And why do they fall into two such distinct classes? We can begin by tentatively ruling out a relationship with adult morphology. Both kinds of changes are scattered across the cladogram, and do not correlate in any understandable way with changes in adult morphology. However, there does seem to be a correlation with life history strategy: all the large delays in expression coincide with the transformation to non-feeding larvae (Fig. 3, black ovals).

To decide whether this association could be the result of chance, we can compute the probability that three independent changes in one feature (timing of gene expression) would by chance change in concert with another feature (life history

Table 1. Reasons why developmental processes do and do not evolve

Reason for change	Example in text
Generate derived larval morphology	Timing of msp130 expression (major changes) Cleavage geometry
Generate derived life history feature	Specification of adult cell fates Morphogenesis of adult imaginal rudiment
Compensate for other changes	Gastrulation cell movements (for yolkier egg) Fate specification (for gastrulation movements)
Drift (nearly neutral changes)	Timing of msp130 expression (minor changes) Position of first cleavage plane (most cases)
Generate derived adult morphology	Mode of rudiment formation ?? Skeletogenic cell movements ??

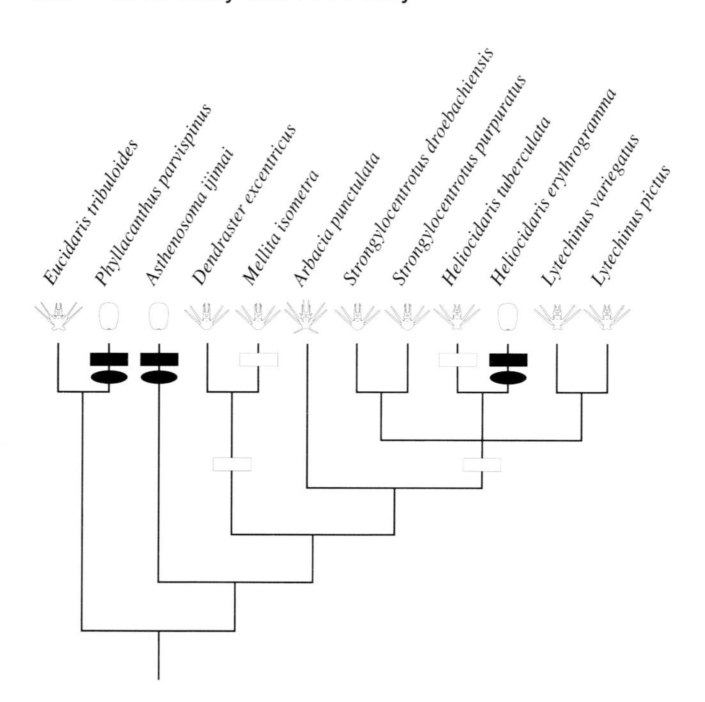

Fig. 3. Evolutionary transformations in gene expression. Changes in timing of *msp130* expression and life history are mapped, using parsimony, onto a cladogram of sea urchin species. Black ovals represent a change from feeding to non-feeding larvae. Large delays in onset of *msp130* expression (black bars) are significantly associated with changes in life history mode (P=0.002; concentrated changes test, Maddison and Maddison, 1992) and may be driven by the switch to non-feeding larvae. Changes to slightly earlier expression of *msp130* (white bars) do not correlate with either morphology or life history, and may represent effectively neutral changes. *Phylogenetic methods*: Phylogenetic relationships among sea urchins are based on parsimony analyses of adult morphology and 18S rRNA sequence data by Smith (1984, 1988, 1993). Topologies based on molecular and morphological data are congruent. None of the characters we map were used in generating the cladogram.

strategy). The concentrated changes test (Maddison, 1990) calculates this probability, given a particular phylogeny and a particular distribution of evolutionary changes in two traits. This statistical test, as well as other analytic tools for addressing similar issues, is implemented in the computer program MacClade (Maddison and Maddison, 1992). The concentrated changes test indicates that the observed association between the two traits would arise less than 1% of the time by chance alone. The correlation is therefore statistically robust.

The large changes in the timing of *msp130* expression might be driven by changes in life history strategy. This interpretation makes sense functionally. msp130 protein is a structural component of the skeleton, which supports larval arms that are used for feeding (Hart, 1991; Wray, 1992). In species with feeding larvae, msp130 protein must be present very early in development in order to build these arms. Species whose larvae do not feed no longer require arms, and most have lost not only the arms but the larval skeleton that supports them (Wray, 1992). In these species, therefore, msp130 protein is not required until later in development when the adult skeleton is made.

The small changes in the onset of *msp130* expression (Fig. 3, white bars) are not associated with life history differences or adult morphology, and must be explained in another way. In most species, msp130 protein appears before skeletogenesis actually begins (Wray and McClay, 1989). Activating *msp130* transcription slightly before it is required may not be detrimental. It is possible, therefore, that these minor differences in the timing of expression are functionally neutral.

There are several ways to test these hypotheses rigorously. The hypothesis that large delays in *msp130* expression are driven by life history can be tested in two ways. First, there is a strong prediction about what *msp130* expression patterns should look like in the other 5,000 or so echinoderm species that have not yet been examined. To the extent that the timing of *msp130* expression continues to correlate with life history strategy as additional species are examined, the hypothesis is increasingly supported. Second, a similar evolutionary correlation should exist for other genes involved in skeletogenesis. Several other structural components of the echinoderm endoskeleton have been characterized molecularly (Harkey and Whiteley, 1983; Benson et al., 1987; Livingston et al, 1991), and their expression could be examined in a comparative way. The hypothesis that the minor timing changes in *msp130* expression are neutral or nearly so could be tested in the following ways. First, it is possible to experimentally manipulate the timing and level of gene expression in echinoderm embryos (e.g., Franks et al., 1990). A close examination of larvae reared from embryos where the onset of *msp130* expression was altered slightly should reveal no effects. Even if this is not the ideal confirmatory test (subtle but functionally significant effects could be overlooked), it is a means of potentially falsifying the hypothesis. Second, there should exist heritable variation for minor differences in the timing of *msp130* expression within natural populations. If the smaller changes in gene expression are indeed evolving neutrally, they should at least occasionally appear as intraspecific variation.

Patterns of early zygotic gene expression are sometimes very highly conserved evolutionarily (Patel et al., 1989, 1994; Püschel et al., 1992; Brown et al., 1994). When the gene being compared across taxa has a direct role in pattern formation, and the study taxa share basic elements of body organization, this is to be expected. Indeed, any other result would suggest that the developmental role of the gene has changed, and a few examples of this are known (e.g., Patel et al., 1992). However, when the gene operates downstream of pattern formation, as most do, evolutionary changes may be much more numerous. As the foregoing example illustrates, these changes can be quite sensitive to evolutionary changes in the ecology of an organism.

Cell fate specification

The second set of examples concerns evolutionary changes in the processes that establish axes and cell fates. Comparative data exist for fate specification mechanisms, cell lineages, and cleavage geometry for several sea urchin and a few starfishes (see Wray, 1994). Echinoderms have stereotypic patterns of cell division and cell lineage segregation during cleavage (Hörstadius, 1973; Cameron and Davidson, 1991; Wray, 1994), as is the case for most metazoans (Davidson, 1990, 1991). Specification begins within the first few cleavage divisions, as demonstrated by clonal establishment of cell fates

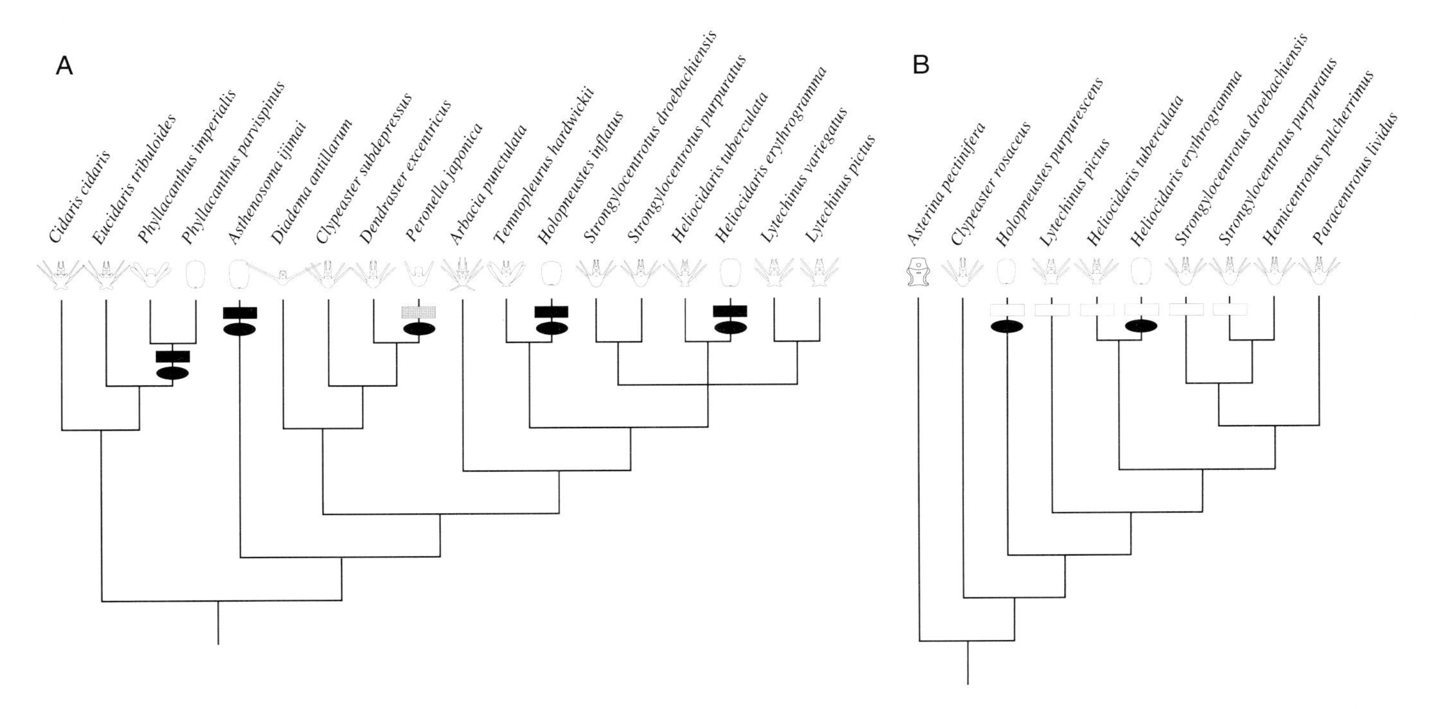

Fig. 4. Evolutionary transformations in cell fate specification. (A) Cleavage geometry. Black bars mark changes from the ancestral cleavage geometry (see text) to a two-tiered, 16-cell embryo with blastomeres of equal size. The grey bar marks a change from the ancestral condition to a 16-cell embryo with enlarged micromeres. These modifications in cleavage geometry are significantly associated with the switch to non-feeding larvae (black ovals) (P=0.00002). (B) Axial establishment. Changes in the relative position of the first cleavage plane to the dorsoventral axis (white bars) show no obvious association with either morphology or derived life history mode (black ovals). These transformations may represent neutral changes. Alternative, equally parsimonious reconstructions of axis formation exist, but do not alter our conclusions. Methods for phylogenetic reconstruction are described in the legend for Fig. 3.

(Cameron et al., 1987; Wray and Raff, 1990) and by direct experimental manipulations (Hörstadius, 1973; Henry et al., 1989; Ransick and Davidson, 1993). Because specification begins so early, most evolutionary changes in cleavage geometry will alter the timing of fate specification, or the number or position of descendent cells.

Mapping overall cleavage geometry onto a cladogram reveals several changes among sea urchins (Fig. 4A). A characteristic, but not universal, feature is a 16-cell embryo with three tiers of cells: mesomeres, macromeres, and micromeres (Hörstadius, 1973). This feature is reconstructed as the ancestral geometry, and has been retained in many phylogenetically diverse species. Since species with this cleavage geometry diverged as long ago as 250 million years (see Fig. 2), it is clearly an ancient feature of sea urchin development. On at least four separate occasions, however, there has been a parallel modification to a 2-tiered, 16-cell embryo with all blastomeres of equal size (Fig. 4A, black bars) (Wray, 1994). An additional case seems to be intermediate, with enlarged micromeres (Fig. 4A, grey bar) (Okazaki, 1975). Each of these modifications in cleavage geometry is associated with the switch to non-feeding larvae. This association is highly significant statistically ($P = 0.00002$) using the concentrated changes test. Because the intermediate case occurs in a species that already has non-feeding larvae, it suggests that the change in cleavage geometry is a consequence, rather than a cause, of the change in larval ecology.

It is not immediately obvious why this particular change in cleavage geometry should be favored in non-feeding larvae. However, a detailed analysis of modified cell lineages and fate specification in *Heliocidaris erythrogramma*, one of the species with non-feeding larvae, provides some clues (Wray and Raff, 1989, 1990; Henry and Raff, 1990). For example, adult ectoderm is specified much earlier and in much larger quantity in *H. erythrogramma* than in species that produce feeding larvae. Instead of involving induction between mesoderm and ectoderm, as is the case in feeding larvae (Okazaki, 1975), specification takes place at the 8-cell stage through interactions among blastomeres (Wray and Raff, 1990). This change in fate specification allows much more rapid development through metamorphosis, by avoiding the prolonged period of proliferation of adult ectodermal cells that occurs in feeding larvae. Accelerated development of non-feeding larvae is probably favored by selection. A non-feeding larva does not need to spend much time in the plankton, where mortality rates are of the order of 10-20% per day (Rumrill and Chia, 1984). Thus, there should be strong selection to minimize the time to metamorphosis. This is borne out by empirical data: all species with non-feeding larvae complete metamorphosis faster than their closest relatives with feeding larvae. Whether other species with non-feeding larvae use the same modifications in fate specification as *H. erythrogramma* in order to achieve this end remains to be seen.

If some aspects of fate specification evolve in response to altered life history strategies, this may not be true of all. The spatial relationship between the first cleavage plane and the dorsoventral axis varies among echinoderms with no obvious relationship to either morphology or life history mode (Fig. 4B; Hörstadius, 1973; Kominami, 1983, 1988; Wray and Raff, 1989; Cameron et al., 1989; Henry et al., 1992). This spatial relationship is variable even within several species. The functional significance of this variation is a change in the timing of dorsoventral axis formation (Henry et al., 1992). Evidently, the exact time of dorsoventral axis specification is not crucial to building particular phenotypes in echinoderms. Therefore, minor differences in the timing of this developmental process may be evolving in a nearly neutral mode.

Other differences in axis formation may, however, be selected for. In *H. erythrogramma*, the dorsoventral axis is not only specified earlier, it is determined much earlier (Henry and Raff, 1990). This earlier determination in a species with non-feeding larvae may have evolved in response to selection for a general acceleration of premetamorphic development. Until comparative data for other species with non-feeding larvae become available, this hypothesis remains untested.

The hypotheses just discussed are all testable. The hypothesis that changes in cleavage geometry and specification of adult ectoderm are driven by changes in life history strategy is testable with further comparative data. There is a strong prediction that the ancestral cleavage geometry will always be retained in species with feeding larvae, but usually modified in those whose larvae do not feed. Further, there should be evolutionary changes in fate specification that reduce the time to metamorphosis in sea urchins with non-feeding larvae. The hypothesis that the position of the first two cleavage planes is nearly neutral can be tested functionally by artificially altering their position (e.g., Hörstadius, 1973), or comparatively, by examining additional species. This hypothesis predicts that changes in the position of the first cleavage planes should neither cause nor correlate with differences in morphology. Finally, the hypothesis that accelerated determination of the dorsoventral axis is driven by the loss of larval feeding can be tested by examining this process in additional species with feeding and non-feeding larvae.

Cell lineage and axis formation are often considered to be evolutionarily conservative developmental processes. As the foregoing examples demonstrate, however, this is not necessarily the case. Further, these are not isolated cases. There are other examples of both minor and major evolutionary changes in cell lineage among related animals, including nematodes (Skiba and Schierenberg, 1992), mollusks (Lillie, 1895; Freeman and Lundelius, 1992), nemerteans (Henry and Martindale, 1994), and ascidians (Berrill, 1931; Jeffery and Swalla, 1991). If studies in these other phyla could be expanded to cover a greater range of related species, it should be possible to begin to understand why these changes have evolved.

Morphogenesis

The third and last set of examples concerns evolutionary changes in the cell movements that drive morphogenesis. Morphogenesis has been extensively studied in echinoderm embryos because of their exceptional clarity, simple organization, and robust constitution (Gustafson and Wolpert, 1967; Wilt, 1987; Ettensohn and Ingersoll, 1992). The comparative data base on echinoderm morphogenesis is rapidly expanding. It is now possible to begin drawing some conclusions about why certain features have changed or remained the same, and to begin framing specific hypotheses for further testing. We consider two distinct morphogenetic processes: gastrulation and formation of the adult body rudiment.

Gastrulation cell movements have been studied in several echinoderms (Fig. 5A). In sea urchins with feeding larvae, gastrulation occurs in three temporally and mechanistically

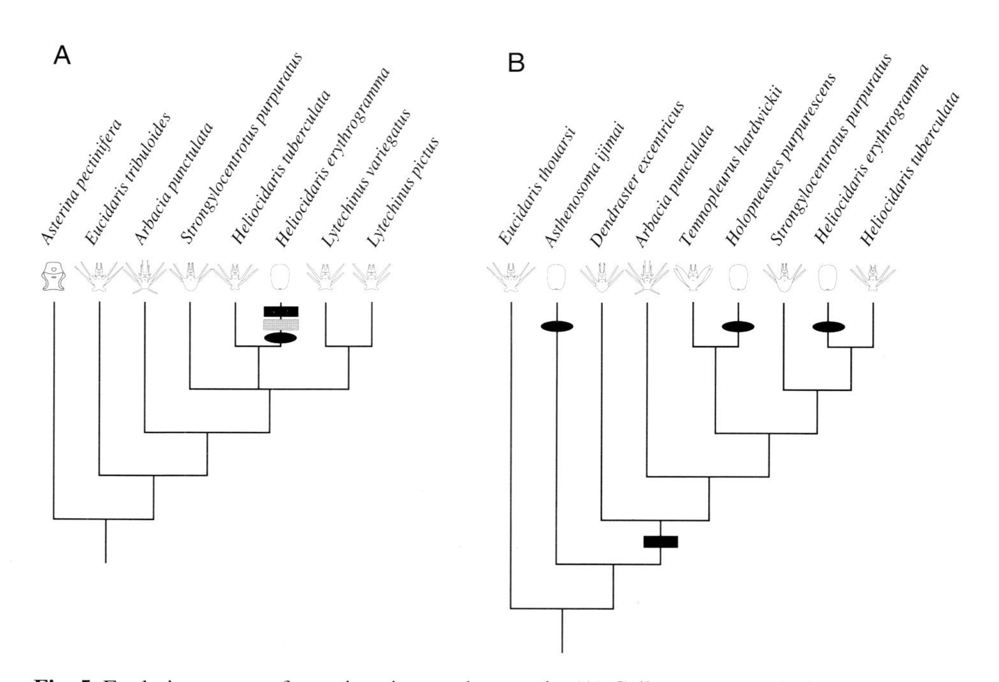

Fig. 5. Evolutionary transformations in morphogenesis. (A) Cell movements during gastrulation. Two significant changes in cell movements are the origin of extensive involution during gastrulation (black bar) and the imposition of a strong dorsoventral asymmetry in cell movements (grey bar). Both correlate with the life history transformation from feeding to non-feeding larvae (black oval). This association is not statistically significant by the concentrated changes test (P=0.07), but suggests that additional species with the derived life history pattern should be examined for parallel changes in morphogenesis. (B) Rudiment formation. During sea urchin development, the adult body forms as an imaginal rudiment on the left side of the larval body. In some species, this rudiment is exposed on the surface of the larva, but in one morphologically derived clade, it is enclosed within an invagination called the vestibule (black bar). Vestibule formation (black bar) does not correlate with a switch in life history (black ovals) but is significantly associated with adult morphology (P=0.03). (Note: some species included in the statistical analysis are not shown here.) It is possible that this developmental change is part of the reason that this derived clade of sea urchins looks different as adults. Methods for phylogenetic reconstruction are described in the legend for Fig. 3.

distinct phases: a brief period of involution, an extended period of convergent extension involving cell rearrangements and shape changes but no further involution, and a final period of "target recognition" during which filopodial traction guides the tip of the archenteron to the site of the future mouth (Ettensohn, 1984, 1985; Hardin and Cheng, 1986; Hardin, 1988; Hardin and McClay, 1990; Burke et al., 1991). Gastrulation has been less thoroughly studied in starfishes; it apparently involves a longer period of involution, but otherwise may be quite similar (Kuraishi and Osanai, 1992). This suggests a very lengthy conservation in the basic cell movements of echinoderm gastrulation, since starfishes and sea urchins diverged over 400 million years ago (Smith, 1988).

In *H. erythrogramma*, which has non-feeding larvae, gastrulation cell movements are modified in several important ways (Wray and Raff, 1991b). First, involution persists throughout gastrulation, helping to drive archenteron extension. Second, cell movements rapidly become asymmetric, with involution occurring primarily over the ventral lip of the blastopore. And third, the last phase of gastrulation is absent, in that no larval mouth is produced. Although gastrulation cell movements have not been studied in any other species with non-feeding larvae, the larval mouth is usually absent, suggesting that the third modification has evolved in parallel several times.

Two distinct evolutionary mechanisms may account for these changes in gastrulation cell movements. Loss of the larval mouth is consistent with generally accelerating premetamorphic development (see previous section), since it is a feeding structure that is no longer needed. The other changes, prolonged and asymmetric involution, may have a very different cause. Sea urchins with non-feeding larvae all have greatly enlarged, lipid-rich eggs. In vertebrates, large increases in egg size require changes in gastrulation cell movements due to mechanical constraints (Ballard, 1976; del Piño and Elinson, 1983), and the same may be true of echinoderms. If this interpretation is correct, it would constitute an example of yet another reason for evolutionary changes in development: to accommodate, or compensate for, another change in development. Indeed, the asymmetry of cell movements during gastrulation in *H. erythrogramma* may in turn account for the dorsoventral asymmetry in cell fate specification present in this species (Fig. 5A). Thus, evolutionary changes in two mechanistically distinct developmental processes, cell fate specification and cell movements, may be evolutionarily tracking changes in egg size that are in turn being driven by ecological factors.

Morphogenesis of the adult rudiment is similarly affected by several evolutionary mechanisms. The adult sea urchin body develops on the left side of the larva from imaginal cell populations (Okazaki, 1975). The arrangement of this adult rudiment varies among sea urchin species (Fig. 5B): ancestrally, the rudiment develops on the surface of the larva (Emlet, 1988; Amemiya and Emlet, 1992), while in a single derived clade (the non-echinothuriod euechinoids) it is enclosed in a vestibule (Bury, 1895; Okazaki, 1975). This change in the morphogenesis of the rudiment could be directly related to producing a derived adult morphology: the statistical correlation is robust ($P = 0.03$). However, there are no replicate origins of the derived adult morphology, making this interpretation difficult to test directly with further comparative data.

A more extensive change in rudiment morphogenesis is, however, amenable to hypothesis testing. In sea urchins with feeding larvae, the rudiment appears as a tiny invagination that grows by cell proliferation over the course of days or weeks (Okazaki, 1975). In *H. erythrogramma*, in contrast, the rudiment forms at very nearly full size (Williams and Anderson, 1975). Instead of a tiny invagination a few tens of cells in diameter, the vestibular invagination is hundreds of cells in diameter. This change in morphogenesis may be the result of selection for a shortened time to metamorphosis, as discussed earlier. If so, other species with non-feeding larvae should have enlarged vestibular invaginations. Because there exist replicates of this ecological shift, it is possible to test this hypothesis with comparative data.

Gastrulation and formation of the adult body plan are among the most fundamental morphogenetic movements during development. As such, they are not usually considered to be evolutionarily labile among closely related species (Thomson, 1988; Hall, 1992). However, the examples just discussed suggest that morphogenetic processes may change for reasons other than producing a different adult morphology. Because mechanistically distinct underlying cell movements can generate very similar phenotypic outcomes, morphological comparisons alone are not sufficient to rule out extensive evolutionary changes in morphogenetic mechanisms. Understanding why morphogenesis evolves will therefore require detailed comparisons of cell movements across a broader array of species.

ELEMENTS OF A COMPARATIVE ANALYSIS OF DEVELOPMENT

The preceding examples suggest several reasons why developmental processes have or have not changed within a clade. These evolutionary mechanisms are summarized in Table 1. The examples also suggest that the evolution of a particular developmental process can be influenced by different factors. Discriminating between competing explanations will not always be easy or even possible. However, comparative and experimental evidence will together often support or reject one or more of these alternatives. Such an analysis requires three elements.

Extensive and judicious sampling of taxa

Many published studies of developmental evolution compare a feature in two species. This approach provides a valuable starting point. When a developmental process is evolutionarily conserved, the divergence time between the two species provides a minimum estimate of its antiquity. When the process differs, it clearly can change without incurring a drastic functional cost.

Sampling additional taxa can provide a great deal more information. First, it becomes possible to reconstruct the polarity of the change: which is the ancestral and which the derived state of the developmental process. This requires sampling at least three species of known phylogenetic relationship, and preferably more (Maddison and Maddison, 1992). Second, sampling additional species provides a measure of how widely conserved the developmental process is within a clade. The more species that are examined, the stronger the support for conservation or variation. And third, discriminat-

Table 2. Probabilities of evolutionary associations*

Hypothesized factor	Developmental correlate	Number of changes†	Number of taxa‡	Probability§ (P - value)
Non-feeding larva	Delay msp130 expression	3	12	0.002
	Loss of larval skeleton	5	19	0.00002
	Alter cleavage geometry	5	18	0.00002
	Alter gastrulation movements	1	8	0.07
	Larger egg	5	19	0.00002
Adult morphology	Skeletogenic cell movements	1	15	0.04
	Mode of rudiment formation	1	19	0.03

*Probabilities < 0.05 suggest that the association between a hypothesized influence and a developmental process is non-random.

†Determined from maximum parsimony reconstructions with MacClade 3.0 (Maddison and Maddison, 1992) using the phylogenetic relationships shown in Figs 3-5.

‡Each analysis is based on a phylogeny containing only those species for which the particular trait has been studied.

§Calculated with MacClade 3.0 (Maddison and Maddison, 1992) using Maddison's (1990) concentrated changes test. Expressed as the probability that the hypothesized association could have arisen by chance.

ing among potential evolutionary mechanisms with any statistical confidence may require sampling on the order of ten species (Fig. 1). In each case, then, conclusions will become more robust when more taxa are examined.

Quantity is not, however, everything. Judicious choices about study taxa can considerably increase the information that can be extracted from any comparative analysis. If the issue is reconstructing the polarity of the change, it is best to choose closely related taxa. If the issue is how widely a particular developmental process is conserved, it is important that a representative of each major subgroup within a clade be examined, as well as any species that are distinguished by unusual morphology or life history. Many supposedly invariant features of development have later been found to vary within a group when the number of examined species rises to three or four. Study taxa are often chosen primarily for convenience; however, the extra effort of studying non-standard species is often well worthwhile, and can provide particularly illuminating comparisons.

A robust phylogenetic framework

Understanding the phylogenetic relationships of study taxa is crucial for two reasons. First, it provides the only rigorous way to reconstruct the evolutionary history of development. By mapping differences in developmental processes onto a cladogram, it is possible to reconstruct the polarity of changes, how many times they have arisen independently, the sequence in which they have occurred, and what clades they define (for examples, see Figs 3, 4, and 5). By combining this information with divergence times between study taxa, it is possible to determine when particular changes occurred or how old a conserved feature is. Simple reconstructions can be done by eye, but large data sets require the assistance of computer-based algorithms. These programs are particularly important for exploring alternative, equally parsimonious reconstructions, and for assessing the impact of uncertainties in cladogram topology on the evolution of the characters being studied. Methods for reconstructing evolutionary changes are reviewed in detail by Harvey and Pagel (1991) and by Maddison and Maddison (1992).

Second, a phylogeny provides a framework for testing associations between developmental changes and other features, such as environment, life history, or morphology. The fact that many species share a derived feature of development and derived life history strategy may be a coincidence if each change has only evolved once and these species form a single clade. An explicit statistical test for associations, such as the concentrated changes test (Maddison, 1990), is therefore essential. Statistical significance will typically require sample sizes of more than six species, and two or more replicate transformations (Fig. 1; Table 2). Brooks and McLennan (1991) and Harvey and Pagel (1991) provide general discussions about using phylogenetic information to frame and test evolutionary hypotheses.

An understanding of the functional consequences of variation

Finally, it is important to understand the consequences of differences in developmental processes, because, ultimately, it is phenotype that interacts with the environment and largely determines which genetic variants become fixed in populations. In some cases, a change in development may have no phenotypic effect: it may be canalized, or may be compensated for by other changes in development, as in the gastrulation example discussed above. However, most changes in development probably do have phenotypic consequences. These may be manifest in the morphology of adults or preadult phases such as larvae, or they may be limited to changing a non-morphological aspect of phenotype such as the rate of development or behavior.

Understanding why a particular developmental process has changed or remained the same requires distinguishing among these alternatives. Developmental processes can often be manipulated experimentally or genetically. Such perturbation experiments provide useful information for understanding not only how a single species develops, but why different species use the same or different developmental processes.

CONCLUSIONS

In 1975, VanValen wrote that "evolution is the control of development by ecology". To a large extent, this may be true. The available empirical evidence from echinoderms, when explicitly analyzed within a phylogenetic framework, suggests that ecology can exert a profound influence on a variety of important developmental processes. In sea urchins, the ecologically significant shift from feeding to non-feeding larvae has evolved on several separate occasions. This change in life history strategy seems to be driving functionally significant changes in the regulation of gene expression, cell fate specification, and cell movements during morphogenesis. A second way in which ecology controls development is through more conventional aspects of phenotype. Some evolutionary changes in development are probably fixed because their phenotypic consequences improve locomotion or feeding efficiency or provide additional protection from predation. However, it is unlikely that all evolutionary changes in development are driven directly by ecology. A third class of modifications appears to be driven indirectly, evolving in order to compensate for other changes in development. And, fourth, some changes may not be related to ecology at all; instead, these changes may evolve simply because they produce neutral or nearly neutral phenotypic effects. This fourth kind of change is possible when developmental processes are canalized and buffer some perturbations.

We thank Chris Lowe and two anonymous reviewers for helpful comments on the manuscript. The National Science Foundation and the National Institutes of Health supported G. A. W.'s research.

REFERENCES

Amemiya, S. and Emlet, R. B. (1992). The development and larval form of an echinothurioid echinoid, *Asthenosomal ijimai*, revisited. *Biol. Bull.* **182**, 15-30.

Anderson, D. T. (1973). *Embryology and Phylogeny in Annelids and Arthropods*. Oxford: Pergamon Press.

Arthur, W. (1988). *A Theory of the Evolution of Development*. Chichester: Wiley.

Ballard, W. W. (1976). Problems of gastrulation: real and verbal. *BioScience* **26**, 36-39.

Benson, S., Sucov, H., Stephens, L., Davidson, E. and Wilt, F. (1987). A lineage-specific gene encoding a major matrix protein of the sea urchin embryo spicule. I. Authentication of the cloned gene and its developmental expression. *Dev. Biol.* **120**, 499-506.

Berrill, N. J. (1931). Studies in tunicate development. II. Abbreviation of development in the Molgulidae. *Phil. Trans. R. Soc. Lond.* **B219**, 281-346.

Brooks, D. R. and McLennan, D. A. (1991). *Phylogeny, Ecology, and Behavior: A Research Program in Comparative Biology*. Chicago: University of Chicago Press.

Brown, S. J., Patel, N. H. and Denell, R. E. (1994). Embryonic expression of the single *Tribolium engrailed* homolog. *Dev. Genet.* **15**, 7-18.

Burke, R. D., Myers, R. L., Sexton, T. L. and Jackson, C. (1991). Cell movements during the initial phase of gastrulation in the sea urchin embryo. *Dev. Biol.* **146**, 542-557.

Bury, H. (1895). The metamorphosis of echinoderms. *Quart. J. Microsc. Sci., N.S.* **38**, 45-135.

Byrne, M. (1991). Developmental diversity in the starfish genus *Patiriella* (Asteroidea: Asterinidae). In *Proceedings of the 7th International Echinoderm Conference* (ed. T. Yanagisawa, I. Yasumasu, C. Oguro, N. Suzuki, and T. Motokawa), pp. 499-508. Rotterdam: Balkema.

Cameron, R. A. and Davidson, E. H. (1991). Cell type specification during sea urchin development. *Trends Genet.* **7**, 212-218.

Cameron, R. A., Fraser, S. E., Britten, R. J. and Davidson, E. H. (1989). The oral-aboral axis of a sea urchin embryo is specified by first cleavage. *Development* **106**, 641-647.

Cameron, R. A., Hough-Evans, B., Britten, R. J. and Davidson, E. H. (1987). Lineage and fate of each blastomere of the eight-cell sea urchin embryo. *Genes Dev.* **1**, 75-84.

Coe, W. R. (1949). Divergent methods of development in morphologically similar species of prosobranch gastropods. *J. Morph.* **84**, 383-399.

Davidson, E. H. (1990). How embryos work: a comparative view of diverse modes of cell fate specification. *Development* **108**, 365-389.

Davidson, E. H. (1991). Spatial mechanisms of gene regulation in metazoan embryos. *Development* **113**, 1-26.

Decker, G. L., Valdizan, M. C., Wessel, G. M. and Lennarz, W. J. (1988). Developmental distribution of a cell surface glycoprotein in the sea urchin *Strongylocentrotus purpuratus*. *Dev. Biol.* **129**, 339-349.

del Piño, E. M. and Elinson, R. P. (1983). A novel development pattern for frogs: gastrulation produces an embryonic disk. *Nature* **306**, 589-591.

Drager, B. J., Harkey, M. A., Iwata, M. and Whiteley, A. H. (1989). The expression of embryonic primary mesenchyme genes of the sea urchin, *Strongylocentrotus purpuratus*, in the adult skeletogenic tissues of this and other echinoderms. *Dev. Biol.* **133**, 14-23.

Elinson, R. P. (1990). Direct development in frogs: wiping the recapitulationist slate clean. *Sem. Dev. Biol.* **1**, 263-270.

Emlet, R. B. (1988). Larval form and metamorphosis of a "primitive" sea urchin, *Eucidaris thouarsi* (Echinodermata: Echinoidea: Cidaroida), with implications for developmental and phylogenetic studies. *Biol. Bull.* **174**, 4-19.

Emlet, R. B. (1990). World patterns of developmental mode in echinoid echinoderms. In *Advances in Invertebrate Reproduction* (ed. M. Hoshi and O. Yamishita), pp. 329-335. Amsterdam: Elsevier.

Ettensohn, C. A. (1984). Primary invagination of the vegetal plate during sea urchin gastrulation. *Amer. Zool.* **24**, 571-588.

Ettensohn, C. A. (1985). Mechanisms of epithelial invagination. *Quart. Rev. Biol.* **60**, 289-307.

Ettensohn, C. A. and Ingersoll, E. P. (1992). Morphogenesis of the sea urchin embryo. In *Morphogenesis: An Analysis of the Development of Biological Form* (ed. E. F. Rossomando and S. Alexander), pp. 189-262. New York: Marcel Dekker Press.

Franks, R. R., Anderson, R., Moore, J. G., Hough-Evans, B. R., Britten, R. J. and Davidson, E. H. (1990). Competitive titration in living sea urchin embryos of regulatory factors required for expression of the CyIIIa actin gene. *Development* **110**, 31-40.

Freeman, G. and Lundelius, J. W. (1992). Evolutionary implications of the mode of D quadrant specification in coelomates with spiral cleavage. *J. Evol. Biol.* **5**, 205-247.

Garstang, W. (1929). The origin and evolution of larval forms. *British Association for the Advancement of Science* 1929, 77-98.

Gould, S. J. (1977). *Ontogeny and Phylogeny*. Cambridge, MA: Belknap Press.

Gustafson, T. and Wolpert, L. (1967). Cellular movement and contact in sea urchin morphogenesis. *Biol. Rev.* **42**, 442-498.

Hall, B. K. (1992). *Evolutionary Developmental Biology*. London: Chapman & Hall.

Hardin, J. D. (1988). The role of secondary mesenchyme cells during sea urchin gastrulation studied by laser ablation. *Development* **103**, 317-324.

Hardin, J. D. and Cheng, L. Y. (1986). The mechanisms and mechanics of archenteron elongation during sea urchin gastrulation. *Dev. Biol.* **115**, 490-501.

Hardin, J. and McClay, D. R. (1990). Target recognition by the archenteron during sea urchin gastrulation. *Dev. Biol.* **142**, 86-102.

Harkey, M. A. and Whiteley, A. H. (1983). The program of protein synthesis during the development of the micromere-primary mesenchyme cell line in the sea urchin embryo. *Dev. Biol.* **100**, 12-28.

Hart, M. (1991). Particle capture and the method of suspension feeding by echinoderm larvae. *Biol. Bull.* **180**, 12-27.

Harvey, P. H. and Pagel, M. D. (1991). *The Comparative Method in Evolutionary Biology*. New York: Oxford University Press.

Henry, J. J., Amemiya, S., Wray, G. A. and Raff, R. A. (1989). Early inductive interactions are involved in restricting cell fates of mesomeres in sea urchin embryos. *Dev. Biol.* **136**, 140-153.

Henry, J. J., Klueg, K. M. and Raff, R. A. (1992). Evolutionary dissociation between cleavage, cell lineage and embryonic axes in sea urchin embryos. *Development* **114**, 931-938.

Henry, J. Q. and Martindale, M. Q. (1994). Establishment of the dorsoventral axis in nemertean embryos: evolutionary considerations of spiralian development. *Dev. Genet.* **15**, 64-78.

Henry, J. J. and Raff, R. A. (1990). Evolutionary change in the process of dorsoventral axis determination in the direct developing sea urchin, *Heliocidaris erythrogramma*. *Dev. Biol.* **141**, 55-69.

Hörstadius, S. (1973). *Experimental Embryology of Echinoderms*. London: Clarendon, Oxford Univ. Press.

Janies, D. A. and McEdward, L. R. (1993). Highly derived coelomic and water-vascular morphogenesis in a starfish with pelagic direct development. *Biol. Bull.* **185**, 56-76.

Jeffery, W. R. and Swalla, B. J. (1991). An evolutionary change in the muscle lineage of an anural ascidian embryo is restored by interspecific hybridization with a urodele ascidian. *Dev. Biol.* **145**, 328-337.

Jeffery, W. R. and Swalla, B. J. (1992). Evolution of alternate modes of development in ascidians. *Bioessays* **14**, 219-226.

Kominami, T. (1983). Establishment of the embryonic axes in larvae of the starfish, *Asterina pectinifera*. *J. Embryol. Exp. Morphol.* **75**, 87-100.

Kominami, T. (1988). Determination of dorso-ventral axis in early embryos of the sea urchin, *Hemicentrotus pulcherrimus*. *Dev. Biol.* **127**, 187-196.

Kumé, M. and Dan, K. (1968). *Invertebrate Embryology*. Belgrade: NOLIT Press.

Kuraishi, R. and Osanai, K. (1992). Cell movements during gastrulation of starfish larvae. *Biol. Bull.* **183**, 258-268.

Lillie, F. (1895). The embryology of the Unionidae. *J. Morph.* **10**, 1-100.

Livingston, B. T., Shaw, R., Bailey, A. and Wilt, F. (1991). Characterization of a cDNA encoding a protein involved in formation of the skeleton during development of the sea urchin *Lytechinus pictus*. *Dev. Biol.* **148**, 473-480.

Maddison, W. P. (1990). A method for testing the correlated evolution of two binary characters: are gains or losses concentrated on certain branches of a phylogenetic tree? *Evolution* **44**, 539-557.

Maddison, W. P. and Maddison, D. R. (1992). MacClade. v.3.01. Sunderland, MA: Sinauer Associates.

McEdward, L. R. and Janies, D. A. (1993). Life cycle evolution in asteroids: what is a larva? *Biol. Bull.* **184**, 255-268.

McGinnis, W. and Krumlauf, R. (1992). Homeobox genes and axial patterning. *Cell* **68**, 283-302.

Okazaki, K. (1975). Normal development to metamorphosis. In *The Sea Urchin Embryo* (ed. G. Czihak), pp. 177-232. Berlin: Springer-Verlag.

Parks, A. L., Bisgrove, B. W., Wray, G. A. and Raff, R. A. (1989). Direct development in the sea urchin *Phyllacanthus parvispinus* (Cidaroidea): phylogenetic history and functional modification. *Biol. Bull.* **177**, 96-109.

Parks, A. L., Parr, B. A., Chin, J., Leaf, D. S. and Raff, R. A. (1988). Molecular analysis of heterochronic changes in the evolution of direct developing sea urchins. *J. Evol. Biol.* **1**, 27-44.

Parr, B. A., Parks, A. L. and Raff, R. A. (1990). Promoter structure and protein sequence of msp130, a lipid-anchored sea urchin glycoprotein. *J. Biol. Chem.* **265**, 1408-1413.

Patel, N. H., Ball, E. E. and Goodman, C. S. (1992). Changing role of *even-skipped* during the evolution of insect pattern formation. *Nature* **357**, 339-342.

Patel, N. H., Condron, B. G. and Zinn, K. (1994). Pair-rule expression patterns of *even-skipped* are found in both short- and long-germ beetles. *Nature* **367**, 429-434.

Patel, N. H., Martin-Blanco, E., Coleman, K. G., Poole, S. J., Ellis, M. C., Kornberg, T. B. and Goodman, C. S. (1989). Expression of *engrailed* proteins in arthropods, annelids, and chordates. *Cell* **58**, 955-968.

Püschel, A. W., Westerfield, M. and Dressler, G. R. (1992). Comparative analysis of Pax-2 protein distributions during neurulation in mice and zebrafish. *Mech. Dev.* **38**, 197-208.

Raff, R. A. (1987). Constraint, flexibility, and phylogenetic history in the evolution of direct development in sea urchins. *Dev. Biol.* **119**, 6-19.

Raff, R. A. (1992). Direct-developing sea urchins and the evolutionary reorganization of early development. *BioEssays* **14**, 1-8.

Ransick, A. and Davidson, E. H. (1993). A complete second gut induced by transplanted micromeres in the sea urchin embryo. *Science* **259**, 1134-1138.

Rumrill, S. and Chia, F. (1984). Differential mortality during the embryonic and larval lives of northeast Pacific echinoids. In *Echinodermata: Proceedings of the Fifth International Echinoderm Conference* (ed. B. F. Keegan and B. D. S. O'Connor), pp. 333-338. Amsterdam: Balkema.

Sander, K. (1983). The evolution of patterning mechanisms: gleanings from insect embryogenesis and spermatogenesis. In *Development and Evolution* (ed. B. C. Goodwin, N. Holder, and C. C. Wylie), pp. 137-159. Cambridge, UK: Cambridge Univ. Press.

Scott, L. B., Lennarz, W. J., Raff, R. A. and Wray, G. A. (1990). The "lecithotrophic" sea urchin *Heliocidaris erythrogramma* lacks typical yolk platelets and yolk glycoproteins. *Dev. Biol.* **138**, 188-193.

Skiba, F. and Schierenberg, E. (1992). Cell lineages, developmental timing, and spatial pattern formation in embryos of free-living soil nematodes. *Dev. Biol.* **151**, 597-610.

Smith, A. B. (1984). *Echinoid Palaeobiology*. London: Allen & Unwin.

Smith, A. B. (1988). Fossil evidence for the relationship of extant echinoderm classes and their times of divergence. In *Echinoderm Phylogeny and Evolutionary Biology* (ed. C. R. C. Paul and A. B. Smith), pp. 85-97. New York: Oxford University Press.

Smith, A. B. (1992). Echinoderm phylogeny: morphology and molecules reach accord. *Trends Ecol. Evol.* **7**, 224-229.

Smith, A. B., Lafay, B. and Christen, R. (1993). Comparative variation of morphological and molecular evolution through geological time: 28S ribosomal RNA versus morphology in echinoids. *Phil. Trans. Roy. Soc. Lond. B* **338**, 365-382.

Strathmann, R. R. (1978). The evolution and loss of feeding larval stages of marine invertebrates. *Evolution* **32**, 894-906.

Strathmann, R. R. (1985). Feeding and nonfeeding larval development and life-history evolution in marine invertebrates. *Ann. Rev. Ecol. Syst.* **16**, 339-361.

Thomson, K. S. (1988). *Morphogenesis and Evolution*. Cambridge, UK: Cambridge University Press.

Williams, D. H. C. and Anderson, D. T. (1975). The reproductive system, embryonic development, larval development and metamorphosis of the sea urchin *Heliocidaris erythrogramma* (Val.) (Echinoidea: Echinometridae). *Australian J. Zool.* **23**, 371-403.

Wilt, F. H. (1987). Determination and morphogenesis in the sea urchin embryo. *Development* **100**, 559-575.

Wray, G. A. (1992). The evolution of larval morphology during the post-Paleozoic radiation of echinoids. *Paleobiol.* **18**, 258-287.

Wray, G. A. (1994). The evolution of cell lineage in echinoderms. *Amer. Zool.* (in press).

Wray, G. A. and McClay, D. R. (1988). The origin of spicule-forming cells in a 'primitive' sea urchin (*Eucidaris tribuloides*) which appears to lack primary mesenchyme cells. *Development* **103**, 305-315.

Wray, G. A. and McClay, D. R. (1989). Molecular heterochronies and heterotopies in early echinoid development. *Evolution* **43**, 803-813.

Wray, G. A. and Raff, R. A. (1989). Evolutionary modification of cell lineage in the direct-developing sea urchin *Heliocidaris erythrogramma*. *Dev. Biol.* **132**, 458-470.

Wray, G. A. and Raff, R. A. (1990). Novel origins of lineage founder cells in the direct-developing sea urchin *Heliocidaris erythrogramma*. *Dev. Biol.* **141**, 41-54.

Wray, G. A. and Raff, R. A. (1991a). The evolution of developmental strategy in marine invertebrates. *Trends Ecol. Evol.* **6**, 45-50.

Wray, G. A. and Raff, R. A. (1991b). Rapid evolution of gastrulation mechanisms in a sea urchin with lecithotrophic larvae. *Evolution* **45**, 1741-1750.

Development 1994 Supplement, 107-116 (1994)
Printed in Great Britain © The Company of Biologists Limited 1994

Evolution of flowers and inflorescences

Enrico S. Coen* and Jacqueline M. Nugent

John Innes Centre, Colney, Norwich NR4 7UH, UK

*Author for correspondence

SUMMARY

Plant development depends on the activity of meristems which continually reiterate a common plan. Permutations around this plan can give rise to a wide range of morphologies. To understand the mechanisms underlying this variation, the effects of parallel mutations in key developmental genes are being studied in different species. In *Antirrhinum*, three of these key genes are: (1) *floricaula* (*flo*) a gene required for the production of flowers (2) *centroradialis* (*cen*), a gene controlling flower position (3) *cycloidea* (*cyc*), a gene controlling flower symmetry. Several plant species, exhibiting a range of inflorescence types and floral symmetries are being analysed in detail. Comparative genetic and molecular analysis shows that inflorescence architecture depends on two underlying parameters: a basic inflorescence branching pattern and the positioning of flowers. The *flo* and *cen* genes play a key role in the positioning of flowers, and variation in the site and timing of expression of these genes, may account for many of the different inflorescence types. The evolution of inflorescence structure may also have influenced the evolution of floral asymmetry, as illustrated by the *cen* mutation which changes both inflorescence type and the symmetry of some flowers. Conflicting theories about the origins of irregular flowers and how they have coevolved with inflorescence architecture can be directly assessed by examining the role of *cyc*- and *cen*-like genes in species displaying various floral symmetries and inflorescence types.

Key words: *Antirrhinum*, *Arabidopsis*, symmetry

INTRODUCTION

The development of plants depends on the activity of their meristems, groups of dividing cells located at the growing points. These meristems can continue to add new structures to the plant throughout its life history, giving it the potential for indeterminate growth. They achieve this by generating two types of structure on their periphery: **primordia** which will grow to form organs such as leaves, petals and stamens; and **secondary meristems**, which will form side branches or flowers. In most cases each primordium has a secondary meristem located in its **axil** (the angle between the base of the primordium and the main stem). By changing a few of the key features of meristems, it is possible to account for much of the evolutionary variation in plant form. Here we present a comparative molecular genetic approach to understanding how such changes may account for variation in inflorescence architecture and floral symmetry.

Plants offer several advantages for comparative analysis. Parallel genetic studies have been carried out in several species, allowing comparisons of mutations in key genes. Several species can be transformed, permitting the transfer of genes between species to assess how the role of these genes may have changed during evolution. The indeterminate growth pattern of plants gives them the potential to produce many different morphologies through limited modifications in the properties of their meristems. Thus, analysing key genes controlling meristem behaviour may give important insights into how diverse plant forms have evolved.

TYPES OF INFLORESCENCE

Most plants have flowers clustered together in a region termed the inflorescence. Inflorescences have been classified according to several criteria (Weberling, 1989). These include whether the inflorescence is: (1) **determinate**, where the main axis of growth ends in a flower (Fig. 1a-d), or **indeterminate**, in which there is no terminal flower (Fig. 1e-i); (2) **racemose**, having a branching pattern with a single main axis of growth (Fig. 1e-h), or **cymose**, having a branching pattern which lacks a main axis (Fig. 1c,d); (3) **simple**, in which no secondary branches are produced (Fig. 1a,b,e,h,i), or **compound**, which produce secondary and sometimes higher order branches (Fig. 1c,d,f,g); (4) **bracteose** which have leaf-like organs (bracts) subtending flowers, or **bractless**.

Most of these criteria depend on two underlying parameters, a basic inflorescence branching pattern and the positions of flowers. Since, by definition, an inflorescence must comprise both flowers and branch points, these two parameters normally cannot be uncoupled and their individual contributions to inflorescence architecture assessed separately. However, by using a molecular genetic approach to analyse inflorescence mutants, we can begin to separate the various parameters and assess their role in the evolution of inflorescence types.

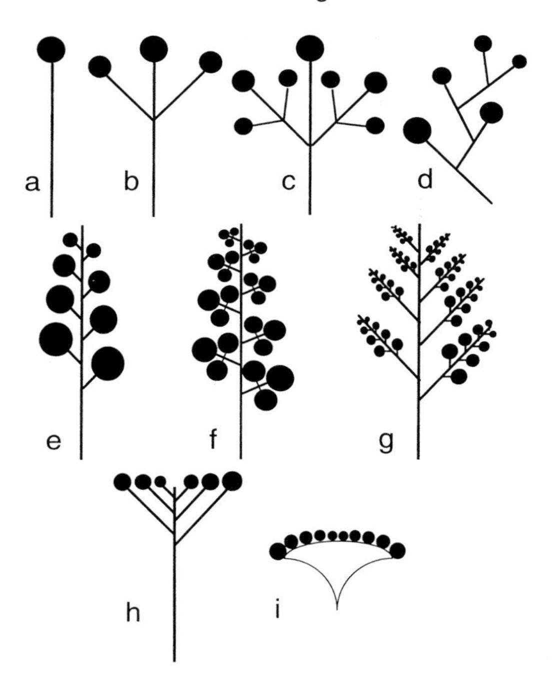

Fig. 1. Diagrammatic illustration of various inflorescence types. (a) Solitary terminal flower; (b) simple dichasium; (c) compound dichasium; (d) cyme; (e) raceme; (f) thyrse; (g) compound raceme; (h) corymb; (i) capitulum. Flowers are indicated by solid circles. Bracts are not shown.

DEFINING THE PRIMARY INFLORESCENCE BRANCHING PATTERN

The first gene affecting inflorescence development to be studied in detail was *floricaula* (*flo*) in *Antirrhinum majus* (Carpenter and Coen, 1990; Coen et al., 1990). Vegetative growth in *Antirrhinum* is characterised by the production of two opposite leaves at each node of the main stem. After the plant has produced several vegetative nodes, it switches to the reproductive phase, characterised by the production of single flowers in the axils of bracts separated by short internodes on a hairy stem (Fig. 2). This type of inflorescence, comprising a series of axillary flowers arranged along a single main axis, is termed a raceme (Fig. 1e). In *flo* mutants, flowers are replaced by shoots (Fig. 2) that have some characteristics of the reproductive phase of growth (bracts, short internodes, hairy stem). These secondary shoots can themselves produce further shoots, termed tertiary shoots, in the axils of their bracts. The overall result is that the raceme has been replaced by a *flo*-independent branching structure that is never terminated by the production of flowers.

One interpretation of the *flo* mutant phenotype is that it reflects an alteration in the identity of meristems. In *Antirrhinum* there are three types of meristem that are distinguished by the types and arrangements of organ primordia and the secondary meristems they produce. The **vegetative** meristem produces leaf primordia with secondary vegetative meristems in their axils. The **inflorescence** meristem normally produces bract primordia with floral meristems in their axils. **Floral** meristems produce floral organ primordia but unlike the other meristem types, do not give rise to secondary meristems. A further distinguishing feature of floral meristems is that after they have produced a defined number of primordia they cease to be active. Thus floral meristems are **determinate** and provide endpoints to the growth of an axis whereas vegetative and inflorescence meristems are **indeterminate**.

In the *flo* mutant, the main inflorescence meristem produces secondary inflorescence meristems, rather than floral meristems, in the axils of its bract primordia. These secondary meristems can in turn produce further meristems on their periphery and this process can continue indefinitely leading to even higher order meristems. The *flo* gene is therefore required to establish floral meristem identity and in its absence, indeterminate branching ensues. This early requirement for *flo* correlates with its expression pattern, as determined by in situ hybridisation (Coen et al., 1990). Transcripts of *flo* are detected in initiating floral meristems and their subtending bract primordia but are not observed in either the main infloresence meristem or in vegetative meristems (Fig. 3). Localisation of *flo* expression within particular meristems is therefore an important determinant of which meristems will form flowers. At later stages of development, *flo* is expressed transiently in all floral organ primordia except stamens.

One way to alter inflorescence architecture might be to change the site or timing of *flo* expression. This can be illustrated by using wild-type *Antirrhinum* as a starting point and predicting the phenotype produced when *flo* expression is altered in several ways. A slight delay in *flo* activation, such that it does not come on immediately in secondary meristems but after some tertiary branching is initiated, would give a cluster of flowers instead of single axillary flowers. This could give rise to a compound inflorescence where each secondary branch comprises a small cyme (thyrse, Fig. 1f). Another possiblity would be changing *flo* expression such that it only came on in tertiary rather than secondary meristems. This would result in each flower being replaced by an indeterminate axillary raceme, giving rise to a compound raceme (Fig. 1g).

DETERMINATE VERSUS INDETERMINATE INFLORESCENCES

Whereas a delay of *flo* expression might be expected to produce a more highly branched inflorescence, ectopic expression of *flo* could be involved in producing more compact or determinate inflorescences. One way to convert an indeterminate inflorescence to a determinate condition would be to replace the main inflorescence apex by a flower. This is illustrated by *centroradialis* (*cen*) mutants of *Antirrhinum* (Stubbe, 1966) which produce a short raceme, terminated by a single flower (Fig. 4, note that the shape of the terminal flower is distinct, see below). Presumably, in wild-type plants this gene prevents expression of *flo* and other genes needed for flower development in the inflorescence apex.

It has been suggested that the indeterminate inflorescence is a derived condition, which required a mechanism for repressing terminal flowers to have evolved (Stebbins, 1974). Furthermore, because species with indeterminate inflorescences are present in a wide range of taxa, interspersed with species having a determinate condition, it is believed that a genetic mechanism for repressing terminal flowers must have arisen independently many times. However, it is equally plausible that a mechanism for repressing terminal flowers was derived at an early stage during the evolution of flowering plants and that in many lineages determinate inflorescences have been secondarily derived through loss of this mechanism. Is it possible to distinguish between these two scenarios? According to the multiple gain hypothesis, the mechanism for establishing indeterminacy might be expected to be different in distantly related species. However, according to the multiple loss theory, many species could share the same underyling mechanism for establishing indeterminacy. One way to distinguish between these possibilities is to establish whether the genes controlling indeterminacy in other species are homologous to *cen*. For example, the *terminal flower* (*tfl*) mutant of *Arabidopsis thaliana* has a comparable phenotype to *cen*. If *tfl* and *cen* encode distinct proteins, it would suggest that *Arabidopsis* and *Antirrhinum* might have acquired indeterminacy through different routes. If the *cen* and *tfl* genes are homologous, it would imply that the indeterminate condition of *Antirrhinum* and *Arabidopsis* may not have been independently acquired but was present in their common ancestor.

RACEMOSE VERSUS CYMOSE BRANCHING

The previous examples illustrate how some inflorescence architectures might arise through changing the expression pattern of floral meristem identity genes such as *flo*. If this were the only way of altering inflorescence type, *flo*-like mutants in all species would reveal essentially the same primary branching pattern. This can be directly assessed in several cases.

The most intensively studied *flo* homologue in another species is the *lfy* gene of *Arabidopsis* (Weigel et al., 1992; Schultz and Haughn, 1991). Vegetative growth of wild-type *Arabidopsis* produces a rosette of leaves separated by short internodes. On entering the reproductive phase, the plant bolts to generate an elongated main stem, the lower part of which has several small leaves (cauline leaves) and the upper part bears flowers arranged as a raceme. Secondary inflorescences are initiated within the axils of the cauline leaves (Fig. 2). In *lfy* mutants, the plant bolts as normal but flowers are replaced by secondary shoots subtended by cauline leaves. In older plants, the secondary shoots produced at the top of the bolt, display some floral characteristics, showing that some features of flower development can be restored, even in the absence of *lfy* activity.

The basic branching pattern underlying the *Arabidopsis* inflorescence, as revealed by the *lfy* mutant, is very similar to that of *Antirrhinum* (Fig. 2). In both cases there is a main axis of growth with secondary shoots arising in axillary positions. This similarity may reflect the fact that wild-type *Arabidopsis* and *Antirrhinum* share the same overall inflorescence architecture, the raceme. A more revealing comparison might be to analyse the *flo*-independent branching pattern in species with markedly different inflorescence types.

The tomato (*Lycopersicon esculentum*) inflorescence has a cymose branching pattern. After vegetative growth, the main apical meristem becomes a terminal floral meristem associated with an adjacent secondary meristem. The secondary meristem terminates with the production of a flower and a tertiary meristem. By repeating this pattern of growth, tertiary and higher order flowers are successively produced to form a cymose inflorescence. Although initially in an apical position, the whole inflorescence eventually becomes displaced by an axillary leafy shoot which continues the main growth of the plant (Fig. 2). After producing about three leaves the axillary shoot repeats the growth pattern of the primary shoot. This growth pattern, termed sympodial growth, could be a consequence of expressing *flo*-like genes in the apical meristem and in secondary, tertiary and higher order meristems derived from it. By committing the apical meristem to producing flowers, the continued growth of the plant would have to occur from an axillary meristem. According to this view, a *flo*-like mutant in tomato should have an indeterminate leafy shoot with a single primary axis of growth, similar to that seen for the *flo* and *lfy* mutants of *Antirrhinum* and *Arabidopsis*. Alternatively, sympodial growth may not be a consequence of flower production but may reflect the *flo*-independent primary branching pattern of tomato. In this case, *flo*-like mutants in tomato should retain sympodial growth.

The *falsiflora* (*fa*) mutant of tomato lacks a flowering inflorescence and produces a highly branched system of shoots instead (Stubbe, 1963, Fig. 2). The mutant still exhibits sympodial growth, showing that this growth pattern is independent of flower production. Preliminary genetic and molecular results indicate that *fa* may be the tomato homologue of *flo*. However, the *fa* mutation reveals a primary branching pattern that is quite distinct from that revealed by *flo* or *lfy*. Unlike *flo* and *lfy*, the *fa* branching system that replaces the inflorescence, lacks a main axis of growth, even though it proliferates extensively. This suggests that the wild-type tomato inflorescence can be distinguished from that of *Antirrhinum* or *Arabidopsis* by two features: a cymose primary branching pattern which is *flo*-independent and the production of terminal flowers which is *flo*-dependent. Both of these features can be seen as reflecting early meristem behaviour.

BRACTEOSE VERSUS BRACTLESS INFLORESCENCE

In addition to overall architecture, another feature that distinguishes inflorescences is the presence or absence of bracts subtending the flowers. *Antirrhinum* has flowers subtended by bracts (bracteose infloresence) whereas *Arabidopsis* flowers are bractless (Fig. 2). The absence of bracts is a general feature of the family Cruciferae, to which *Arabidopsis* belongs and is thought to be a derived condition. In *lfy* mutants, shoots or abnormal flowers often with subtending bracts, are produced where bractless flowers would normally arise (Schultz and Haughn, 1991; Weigel et al., 1992). This demonstrates that the bractless condition is to a large extent *lfy*-dependent. The analysis of *flo* and *lfy* expression indicates how the bractless condition may have evolved.

In *Antirrhinum*, *flo* is expressed in floral meristems and their subtending bract primordia. In *Arabidopsis*, *lfy* is only expressed in floral meristems. However, when bract primordia are allowed to form in *Arabidopsis*, as in *lfy* mutants, they accumulate *lfy* RNA, giving an expression pattern similar to that of wild-type *Antirrhinum* (Fig. 3).

One interpretation of these results is that originally *lfy* was

Antirrhinum
floricaula
Arabidopsis
leafy
tomato
falsiflora

Fig. 2

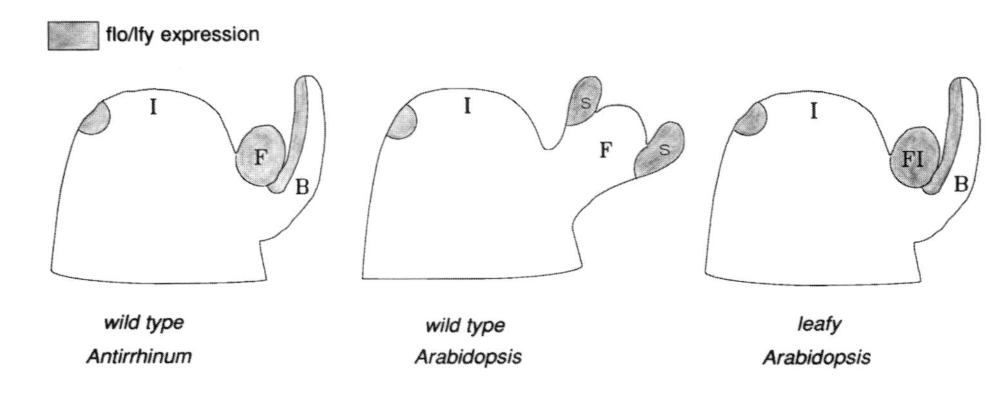

Fig. 3. Expression pattern of *flo* in wild-type *Antirrhinum* and *lfy* in wild-type and mutant *Arabidopsis*. Abbreviations: I, inflorescence meristem; F, floral meristem; B, bract primordium; s, sepal primordium; FI, meristem with both floral and inflorescence character.

expressed in flower meristems and bract primordia, as in *Antirrhinum*. Subsequently, in a common ancestor of the Cruciferae, *lfy* acquired the role of incorporating cells that would normally go on to form the bract, into the floral meristem. The bracts are therefore suppressed, not by inhibiting the activity of a bract-forming region, but by incorporating the cells from such a region into the flower. The advantage of recruiting all cells to form a flower without the attending bract might be to accelerate flower development. This is illustrated in *lfy* mutants, where the development of floral meristems subtended by bracts is retarded relative to wild-type, presumably because they start off with fewer cells (Fig. 3).

In some members of the Cruciferae, the role of *lfy* in cell recruitment may be partly separated from its role in flower production. For example, in *Alyssum*, the secondary inflorescences that are produced just below the main raceme, are not subtended by cauline leaves (Fig. 5). One possible explanation is that

Fig. 2. Diagrammatic illustrations and photographs comparing wild-type and *flo* mutant of *Antirrhinum* (top panel), wild-type and *lfy* mutant of *Arabidopsis* (middle panel) and wild-type and *fa* mutant of tomato (bottom panel). In all cases, wild type is shown on the left. In the diagrammatic illustrations, leaves are shown in brown, bracts or cauline leaves in green, flowers as blue circles and abnormal "flowers" as green circles. In the photographs, only the upper part of the plant is shown for *Antirrhinum* and *Arabidopsis* and only single inflorescences are shown for tomato.

Fig. 4. Photograph of the inflorescence of an *Antirrhinum cen* mutant shown from the side. Note the symmetrical terminal flower.

in these cases, *lfy* is acting to incorporate cells, that would normally go on to form the cauline leaf, into the meristem of the secondary inflorescence. Within the main raceme of *Alyssum*, the *lfy* gene acts as it does in *Arabidopsis*, recruiting cells into the secondary meristem and switching its identity to that of a floral meristem.

The emergence of the bractless condition may reflect an alteration in the *lfy* itself and/or alterations in target genes acting downstream of *lfy*. One way to test this would be to transform *lfy* mutants with *flo* and observe its expression pattern and phenotypic consequences. Preliminary results indicate that *flo* does not complement the *lfy* mutant, suggesting that some functionally important differences exist between these genes (R. Elliott, J. Nugent, R. Carpenter and E. Coen, unpublished results). Further experiments, such as swapping domains between *lfy* and *flo*, may shed light on where the key differences reside.

EVOLUTION OF FLORAL SYMMETRY

Flowers are classified as being either **irregular**, having only one plane of mirror symmetry or **regular**, having more than one plane of symmetry. The most intensive genetic analysis of floral symmetry has been carried out in *Antirrhinum*, which has irregular flowers that are markedly asymmetrical along their dorsoventral axis. Wild-type *Antirrhinum* flowers consist of five petals that are united for part of their length to form a tube ending with five lobes. The petal lobes are of three types: two large dorsal (upper) petals, two side petals and a ventral (lowest) petal. The flower is also irregular with respect to the stamens. Five stamens are initiated, alternate with the petals and are also of three types: the dorsal stamen is aborted and the two side stamens are shorter than the two ventral stamens. Mutations in *cycloidea* (*cyc*) give regular flowers with five-fold symmetry in certain genetic backgrounds (Fig. 6). All five petals and stamens resemble the ventral petal and stamens of the wild-type. It has been proposed that the irregular phenotype of wild-type flowers is dependent on *cyc* activity establishing an axis of dorsoventral asymmetry. The activity of *cyc* is predicted to be greatest in the dorsal regions of the flower meristem and to decline towards the more ventral regions. This would account for the ventralised phenotype of *cyc* mutants (Carpenter and Coen, 1990).

Early flowering plants are thought to have had regular flowers and irregularity is considered to be a derived condition. The analysis of genes like *cyc* allows us to address several important questions about the genetic basis of this evolutionary change: what role might *cyc* have played in the ancestral species with regular flowers and how was *cyc* subsequently recruited to establish dorsoventral asymmetry? How many times has irregularity evolved and has it always involved *cyc*? Before considering these questions it is important first to consider the current view of how irregularity has evolved, largely based on comparative morphology.

Irregularity is thought to have evolved independently many times, perhaps arising on as many as 25 separate occasions (Stebbins, 1974). The alternative to this multiple gain hypothesis, is that irregularity arose only a limited number of times and was subsequently lost several times in independent lineages. According to this view, irregularity may be much more ancient than is commonly believed. Three types of argument can be used to evaluate these two hypotheses:

(1) The multiple gain hypothesis is claimed to be the most parsimonious way to explain the broad phylogenetic distribution of regularity as compared to the more sporadic occurrence of irregularity. However, this is partly a circular argument because the phylogenies it is based upon have been constructed using morphological data that includes floral symmetry and

Fig. 5. Photograph of the *Alyssum* inflorescence. Arrow shows a secondary inflorescence not subtended by a cauline leaf.

Fig. 6. Photographs of wild-type and *cyc* mutant of *Antirrhinum* (A) and comparable peloric mutants from *Linaria vulgaris* (B), *Saintpaulia* (C) and *Sinningia speciosa* (D). For all species, the wild type is on the left and mutant on the right.

correlated characters. An objective evaluation can only be made if the phylogenies are independent of the morphological character being assessed.

Even with a more objective phylogeny, arguments based on parsimony are problematical because they depend on knowing the relative probabilities of gaining or losing irregularity. This is illustrated in Fig. 7, which shows the relationships between families in a portion of the subclass Asteridae and is based on molecular data (Olmstead et al., 1993). A traditional view would be that the common ancestor of these species was regular, implying that irregularity evolved a minimum of 3 times independently (see (+) in Fig. 7). The alternative, is that the ancestor was irregular and implies that regularity was derived a minimum of 4 times (see (–), Fig. 7). The two explanations are about equally parsimonious, assuming that the probabilities of gaining or losing irregularity are equal. However, this is unlikely to be true because the probabilities depend on two biological variables: the genetic facility with which irregularity can be gained or lost and the adaptive consequences of the change. The *cyc* mutation of *Antirrhinum* illustrates an alteration from the irregular to the regular condition. To our knowledge, no mutations have been described in species with regular flowers that render them irregular. This would argue that it is easier to lose irregularity than to gain it. However, this must be tempered by adaptive considerations. The *cyc* mutation would clearly confer a selective disadvantage because bees, the primary pollinators of *Antirrhinum*, would be unable to enter and cross-pollinate the mutant flowers efficiently. It is therefore very difficult to establish overall whether gain or loss is the more likely explanation for the phylogeny in Fig. 7.

(2) Another argument used in favour of the multiple gain hypothesis is that irregularity is a more specialised adaptation for pollination and is therefore more likely to be derived. However, although irregularity may have originally evolved from the regular state, it need not always be the derived condition. Evolution is not a unidirectional process and specialised characters can be lost as well as gained.

(3) Irregular flowers occur in many different guises. For example, the irregular flowers of mint, pea and *Antirrhinum*, are all structurally very different. This has been taken as evidence in support of the multiple gain hypothesis as it seems to suggest that many independent mechanisms for generating irregularity have evolved. However, it is possible that some of the different types of irregular flowers share the same underlying mechanism for generating asymmetry. The differences may simply reflect the imposition of irregularity on different frameworks of floral development.

It is impossible to resolve these issues purely on the basis of taxonomic information. Isolating genes like *cyc* that are involved in controlling floral symmetry allows a more direct approach to addressing these problems. The molecular basis of both the irregular and the regular condition could be addressed by comparing the activity of *cyc*-like genes in a range of species using a combination of genetic, molecular and transgenic techniques.

Mutants that give regular instead of irregular flowers, termed peloric mutants, have been described in several species in

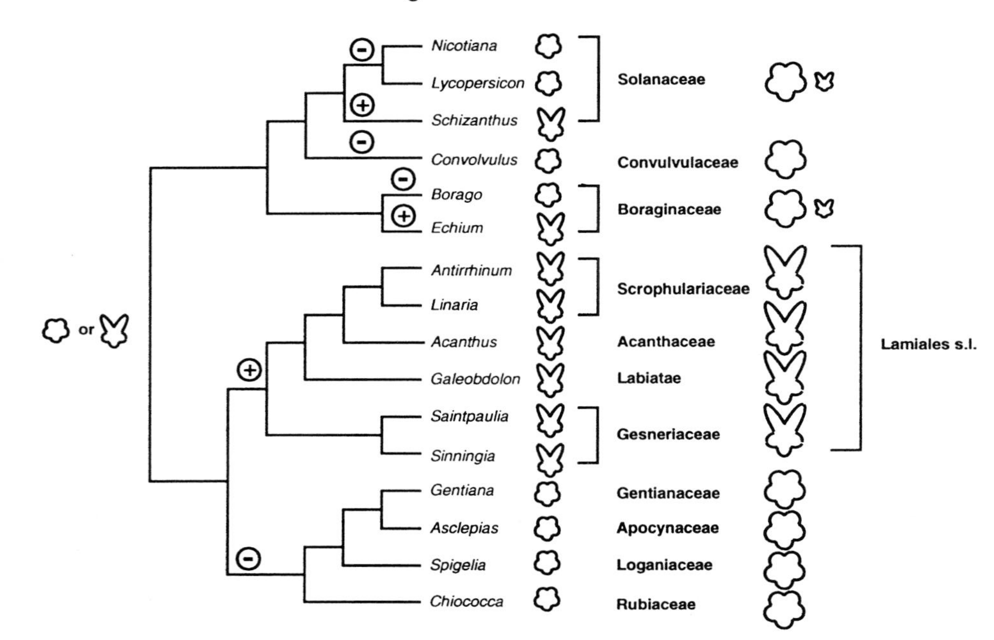

Fig. 7. Phylogenetic tree showing relationships between species and families of a portion of the subclass Asteridae. Examples of species and families having regular flowers are indicated by symmetrical symbols whereas those with irregular flowers are indicated by asymmetrical symbols. In families having a mixture of species with irregular and regular flowers, both symbols are shown, the larger symbol indicating the predominant condition within the family. If the ancestor of the group was regular, irregularity was derived independently at least 3 times in the group (indicated by +). If the ancestor of the group was irregular, this condition was lost at least 4 times in the group (indicated by –).

addition to *Antirrhinum*. The first peloric mutant was described by Linnaeus in *Linaria vulgaris* (toadflax), a close relative of *Antirrhinum* (Fig. 6B, Linnaeus, 1749). As with *cyc*, this mutant confers a ventralised phenotype, all petals resemble the lowest petal of the wild type, which is easily distinguished in *Linaria* by the presence of a spur. Peloric mutants with ventralised phenotypes have also been described for *Saintpaulia* (African violets, Fig. 6C) and *Sinningia* (gloxinias, Fig. 6D), both members of the family Gesneriaceae (Fig. 7). Although genetic analysis of peloric mutants has not yet been carried out in species other than *Antirrhinum*, it is tempting to speculate that they reflect alterations in *cyc*-like genes. This can be tested by analysing the structure and expression of *cyc* homologues in these mutants.

All of these examples of peloric mutants are in species that are quite closely related. According to the phylogeny presented in Fig. 7 these species fall within a monophyletic group (the Lamiales s.l.; Olmstead et al., 1993) that includes only species with irregular flowers. It is likely that the common ancestor of this group had irregular flowers that depended on *cyc*-like gene activity. However, it is unclear whether irregularity in species that lie outside the Lamiales is also *cyc*-dependent. According to the multiple gain hypothesis, the mechanism for establishing irregularity might be expected to be different outside this group. The alternative hypothesis, that irregularity is more ancient, predicts that irregularity in some species outside the Lamiales should also be *cyc*-dependent. One way to distinguish between these possibilities is to determine the role of *cyc* homologues in species with irregular flowers, such as *Schizanthus* (butterfly flower), *Echium* (Fig. 7) or even more distantly related species. If *cyc*-like genes are involved in controlling irregularity in these species, it then raises the question of whether *cyc* was recruited once in a common ancestor of all these species or whether it was recruited independently several times. This might be resolved by studying the role, if any, of *cyc*-like genes in species with regular flowers and how this compares with its role in irregular species.

COEVOLUTION OF INFLORESCENCE STRUCTURE AND FLORAL SYMMETRY

There is a strong correlation between floral asymmetry and inflorescence architecture: irregular flowers are commonly associated with indeterminate racemose inflorescences whereas regular flowers can occur in both racemose and cymose inflorescences (Stebbins, 1974). This correlation may be a consequence of either selective or developmental constraints. For example, if the adaptive value of the irregular condition depends on the flowers being presented to pollinators on a racemose inflorescence, this would mean that selection was involved. Alternatively, if the genetic mechanisms for generating asymmetry are dependent on the racemose condition, a developmental constraint would be involved.

The *cen* mutant of *Antirrhinum*, which has a determinate inflorescence, provides strong evidence for developmental constraints. The *cen* mutant produces two types of flower: axillary flowers that are irregular and a terminal flower that is regular (Fig. 4). This appears to be a general property of *cen*-like mutants in a wide range of species with irregular flowers, such as *Digitalis* (Scrophulariaceae), *Salvia grandiflora* and *Galeobdolon luteum* (Labiatae), *Delphinium elatum* and *Aconitum variegatum* (Ranunculaceae); in all of these cases the terminal flower produced in the mutants is regular (Peyritsch, 1870, 1872). The terminal flower of *cen* resembles the axillary flowers of *cyc* mutants (Fig. 6), suggesting that *cyc* is not active in the apical meristem. The production of *cyc*-dependent asymmetry seems to require that floral meristems are in axillary positions, and have a particular cellular environment. Axillary meristems are in an asymmetric environment, with the inflorescence apex above them and a bract below. This polarised environment could provide necessary cues for establishing the *cyc* system. By contrast, a terminal flower meristem is in a symmetrical environment and may lack the cues required to activate *cyc* in an asymmetrical manner. Accordingly, species producing only terminal flowers might be unable to make them irregular. However, there are apparent exceptions to this view because some species, such as *Schizanthus* (Fig. 7) have irregular flowers occupying terminal positions. Molecular and phenotypic analysis of these species may further our understanding of the relationship between flower position and symmetry.

Other aspects of the coevolution of inflorescence type and floral symmetry appear to reflect selective constraints. Most members of the Cruciferae have regular flowers borne on extended racemes, as shown for *Arabidopsis* and *Alyssum* (Figs 2 and 5). However, in some members of this family, such as *Iberis*, the inflorescence axis is shorter and produces a broad dome of flowers (corymb, Figs 1h and 8). The overall effect of the corymb is to simulate a single large regular flower. Within the corymb flowers are irregular, the two ventral petals of each flower being much larger than the dorsal ones. This is presumably an adaptation that prevents the petals from overlapping and protruding into the centre of the inflorescence. Thus selection may have played a role in the coevolution of the compressed inflorescence axis and the irregularity of the flowers.

The coevolution of inflorescences and flowers is further illustrated by the mixed inflorescences of some of the Compositae. The inflorescence axis within this group is highly compressed to form a dense cluster of flowers (florets), termed a capitulum (Fig. 1i). In some species, such as daisy or chrysanthemum it consists of two types of florets: ray florets that are highly irregular and occupy the periphery and disc florets that are regular and occupy the central dome of the capitulum (Fig. 8). This is reminiscent of the *cen* phenotype of *Antirrhinum*, where the central terminal flower is regular and the axillary flowers are irregular. In both cases, the activation of genes controlling irregularity seems to be restricted to a peripheral zone around the inflorescence apex. In the case of a daisy, this zone gives rise only to the outermost florets whereas in *Antirrhinum* it produces all of the axillary flowers. Nevertheless, the production of inflorescences with a mixture of regular and irregular flowers is not found as a wild-type condition in plants with extended racemes like *Antirrhinum* but is common in species with a capitulum. This may reflect a selective advantage of mixed inflorescences (such as mimicking a large flower) when the flowers are tightly clustered within a capitulum.

The similarity between *cen* and daisy-like inflorescences raises the question of whether *cyc*-like genes are involved in the control of irregularity within the Compositae, a family that is quite distantly related to *Antirrhinum* (e.g. it lies outside the phylogeny shown in Fig. 7). It is possible to test this by looking for, and analysing the expression of, *cyc*-like genes in species from this group. This analysis may be helped by exploiting mutants described in some species of the Compositae that affect floret development. Mutants in species such as Chrysanthemum have been described that result in a capitulum comprising only irregular flowers (Fig. 8). These might be explained in terms of an extension of *cyc*-like gene activity into the central dome. Similarly, mutants that give only regular flowers might result from a loss of *cyc*-like gene activity.

Fig. 8. Photographs of the corymb inflorescence from *Iberis* (top panel) and the capitulum from wild-type *Chrysanthemum* (bottom left) with outer irregular florets (white) and inner regular florets (yellow). A mutant *Chrysanthemum*, with only irregular florets is shown for comparison (bottom right).

These examples illustrate the importance of studying the evolution of symmetry and inflorescence architecture together. The use of comparative molecular genetic analysis of genes such as *flo*, *cen* and *cyc* should start to reveal how this coevolution may have occurred.

We would like to thank Rosemary Carpenter, Desmond Bradley, Pilar Cubas and Richard Olmstead for critical comments and discussions of the manuscript and researchers at Kew Gardens for helpful discussions.

REFERENCES

Carpenter, R. and Coen, E. S. (1990). Floral homeotic mutations produced by transposon-mutagenesis in *Antirrhinum majus*. *Genes Dev.* **4**, 1483-1493.

Coen, E. S., Romero, J. M., Doyle, S., Elliott, R., Murphy, G. and Carpenter, R. (1990). *floricaula*: a homeotic gene required for flower development in *Antirrhinum majus*. *Cell* **63**, 1311-1322.

Linnaeus, C. (1749). *De Peloria. Diss.* Uppsala:Amoenitates Acad.

Olmstead, R. G, Bremer, B., Scott, K. M. and Palmer, J. D. (1993). A parsimony analysis of the Asteridae sensu lato based on *rbc*L sequences. *Ann. Missouri Bot. Gard.* **80**, 700-722.

Peyritsch, J. (1870). Uber pelorienbildungen. *Sitzunger Akad. Wiss* **62**, 1-27.

Peyritsch, J. (1872). Uber Pelorienbildungen. *Sitzunger Akad. Wiss* **66**, 1-35.

Schultz, E. A. and Haughn, G. W. (1991). LEAFY a homeotic gene that regulates inflorescence development in *Arabidopsis*. *Plant Cell* **3**, 771-781.

Stebbins, G. L. (1974). *Flowering Plants, Evolution above the Species Level.* MA: Harvard University Press.

Stubbe, H. (1963). Mutanten der Kulturtomate *Lycopersicon esculentum* Miller IV. *Die Kulturpflanze, II*, 603-644.

Stubbe, H. (1966). *Genetik und Zytologie von Antirrhinum L. sect. Antirrhinum Veb.* Jena: Gustav Fischer.

Weberling, F. (1989). *Morphology of Flowers and Inflorescences.* Cambridge: Cambridge University Press.

Weigel, D., Alvarez, J., Smyth, D. R., Yanofsky, M. F. and Meyerowitz, E. M. (1992). LEAFY controls floral meristem identity in *Arabidopsis*. *Cell* **69**, 843-859.

Development 1994 Supplement,117-124 (1994)
Printed in Great Britain © The Company of Biologists Limited 1994

The evolution of vertebrate gastrulation

E. M. De Robertis*, A. Fainsod, L. K. Gont and H. Steinbeisser

Molecular Biology Institute, Department of Biological Chemistry, University of California, Los Angeles, CA 90024-1737, USA

*Author for correspondence

SUMMARY

The availability of molecular markers now permits the analysis of the common elements of vertebrate gastrulation. While gastrulation appears to be very diverse in the vertebrates, by analyzing a head-organizer marker, *goosecoid*, and a marker common to all forming mesoderm, *Brachyury*, we attempt to identify homologous structures and equivalent stages in *Xenopus*, zebrafish, chick and mouse gastrulation. Using a tail-organizer marker, *Xnot-2*, we also discuss how the late stages of gastrulation lead to the formation of the postanal tail, a structure characteristic of the chordates.

Key words: Spemann organizer, tailbud, *goosecoid, Brachyury, Xnot-2*, gastrulation

INTRODUCTION

Evolutionary and embryological studies have been intertwined throughout their history. The common elements of body plans can be visualized best at mid embryogenesis, the phylotypic stage, as noted initially for the vertebrates by von Baer (1828; Gould, 1977). At an earlier stage, gastrulation, the process by which morphogenetic movements of cell layers produce an embryo consisting of endoderm, mesoderm and ectoderm, appears to be very different in the various vertebrate classes.

In the zebrafish the most striking gastrulation movement is epiboly, in which cells of the embryo proper envelop a spherical yolk mass. In *Xenopus*, which has holoblastic cleavage, the endomesoderm invaginates through a circular blastopore. In the chick the main gastrulation movement is the invagination of the mesoderm and endoderm through a linear primitive streak. In the mouse the gastrula has the shape of a cup, whereas in the human, gastrulation takes place in a flat epithelial epiblast layer and is very similar to that of chick. Mammalian gastrulation is thought to have evolved from ovo-viviparous ancestors that lost the yolk after nutrients could be obtained from the mother (Kerr, 1919), thus explaining the similarities observed among the amniotes.

Because the outcome of gastrulation in all the animals is a very similar body plan at mid embryogenesis, it is reasonable to assume that the underlying mechanisms are similar in all vertebrates. The availability of molecular markers now permits the identification of the common elements in vertebrate gastrulation.

In this paper we will compare the distribution of two gene markers, *goosecoid* and *Brachyury*, in *Xenopus*, zebrafish, chick and mouse embryos, which shed light on the evolution of the organizer and of mesoderm precursor cells. In addition we will examine the mechanism of development of the postanal tail, which arises as a continuation of gastrulation in *Xenopus*, and perhaps in other vertebrates as well.

ORGANIZER-SPECIFIC GENE MARKERS

An experiment that has had a permanent influence in the way we think about embryos was carried out by Spemann and Mangold in 1924. When the region at which involution starts in the amphibian gastrula, the dorsal lip of the blastopore, was transplanted to the opposite (ventral) side of a host embryo, a twinned embryo resulted. Because the transplanted tissue recruited cells from the host into the secondary axis, this region was called the organizer (Spemann, 1938; Hamburger, 1988). The organizer was able to induce neural tissue and segmented mesoderm such as somites and pronephros, while the graft itself differentiates predominantly into prechordal plate, notochord and part of the tailbud (Hamburger, 1988; Gont et al., 1993).

One of the recent advances has been the isolation of genes that mark the organizer. These include the genes for transcription factors such as *goosecoid* (Blumberg et al., 1991; Cho et al., 1991), *Xlim-1* (Taira et al., 1992), *FKHD-1* (Dirksen and Jamrich, 1992) and *Xnot* (von Dassow et al., 1993; Gont et al., 1993) as well as those for the secreted protein *noggin* (Smith and Harland, 1992) and the membrane bound molecule *integrin* $\alpha 3$ (Whittaker and DeSimone, 1993).

The homeobox-containing gene *goosecoid* has been studied extensively and its expression pattern in *Xenopus*, zebrafish, chick and mouse is reviewed below. From a functional point of view, ectopic expression of *goosecoid* in ventral blastomeres of *Xenopus* mimics many of the functions of organizer grafts, such as formation of secondary axes (Cho et al., 1991; Steinbeisser et al., 1993), recruitment of neighboring uninjected cells into new cell fates, and triggering of cell migration in the dorsoanterior direction (Niehrs et al., 1993). It also causes dose-dependent formation of dorsal tissues (Niehrs et al., 1994). While *goosecoid* has been studied most intensively, the other dorsal markers are also expected to play roles in the organizer phenomenon. In the case of *noggin* it has been shown

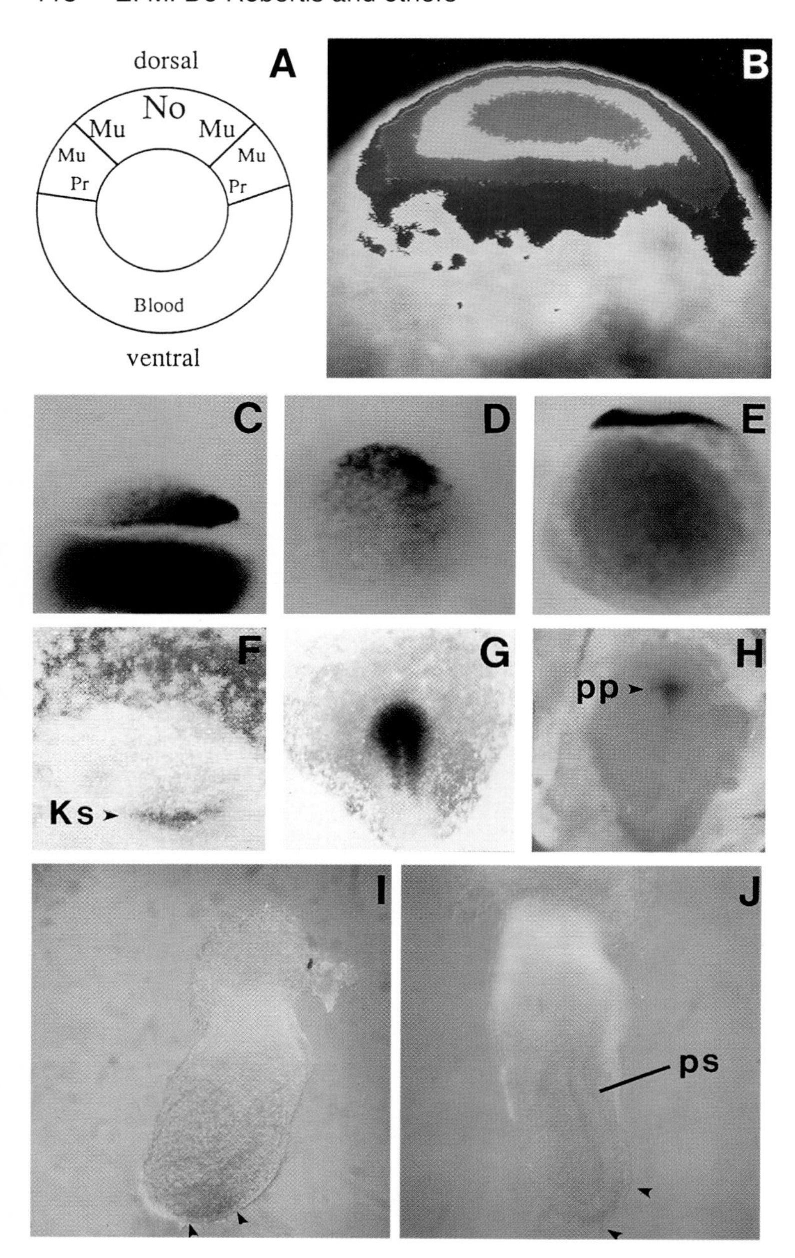

Fig. 1. Expression of *goosecoid* mRNA in *Xenopus*, zebrafish, chick and mouse gastrulae. (A) Specification map of the early *Xenopus* gastrula as determined by explantation experiments (Dale and Slack, 1987). No, Notochord; Mu, Muscle; Pr, Pronephros. (B) Graded distribution of *Xenopus goosecoid* in the marginal zone of the stage 10 dorsal marginal zone. Red is maximal staining in this image analysis of a whole-mount in situ hybridized embryo. (C) Zebrafish blastula (3.5 hours) showing graded distribution of *goosecoid* in the cells overlying the yolk mass, side view. (D) Same embryo as C, showing the graded distribution in top view. (E) Zebrafish embryo showing *goosecoid* expression in the prechordal plate at the early tailbud stage. (F) Chick embryo showing *goosecoid* expression in Koller's sickle (Ks) and in cells immediately anterior to it. At this stage (XII) the forming hypoblast has covered 50% of the area pellucida, which remains transparent at the top of the panel. (G) Maximal expression of chick *goosecoid* at stage 3^+. (H) Expression in cells that have left the node to form the prechordal plate (pp) at stage 4^+. (I) Mouse gastrula showing *goosecoid* expression (arrowheads) in the anterior region of the early primitive streak. (J) *goosecoid* expression (arrowheads) in a mouse gastrula at the anterior of the primitive streak (ps) at the stage in which *goosecoid* is down-regulated. Panels are reproduced, with permission, from the following publications. (A, B) Niehrs et al. (1994) *Science* **263**, 817-820; (C,D,E) Schulte-Merker et al. (1994) *Development* **120**, 843-852; (F,G,H) Izpisúa-Belmonte et al. (1993) *Cell* **74**, 645-659; (I,J) Blum et al. (1992) *Cell* **69**, 1097-1106 .

that its ectopic expression produces complete axial rescue of *Xenopus* embryos ventralized by UV light (Smith and Harland, 1992). The analysis of many of the other genes is in progress, particularly by targeted inactivation of their mouse homologues, and one can expect that much information will be forthcoming. In this paper we concentrate on *goosecoid*, which is a marker of the head organizer region.

The *Brachyury* gene has also been intensively studied (reviewed by Beddington and Smith, 1993). Named after its short tailed phenotype in heterozygous mice, it was one of the first embryonic lethal mutations to be identified (Dobrovolskaia-Zavadskaia, 1927; Gluecksohn-Schoenheimer, 1938). The expression pattern of *Brachyury* has been studied in mouse (Herrmann, 1991), *Xenopus* (Smith et al., 1991) and zebrafish (Schulte-Merker et al., 1992). *Brachyury* provides a marker for both ventral and dorsal mesoderm. At mid gastrula *Brachyury* marks the notochord and uncommitted mesoderm and, together with *goosecoid*, which marks the prechordal plate at this stage, permits one to identify the main territories of the gastrula. At later stages *Brachyury* is expressed in the tailbud and, together with *Xnot-2* (Gont et al., 1993), permits the analysis of the developing tail.

goosecoid AND *Brachyury* IN *XENOPUS* GASTRULATION

goosecoid mRNA is present at low levels in the unfertilized egg (De Robertis et al., 1992; Lemaire and Gurdon, 1994), but shortly after mid blastula the levels of its transcripts increase, reaching a maximum just before the start of gastrulation at stage 10. At this stage *goosecoid* mRNA is localized to the dorsal side of the marginal zone, forming a gradient over tissue that at this stage is already fated to form dorsal mesoderm (Fig. 1A,B). By culturing fragments of marginal zone (Dale and Slack, 1987) it has been shown that dorsal mesodermal tissues (notochord and muscle) are formed by the organizer (dorsal lip) region, that intermediate regions give rise to pronephros and muscle, and that the rest of the marginal zone produces ventral mesoderm (mesothelium, mesenchyme and blood), as indicated in Fig. 1A. At later stages the more lateral regions of the marginal zone become induced to become muscle under the influence of a 'horizontal' signal emanating from the organizer (reviewed by Slack, 1991), but at the early gastrula only the *goosecoid*-expressing region is specified to form dorsal mesoderm. This early gradient of expression is thought to be involved in mesodermal patterning, because two-fold increments in the amounts of microinjected synthetic *goosecoid* mRNA can trigger the expression of increasingly dorsal molecular markers, forming several sharp thresholds in explanted ventral marginal zones (Niehrs et al., 1994).

Once gastrulation starts, *goosecoid* is down-regulated, and the cells that continue to express it leave the dorsal lip with the involuting endomesoderm (Fig. 2A), eventually becoming restricted to the small group of cells that form the prechordal plate or head mesoderm (Steinbeisser and De Robertis, 1993), as shown in Fig. 2B. Not all genes expressed in the dorsal lip follow this pattern. For example, the homeobox gene *Xnot-2* is also expressed dorsally, but its expression stays localized in the dorsal lip, and at neurula stages is expressed in the notochord and eventually becomes restricted to the tail of the embryo (Fig. 2C,D).

At the early gastrula stage *Xenopus Brachyury (Xbra)* is expressed in a band or ring encompassing the entire marginal zone (Fig. 2E). In other words, *Xbra* is expressed in the progenitors of both dorsal and ventral mesoderm (Smith et al., 1991). Once cells involute through the circular blastopore, they lose *Xbra* expression (except in the notochord which is transiently positive), so that at later stages the only region expressing *Xbra* is the tailbud, as shown in Fig. 2F. Two regions of the tailbud express *Brachyury*, one of dorsal origin (the chordoneural hinge, which also expresses *Xnot-2*) and the other deriving from the lateral marginal zone (the posterior wall). These two populations are separated by the neurenteric canal (indicated by an arrowhead in Fig. 2F). The role of these two regions in late gastrulation (Gont et al., 1993) will be discussed below.

We conclude that in the early *Xenopus* gastrula *goosecoid* marks most cells specified to become dorsal mesoderm, including notochord and muscle, but that as gastrulation proceeds only those cells forming the prechordal plate continue to express *goosecoid*. *Xbra* is expressed in all mesoderm progenitors at early gastrula, but at later stages becomes restricted to posterior notochord and the tailbud.

goosecoid IN THE ZEBRAFISH

In zebrafish *goosecoid* expression starts at the blastula stage, where it forms a striking dorsoventral gradient (Schulte-

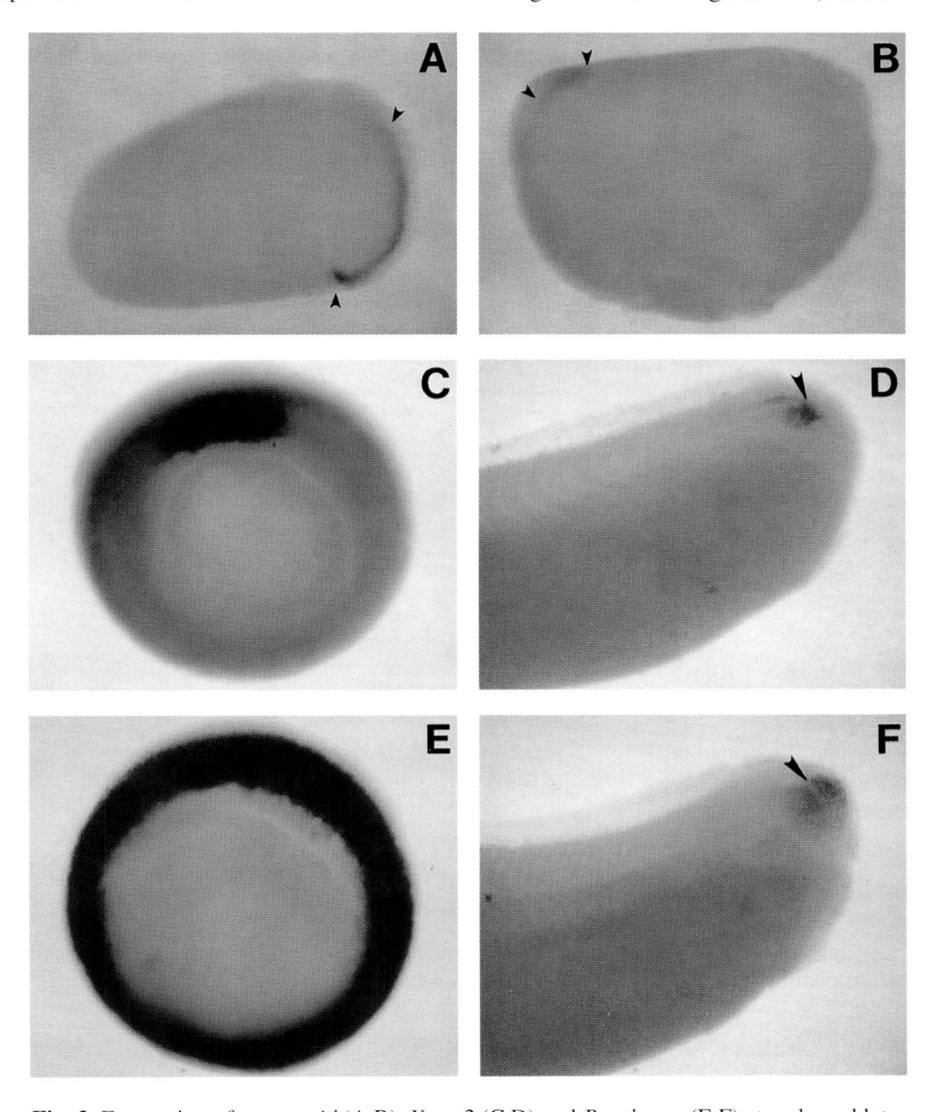

Fig. 2. Expression of *goosecoid* (A,B), *Xnot-2* (C,D) and *Brachyury* (E,F) at early and late stages. (A) *goosecoid* at mid-gastrula (stage 11) has left the dorsal lip and involutes with the anterior endomesoderm (arrowheads). (B) At the neurula stage, expression can be seen in the prechordal plate or mesoderm of the head. (C) *Xnot-2* at early gastrula, note expression in the dorsal lip. (D) *Xnot-2* at the tailbud stage, note expression at the chordoneural hinge (arrowhead). The floor plate is seen as a thin trail of expression. (E) *Brachyury* at early gastrula, the entire marginal zone is marked, forming a ring. (F) *Brachyury* at the tailbud stage, two areas of expression can be seen. The arrowhead indicates the neurenteric canal, which separates the chordoneural hinge from the posterior wall area of expression.

Merker et al., 1994), as shown in Fig. 1C,D. These cells then cover the yolk through the movement of epiboly, which is accompanied by convergence towards the dorsal midline (forming the 'embryonic shield' on the dorsal side). Zebrafish *goosecoid* transcripts are maximal at 50% epiboly (Stachel et al., 1993; Schulte-Merker et al., 1994), the stage at which involution of the endomesoderm starts; expression is then down-regulated and follows the involuting cells in the anterior-most regions. The notochord is negative at all times, and by the end of gastrulation expression can be seen in the endomesoderm of the prechordal plate (Schulte-Merker et al., 1994), as shown in Fig. 1E. Patches of expression can be seen in the CNS and branchial arches at later stages (Schulte-Merker et al., 1994).

Zebrafish *Brachyury* (also called *notail* due to the phenotype of its mutant) has been studied in considerable detail, as an excellent antibody is available (Schulte-Merker et al., 1992). Expression starts on the dorsal side and forms a ring along the entire marginal zone, as in *Xenopus*. Interestingly, double staining experiments using *goosecoid* RNA probes and Brachyury antibodies have shown that both genes are expressed in the same cells of the dorsal side before involution starts. As gastrulation proceeds, the *goosecoid* signal follows the prechordal plate, which becomes negative for *Brachyury*. *Brachyury* is found in the notochord and tailbud (Schulte-Merker et al., 1992, 1994). Although both genes are expressed initially in the same cells, in *Brachyury (notail)* mutants the expression of *goosecoid* is normal (Schulte-Merker et al., 1994).

We conclude that despite the predominance of epiboly movements in zebrafish gastrulation, the expression of *goosecoid* is not unlike that of its *Xenopus* homologue: it starts as a dorso-ventral gradient, has maximal expression at the start of involution of the endomesoderm and is then down-regulated, following the prechordal plate.

goosecoid IN THE CHICK

The chick embryo offers many advantages to the experimentalist, one of which is that the embryo develops from a flat epithelial sheet, the epiblast, from which all germ layers derive. This facilitates the study of pre-gastrulation stages. Chick *goosecoid* (Izpisúa-Belmonte et al., 1993) is first detectable in the unincubated egg, which as it is laid by the hen already contains several thousand cells. Expression starts in a thickening of the posterior marginal zone called Koller's sickle, where it is confined to a group of cells located in a middle layer of cells between epiblast and the forming hypoblast. The existence of this cell population had been overlooked, although Koller's sickle had been noted almost 100 years ago, but was revealed by the *goosecoid* marker. It is a very interesting group of cells, for lineage tracing with the hydrophobic dye DiI showed that they migrate anteriorly to the *goosecoid*-expressing region of Hensen's node, indicating that this group of cells constitutes the earliest mesendodermal cells of the chick gastrula (Izpisúa-Belmonte et al., 1993). The Koller sickle region can also induce secondary axes when grafted. The main conclusion from these studies was that in the chick gastrulation starts much earlier than previously thought, in the unincubated egg (Izpisúa-Belmonte et al., 1993). Fig. 1F shows *goosecoid* expression in Koller's sickle in an embryo that has been incubated for a few hours (stage XII, when the forming sheet of hypoblast covers 50% of the area pellucida).

As development continues, the primitive streak forms in the posterior end. As the extension of the streak progresses in the anterior direction, *goosecoid* is expressed at the tip of the primitive streak. Transcripts become more abundant as the streak elongates, and by stage 3^+, when the streak develops a groove but has not yet reached its maximal extension, *goosecoid* reaches its maximal expression (Fig. 1G). Transcripts are maximal anteriorly and decrease gradually, like the tail of a comet, towards the posterior primitive streak (Izpisúa-Belmonte et al., 1993). This stage may be considered homologous to stage 10 (beginning of dorsal lip formation) in *Xenopus* (Fig. 1B). This is also the stage in which the anterior end of the primitive streak, the young Hensen's node, has its maximal inductive activity. Hensen's node is classically considered to be the equivalent of the amphibian organizer (Waddington, 1933). Once the primitive streak reaches its full extension at stage 4, *goosecoid* is down-regulated, and this correlates with a decrease in inducing activity by the node (Storey et al., 1992). By stage 4^+, *goosecoid* RNA is found in the cells that fan out from the node ingressing to form the endomesoderm of the head, as shown in Fig. 1H. At later stages (not shown) *goosecoid* expression is found in foregut endoderm, prechordal plate, and ventral diencephalon. The expression pattern of *Brachyury* in the chick has not yet been reported, but from what is known from *Xenopus* and mouse one would expect this gene to be active throughout the length of the primitive streak, where the forming mesoderm is located.

We conclude that while two of the regions in which chick *goosecoid* is found, Hensen's node and prechordal plate, have clear homologues in *Xenopus* and zebrafish, the earliest one, Koller's sickle, is more difficult to fit into the picture of a generalized gastrulation mechanism.

goosecoid IN THE MOUSE

The mouse gastrula develops from a cup-shaped epiblast, the egg cylinder. Expression is first detected at the posterior end of the egg cylinder at 6.4 days after fertilization. It is seen as a spot on the side of the egg cylinder on the initial group of cells that leave the epiblast in the initial epithelial-mesenchymal transition that starts formation of the primitive streak (Blum et al., 1992). This early expression may be homologous to that found in Koller's sickle in the chick. Once the primitive streak is formed, *goosecoid* expression is found at its anterior end (Fig. 1I,J). Expression is maximal at day 6.7, also at the anterior of the streak. This early phase of expression is brief, lasting for about 10 hours. Transcripts were not detectable in the prechordal plate or other regions. The whole-mount in situ hybridization signal of mouse *goosecoid* is less intense than in other organisms, perhaps explaining why the signal is lost. After day 10.5 mouse *goosecoid* has a late phase of expression that has been described in some detail (Gaunt et al., 1993). *goosecoid* transcripts are found in limb buds (in the proximal, ventral and anterior region), neural crest of pharyngeal arch 1 (mandibular and maxillary processes) and anterior third of arch 2 (hyoid), the floor of the diencephalon, and other sites (Gaunt et al., 1993). The late expression pattern of *goosecoid* might be similar in other organisms, but systematic studies have not been carried out.

The studies in the mouse suggest that the anterior primitive streak at day 6.7, when *goosecoid* expression is maximal, should be homologous to the early dorsal lip of the *Xenopus* early gastrula. Mouse embryo fragments containing this region have inducing activity when transplanted into *Xenopus* embryos by the Einsteck procedure (Blum et al., 1992). The mouse node, which at 7.5 days can induce trunk structures when grafted to mouse embryos (Beddington, 1994) would be the equivalent to the late (regressing) Hensen's node of the chick or the late dorsal lip in amphibians, which have trunk-tail organizer activities but do not express *goosecoid.*

The *Brachyury* gene was initially isolated from the mouse, and it has been studied in detail. It is expressed first in the entire primitive steak and then in the notochord and tailbud (Herrmann, 1991; Beddington and Smith, 1993). In mouse, expression of *Brachyury* in the notochord is more persistent than in *Xenopus*.

TAIL FORMATION IS A CONTINUATION OF GASTRULATION

It has been observed recently that the formation of the postanal tail results from a continuation of gastrulation in *Xenopus* and is induced by cells originating from the late dorsal lip (Gont et al., 1993). These findings will be summarized here and then discussed in the context of the evolution of vertebrate gastrulation. Three lines of evidence support the view that tail formation is a continuation of gastrulation: (1) studies with the gene markers *Xnot-2* and *Brachyury*, (2) lineage tracing studies and (3) tail-organizer transplantations (Gont et al., 1993).

The gene marker *Xnot-2* is found first in the dorsal lip (Fig. 2C) and then in a part of the tailbud that forms a hinge between the notochord and floor plate (Fig. 2D). *Brachyury* is expressed as a ring in the marginal zone of the early gastrula (Fig. 2E) but is also found in the tailbud at late stages, although not exclusively in the hinge region, but also in cells located more posteriorly (Fig. 2F). The identification of these two cell populations, called the chordoneural hinge and the posterior wall cells, was our first indication that the tailbud is indeed heterogeneous. Both cell populations are separated by the neurenteric canal. Fifty years earlier Pasteels had reached similar conclusions based on careful examination of histological sections (Pasteels, 1943) but we were not aware of this at the time.

Lineage tracing of the late blastopore lip revealed that the chordoneural hinge is derived from the organizer. As shown in Fig. 3A, after the *goosecoid*-expressing cells that will form the prechordal plate leave the organizer, the predominant movement at mid-gastrula is that of involution (driven by convergence and extension), which leads to the formation of the trunk notochord and somites. At the early neurula (stage 13), the ectoderm and mesoderm attach to each other and the two layers undergo posterior extension movements instead of involution (Fig. 3B). It is these posterior movements of the organizer region (driven by continuing cell intercalations in the neural plate and notochord) that cause tail formation. When marks of DiI were placed in the closing blastopore at the slit stage (Fig. 4) and the embryos were left to develop, it was observed that: (1) the dorsal lip becomes the chordoneural

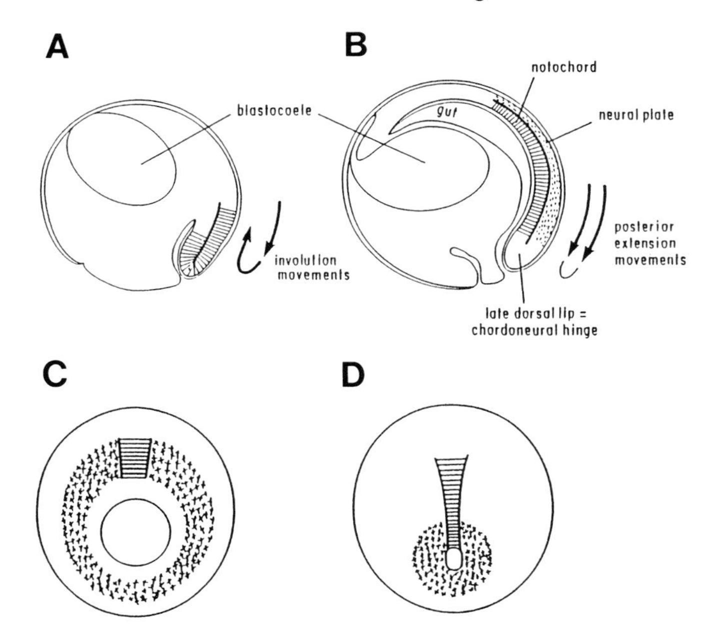

Fig. 3. Diagrams of mid and late gastrulation in *Xenopus*. (A) At the mid gastrula the main movement is involution, which forms the trunk axial structures. (B) At late gastrula/early neurula the main movement is posterior extension, which drives tail formation. Note that the neural plate and notochord are connected at the hinge, which descends from the organizer. (C) Mid-gastrula in vegetal view, the prospective notochord is hatched, the *Xbra*-positive region of the marginal zone is indicated by small crosses. (D) Early neurula, ventral view, the *Xbra*-positive cells from a thick circumblastoporal collar.

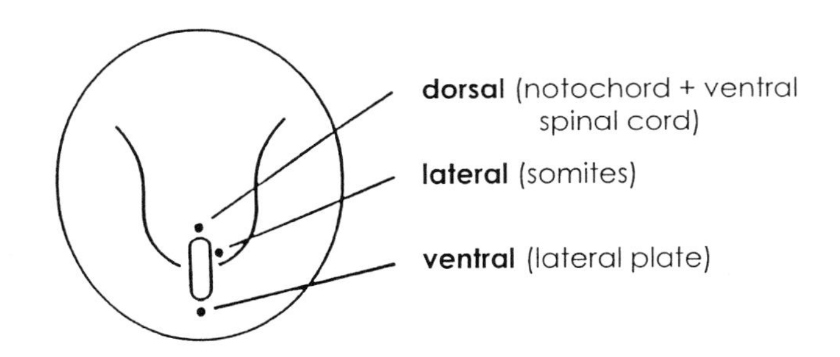

Fig. 4. Fate map of the late blastopore lip. DiI marks were placed in the slit blastopore of an early neurula (note neural folds) as indicated, and embryos allowed to develop to the tailbud stage. Each mark gives rise to different tissues of the tail. The chordoneural hinge, which gives rise to notochord and ventral spinal cord, derives from the late dorsal lip (from experiments of Gont et al., 1993).

hinge and gives rise to tail notochord and ventral spinal chord, (2) the lateral lip gives rise mostly to posterior wall and the tail somites that derive from it and, (3) the ventral lip gives rise to lateral plate mesoderm as well as tissue spanning the region between the anus and the tailbud. The neurenteric canal, which connects the gut and the spinal cord cavity, is formed by the fusion of the lateral lips and the closure of the neural plate so that the lateral lips become the posterior wall (Fig. 4; Gont et al., 1993). As shown in Fig. 5, at tailbud stages the neurenteric canal separates the chordoneural hinge (which expresses *Xnot-*

2 and *Brachyury*) from the posterior wall (which expresses only *Brachyury*).

Transplantation of the chordoneural hinge into host gastrulae by the Einsteck procedure showed that it has potent organizing activity, inducing tails with structures such as somites, spinal cord and fins, while the graft itself gives rise mostly to chordoneural hinge and notochord (Gont et al., 1993).

The main conclusion is that the organizer region retains its activity at tailbud stages and occupies a distinct region of the tailbud. This has implications on the way we think about *Xenopus* gastrulation. For example, the 'horizontal' signal emanating from the organizer that induces somite formation (Slack, 1991) might continue well into tailbud stages. The source of inductive signals would be the chordoneural hinge, which as it develops could recruit uncommitted cells from the *Brachyury*-positive posterior wall into the myotome pathway. In addition, the experiments in *Xenopus* tell us that the gastrulation process continues much longer than previously thought in the tail region (Gont et al., 1993). We will discuss below whether other vertebrates might share a similar mechanism of late gastrulation.

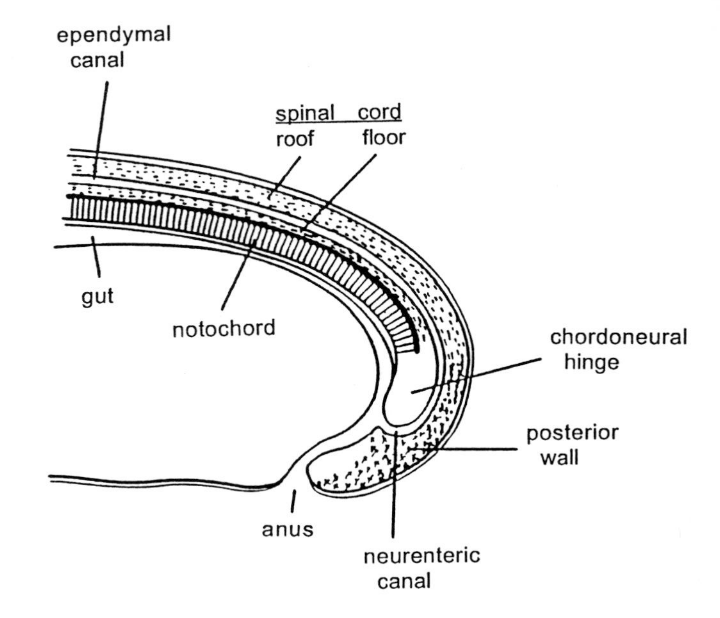

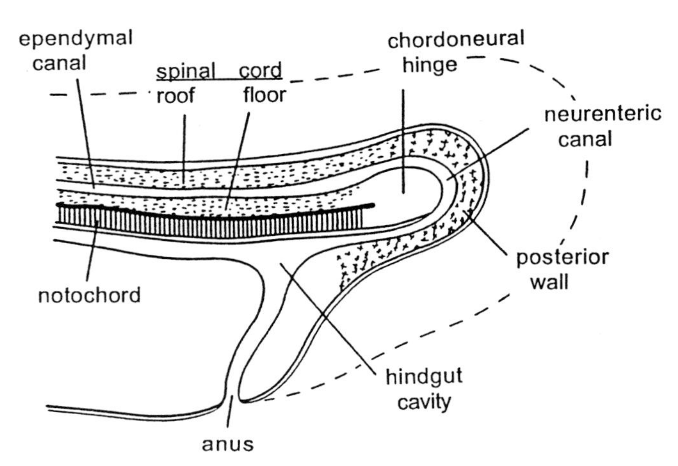

Fig. 5. Anatomy of the *Xenopus* tailbud. (A) Stage 23. (B) Stage 28. The neurenteric canal connects the ependymal and gut cavities. The chordoneural hinge gives rise to notochord, floor plate and roof of the postanal gut. The notochord is hatched, the *Xbra*-positive cells of the posterior wall are indicated by small crosses, and the spinal chord is stippled. All these structures form as a consequence of late gastrulation movements.

COMPARATIVE MOLECULAR ANATOMY OF VERTEBRATE GASTRULATION

Despite their considerable anatomical differences, the common elements of gastrulation in the various vertebrate classes are highlighted by analyzing the expression of two markers, *goosecoid* and *Brachyury*. The organizer is marked by *goosecoid*. The period of highest *goosecoid* expression corresponds to that of maximal organizer activity in *Xenopus* and chick, the two organisms in which detailed transplantation analyses have been possible.

The future mesoderm, both dorsal and ventral, is marked by *Brachyury*. In *Xenopus* and zebrafish, this can be seen as a ring encircling the spherical embryo, the marginal zone. In the mouse embryo, it is expressed in the entire primitive streak. Thus, the marginal zone ring has become flattened, forming two parallel accumulations of cells in the primitive streak in the amniote embryo. At the anterior end of the early primitive streak lies the organizer, which expresses *goosecoid*. We conclude that the primitive streak is homologous to the circumblastoporal marginal zone present in *Xenopus* and zebrafish.

It might be worthwhile to attempt to identify the equivalent stages of gastrulation from the expression patterns. The maximal expression of *goosecoid* correlates with maximal inducing activity, tested by grafting, in *Xenopus* and chick. These stages may therefore be considered truly homologous. Furthermore, maximal expression of *goosecoid* is followed by the migration of cells that form the endomesoderm of the head in *Xenopus*, zebrafish, chick and mouse. These anterior cells continue to express *goosecoid*, while the gene is down-regulated in other cells that will become a number of other dorsal tissues. The region of maximal expression corresponds to the dorsal marginal zone in *Xenopus*, the embryonic shield in zebrafish, the early Hensen's node in chick and the anterior primitive streak in the mouse. Maximal expression of *goosecoid* is achieved just before dorsal lip formation in *Xenopus* (stage 10, Fig. 1B), before the primitive streak is fully extended in the chick (stage 3^+, Fig. 1G), at 50% epiboly in zebrafish (Stachel et al., 1993; Schulte-Merker et al., 1994), and at day 6.7 in the mouse (Blum et al., 1992). We propose that these stages are equivalent in the vertebrate gastrula.

Xenopus embryologists usually consider the beginning of gastrulation as the start of involution of cells through the dorsal lip at stage 10. However, if we consider this stage to be homologous to the chick embryo as it reaches full extension of the primitive streak (stage 3^+ to 4), it is evident that in the chick gastrulation starts much earlier. The extensive morphogenesis involved in the formation of the primitive streak, starting from Koller's sickle, is being ignored in most current *Xenopus* work. However, there are indications that also in *Xenopus*, pregastrulation movements are much more extensive than commonly thought (Hausen and Riebsell, 1991; Bauer et al., 1994). The challenge for the future lies in discovering whether homologous events exist at the very early stages of vertebrate gastrulation.

THE EARLIEST ORGANIZER CELLS

In *Xenopus* inducing activity is first found in dorsal-vegetal blastomeres, in a region designated the Nieuwkoop center (Gerhart et al., 1989). As early as the 32- to 64-cell stage the vegetal cells are thought to emit a signal that induces organizer activity in the overlying marginal zone. Nieuwkoop center cells do not themselves participate in the axis, but remain as yolky endodermal cells (Smith and Harland, 1991). The next step for those working in zebrafish, chick or mouse would be to identify the source of the equivalent Nieuwkoop center signal in these animals, if one exists. In zebrafish we may assume that the early gradient of *goosecoid* transcripts that is observed at blastoderm (Fig. 1C,D) reflects the intensity of the inductive signal from the Nieuwkoop center equivalent. This signal could emanate for example from the yolk region.

In chick, the *goosecoid*-positive cells of Koller's sickle have been shown by fate mapping to contribute to the main body axis, in particular Hensen's node and its derivatives (Izpisúa-Belmonte et al., 1993). Therefore these cells are not homologous to the Nieuwkoop center, whose cells can induce but do not themselves form part of the axis. If a Nieuwkoop center exists in the chick, one would venture that it should signal before the chick egg is laid and that it would be located in the posterior marginal zone, in the vicinity of the region where Koller's sickle is formed. Another possibility is that Koller sickle cells may not require a separate localized inducing center.

In *Xenopus* there is recent evidence suggesting that the redistribution of egg cytoplasmic determinants may be all that is required for the activation of *goosecoid*. Lemaire and Gurdon (1994) marked eggs on the dorsal or ventral side with vital dyes and then cultured dissociated blastomeres. Even in the absence of any cell-cell interactions, *goosecoid* was turned on in dorsal blastomeres and *Xwnt-8*, a ventral marker, was activated in ventral blastomeres. One possibility worth exploring is that perhaps the inductive activity found in vegetal-dorsal blastomeres (Nieuwkoop center) is just a manifestation of a diffuse organizer present from the earliest stages in the dorsal side of the *Xenopus* embryo. Perhaps the vegetal cells do not participate in the axis proper because they are too bulky to migrate, while smaller more dorsal cells might be able to migrate and still have similar inductive properties. While the possibilities mentioned in the last two paragraphs are entirely hypothetical, they were raised here to illustrate how a comparative approach to the Nieuwkoop center/Koller's sickle problem helps us think about how the vertebrate embryo is formed.

COMMON MECHANISMS IN LATE GASTRULATION?

Unlike the studies on *goosecoid* and *Brachyury*, the experiments reviewed here on the development of the tail apply only to *Xenopus*. Given the conservation of developmental mechanisms, we would propose that the mechanism of tail formation will turn out to be very similar in all vertebrates. However, it could be argued that what happens in *Xenopus* might be the exception, because the tadpole is under strong selective pressure to develop a tail rapidly in order to swim for survival. The currently prevailing view states that the tailbud of a chick or a mouse is a proliferation blastema of undifferentiated cells, from which the various tissues of the tail arise (Griffith et al., 1992).

The comparative approach could be very productive in advancing knowledge in this case. If the *Xenopus* tail formation mechanism (Gont et al., 1993) operates in amniotes, several strong predictions can be made. For example, a population of cells derived from the organizer should regress through the primitive streak (the region equivalent to the lateral blastopore lip and the posterior wall of *Xenopus*). This population of cells would be homologous to the chordoneural hinge of *Xenopus* and should be endowed with potent organizing properties, inducing somites and other tail structures as it comes in contact with pluripotent primitive streak cells. Some observations on mouse somite formation (Tam and Tan, 1992) may be considered to be at least consistent with the mechanism proposed here. Gene markers will be useful, but ultimately definitive answers will require the traditional transplantation and lineage tracing tools available to the embryologist.

In the case of chick and mouse, the tail formation mechanism might be masked by the extensive cell proliferation required for the growth of the embryo which, unlike *Xenopus*, greatly increases in size at late gastrulation. Perhaps the largest obstacle is the lack of a visible neurenteric canal, which serves as a useful landmark, in chick and mouse. However, this canal must be an ancestral chordate characteristic, for it is present in *Amphioxus*, shark, turtle and human embryos (see Gont et al., 1993).

Thus in the immediate future the task will be to establish whether a common mechanism exists in the development of the postanal tail, which is a structure found only in the chordates. By focusing attention on this problem it is hoped that more will be learnt about the mechanisms that control late gastrulation in the vertebrates.

CONSERVATION VERSUS VARIATION

We have emphasized throughout this paper how developmental mechanisms that control gastrulation in the vertebrates appear to be conserved when examined through the narrow prism of gene markers. One might argue that a productive area of research might be to extend these comparisons by studying homologues of organizer-specific genes in invertebrates, e.g., hemichordates, echinoderms and even in protostomes. Such studies might help answer the question of whether the organizer phenomenon is restricted only to the vertebrates. On the other hand, one could also argue that understanding how differences in body shape arise will be of greater importance in evolutionary studies. After all, mutations in the genes that regulate development may provide one of the main sources of the variation upon which natural selection acts in metazoans (reviewed by De Robertis, 1994).

The choice of emphasizing either similarities or differences can be seen in articles throughout this volume. But this is not something new in evolutionary biology. In 1830 a famous debate took place in the French Academy between Cuvier and Geoffroy Saint-Hilaire (see Appel, 1987). The latter defended the principle of unity of composition among animals. Those who today choose to emphasize homologies marvel at the unity of composition of the mechanisms of development. But others marveled before. One of the observers of the 1830 debate summarized his views in the following way:

"Il n'y a qu'un animal. Le créateur ne s'est servi que d'un seul et même patron pour tous les êtres organisés."

"There is but one animal. The Creator used only a single pattern for all organisms."

Honoré de Balzac
in "La Comédie Humaine", 1842

We are indebted to Stephan Schulte-Merker, Martin Blum, Juan Carlos Izpisúa-Belmonte and Claudio Stern for their contributions to the study of expression patterns of *goosecoid*. H. S. was supported by a DFG postdoctoral fellowship and A. F. was recipient of an American Cancer Society International Cancer Research Fellowship. This work was funded by grants of the NIH (HD21502-09) and HFSPO.

REFERENCES

Appel, T. A. (1987). *The Cuvier-Geoffroy Debate.* Oxford: Oxford University Press.

Bauer, D. V., Huang, S. and Moody, S. A. (1994). The cleavage stage origin of Spemann's Organizer: analysis of the movements of blastomere clones before and during gastrulation in *Xenopus. Development* **120**, 1179-1189.

Beddington, R. S. P. and Smith, J.C. (1993). The control of vertebrate gastrulation: inducing signals and responding genes. *Curr. Opin. Genet. Dev.* **3**, 655-600.

Beddington, R.S.P. (1994). Induction of a second neural axis by the mouse node. *Development* **120**, 613-620.

Blum, M., Gaunt, S. J., Cho, K. W. Y., Steinbeisser, H., Blumberg, B., Bittner, D. and De Robertis, E. M. (1992). Gastrulation in the mouse: The role of the homeobox gene *goosecoid*. *Cell* **69**, 1097-1106.

Blumberg, B., Wright, C. V. E., De Robertis, E. M. and Cho, K. W. Y. (1991). Organizer-specific homeobox genes in *Xenopus laevis* embryos. *Science* **253**, 194-196.

Cho, K. W. Y., Blumberg, B., Steinbeisser, H. and De Robertis, E. M. (1991). Molecular nature of Spemann's organizer: the role of the *Xenopus* homeobox gene *goosecoid*. *Cell* **67**, 1111-1120.

Dale, L. and Slack, J. M. W. (1987). Regional specification within the mesoderm of early embryos of *Xenopus laevis*. *Development* **100**, 279-295.

De Robertis, E. M., Blum, M., Niehrs, C. and Steinbeisser, H. (1992). *goosecoid* and the organizer. *Development* **Supplement**, 167-171.

De Robertis, E. M. (1994). The homeobox in cell differentiation and evolution. In *Guidebook of Homeobox Genes*. (ed. D. Duboule), Oxford, IRL press.

Dirksen, M. L. and Jamrich, M. (1992). A novel, activin-inducible, blastopore lip-specific gene of *Xenopus laevis* contains a *fork head* DNA-binding domain. *Genes Dev.* **6**, 599-608.

Dobrovolskaia-Zavadskaia, N. (1927). Sur la mortification spontanee de la queue chez la souris nouveau-nee et sur l'existence d'un caractere heriditaire "non-viable". *C. R. Soc. Biol.* **97**, 114-116.

Gaunt, S. J., Blum, M. and De Robertis, E. M. (1993). Expression of the mouse *goosecoid* gene during mid-embryogenesis may mark mesenchymal cell lineages in the developing head, limbs and body wall. *Development* **117**, 769-778.

Gerhart, J., Danilchik, M., Doniach, T., Roberts, S., Rowning, B. and Stewart, R. (1989). Cortical rotation of the *Xenopus* egg: consequences for the anteroposterior pattern of embryonic dorsal development. *Development* **107 Supplement**, 37-51.

Gluecksohn-Schoenheimer, S. (1938). The development of two tailless mutants in the house mouse. *Genetics* **23**, 573-584.

Gont, L. K., Steinbeisser, H., Blumberg, B. and De Robertis, E. M. (1993). Tail formation as a continuation of gastrulation: the multiple cell populations of the *Xenopus* tailbud derive from the late blastopore lip. *Development* **119**, 991-1004.

Gould, S. J. (1977). *Ontogeny and Phylogeny.* Cambridge, Mass: Harvard University Press.

Griffith, C. M., Wiley, M. J., and Sanders, E. J. (1992). The vertebrate tail bud: three germ layers from one tissue. *Anat. Embryol.* **185**, 101-113.

Hausen, P. and Riebesell, M. (1991). The early development of *Xenopus laevis*. Berlin: Springer.

Hamburger, V. (1988). *The Heritage of Experimental Embryology: Hans Spemann and the Organizer.* Oxford: Oxford University press.

Herrmann, B. G. (1991). Expression pattern of the *Brachyury* gene in whole-mount T^{wis}/T^{wis} mutant embryos. *Development* **113**, 913-917.

Izpisúa-Belmonte, J. C., De Robertis, E. M., Storey, K. G. and Stern, C. D. (1993). The homebox gene *goosecoid* and the origin of organizer cells in the early chick blastoderm. *Cell*, **74**, 645-659.

Kerr, J. G. (1919). *Textbook of Embryology* Vol. II. London: Macmillan and Co.

Lemaire, P. and Gurdon, J. B. (1994). A role for cytoplasmic determinants in mesoderm patterning: cell-autonomous activation of the *goosecoid* and *Xwnt-8* genes along the dorsoventral axis of early *Xenopus* embryos. *Development* **120**, 1191-1199.

Niehrs, C., Keller, R., Cho, K. W. Y. and De Robertis, E. (1993). The homeobox gene *goosecoid* controls cell migration in Xenopus embryos. *Cell* **72**, 491-503.

Niehrs, C., Steinbeisser, H. and De Robertis, E. M. (1994). Mesodermal patterning by a gradient of the vertebrate homeobox gene *goosecoid. Science* **263**, 817-820.

Pasteels, J. (1943). Proliférations et croissance dans la gastrulation et la formation de la queue des Vertébrés. *Archives de Biologie* **54**, 1-51.

Schulte-Merker, S., Ho, R. K., Herrmann, B. G., and Nüsslein-Volhard, C. (1992). The protein product of the zebrafish homologue of the mouse *T* gene is expressed in nuclei of the germ ring and the notochord of the early embryo. *Development* **116**, 1021-1032.

Schulte-Merker, S., Hammerschmidt, M., Beuchle, D., Cho, K. W., De Robertis, E. M. and Nüsslein-Volhard, C. (1994). Expression of zebrafish *goosecoid* and *no tail* gene products in wild-type and mutant *no tail* embryos. *Development* **120**, 843-852.

Slack, J. M. W. (1991). *From Egg to Embryo.* Cambridge University Press, Cambridge, UK.

Smith, J. C., Price, B. M. J., Green, J. B. A., Weigel, D. and Herrmann, B. G. (1991). Expression of a *Xenopus* homolog of *Brachyury* (*T*) is an immediate-early response to mesoderm induction. *Cell* **67**, 79-87.

Smith, W. C. and Harland, R. M. (1991). Injected *Xwnt-8* RNA acts in early *Xenopus* embryos to promote formation of a vegetal dorsalizing center. *Cell* **67**, 753-765

Smith, W. C. and Harland, R. M. (1992). Expression cloning of *noggin*, a new dorsalizing factor localized to the Spemann organizer in Xenopus embryos. *Cell* **70**, 829-840.

Spemann, H. (1938). *Embryonic Development and Induction.* New Haven, Conn: Yale University Press.

Stachel, S. E., Grunwald, D. J. and Myers, P. (1993). Lithium perturbation and *goosecoid* expression identify a dorsal specification pathway in the pregastrula zebrafish. *Development* **117**, 1261-1274.

Steinbeisser, H. and De Robertis, E. M. (1993). *Xenopus goosecoid*: a gene expressed in the prechordal plate that has dorsalizing activity. *Compt. Rend. Academ. Scienc., Paris* **316**, 966-971.

Steinbeisser, H., De Robertis, E. M., Ku, M., Kessler, D. S. and Melton, D. A. (1993). *Xenopus* axis formation: induction of *goosecoid* by injected *Xwnt-8* and activin mRNA. *Development* **118**, 499-507.

Storey, K. G., Crossley, J. M., De Robertis, E. M., Norris, W. E. and Stern, C. D. (1992). Neural induction and regionalisation in the chick embryo. *Development* **114**, 729-741.

Taira, M., Jamrich, M., Good, P. J. and Dawid, I. B. (1992) The LIM domain-containing homeo box gene *Xlim-1* is expressed specifically in the organizer region of *Xenopus* gastrula embryos. *Genes Dev.* **6**, 356-366.

Tam, P. P. L. and Tan, S. S. (1992). The somitogenetic potential of cells in the primitive streak and the tail bud of the organogenesis-stage mouse embryo. *Development* **115**, 703-715.

von Baer, K. E. (1828). *Entwicklungsgeschichte der Thiere: Beobachtung und Reflexionen.* Königsberg: Bornträger.

von Dassow, G., Schmidt, J. E. and Kimelman, D. (1993). Induction of the *Xenopus* organizer: expression and regulation of *Xnot*, a novel FGF and activin-regulated homeobox gene. *Genes Dev.* **7**, 355-366.

Waddington, C. H. (1933). Induction by the primitive streak and its derivatives in the chick. *J. Exp. Biol.* **10**, 38-46.

Whittaker, C. A. and DeSimone, D. W. (1993). Integrin α subunit mRNAs are differentially expressed in early *Xenopus* embryos. *Development* **117**, 1239-1249.

Development 1994 Supplement, 125-133 (1994)
Printed in Great Britain © The Company of Biologists Limited 1994

Gene duplications and the origins of vertebrate development

Peter W. H. Holland[1,*], Jordi Garcia-Fernàndez[1,†], Nic A. Williams[1,*] and Arend Sidow[2]

[1]Department of Zoology, University of Oxford, South Parks Road, Oxford, OX1 3PS, UK
[2]Department of Molecular and Cellular Biology, 401 Barker Hall, University of California, Berkeley, CA 94720, USA

*Present address: Department of Pure and Applied Zoology, University of Reading, Whiteknights, PO Box 228, Reading, RG6 2AJ, UK
†Present address: Departament de Genètica, Facultat de Biologia, Universitat de Barcelona, Avda. Diagonal 645, 08071 Barcelona, Spain

SUMMARY

All vertebrates possess anatomical features not seen in their closest living relatives, the protochordates (tunicates and amphioxus). Some of these features depend on developmental processes or cellular behaviours that are again unique to vertebrates. We are interested in the genetic changes that may have permitted the origin of these innovations. Gene duplication, followed by functional divergence of new genes, may be one class of mutation that permits major evolutionary change. Here we examine the hypothesis that gene duplication events occurred close to the origin and early radiation of the vertebrates. Genome size comparisons are compatible with the occurrence of duplications close to vertebrate origins; more precise insight comes from cloning and phylogenetic analysis of gene families from amphioxus, tunicates and vertebrates. Comparisons of Hox gene clusters, other homeobox gene families, Wnt genes and insulin-related genes all indicate that there was a major phase of gene duplication close to vertebrate origins, after divergence from the amphioxus lineage; we suggest there was probably a second phase of duplication close to jawed vertebrate origins. From amphioxus and vertebrate homeobox gene expression patterns, we suggest that there are multiple routes by which new genes arising from gene duplication acquire new functions and permit the evolution of developmental innovations.

Key words: gene duplication, evolution, amphioxus, tunicate, homeobox, Wnt genes

INTRODUCTION

The origin of vertebrates has been the subject of conjecture and debate for over a century. Discussion has centred on the affinities of the vertebrates, the nature of their ancestors and the anatomical changes that must have occurred during vertebrate evolution. (There is also disagreement concerning usage of the term 'vertebrate'; here we include mammals, birds, reptiles, amphibians, true fish, lampreys and, unlike some authors, hagfish). Many attempts have been made to derive vertebrates from either extant invertebrate taxa or hypothetical ancestral forms; each scenario suggests various morphological changes to the body plan, but only rarely have authors considered the underlying genetic causes or the plausibility in a developmental context. In this regard, comparative genome analysis, phylogenetic studies of developmental regulator genes and comparative developmental biology of vertebrates and their extant relatives has much to offer, since it could reveal how genes and developmental processes have changed in evolution.

One influential hypothesis for vertebrate origins, which did take a developmental perspective, was proposed ten years ago by Gans and Northcutt (for a recent review see Gans, 1993). These authors proposed an evolutionary scenario in which the origin of a suite of novel vertebrate characters (including the sensory and cranial ganglia, three paired special sense organs, sensory capsules, and cartilaginous gill arches), were dependent on the origin of neural crest cells and ectodermal placodes. These have important developmental roles in the cranial region of vertebrates, and the structures derived from them (and through their interactions with other cells) dominate the vertebrate head. In this sense, much or all of the vertebrate head was proposed to be an evolutionarily new structure (or neomorph): an innovation of the vertebrates. Other significant morphological changes are proposed to have occurred earlier or later in chordate evolution; for example, the origin of segmentation within early chordates, and the origin of vertebrae and jaws during early vertebrate radiation (Fig. 1).

How did each of these developmental changes occur in evolution? What kinds of genetic changes allowed the origin of the vertebrate developmental program? For example, did new genes permit the evolution of new cell behaviours (seen in neural crest cell migration and differentiation)? Answers to questions such as this may come from considering the genetic basis for evolutionary changes in development. Specifically, we must ask what sort of mutations were potentially, and actually, responsible for particular changes in developmental control during vertebrate origins.

One type of mutation that may have played an important role is gene duplication. The great potential of gene duplication in the evolution of increasing complexity was discussed by Susumu Ohno in his classic book (Ohno, 1970). He argued that tandem duplication of genes, and polyploidy, could create

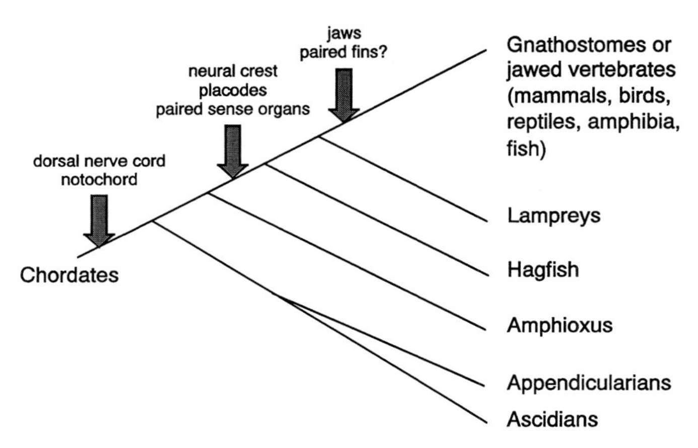

Fig. 1. Possible phylogenetic relationships between chordates showing origin of developmental and anatomical innovations.

redundant genes that were then able to diverge, relatively unchecked by purifying selection, until co-opted for new functions. In Ohno's words "natural selection merely modified, while redundancy created". Since Ohno's insight, the hypothesis that gene duplications are a major force in the generation of organismal complexity has been put on a sound population genetic basis (Ohta, 1989).

With respect to the origin and radiation of the vertebrates, little data on gene duplications were available at the time of Ohno's book. Even so, he was able to make some speculations, based on allozyme data and genome sizes within the chordates. Ohno suggested that at least one round of tetraploidization occurred in the lineage leading to amniotes (reptiles, birds and mammals), probably in our Devonian fish or amphibian ancestors, and that independent tetraploidization events occurred in other fish and amphibian lineages (see also Ohno, 1993). He also suggested that, much earlier in evolution, genome expansion (by either tetraploidy or tandem gene duplication) occurred in the common ancestor of amphioxus and vertebrates (after divergence from tunicates); he did not explicitly propose significant genome changes at the origin of vertebrates. More recently, Holland (1992) speculated that multiple gene duplications may have occurred at vertebrate origins; new genes could then have been co-opted to new roles, facilitating the evolution of new developmental pathways.

These hypotheses make predictions concerning the diversification of gene families that are testable. For example, the number of related genes in a particular gene family can be estimated by application of the polymerase chain reaction (PCR) using degenerate primers; although this technique may not detect all related genes, it does have the advantage of being applicable to multiple species (essential for the comparative approach needed). Furthermore, even if only a subset of genes within a gene family are cloned, from a few key species, molecular phylogenetic analyses can reveal relationships between genes and hence the pathways and timings of gene duplication. Linkage analysis by genomic walking or chromosome in situ hybridization is also now widely applicable, and can be used to distinguish tandem duplication from polyploidy. Of course, it should not be expected that all duplicated genes are retained in the genome after duplication; unused genes could be deleted or scrambled during evolution with little consequence. In addition, it seems possible that even genes that were once essential could be secondarily lost. Nonetheless, applying PCR, genomic library screening and molecular phylogenetic analysis to multiple gene families, in multiple chordate species, should allow the general patterns of genome evolution to be elucidated.

Here we examine the evidence for gene duplications during chordate evolution, comparing the conclusions drawn from genome size comparisons to the insights now possible from gene cloning. All protein-coding genes reported to date from amphioxus are reviewed in an evolutionary context, and two amphioxus *Wnt* genes are reported. Tunicate genes are compared where relevant; we also report the PCR cloning of a Hox gene from an appendicularian. We then consider alternative ways in which duplication of developmental control genes could contribute to the evolution of vertebrate development, and assess these alternatives in the light of in situ hybridization analyses of amphioxus homeobox gene expression.

CAN GENOME SIZE GIVE ANY EVOLUTIONARY INSIGHTS?

Since the cephalochordates (amphioxus) are generally thought to be the closest extant relatives of the vertebrates (Fig. 1), genome comparisons between amphioxus and vertebrates may yield clues to the genetic events that accompanied the evolution of developmental innovations at the origin of vertebrates. Atkin and Ohno (1967) reported the haploid genome of the amphioxus *Branchiostoma lanceolatum* to be approximately 0.6 pg, about 17% of the value for placental mammals. This is considerably larger than seen in many tunicates (for example, *Ciona* at 0.2 pg), but similar to the smallest vertebrate genomes (for example, puffer fish at 0.5 pg; see also Brenner et al., 1993). This led Ohno (1970) to suggest that genome enlargement by tandem gene duplications and/or polyploidy occurred in a common ancestor of amphioxus and vertebrates, but not significantly in the immediate vertebrate ancestors. Taking into account the genome sizes of mammals, birds and reptiles, he also suggested the occurrence of one or more additional rounds of tetraploidy in our Devonian fish or amphibian ancestors (Ohno, 1970, 1993).

These proposed timings of genome expansion do not correlate with vertebrate origins. Does this mean that gene duplications did not play an important role in the origin of the complex vertebrate body plan? Not necessarily, since genome size may be only a very approximate indicator of gene number: for example, repetitive DNA comprises from 20% to over 50% of metazoan genomes, and this fraction is prone to dramatic changes in evolution, probably without concomitant changes in gene number (Lewin, 1990). Furthermore, the distribution of genome sizes within the fishes, together with phylogenetic considerations, make it very unlikely that the extremely compact puffer fish genome is representative of early vertebrates. Puffer fish, being members of the order Tetraodontiformes, occupy a very derived phylogenetic position within the ray-finned fish, and have genome sizes well below the modal value for fishes; this unusually small genome size must be secondarily derived, unless one is willing to accept the occurrence of very frequent, but independent, genomic expansion events in many divergent fish lineages (P. E. Ahlberg, unpublished analyses). If amphioxus is compared instead to the living members of the ***earliest*** vertebrate lineages, hagfish and lampreys, significantly larger genomes are indeed seen in vertebrates (haploid values 1.4-2.8 pg). Furthermore, it has been shown that the brook lamprey genome (at 1.4 pg) is ***not*** complicated by very recent tetraploidy (Ward et al., 1981); hence, it may be valid to use it as an approximate guide to early vertebrate genome size. Of course, modern lampreys could have secondarily expanded or compacted genomes, in which case it would not be valid to infer early vertebrate genome size from them.

The assumption would be testable if genome sizes could be measured from representatives of other (now extinct) jawless vertebrate lineages. Perhaps surprisingly, this may be feasible since the outlines of cells are preserved in some fossils. Cell outlines give an estimate of cell volume, which in turn is an approximate indicator of genome size within vertebrates (if the same cell type is compared between species). The feasibility of this approach was demonstrated by Conway Morris and Harper (1988), who estimated genome size in extinct conodonts (thought to be an ancient lineage of jawless vertebrate; Sansom et al., 1992). These analyses need extending to other lineages; at present, however, the data from both living and fossil jawless vertebrates support the contention that significant genome enlargement occurred at vertebrate origins (after divergence of the amphioxus lineage).

EVOLUTIONARY INSIGHTS FROM AMPHIOXUS HOMEOBOX GENES

More accurate insight into the evolution of vertebrate genome organization will undoubtedly come from cloning and phylogenetic analysis of gene families in representatives of several protochordate and vertebrate lineages. Phylogenetic considerations make amphioxus a particularly important protochordate for gene family analysis, since its lineage diverged after the urochordates but before the diversification of vertebrates (Fig. 1). Of particular interest in these analyses will be multigene families implicated in the control and coordination of developmental processes (for example homeobox and Wnt genes), since their molecular evolution may give insight into the origin of vertebrate developmental control. Relatively few protein-coding genes have been cloned from amphioxus, but these include genes related to vertebrate transcription factors, growth factors, signalling molecules, structural proteins and enzymes. In this section and the next, we look at every example published to date.

One group of homeobox genes for which comparative surveys have been undertaken in the chordates is the Msx gene family. Three members of this gene family were cloned from the mouse genome by PCR (Holland, 1991b); two of these are known to be expressed in cranial neural crest-derived mesenchymal tissue and in complementary patterns at many sites of tissue interaction during development (including during branchial arch development, palate development, tooth morphogenesis, and development of the paired eyes; Davidson and Hill, 1991). Aspects of the gene expression patterns are certainly functional; for example, a point mutation in the homeobox of the human *MSX2* gene is thought be one cause of a skull morphology abnormality, craniosynostosis (Jabs et al., 1993), whilst deletion of mouse *Msx1* by gene targeting causes a range of cranial defects (Satokata and Maas, 1994). Many of the expression sites of the *Msx-1* and *Msx-2* genes, although not all, are vertebrate-specific features (Holland, 1992); hence it is intriguing to ask whether amphioxus has homologues of these genes.

To date, we have succeeded in isolating only a single member of the Msx homeobox gene family from the genome of *Branchiostoma floridae*, both by PCR (Holland et al., 1994) and by genomic library screening (A. Sharman and P. W. H. H., unpublished data). This parallels the results from an ascidian, but contrasts with the multiple Msx genes present in a teleost fish, *Brachydanio rerio* (Holland, 1991b). These results are consistent with the hypothesis that gene duplications in this gene family occurred in the vertebrate lineage after divergence of amphioxus; however, a wider survey of vertebrates must be completed before the timing of duplication can be ascertained.

Comparative data are more sparse for three other homeobox gene families analyzed in amphioxus: the En, Cdx and the XlHbox8-related genes. In the latter two cases, PCR has identified a single homologue to date in *B. floridae* (Holland et al., 1994); the Cdx genes, at least, form a multigene family in mammals (Gamer and Wright, 1994). The size of the XlHbox8 gene family is unknown in any species; within vertebrates, representatives have been cloned from *Xenopus* (Wright et al., 1989), mouse (Ohlsson et al., 1993) and rat (Miller et al., 1994). For both the Cdx and XlHbox8 gene families, additional species need to be analyzed (including jawless vertebrates) before the timing and extent of gene duplications can be ascertained.

For the largest homeobox gene family, the Hox genes, more extensive cloning and phylogenetic surveys have been undertaken. Mammals (and probably all higher vertebrates) have four similar clusters of Hox genes, homologous to the single Hox or HOM-C cluster of arthropods and nematodes (reviewed by Holland, 1992; Burglin and Ruvkun, 1993). Elucidating the number of Hox clusters in amphioxus and lower vertebrates is crucial to determining the time of Hox cluster duplication. In addition to cluster duplication, there is the question of tandem duplications within a cluster. The 38 mammalian Hox genes are divisible between 13 paralogous groups (containing genes related by the cluster duplication events); many of these groups are not present in arthropods and nematodes (Holland, 1992; Burglin and Ruvkun, 1993). Phylogenetic reconstructions suggest that the pre-duplication Hox cluster organization was not identical to any of the clusters of mouse or human (Kappen and Ruddle, 1993); hence, tandem duplications and/or gene losses must have occurred after cluster duplication. Amphioxus Hox genes could give clues to the timing of these events.

The first amphioxus Hox gene published was *AmphiHox3* (Holland et al., 1992) from *Branchiostoma floridae*; complete gene sequence showed this gene is homologous to paralagous group 3 of mammalian genes. This assignment suggests that the tandem duplication event that yielded paralogous groups 2 and 3 (both related to the *Drosophila pb* gene) predated the divergence of amphioxus and vertebrates; it cannot be dated more accurately at present. The number of Hox clusters in the amphioxus genome has been estimated in two studies using PCR (Pendleton et al., 1993; Holland et al., 1994). Both studies utilized the same species (*B. floridae*) and identified multiple Hox genes. From analysis of the deduced translation products of short Hox clones, Pendleton et al. (1993) conclude that "the amphioxus data are in good agreement with a two cluster model"; however, from similar data Holland et al. (1994) conclude that there is "probably a single Hox cluster". The difficulty in determining the number of clusters stems partly from the fact that PCR primers capable of amplifying a broad spectrum of Hox genes can only yield up to 82 nucleotides of unique sequence from each gene. This is often insufficient to assign a gene accurately to a paralogous group (Garcia-Fernàndez and Holland, unpublished data). To overcome this problem, and resolve the discrepancy, we have isolated genomic clones of ten amphioxus Hox genes and mapped their genomic organisation. We find there is a single cluster of Hox genes in the amphioxus genome (Garcia-Fernàndez and Holland, unpublished data).

It is interesting to compare our one cluster model for amphioxus Hox genes (Holland et al., 1994) with PCR results obtained for a lamprey (Pendleton et al., 1993). Despite the difficulty in assigning PCR clones to paralogous groups, the 19 Hox genes identified in *Petromyzon marinus* are consistent with lampreys having at least two, and perhaps three or four, Hox clusters. This suggests the initial Hox cluster duplication(s) occurred in the lineage leading to the first vertebrates, after the divergence of amphioxus.

EVOLUTIONARY INSIGHTS FROM OTHER AMPHIOXUS GENES

The first clues to gene family complexity in amphioxus, predating the homeobox results discussed above, came from Chan et al. (1990). These authors reported that *Branchiostoma californiensis* has a single insulin-like gene (*ILP*), homologous to three gene family members in mammalian genomes (*insulin, IGF-1, IGF-2*); the deduced mature protein sequence shares equal identity with each of the three human proteins. The simplest explanation is that amphioxus retains a single member of this gene family, and that insulin gene duplications occurred on the vertebrate lineage, ***after*** divergence of amphioxus and vertebrates. One of the duplication events occurred very early on the vertebrate (or pre-vertebrate) lineage, since both hagfish and lampreys possess an insulin gene and at least one IGF gene (Nagamatsu et al., 1991). Remarkably, evidence for this very ancient duplication may still be present in the human genome: *IGF-1* maps to chromosome 12, within a region of paralogy to chromosome 11 that contains the *insulin* and *IGF-2* genes (Brissenden et al., 1984; Lundin, 1993). It should be possible to test if this paralogy is genuinely the result of a very early duplication event (of a chromosome, chromosomal region or entire genome) by examination of the genes linked to the *ILP* gene in amphioxus.

The Mn superoxide dismutase (*Mn SOD*) gene and an intermediate filament (*IF*) gene have also been cloned from amphioxus (Smith and Doolittle, 1992; Riemer et al., 1992). The former appears to be a single copy gene in all animals studied, implying that gene duplications during vertebrate ancestry did not affect every gene (or subsequent gene loss has returned some gene families to singletons). The intermediate filament genes could be an informative source of data on duplications, since in mammals they form five subfamilies (types I to V), each containing multiple genes. Exhaustive surveys have not been carried out in amphioxus; the one gene reported to date is clearly a type III gene (vimentin/desmin family), consistent with the idea that initial subdivision of the IF gene superfamily predated the divergence of amphioxus and vertebrates (Riemer et al., 1992).

It would be interesting to know the number of amphioxus genes in each IF gene subfamily, particularly since mammalian gene mapping studies reveal that some of the 'within group' duplications almost certainly coincided with Hox cluster duplications. For example, within both the type I (acidic cytokeratin) and the type II (basic cytokeratin) gene families, there are related genes very closely linked to the HOXB and HOXC gene clusters on human chromosomes 12 and 17 (Bentley et al., 1993; Lundin, 1993).

There are several other cases where members of a gene family are chromosomally linked to more than one mammalian Hox cluster; in each case, their origin by chromosomal or genome duplication ***may*** have coincided with Hox cluster duplication. Possible examples of 'co-duplicated' gene families include (in addition to the cytokeratins): collagen genes, retinoic acid receptor genes, Evx homeobox genes, erythrocyte band 3 related genes, glucose transporter genes, actin genes, GLI/ci^D zinc finger genes, myosin light chain genes, some Wnt genes (but see below) and the neuropeptide Y/pancreatic polypeptide genes (gene mapping data from Bentley et al., 1993; Lundin, 1993). Extrapolating from data on the timing of Hox cluster duplications (Pendleton et al., 1993; Holland et al., 1994; Garcia-Fernàndez and Holland, unpublished data), we suggest that expansion of many of these gene families occurred close to vertebrate origins. We do not, however, discount the possibility that additional duplication events occurred in these gene families during the subsequent evolutionary radiation of the vertebrates.

The Wnt gene family is an interesting case from the perspective of gene duplications. These genes encode an extensive family of secreted proteins implicated in cell-cell signalling during vertebrate and invertebrate embryogenesis (Nusse and Varmus, 1992). Sidow (1992) investigated the diversification and molecular evolution of the Wnt gene family, by phylogenetic analysis of 72 partial Wnt gene sequences isolated from a diversity of vertebrates, echinoderms and *Drosophila*. The results suggested that *Wnt-1*, *-3*, *-5*, and *-7*, and one or more ancestors of *Wnt-2*, *-4*, *-6*, and *-10* were probably present in the genome of the last common ancestor of arthropods and vertebrates. Later duplications of *Wnt-3*, *-5*, *-7*, *-8* and *-10* (giving rise to, for example, *Wnt-3a* and *-3b*) occurred before the diversification of jawed vertebrates, perhaps after divergence of the hagfish lineage.

We used PCR to search for Wnt genes in amphioxus genomic DNA, since no protochordate genes were included in the original analysis. After cloning of the amplified band, sequence analysis of 12 recombinants revealed just two amphioxus Wnt genes (Fig. 2). The phylogenetic position of amphioxus and the history of gene duplications in the Wnt family (Sidow, 1992) imply that amphioxus should have additional Wnt genes, unless they were lost during its evolution. It will be particularly interesting to determine if amphioxus has homologues of those genes that are duplicated in jawed vertebrates (*Wnt-3, -5, -7, -8,* and *-10*).

EVOLUTIONARY INSIGHTS FROM TUNICATE GENES

In the examples discussed above, the assumption is made that

Amphioxus *Wnt-A*

```
1     GGCGTGTCGG GATCCTGCGA GCTCAAGACC TGCTGGCGGG CCATGCCGCC
51    TTTCCGGGAG GTCGGGGCGA GGCTGAAGGA AAAATTCGAC GGCGCCACCG
101   AGGTGCAACA GAAAAAGATC GGCAGCAGGA GAGAACTCGT GCCGCTCAAT
151   TCTGACTTCA AGCCGCACTC GAGTTCCGAC CTGGTGTATC TGGATGCTTC
201   CCCAGACTTT TGCGTGCGGG ACACCAAGGT GGGGTCGATG GGTACAGTCG
251   GGAGGGTGTG CAACAAGACT TCCAAGGCCA TCGATGGCTG CGAACTTCTG
301   TGCTGTGGGA GAGGGTACAA CACCCATACC CGCGAAGTAG TGGAGAGATG
351   TAGCTGTAAG
```

translation

```
      GVSGSCELKTCWRAMPPFREVGARLKEKFDGATEVQQKKIGSRRELVPLNSDFK
      PHSSSDLVYLDASPDFCVRDTKVGSMGTVGRVCNKTSKAIDGCELLCCGRGYNT
      HTREVVERCSCK
```

Amphioxus *Wnt-B*

```
1     GGACTGTCTG GCTCATGCGC AGTAAAGACG TGTTGGAAAA AGATGCCGAT
51    ATTCCGGGAG GTCGGAGTTC GGCTAAAGGA GAGGTTTAAC GGTGCCTTCC
101   AAGTCATGGG CTCCAACAAC GGCAAATATC TCATACCCGT CGGGGACACT
151   ATCAAAGCTC CTACGGCAGA GGACCTCGTG TATACGAACG AGTCGCCGAA
201   TTTTTGCAAA AGGAACAGAA AAACAGGGTC GCAAGGGACC AAAGGGCGGG
251   CCTGTAACGC CACGTCCATG GGGATTGGCG GCTGTGACTT GTTGTGTTGT
301   GGGAGAGGGT ACAAGGAGAG ACAGGTGGTC GTGGAGGAGA ACTGCAAGTG
351   TCGC
```

translation

```
      GLSGSCAVKTCWKKMPIFREVGVRLKERFNGAFQVMGSNNGKYLIPVGDTIKAP
      TAEDLVYTNESPNFCKRNRKTGSQGTKGRACNATSMGIGGCDLLCCGRGYKERQ
      VVVEENCKCR
```

Fig. 2. Nucleotide and deduced amino acid sequences of two partial Wnt genes cloned by PCR from amphioxus (*Branchiostoma floridae*). PCR primers were as described by Gavin et al. (1990); thermal cycling and recombinant cloning was performed as described by Holland (1993). Molecular phylogenetic analyses suggest amphioxus *Wnt-A* is orthologous to vertebrate *Wnt-4*. EMBL/GenBank/DDBJ accession numbers Z34273 and Z34274.

gene family organization in amphioxus is primitive with respect to the vertebrate condition. To test this assumption, comparison can be extended to an outgroup. The Urochordates (tunicates) may be useful for this comparison, since they are thought to be the sister group to the clade comprising amphioxus plus vertebrates (Fig. 1). In addition, this comparison could help evaluate the hypothesis that, after divergence of the tunicates, substantial gene duplications occurred on the lineage leading to amphioxus plus vertebrates (Ohno, 1970).

The majority of tunicates belong to the Ascidiacea: a group of animals widely used in developmental studies. Consequently, many genes and gene families have been cloned from ascidians. For the present purposes, however, we are only concerned with members of those gene families also analyzed in amphioxus and several vertebrates. One example alluded to above was the Msx homeobox gene family; PCR analyses suggest that the ascidian *Ciona intestinalis* probably has a single member of this gene family (Holland, 1991b), as also found for amphioxus. Hence in this example, Msx gene duplications postdated the amphioxus-vertebrate divergence.

Perhaps surprisingly, at the time of writing, few homeobox genes from the Hox family have been reported from ascidians. Single Hox genes have been isolated from *Phallusia mammilata* (W. Gehring and Paul Baumgartner, personal communication) and, by PCR, from several other ascidians (Ruddle et al., this volume). In addition, in the ascidian *Halocynthia roretzi*, screening of genomic Southern blots and a cDNA library using *Antp* as a probe yielded only the divergent homeobox gene, *AHox1* (Saiga et al., 1991)

We decided to examine complexity of the Hox gene family in a group of tunicates related to the ascidians, the appendicularians. These are small (1.5 mm) pelagic tunicates with an adult morphology similar to ascidian tadpole larvae (Fig. 3A);

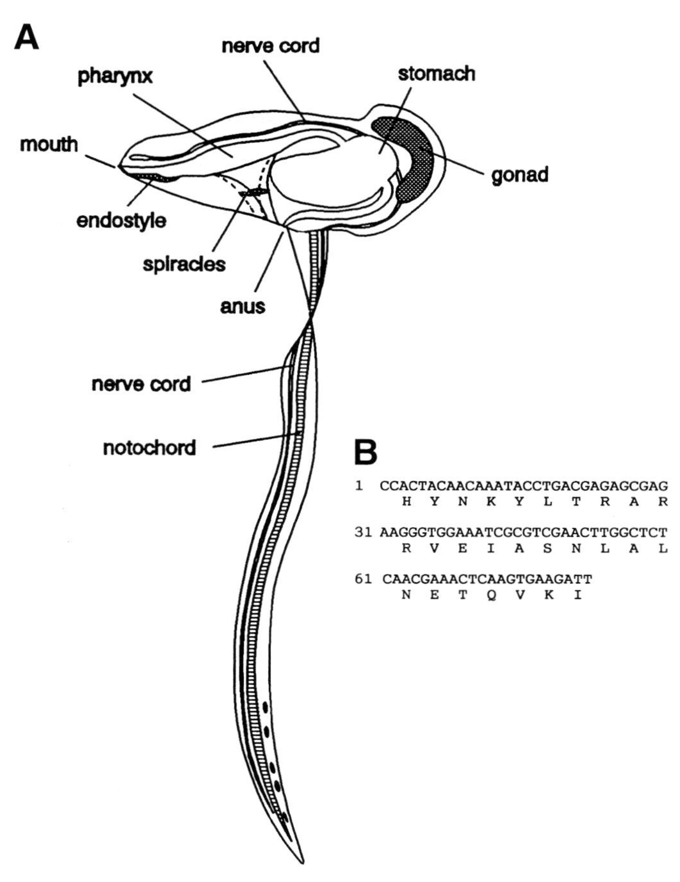

Fig. 3. (A) Schematic diagram of an appendicularian (modified from Alldridge, 1976), plus (B) the nucleotide and deduced amino acid sequence of a partial Hox gene cloned by PCR from *Oikopleura dioica*. PCR primers were based on those of Frohman et al. (1990). Diagnostic amino acids suggest affinity to the lab group of Hox genes. EMBL/GenBank/DDBJ accession number Z34284.

they do not metamorphose into a sessile stage. We reasoned that these animals may be a good outgroup for comparison to amphioxus and vertebrates, since recent studies of sperm morphology suggest they derive from a more basal lineage within the tunicates than do the ascidians (Fig. 1), and they possibly have a less highly modified morphology and life cycle (Holland et al., 1988; Holland, 1991a).

Using degenerate primers (complementary to *Hox* genes from paralagous groups 1 to 10), we employed PCR to search for Hox genes in genomic DNA of *Oikopleura dioica*. After cloning of the amplified band, we determined the DNA sequence of 19 recombinant clones. All clones were identical (or with up to one nucleotide difference), and presumed to derive from the same *Oikopleura* Hox gene (Fig. 3B). Failure to clone additional genes does not disprove their presence in the genome, but similar PCR conditions have yielded multiple Hox genes in many other metazoan species (Averof and Akam, 1993; Pendleton et al., 1993; Holland et al., 1994). It would be surprising if *Oikopleura dioica*, or other tunicates, have only a single Hox gene. An alternative possibility is that some aspect of genome organisation, codon usage or sequence divergence in tunicates causes inefficient cloning or PCR amplification. Further work is necessary to resolve these alternatives. Possession of a single Hox gene cannot be the primitive state within the chordates, since wider comparisons to arthropods, echinoderms and nematodes indicate that a cluster of at least five Hox genes predated the origin of the chordates (Burglin and Ruvkun, 1993). Hence, despite the phylogenetic position of tunicates, it may be difficult to address satisfactorily the question of exactly when genome duplications occurred during the very early phases of chordate radiation.

TIMING OF GENE DUPLICATIONS

Table 1 summarizes the data on timing of gene duplications directly inferred from cloning of amphioxus genes. These data, and the above discussions, suggest that different gene families followed different patterns of diversification during the evolutionary radiation of chordates. Some gene families apparently showed stability without duplication (*Mn SOD* gene); whilst some gene duplications occurred after vertebrate origins (eg: some Wnt gene duplications; one IGF gene duplication). A common theme found is expansion of gene families on the ancestral lineage of vertebrates, occurring ***after*** the lineage leading to amphioxus had diverged. Examples include duplication of the Hox cluster, *Msx* gene, *Cdx* gene and the ancestral insulin/IGF gene, plus probably duplication of several genes chromosomally linked to the Hox clusters.

These gene family analyses can be used to evaluate previous hypotheses concerning the evolution of vertebrate genomes. Ohno suggested that there were two principal phases of gene duplication on the lineage leading to higher vertebrates (Ohno, 1970, 1993; see also Lundin, 1993). He postulated that genome expansion occurred (a) before vertebrate origins (predating divergence of amphioxus and vertebrates), and (b) during vertebrate radiation (in Devonian fish or amphibia).

We believe that the available comparative data from amphioxus, tunicates and jawless vertebrates (see above) suggest either of two different scenarios. Either (1) there was one major phase of genome expansion, involving two rounds of extensive gene duplication (perhaps by complete or partial tetraploidization), close to the origin of the vertebrates; or (2) there were indeed two phases, but the first was close to the origin of the vertebrates and the second was close to the origin of the jawed vertebrates or gnathostomes (Fig. 4). At present we favour the second of these models, since it is compatible with data from the Hox, Wnt, insulin/IGF and En gene families.

We propose that the first phase of gene duplication, close to vertebrate origins, included the initial Hox cluster duplication and an insulin gene duplication. Evidence for the former comes from our demonstration of a single Hox gene cluster in amphioxus (Holland et al., 1994; Garcia-Fernàndez and Holland, unpublished data), coupled with PCR data suggesting more than one cluster in lampreys (Pendleton et al., 1993). Together, these date the initial Hox cluster duplication to after divergence of the amphioxus lineage, but before emergence of lampreys. Suggestive evidence for similar timing of the first insulin gene duplication stems from the single *ILP* gene in amphioxus (Chan et al., 1990), but two genes (insulin and IGF) in lampreys and hagfish (Nagamatsu et al., 1991).

We suggest the second phase of duplication was close to gnathostome origins, and probably included further Hox cluster duplication, a second IGF duplication, duplication of the En gene, and expansion of the Wnt gene family. Pendleton et al. (1993) suggest lampreys have three Hox clusters; we believe their data could reflect a two-cluster model. Either way, a second Hox cluster duplication event is implied, after divergence of the lamprey lineage, possibly around the origin of jawed vertebrates. The second IGF duplication is implied by the existence of a single IGF gene in lampreys and hagfish (Nagamatsu et al., 1991), but two in jawed vertebrates analyzed. Evidence that several Wnt gene duplications may have also occurred close to jawed vertebrate origins comes from PCR surveys of hagfish and gnathostomes (Sidow, 1992). The best placement of duplications affecting the *Wnt-3, -5, -7,* and *-10* genes (in each case giving a and b paralogues) was after the divergence of the hagfish lineage on the ancestral lineage of jawed vertebrates, although this is not at statistical significance. Timing of an En homeobox gene duplication is inferred by the isolation of one En gene in lampreys, but two to three in teleosts, amphibia and mammals (Holland and Williams, 1990). The two En genes in hagfish could be the result of a separate duplication event, as suggested by Holland and Williams (1990); this may be related to recent, independent, tetraploidy in this lineage (see Ward et al., 1981). The Cdx and Msx homeobox gene families also expanded on the ancestral lineage of the vertebrates, but there are fewer clues to timing (see above).

This two-phase model for gene family expansion during early vertebrate evolution is testable by analysis of other gene families. Resolution will require a combination of careful phylogenetic analysis of extant gene sequences, combined with gene family analysis and genome analysis in protochordates, jawless vertebrates, teleost fish and other chordate taxa.

Table 1. Gene number within gene families in amphioxus and selected vertebrates

	Msx genes	Hox clusters	Cdx genes	Mn SOD	Insulin/ IGF
Amphioxus	1	1	1	1	1
Agnatha	n/t	≥2	n/t	1	2
Mammal	3	4	≥4	1	3

Results from hagfish and lampreys are amalgamated under agnatha, although we do not suggest this is a monophyletic group. Not shown are the En, XlHbox8, IF and Wnt gene families, for which insufficient data are available. n/t, not tested. See text for further details and references.

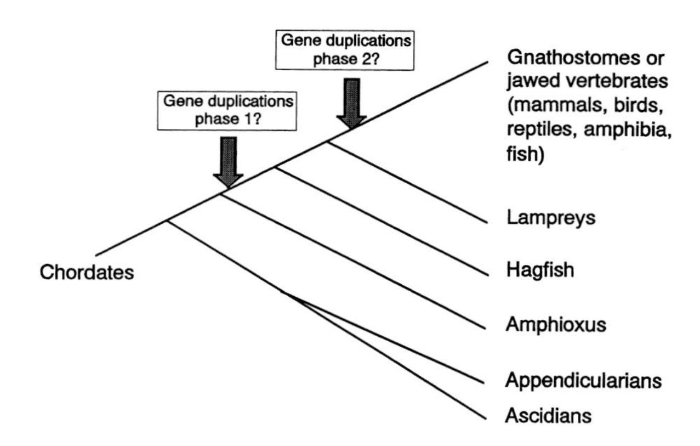

Fig. 4. Phylogenetic tree of the chordates showing the timing of multiple gene duplications proposed in this paper. See text for details and alternative scenarios.

FROM DUPLICATION TO INNOVATION

The hypothesis that mutations in regulatory genes could underlie evolutionary change in embryonic development is now widely accepted (Ohno, 1970; Raff and Kaufman, 1983; Arthur, 1988; Holland, 1992). But different types of mutation could play different roles in developmental evolution. For example, it seems unlikely that minor alterations to the coding sequence or expression pattern of a regulatory gene could allow the origin of completely new morphological features or developmental processes, at least in the majority of cases. Duplication of regulatory genes (followed by divergence of one or both daughter genes), seems more likely to precede the origin of such developmental innovations. Gene duplication would not necessarily ***cause*** major developmental alteration; rather, it is considered ***permissive*** to subsequent phenotypic change. This hypothesis, therefore, does not propose the creation of radically altered 'hopeful monsters' in a single generation, but envisages new genes being made available to the gradual modifying effects of natural selection and genetic drift.

The hypothesis predicts a correlation between the origin of new regulatory genes and the emergence of new cell behaviours, body regions or structures (Holland, 1992); furthermore, significant changes to body organisation may correlate with significant genome changes simultaneously affecting several gene families. The scenario presented in the previous section for the timing of gene duplications during vertebrate evolution suggested there may have been two phases of gene duplication: one close to vertebrate origins and a second close to jawed vertebrate origins. Both stages of chordate evolution involved significant developmental changes; in addition, both phases of genome expansion involved duplication of developmental regulatory (and other) genes.

The origin of vertebrates involved the evolution of several innovations in developmental strategy, notably the involvement of neural crest and placodes in craniofacial morphogenesis, elaboration of the brain (origin of the midbrain?), and specialisation of the segmented mesoderm (Holland, 1992; Gans, 1993; Holland and Graham, 1994). In terms of anatomy, the differences between extant jawed vertebrates and jawless vertebrates are less dramatic than between vertebrates and protochordates, but important developmental transformations can be inferred. These include the origin of paired appendages and a remodelling of the anterior branchial arches to form the jaws and jaw support apparatus (these transformations need not have occurred simultaneously).

It seems at least plausible, therefore, that multiple new genes originating close to vertebrate origins, and close to jawed vertebrate origins, permitted the evolution of these developmental innovations. Without new sets of genes, developmental control may have been constrained from further elaboration; significant gene duplication may have allowed release from these genetic constraints, allowing rapid adaptive radiation of the first vertebrates and, later, the jawed vertebrates.

These hypotheses require new genes to be recruited to new roles after duplication. How could this occur? One possible route would be evolutionary modification of the coding sequence of a gene, such that it is optimized for a different function to that of the ancestral gene. This may be accompanied by changes in gene regulation to allow deployment at a new site or new time. An example that shows the feasibility of this pattern of evolution (although not relating to the early vertebrates) is the lysozyme gene family in mammals. Lysozyme gene duplication in the ruminant mammal lineage was followed by changes in protein sequence and expression, allowing lysozyme to be co-opted to a digestive role in the foregut of cows, sheep and relatives (Irwin et al., 1992). In contrast, a recent experimental analysis of the prd gene family in *Drosophila*, demonstrates that adaptive divergence of protein-coding sequences does not always accompany functional divergence after duplication (Li and Noll, 1994). The related genes *prd*, *gsb* and *gsbn* apparently encode functionally equivalent proteins; their divergent roles may have evolved solely by changes in deployment. We suggest this latter route to functional diversification may have occurred frequently.

At present, there are limited clues to the mechanisms that allowed functional divergence within regulatory gene families during vertebrate evolution. From the inferred timing of gene duplication, and the expression patterns of the mammalian gene family members, a hypothesis can be proposed regarding functional evolution of the vertebrate *Msx* homeobox genes (Holland, 1991b, 1992). As described above, two of the three mammalian Msx genes resultant from duplication are predominantly expressed in vertebrate-specific tissues, including craniofacial neural crest derivatives, developing teeth and eyes. We suggest that these expression characteristics reflect co-option to new functions at vertebrate origins; the origin of *Msx-1* and *Msx-2* might even have permitted the evolution of new patterns of cell behaviour and differentiation and new developmental processes. This hypothesis implies that *Msx-1* and *Msx-2* acquired new control elements after gene duplication, leaving *Msx-3* to persist with an ancestral function. Further insight will come from analysis of the expression pattern of the third mammalian paralogue, *Msx-3*, and comparison to the single amphioxus *Msx* gene. These analyses are in progress (S. Shimeld and P. Sharpe, personal communication; A. Sharman and P. W. H. H., unpublished data).

The evolution of Hox gene function following cluster duplication is discussed in detail by Gaunt (1991) and Holland (1992). The expression patterns of mouse (and other vertebrate) Hox genes suggests that Hox and Msx genes followed different courses of functional diversification. Each mammalian Hox gene is expressed within a precise, regionally restricted, domain along the anteroposterior axis of the developing neural tube, plus often in a subset of tissues from somitic or lateral mesoderm and/or neural crest cells (for reviews, see Holland and Hogan, 1988; Shashikant et al., 1990; Gaunt, 1991). Hox genes related by cluster duplication have similar (but not always identical) expression patterns in the developing neural tube, but there are dramatic differences in which mesodermal and neural crest derivatives express these paralogues. Furthermore, gene targeting of Hox genes often causes more severe disruption in mesodermal and neural crest derivatives than in the neural tube, suggesting partial functional overlap in the latter. This suggests that most Hox genes retained ancestral roles in neural patterning after Hox cluster duplication, but added to these roles were secondary expression sites and functions (perhaps by acquisition of additional *cis*-regulatory elements; Holland, 1992).

This evolutionary scenario, which was based primarily on data from mouse Hox genes, made testable predictions

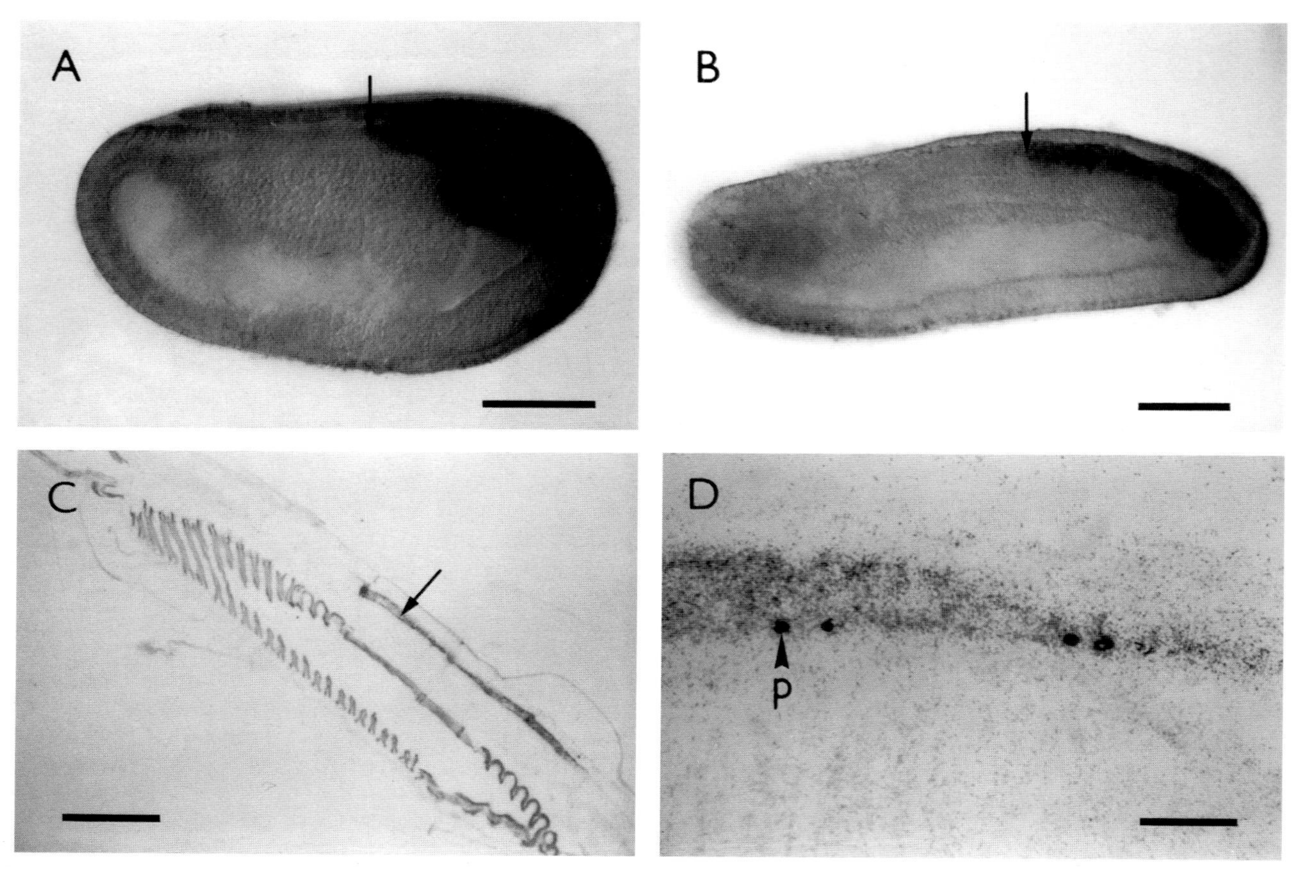

Fig. 5. Distribution of *AmphiHox3* RNA in the developing neural tube of *B. floridae*. Expression in 13 hour (A) and 20 hour (B) embryos was visualised by whole-mount in situ hybridization using a digoxigenin-labelled riboprobe (Holland et al., 1992, 1994); embryos were obtained by in vitro fertilization (Holland and Holland, 1993). (A) Reproduced from Holland et al. (1992). Juvenile amphioxus were collected by fine sieving of sand from Old Tampa Bay, Florida; *AmphiHox3* expression in juveniles (C,D) was examined using radioactive in situ hybridization to wax sections, following the protocol of Wilkinson (1993). C and D are bright-field photographs. Arrows in A and B indicate the anterior expression limit in the neural tube; in C the arrow simply points to the neural tube. Pigment granules (p), ventrally located in the neural tube, are visible in D. Scale bars: 50 µm (A,B,D) and 500 µm (C).

regarding the expression of amphioxus Hox genes. If the ancestral function of chordate Hox genes, prior to cluster duplication, was to encode positional information within the developing neural tube, then this should be the predominant (or only) expression site in amphioxus, which retains a single Hox cluster. Consistent with this prediction, the *AmphiHox3* gene was found to be expressed predominantly in the developing amphioxus neural tube, where it respects a stable anterior boundary at the level of somite five. Expression is also seen in posterior mesoderm, but this does not respect a stable boundary and remains posteriorly localised through development (Holland et al., 1992, 1994; Fig. 5A).

Signals were not detected by whole-mount in situ hybridization on amphioxus larvae older than 5 days, perhaps due to technical problems relating to probe penetration (Holland et al., 1992). We therefore used radioactive in situ hybridization to histological sections to assess if additional, secondary expression sites appear later in development. These experiments revealed that the predominant site of expression remains the dorsal nerve cord in adult and juvenile amphioxus; we find no consistent evidence for secondary expression sites (Fig. 5B).

Acquisition of completely new roles (as proposed for Msx genes) or the supplementation of ancestral roles with secondary roles (Hox genes) are just two of many routes possible for functional diversification of duplicated genes. It seems likely that, even if gene duplication events affected many (or all) gene families simultaneously in evolution, different gene families will have followed quite different routes of functional diversification. These patterns of evolution need to examined in much more detail (including analysis of coding sequences, regulatory elements and function), and in many more gene families, if strong correlations are to be found between particular genetic and phenotypic changes in vertebrate evolution. Correlations should be tested by examination of multiple taxa, but cannot be considered proof of causality in evolution. Even so, by analysis of multiple gene families in many taxa, it should be possible to assess the hypothesis that gene duplications have played an important permissive role in the evolution of vertebrate development. Is it unrealistic to hope that insight will eventually be gained into the mutations that permitted the evolution of specific innovations, such as the origin of neural crest cells and placodes, or the transition from branchial arches to jaws?

We thank Walter Gehring, Frank Ruddle, Anna Sharman, Paul Sharpe and Seb Shimeld for communication of results prior to publication; Per Ahlberg for discussions; and Linda Holland, Nick Holland

and the rest of Team Amphioxus for help with specimen collection. The authors' research in this field was supported by the SERC (P. W. H. H., N. A. W.), a Royal Society Research Fellowship to P. W. H. H. and a Human Frontiers Fellowship to J. G. F.

REFERENCES

Alldridge, A. (1976) Appendicularians. *Scientific American* **235** (1), 95-102.

Atkin, N. B. and Ohno, S. (1967) DNA values of four primitive chordates. *Chromosoma* **23**, 10-13.

Arthur, W. (1988) *A Theory of the Evolution of Development*. Chichester: Wiley.

Averof, M. and Akam, M. E. (1993) HOM/Hox genes of *Artemia*: implications for the origin of insect and crustacean body plans. *Current Biology* **3**, 73-78.

Bentley, K. L., Bradshaw, M. S. and Ruddle, F. H. (1993) Physical linkage of the murine Hox-b cluster and nerve growth factor receptor on yeast artificial chromosomes. *Genomics* **18**, 43-53.

Brenner, S., Elgar, G., Sandford, R., Macrae, A., Venkatesh, B. and Aparicio, S. (1993). Characterisation of the pufferfish (*Fugu*) genome as a compact vertebrate genome. *Nature* **366**, 265-268.

Brissenden, J. E., Ullrich, A. and Francke, U. (1984) Human chromosomal mapping of genes for insulin-like growth factors I and II and epidermal growth factor. *Nature* **310**, 781-784.

Burglin, T. R. and Ruvkun, G. (1993) The *Caenorhabditis elegans* homeobox gene cluster. *Current Opinion in Genetics and Devel.* **3**, 615-620.

Chan, S. J., Cao, Q. -P. and Steiner, D. F. (1990) Evolution of the insulin superfamily: cloning of a hybrid insulin/insulin-like growth factor cDNA from amphioxus. *Proc. Natl. Acad. Sci. USA* **87**, 9319-9323.

Conway Morris, S. and Harper, E. (1988) Genome size in conodonts (Chordata): inferred variations during 270 million years. *Science* **241**, 1230-1232.

Davidson, D. R. and Hill, R. E. (1991) *Msh*-like genes: a family of homeobox genes with wide-ranging expression during vertebrate development. *Seminars Dev. Biol.* **2**, 405-412.

Frohman, M. A., Boyle, M. and Martin, G. R. (1990) Isolation of the mouse *Hox-2. 9* gene; analysis of embryonic expression suggests that positional information along the anteroposterior axis is specified by mesoderm. *Development* **110**, 589-607.

Gamer, L. W. and Wright, C. V. E. (1994). Murine *Cdx-4* bears striking similarities to the *Drosophila caudal* gene in its homeodomain sequence and early expression pattern. *Mech. Dev.* **43**, 71-81.

Gans, C. (1993) Evolutionary origin of the vertebrate skull. In *The Skull* Vol. 2 (ed. J. Hanken and B. K. Hall), pp. 1-35. Chicago: University of Chicago Press.

Gaunt, S. J. (1991) Expression patterns of mouse *Hox* genes: clues to an understanding of developmental and evolutionary strategies. *BioEssays* **13**, 505-513.

Gavin, B. J., McMahon, J. A. and McMahon, A. P. (1990) Expression of multiple *Wnt-1/int-1*-related genes during fetal and adult mouse development. *Genes Dev.* **4**, 2319-2332.

Holland, L. Z. (1991a) The phylogenetic significance of tunicate sperm morphology. In *Comparative Spermatology, 20 years after* (ed. B. Baccetti). *Serono Symposium* **5**, 961-965.

Holland, P. W. H. (1991b) Cloning and evolutionary analysis of *msh*-like homeobox genes from mouse, zebrafish and ascidian. *Gene* **98**, 253-257.

Holland, P. W. H. (1992) Homeobox genes in vertebrate evolution. *BioEssays* **14**, 267-273.

Holland, P. W. H. (1993) Cloning genes using the polymerase chain reaction. In *Essential Developmental Biology: A Practical Approach*. (ed. C. D. Stern and P. W. H. Holland), pp. 243-255. Oxford: IRL Press at Oxford University Press.

Holland, P. W. H. and Graham, A. (1994) Evolution of regional identity in the vertebrate nervous system. *Perspectives on Developmental Neurobiology*. In press.

Holland, P. W. H. and Hogan, B. L. M. (1988) Expression of homeo box genes during mouse development: a review. *Genes Dev.* **2**, 773-782.

Holland, N. D. and Holland, L. Z. (1993) Embryos and larvae of invertebrate deuterostomes. In *Essential Developmental Biology: A Practical Approach*. (ed. C. D. Stern and P. W. H. Holland), pp. 21-32. Oxford: IRL Press at Oxford University Press.

Holland, P. W. H. and Williams, N. A. (1990) Conservation of *engrailed*-like homeobox sequences during vertebrate evolution. *FEBS Lett.* **277**, 250-252.

Holland, L. Z., Gorsky, G. and Fenaux, R. (1988) Fertilization in Oikopleura dioica (Tunicata, Appendicularia): acrosome reaction, cortical reaction and sperm-egg fusion. *Zoomorphology* **108**, 229-243.

Holland, P. W. H., Holland, L. Z., Williams, N. A. and Holland, N. D. (1992) An amphioxus homeobox gene: sequence conservation, spatial expression during development and insights into vertebrate evolution. *Development* **116**, 653-661.

Holland, P. W. H., Garcia-Fernàndez, J., Holland, L. Z., Williams, N. A. and Holland, N. D. (1994) The molecular control of spatial patterning in amphioxus. *J. Marine Biol. Assoc. UK* **74**, 49-60.

Irwin, D. M., Prager, E. M. and Wilson, A. C. (1992) Evolutionary genetics of ruminant lysozymes. *Animal Genetics* **23**, 193-202.

Jabs, E. W., Müller, U., Li, X., Ma, L., Luo, W., Haworth, I. S., Klisak, I., Sparkes, R., Warman, I., Mulliken, J. B., Snead, M. L. and Maxson, R. (1993) A mutation in the homeodomain of the human *MSX2* gene in a family affected with autosomal dominant craniosynostosis. *Cell* **75**, 443-450.

Kappen, C. and Ruddle, F. H. (1993) Evolution of a regulatory gene family: *HOM/Hox* genes. *Current Opinion in Genetics and Devel.* **3**, 931-938.

Lewin, B. (1990) Genes IV. Oxford University Press, Oxford.

Li, X. and Noll, M. (1994) Evolution of distinct developmental functions of three *Drosophila* genes by acquisition of different *cis*-regulatory regions. *Nature* **367**, 83-87.

Lundin, L. (1993) Evolution of the vertebrate genome as reflected in paralogous chromosomal regions in man and the house mouse. *Genomics* **16**, 1-19.

Miller, C. P., McGehee, R. E. and Habener, J. F. (1994) IDX-1: a new homeodomain transcription factor expressed in rat pancreatic islets and duodenum that transactivates the somatostatin gene. *EMBO J.* **13**, 1145-1156.

Nagamatsu, S., Chan, S. J., Falkmer, S. and Steiner, D. F. (1991) Evolution of the insulin gene superfamily: sequence of a preproinsulin-like growth factor cDNA from the atlantic hagfish. *J. Biol. Chem.* **266**, 2397-2402.

Nusse, R. and Varmus, H. E. (1992) *Wnt* genes. *Cell* **69**, 1073-1087.

Ohlsson, H., Karlsson, K. and Edlund, T. (1993) IPF1, a homeodomain-containing transactivator of the insulin gene. *EMBO J.* **12**, 4251-4259.

Ohno, S. (1970) *Evolution by Gene Duplication*. Heidelberg: Springer-Verlag.

Ohno, S. (1993) Patterns in genome evolution. *Current Opinion in Genetics and Devel.* **3**, 911-914.

Ohta, T (1989) Role of gene duplication in evolution. *Genome* **31**,304-310.

Pendleton, J. W., Nagai, B. K., Murtha, M. T. and Ruddle, F. H. (1993) Expansion of the *Hox* gene family and the evolution of chordates. *Proc. Natl. Acad. Sci USA* **90**, 6300-6304.

Raff, R. A. and Kaufman, T. C. (1983) *Embryos, Genes and Evolution*. New York: Macmillan.

Riemer, D., Dodemont, H. and Weber, K. (1992) Analysis of the cDNA and gene encoding a cytoplasmic intermediate filament (IF) protein from the cephalochordate *Branchiostoma lanceolatum*; implications for the evolution of the IF protein family. *Eur. J. Cell Biol.* **58**, 128-135.

Saiga, H., Mizokami, A., Makabe, K. W., Satoh, N. and Mita, T. (1991) Molecular cloning of a novel homeobox gene *AHox1* of the ascidian, *Halocynthia roretzi*. *Development* **111**, 821-828.

Sansom, I. J., Smith, M. P., Armstrong, H. A. and Smith, M. M. (1992) Presence of the earliest vertebrate hard tissues in conodonts. *Science* **256**, 1308-1311.

Satokata, I. and Maas, R. (1994) *Msx1* deficient mice exhibit cleft palate and abnormalities of craniofacial and tooth development. *Nature Genetics* **6**, 348-355.

Shashikant, C. S., Utset, M. F., Violette, S. M., Wise, T. L., Eimat, P., Pendleton, J. W., Schughart, K. and Ruddle, F. H. (1990) Homeobox genes in mouse development. *Critical Reviews in Eukaryotic Gene Expression* **1**, 207-245.

Sidow, A. (1992) Diversification of the *Wnt* gene family on the ancestral lineage of vertebrates. *Proc. Natl. Acad. Sci. USA* **89**, 5098-5210.

Smith, M. W. and Doolittle, R. F. (1992) A comparison of evolutionary rates of the two major kinds of superoxide dismutase. *J. Molec. Evol.* **34**, 175-184.

Ward, R. D., McAndrew, B. J. and Wallis, G. P. (1981) Enzyme variation in the brook lamprey, *Lampetra planeri* (Bloch), a member of the vertebrate group Agnatha. *Genetica* **55**, 67-73.

Wilkinson, D. G. (1993) In situ hybridization. In *Essential Developmental Biology: A Practical Approach*. (ed. C. D. Stern and P. W. H. Holland), pp. 257-274. Oxford: IRL Press at Oxford University Press.

Wright, C. V. E., Schnegelsberg, P. and DeRobertis, E. M. (1989) XlHbox 8 - a novel *Xenopus* homeoprotein restricted to a narrow band of endoderm. *Development* **105**, 787-794.

Development 1994 Supplement, 135-142 (1994)
Printed in Great Britain © The Company of Biologists Limited 1994

Temporal colinearity and the phylotypic progression: a basis for the stability of a vertebrate Bauplan and the evolution of morphologies through heterochrony

Denis Duboule

Department of Zoology, University of Geneva, Sciences III, Quai Ernest Ansermet 30, 1211 Geneva 4, Switzerland

SUMMARY

Vertebrate Hox genes are essential for the proper organization of the body plan during development. Inactivation of these genes usually leads to important alterations, or transformations, in the identities of the affected developing structures. Hox genes are activated in a progressive temporal sequence which is colinear with the position of these genes on their respective complexes, so that 'anterior' genes are activated earlier than 'posterior' ones (temporal colinearity). Here, an hypothesis is considered in which the correct timing of activation of this gene family is necessary in order to properly establish the various expression domains. Slight modifications in the respective times of gene activation (heterochronies) may shift expression domains along the rostrocaudal axis and thus induce concurrent changes in morphologies. It is further argued that temporal colinearity only occurs in cells with high mitotic rates, which results in a strong linkage between patterning and growth control and makes the patterning process unidirectional, from anterior, proximal and early, to posterior, distal and late, a model referred to as the 'Einbahnstrasse'. While the nature of the mechanism(s) behind temporal and spatial colinearities is unknown, it is proposed that such a mechanism relies on meta-*cis* interactions, that is it may necessitate gene contiguity. Such a mechanism would be based on DNA-specific, rather than gene-specific, features such as chromatin configurations or DNA replication. The existence of such a meta-*cis* mechanism would explain the extraordinary conservation of this genetic system during evolution as its basic properties would be linked to that of the genetic material itself. Consequently, it is hypothesized that, in vertebrates, the resistance of this mechanism to evolutionary variations may be the reason for the existence of a short developmental window of morphological invariance (the phylotypic progression).

Key words: Hox genes, vertebrate evolution, gene activation, expression domains

INTRODUCTION

In vertebrates, 38 genes contain a homeobox sequence related to that of the *Drosophila Antp* gene and encode sequence-specific DNA binding transcription factors (reviewed by McGinnis and Krumlauf, 1992). These genes are clustered in four complexes and are members of the so-called Hox gene family. Each complex contains 9-11 genes, regularly spaced over about 200 kb, and transcribed from the same DNA strand. Sequence analyses have revealed that vertebrate HOX complexes can be aligned with the *Drosophila* Antennapedia and Bithorax complexes (ANT-C; BX-C) of homeotic genes when these latter two clusters are conceptually linked together, indicating a common phylogenetic origin of the complexes (Duboule and Dollé, 1989; Graham et al., 1989). This led to the conclusion that, during evolution, an ancestral complex (Hox/HOM-C) was split, in Diptera, and also amplified along the lineage leading to vertebrates. As a consequence of these cluster duplications, genes located at similar relative positions within the HOX complexes (paralogs) show high sequence similarities. The relationships between this structural organization and the expression patterns of the Hox genes during fetal development have been extensively studied and reviewed (e.g. Gaunt, 1991). Briefly, genes located at the 3′ extremities of the complexes (such as group 1 or 2 genes) are expressed starting at anterior positions within the hindbrain, while genes located at 5′ positions (e.g. group 12 or 13 genes) are expressed in progressively more restricted posterior areas (e.g. the genitalia). The same rule can be applied to the expression patterns of Hox genes during limb development (Dollé et al., 1989) and is referred to as 'spatial colinearity' (Gaunt et al., 1988). This type of colinearity was first observed in *Drosophila* by Lewis (1978) and subsequently documented at the molecular level (e.g. Harding et al., 1985).

TEMPORAL COLINEARITY

The anterior-posterior succession in the topography of the Hox gene expression domains may depend on another type of colinearity, which is concerned with the time of activation of these series of genes during development. A delay is observed in the appearance of the transcripts encoded by the more 5′-located genes; for example, *Hoxd-13* transcripts appear after those encoded by *Hoxd-10*. The rule is that one can never detect transcripts from a given Hox gene before transcripts are produced by its 3′-located neighbour in the complex. This implies that the physical ordering of the genes along their complexes reflects the temporal sequence of their activation (Dollé et al., 1989; Izpisúa-Belmonte et al., 1991). This phenomenon, referred to as temporal colinearity, is most clearly visible for those genes located at 5′ positions but can most likely be extended to more 'anterior' genes, as suggested by work carried out in cultured EC cells (Boncinelli et al., 1991) or in *Xenopus* (Dekker et al., 1992). The onset of activation occurs during early gastrulation (Gaunt et al., 1986; Gaunt, 1987; Deschamps and Wijgerde, 1993), at a stage when the embryo establishes its major body axis, and the process complete, in mice, about 2 days later, at the late tail bud stage (Dollé et al., 1991b).

The key role of vertebrate Hox genes in the proper organization of the body plan is now well documented. Experiments involving either gain of function or loss of function have revealed that these genes are required to build (identify) structures properly, usually within a rostro-caudal window that corresponds to the anterior part of their expression domains (see below). The absence of a given Hox product will often result in the transformation of a structure into a similar, but different, structure from the same anatomical series (e.g. a lumbar into thoracic-like vertebra), transformations that are often explained in terms of homeosis (e.g. LeMouellic et al., 1992; Ramirez-Solis et al., 1993; Rijli et al., 1993). If we consider, as an example, the morphogenesis of the vertebral column, it is thus fair to speculate that the coupling between the anterior-posterior (AP) progression in somite formation and the sequential activation of the Hox genes (in presomitic mesoderm) determines the combination of Hox genes expressed at a given AP level of the sclerotome and, consequently, the shape of the future vertebra. In this view, it is not because more posterior metameres are sequentially produced that posterior Hox genes become successively active, but instead, because subsequent Hox genes are turned on that the newly appearing structures can acquire distinct identities. Hence, any variation in the relative speeds of the two processes would lead to mis-identifications of structures. It is, therefore, of great importance to understand what are the mechanistic and molecular bases of temporal colinearity. I would like to argue first, that temporal colinearity is linked to a particular type of clustered organization and second, that it may be dependent on cellular proliferation.

Circumstantial evidence suggests that proper timing of Hox gene expression may be linked to a highly organized type of clustered organization (Duboule, 1992). In *Drosophila*, an animal that does not appear to use temporal colinearity, the homeotic genes are split in two complexes. Furthermore, in this system, the structural organization of ANT-C is quite different from that observed in both BX-C and the homologous gene complexes in vertebrates, where a higher degree of organization is achieved (Kaufman et al., 1990). In mice, expression patterns resembling the endogenous domains have been obtained with a number of Hox transgenes containing different amounts of DNA sequence around particular transcription units (e.g. Whitting et al., 1991; Sham et al., 1992; Gérard et al., 1993). However, while these results demonstrate that *cis*-acting elements contribute to the spatiotemporal specificity of expression, the large majority of transgene expression patterns do not exactly reflect the endogenous situations, in particular with respect to the positions of the cranio-caudal boundaries. Therefore, while these data seem to contradict the above hypothesis (that demands that the onset of transcription is determined entirely by the position of a given gene within the complex), they can be reconciled with a model where the complex would be required to refine and coordinate a preexisting, gene-specific, temporal control.

HOX PATTERNING AND GROWTH CONTROL

The distribution of Hox gene expression domains during development, in particular for the 'posterior' genes, suggests that they are activated in regions of cellular proliferation, as illustrated by the enhanced expression of some *Hoxd* genes in the genital tubercle, in limb buds and in the tail bud (Dollé et al., 1991a,b). Experiments involving manipulation of the mouse or chick limb buds led to the conclusion that the reactivation of *Hox* gene transcription, which always obeys temporal colinearity, can only occur within cells that proliferate (i.e. within those cells that proliferate in order to produce supernumerary structures; Izpisúa-Belmonte et al., 1992; Riddle et al., 1993). Convincing evidence was obtained in the chick wing bud by manipulating both the reactivation of Hox genes and the growth of the additional structure through local release of retinoic acid and subsequent removal of the ectoderm layer. This clearly showed that the process of sequential reactivation was interrupted when proliferation was stopped due to the absence of signaling from the surrounding ectoderm (Izpisúa-Belmonte et al., 1992).

Relationships between growth control and pattern formation is a general feature of epimorphic systems. A mechanistic linkage between the growth of a structure and the processing of its patterning system (like the one postulated here) would help prevent these two aspects of morphogenesis from becoming uncoupled, i.e. prevent growth from occurring faster than patterning or vice-versa. This direct linkage would imply that the HOX complexes act as a mechanism to translate a recurrent process (proliferation) into a linear progression (morphogenesis). Such a relationship is schematized in Fig. 1. While the mechanistic bases of this potential association are unknown, one could imagine that progression in the activation of Hox genes along their complexes would be a function of the rate of proliferation of a given cellular population. In the case of the developing limbs, for example, cells in the progress zone would continually allow further Hox genes to be activated until the time when cells would leave this zone. At this point, the state of activation achieved (e.g. up to *Hoxd-10*) will be maintained in all daughter cells. If a higher rate of proliferation of these cells is resumed, later in development, by producing an extra structure from the anterior margin of the wing bud (e.g. Summerbell, 1981; Cooke and Summerbell, 1981) or in a regeneration blastema (Rose, 1962), the cells will continue to progress on their HOX complexes starting from the point at which they had been stopped (i.e. making *Hoxd-11* available).

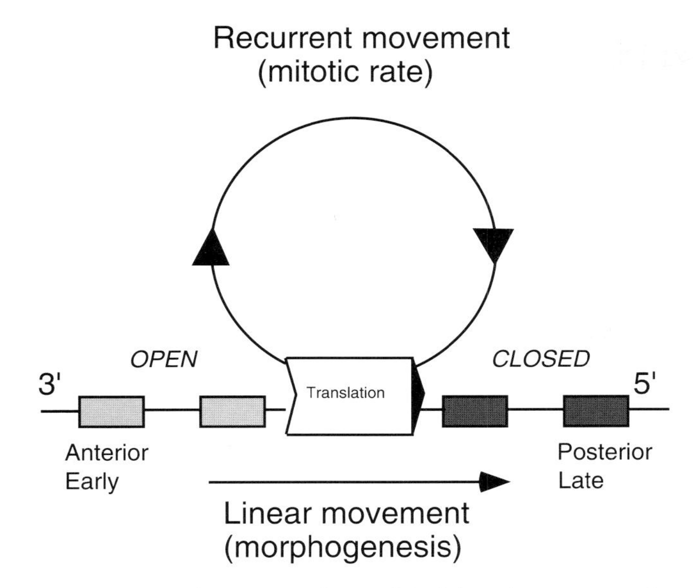

Fig. 1. Schematic illustration of a possible linkage between growth and patterning through the Hox gene complexes. Cells with a high (upper-threshold) mitotic index can 'proceed' along their HOX complexes so that more genes become available (open) for progressively more posterior (distal) patterning. Cells with lower (below-threshold) indexes maintain their state of activation without further opening. In these latter cells, progression can continue if high proliferation is resumed. In this schematic view of one piece of a prototype HOX complex, dark stippled rectangles represent 'closed' genes while light stippled rectangles are 'open' genes. Such open genes may not necessarily be transcribed. The box 'translation' represents the meta-*cis* mechanism referred to in the text. This mechanism may progressively allow more posterior genes to become available in those cells having a high mitotic rate.

This would explain the continuity in patterning observed in the above mentioned experiments, and give a molecular basis to the rule of distal transformation (e.g. Stocum, 1981), the progress zone model (Summerbell et al., 1973) and the acquisition and fixation of 'positional identities' (see Wolpert, 1989).

THE EINBAHNSTRAßE

Such a linkage between growth control and patterning through the HOX complexes would result in the use of series of genes as a system of coordinates; it thus has some traits of the polar coordinate model (French et al., 1976; see also Holder, 1981). However, in marked contrast to this latter model, this system would constrain patterning to be unidirectional and therefore make Hox-dependent intercalation strictly impossible, a proposal previously referred to as the *Einbahnstrasse* (Izpisúa-Belmonte and Duboule, 1992). This model states that the patterning system progresses from early to late, from anterior to posterior, from proximal to distal. It also implies that it can stop and start again (at the same position along the complex) but that it is impossible to come back (i.e. to produce 'anterior' or 'proximal' cells with the progeny of 'posterior' or 'distal' ones, respectively). To achieve the latter transitions, cells would need to de-proliferate (i.e. to go backwards in time), which does not appear to be an acceptable possibility. It means that the process is irreversible and that the designations 'anterior' and 'posterior' are two faces of the same mechanism, acting at different times and always in the same sequence.

At first, such a model seems to be in contradiction to some observations that suggest that, in some systems, posterior body parts can regenerate anterior portions (e.g. Slack, 1980). However, it is not clear whether, in these particular cases, anterior regeneration does require posterior cells (as defined by the expression of posterior Hox genes) to become anterior. It is conceivable that those cells engaged in such 'anterior' regeneration did not previously express any Hox genes or, alternatively, that one of the first responses to experimental injury is to erase any Hox expression, thus allowing cells to progress again from anterior to posterior.

To some extent, the hypothesized relationship between cell proliferation and Hox genes can also be assessed from gene targeting experiments, as many of the defects observed in mice that lack a given Hox gene can be attributed to modifications of local growth rates. Homeotic transformations of vertebrae, for example, are often defined as such because of the presence or absence of morphological traits specific for a neighbouring vertebra (e.g. Jeannotte et al., 1993; Kostic and Capecchi, 1994). This can be best explained in terms of either differential recruitment of cells in the cartilage precursors (e.g. by changing adhesive properties of sub-cellular populations), or differential growth and extension of ossification centres within a vertebra, resulting from local changes in the content of HOX proteins (e.g. Kostic and Capecchi, 1994). Loss or reduction of structures (or part of structures) has been observed in the case of internal organs or the hindbrain of mice lacking the activity of a particular Hox gene (Chisaka and Capecchi, 1991; Dollé et al., 1993a). In this respect, the inactivation of the *Hoxd-13* gene, which is the last gene of the HOXD complex (the last and more posteriorly expressed) is revealing. *Hoxd-13*-deficient mice exhibit limbs that seem to have suffered from a developmental arrest. The defect must occur at an early stage of limb development, since a reduction in the extent of the prechondrogenic condensations is observed. This developmental block is subsequently evidenced by a dramatic delay in the ossification of the limbs; the last skeletal elements to be added in normal development will be missing eventually, leading to animals with neotenic limbs (Dollé et al., 1993a; Fig. 2).

ATAVISTIC TRANSFORMATIONS

In *Hoxd-13*-deficient mice, some phenotypic traits are suggestive of atavism; for example, an increase in the number of carpal bones of the distal row (Dollé et al., 1993a). Another example of atavism in Hox-deficient mice is the phenotype observed in mice lacking a functional *Hoxa-2* gene (Rijli et al., 1993; Gendron-Maguire et al., 1993). In these animals, homeotic-like transformations of the skeletal elements derived from the second branchial arch are observed and remodelling of bones in the middle ear has led to the appearance of a bone resembling a pterygoquadrate element, a bone that is found in reptiles (Rijli et al., 1993). Atavistic transformations have also been observed in gain of function experiments in which Hox gene expression was modified using either ubiquitous (Kessel et al., 1990) or more specific promoters, in transgenic mice. A transformation of the occipital bones to vertebra-like structures (appearance of neural arches) was reported by Lufkin et al. (1992), caused by expression of the *Hoxd-4* gene under the control of the *Hoxa-1* promoter, which was able to drive ectopic expression up to more anterior regions. One could thus imagine that slight modifications to the domains of Hox gene expression may have made important contributions to the evolution of vertebrate morphologies. However, as the overall structure of the Hox gene family seems to be highly conserved between higher and lower vertebrates (F. VanderHoeven and D. D., unpublished data), it is unlikely that many interspecies differences in Hox expression domains derived from loss of Hox genes or significant re-arrangement. Consequently, the approaches described above may not be strictly illustrative of the mechanisms that may have been acting during vertebrate evolution.

POSTERIOR PREVALENCE

One possible way of accommodating these experimental results into a model of how evolution has actually occurred in vertebrates, may be provided by the rule of posterior prevalence. Hox gene inactivation experiments usually lead to phenotypic effects that are restricted to the most anterior part of the expression domain. For example, mice deficient for the *Hoxa-1* gene have no visible alterations in the limbs (Lufkin et al., 1991; Chisaka et al., 1992), even though the gene is expressed in limb buds (Duboule and Dollé, 1989). Similarly, mice lacking the *Hoxb-4* gene show homeotic transformations of most anterior vertebrae only, despite the expression of this gene in more posterior vertebrae (Ramirez-Solis et al., 1993). Thus, the inactivation of a given *Hox* gene generally affects only those structures located at the extreme anterior border of its expression domain, while overexpression seems to induce defects only in those anterior parts of the body where more posterior Hox genes are not expressed. This is in agreement with posterior prevalence, which says that a given Hox gene will exert its function essentially in the domain where this gene is the most posterior of the Hox genes expressed (Duboule, 1991). Most gain of function and loss of function experiments provide some support for the existence of this phenomenon which, to some extent, resembles phenotypic suppression in flies (see Morata et al., 1990; Bachiller et al., 1994). However, as this functional hierarchy must operate exclusively in cells that co-express Hox genes, the phenotype observed when a given Hox gene is knocked out can be more extensive (ie. derived from more posterior cells) if there is a certain degree of mosaicism in the expression of these genes.

In principle, the result of functional hierarchy among Hox gene products would be to restrict potential phenotypic alterations to a reduced area of the embryo; such restriction may prevent too extensive and deleterious transformations from occurring. In this model, minor changes in the temporal control of Hox gene activation could generate evolutionary transformations of structures very similar in outcome to those reported in the experimental approaches described above. For example, a loss of function mutation would correspond, in phenomenological terms, to a situation in which the expression of a given Hox gene was delayed until a stage when its expression would occur too late to compensate for the retardation, or a stage close to the activation of the next posterior (prevalent) gene. Conversely, a gain of function mutation (causing expression of a posterior gene in more anterior domains) would correspond to an advance in the time of expression of this 'posterior' gene, since Hox transcript domains may be shifted anteriorly if activated too early. It is therefore tempting to propose that loss and gain of function mutations of Hox genes may represent the experimental alternatives to heterochronic variations in temporal colinearity that may have occurred during evolution. Minor modifications in timing may have been at the origin of a wide variety of morphological transformations.

In a genetic system such as the Hox gene network, where parts of the same meristic series (e.g. the vertebrae) are individualized by the progressive temporal expression of a series of physically linked genes, variations of the relative times of activation of these genes are likely to produce structures cor-

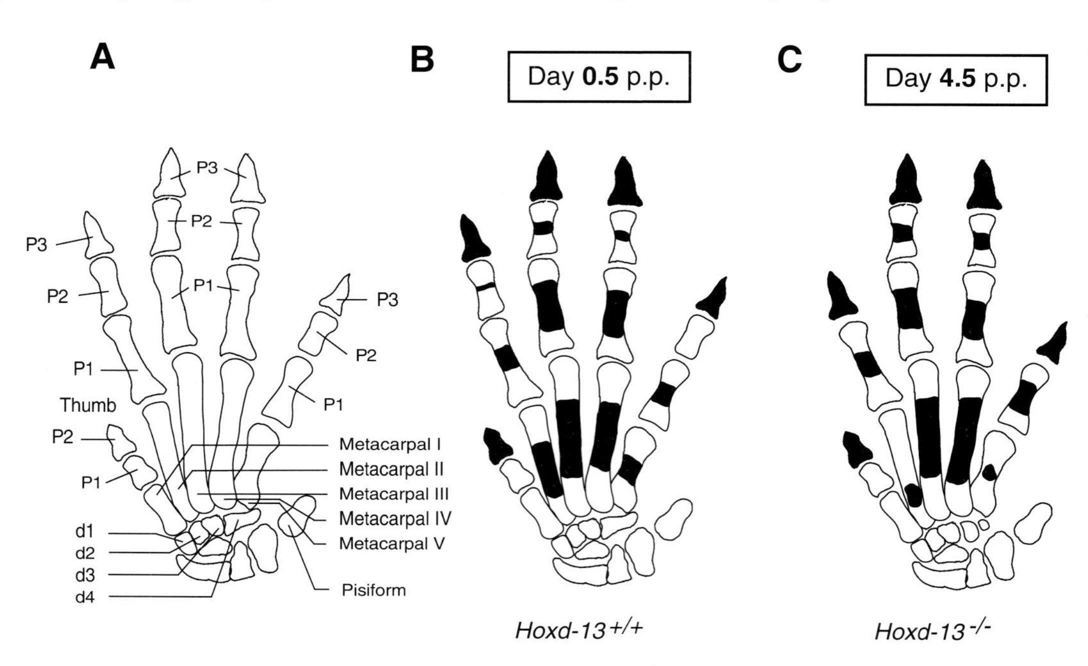

Fig. 2. Illustration of the ossification delay observed in *Hoxd-13*$^{-/-}$ mice. A shows a scheme of a normal hand. The extent of ossification of the various bones is shown in black. B and C show a similar degree of ossification even though the mutated hand (C) is four days older than that of the control littermate (B). This conspicuous delay will result, in adulthood, in hands which will lack some bony elements while the remaining bones will be slightly shorter. In adults, the elements that will be lacking are those that, in normal hands, ossify the last (adapted from Dollé et al., 1993a).

responding to those which are normally produced before or after, within the same series. It is therefore conceivable that heterochronic variations of temporal colinearity make homeotic transformations obligatory rather than exceptional.

A PHYLOTYPIC EGG-TIMER

It is striking that the developmental stage at which Hox genes are sequentially activated corresponds to the so-called phylotypic stage (Sander, 1983): the stage at which vertebrates (for example) express the archetype of the vertebrate body plan (see Wolpert, 1991; Slack et al., 1993, for a discussion and references). This stage (about the early somites stage, but see below) follows the early phase of development and the onset of gastrulation, which, among different species, can be achieved through rather diverse processes, even though molecular analyses suggest a unity of the underlying molecules (e.g. Beddington and Smith, 1993). It precedes the acquisition of species-specific morphological traits which become visible at completion of morphogenesis and organogenesis and thus illustrate that stages of development can exhibit different 'amounts' of evolution (discussed in Raff, 1992). In other words, this stage can be seen as a transition between two phases of development which appear to display some evolutionary variability. Consequently, the phylotypic point, which is (by definition) a stage of high morphogenetic resemblance, can be seen as the narrow point of an ontogenetic egg-timer, a neck of phylogenetic similarity and obligatory passage between two states of higher tolerance for variability (Fig. 3).

In this view, vertebrate embryos, regardless of their early developmental strategies or the way they achieved gastrulation, must converge towards this narrow point to acquire the basic scheme on which subsequent differences will emerge. The fact that Hox genes are expressed at the phylotypic stages of all animal phyla analyzed so far was recently proposed to be a universal trait of animals, hence a basis for the definition of a zootype (Slack et al., 1993). I would like to argue that, in vertebrates, the phylotypic point is neither a point, nor a stage but rather, a *succession* of stages, and propose that the concomitant activation of the Hox gene family is neither a coincidence, nor a consequence of this event, but instead is the *cause* of the apparent invariance of this developmental progression.

There is some difficulty in precisely defining the vertebrate phylotypic stage. It certainly occurs after completion of the major morphogenetic movements, between the head process/early somites stage and the tail bud stage. At these stages, the rostro-caudal progression of morphogenesis is clearly observable through the sequential condensations of somites from pre-somitic mesoderm and progressive closure of the neural tube. While it is clear that a general inspection of all types of vertebrate embryos reveals the existence of a stage at which they all look globally quite similar to each other, the same degree of similarity can be observed in parts of the same embryos before or after this stage. For example, at the time when rostral structures will start to express some morphological differences amongst classes (e.g. a late somite stage), the gastrulating tail bud will still show similarities in all classes. The definition of a phylotypic point is, therefore, a compromise; the similarity is essentially concerned with a morphogenetic progression, in time, rather than with a punctual state. This may be understood in terms of the obligation, for all vertebrate embryos, to use a developmental mechanism that is intrinsically linked to the temporal sequence corresponding to the developmental stages of the phylotypic progression.

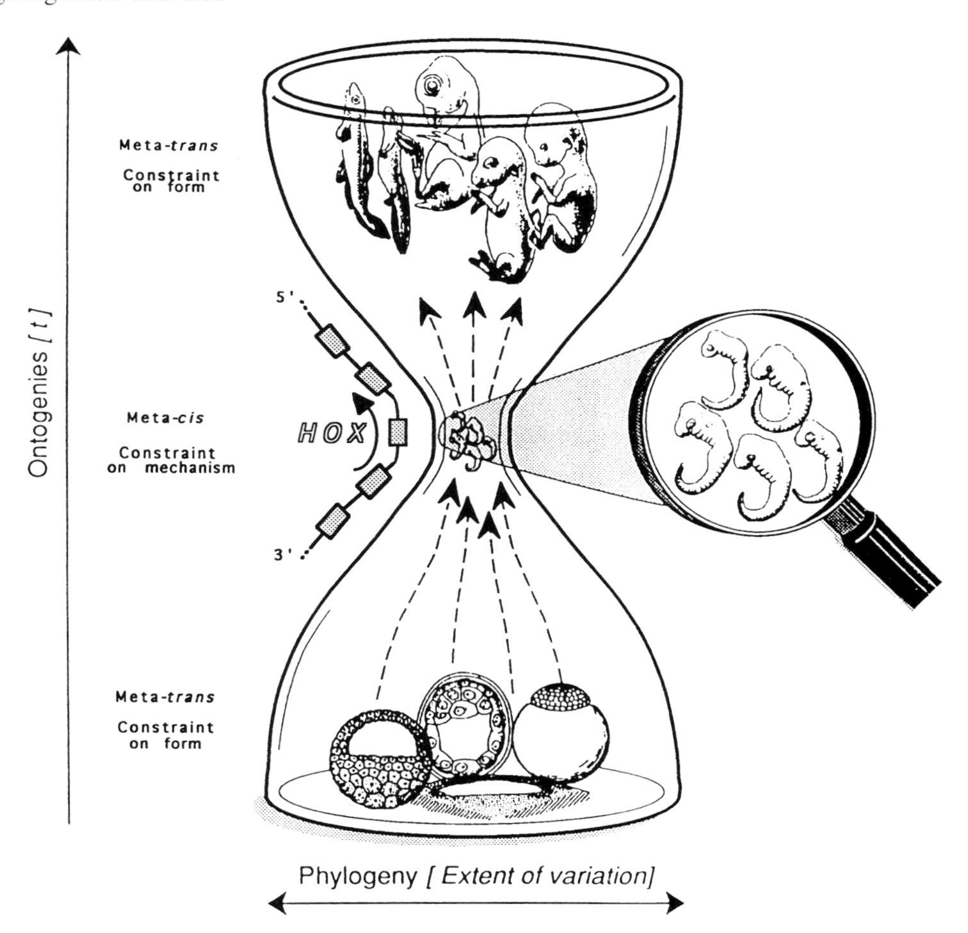

Fig. 3. The phylotypic egg-timer. This scheme illustrates the convergence of vertebrate developmental strategies towards the phylotypic progression, the acquisition of a stable Bauplan. This convergence may be imposed by the necessity to use a meta-*cis* regulatory mechanism whose intrinsic properties, linked to those of the genetic material, may considerably reduce the amount of evolution allowed within this narrow point (see the text). Such a mechanism may result in the progressive temporal activation of Hox genes along their complexes. The diagrams of embryos are schematic to illustrate the concept; it is not suggested that they give an exact description of vertebrate morphologies.

A META-*CIS* REGULATORY MECHANISM

Hox genes are turned on sequentially in time and space; for example, in presomitic mesoderm, in parallel with the craniocaudal morphogenetic progression; this leads to the establishment of a molecular non-equivalence at various levels of the major body axes (e.g. Kessel and Gruss, 1991; Dollé et al., 1989). The pleiotropy of this process (active along all body axes and in many cell types, see above discussion of gene inactivation experiments) and/or a particular mechanistic feature could underlie the observed evolutionary stability of the phylotypic progression. While the pleiotropic effects of Hox gene inactivations are documented (e.g. Dollé et al., 1993a; Small and Potter, 1993), the mechanistic bases of colinearities are as yet unknown. Nevertheless, it appears that the mechanism involved must rely, at least in part, on the physical ordering of genes along segments of DNA, which suggests that the linear structure of our genetic material could be used as a determinant of differences along the linearity of our body axis. The existence of a system of coordinates to encode linear representations of our future body axes within our genetic material itself would explain the invariance of the mechanism involved in the processing of this information. Such a system would be, by essence, invariant as it would rely on the linearity of DNA itself, a feature that may be defined as a meta-*cis* regulation. This regulation would apply to a supra-genic organization, in contrast to regulatory interactions in *trans* between networks of genes (meta-*trans* regulation); the latter should be prone to a higher evolutionary flexibility (acting upstream and downstream the phylotypic progression).

Possible mechanisms underlying meta-*cis* regulation (based on the spatial contiguity of genes; discussed by Horder, 1991) are difficult to conceive, but processes involving spreading of chromatin configurations or DNA replication would fit into this category. Such mechanisms impose significant mechanistic constraints on a developmental process and thus make the evolution of morphologies at these developmental stages unlikely; earlier or later stages could accommodate more 'degeneracy', constraints being imposed by the final form rather than by the mechanisms themselves (Edelman, 1988). In vertebrates, the results of such an invariant mechanism underlying colinearities may be primarily observed in anterior or median parts of the body, i.e. those regions where homologies are more easily recognized. In contrast, more posterior regions (in the trunk) or distal regions (in the limbs), under the control of the last Hox genes at the end of the progression, may suffer from both a lack of precision of the mechanism and an increased tolerance for variations and, consequently, be the source of a higher variability. This could account for the 'distal variability versus proximal stability' observed amongst tetrapod limbs (see Hinchliffe, 1991) or for the observation that: "...Nature got tired of counting towards the tail end of a developing animal..." (Goodrich, 1913).

CHANGING SPEED

In vertebrates, not only can the duration of ontogenesis vary considerably, but also the time required for the phylotypic progression. Thus, fishes reach gastrulation well before rodents (in absolute time) and the processing of the fish Hox network appears to occur faster than in rodents (P. Sordino and D. D., unpublished). This reflects the fact that, whatever mechanism is involved, it can be subject to some variations in its processing speed (heterochronic variations). However, the biochemical nature of a meta-*cis* regulatory mechanism probably imposes a limit on the extent of possible time compression. It is therefore probable that animals developing faster than the minimum time required for temporal colinearity to proceed had to design different developmental strategies. *Drosophila* may represent an example where the reduction in the time needed to reach the fully segmented germ band stage was too important, and where temporal colinearity has been overridden. Interestingly, in this case the design of an alternative developmental system, plus a corresponding innovation in the segmentation mechanism, correlates with a break of the original HOM complex and the disorganization of ANT-C which, in the context of this model, could be seen as a terminal step in heterochrony (the disappearance of a colinear sequence in the time of activation; Izpisúa-Belmonte et al., 1991). As far as the Bithorax complex is concerned, its rather tight organization (which resembles that of the vertebrate HOX complexes; see Duboule, 1992) may simply reflect the fact that most insects develop their abdomen following a time sequence in the addition of segments and may therefore still use temporal colinearity as a mechanism. Thus, it is probable that a correlation exists between a particular mechanism of segmentation (progressive, epimorphic-like), a tight linkage of the corresponding HOM genes within the same complex and the existence of temporal colinearity (a concomitant progressive activation of these genes).

In *Drosophila*, temporal colinearity does not seem to occur anymore, in any part of the body. Accordingly, one might expect the same process that followed the release of this constraint on the ANT-C (the split and disorganization of the complex) should currently be affecting BX-C. In other words, perhaps BX-C can tolerate a degree of disorganization that orthologous pieces of, for example, rodent complexes cannot accept. If this is the case, the differences in organization of the two homeotic complexes in *Drosophila* may simply reflect the fact that an important constraint was released from ANT-C earlier in evolution than its release from BX-C. It is therefore conceivable that the different organization of ANT-C and BX-C reflect two different time points in the evolution of this system in insects. A 'loose' organization (see Duboule, 1992) of the HOM complex is also seen in animals that do not develop according to a segmented body plan, such as *C. elegans* (Bürglin and Ruvkun, 1993), an animal where HOM genes probably specify positions and cell fate but need not be activated in a collinear temporal sequence.

I wish to thank all former and present colleagues in my laboratory or in collaborative teams. I am also grateful to Drs J. Zakany and P. Holland as well as to some referees for their comments and suggestions on the manuscript. This laboratory is supported by funds from the Swiss National Science Foundation, the HFSPO, the University of Geneva and the Claraz Foundation.

REFERENCES

Bachiller, D., Macias, A., Duboule, D. and Morata, G. (1994). Conservation of a functional hierarchy between mammalian and insect Hox/HOM genes. *EMBO J.* **13**, 1930-1941.

Beddington, R. S. P. and Smith, J. C. (1993). The control of vertebrate gastrulation: inducing signals and responding genes. *Current Opin. Genet. Dev.* **3,** 655-661.

Boncinelli, E., Simeone, A., Acampora, D., and Mavilio, F. (1991). HOX gene activation by retinoic acid. *Trends Genet.* **7,** 329-334.

Bürglin, T. R. and Ruvkun, G. (1993). The *Caenorhabditis elegans* homeobox gene cluster. *Current Opin. Genet. Dev.* **3,** 655-661.

Chisaka, O., and Capecchi, M. R. (1991). Regionally restricted developmental defects resulting from targeted disruption of the mouse homeobox gene hox-1.5. *Nature* **350,** 473-479.

Chisaka, O., Musci, T. S., and Capecchi, M. R. (1992). Developmental defects of the ear, cranial nerves and hindbrain resulting from targeted disruption of the mouse homeobox gene Hox-1.6. *Nature* **355,** 516-521.

Cooke, J. and Summerbell, D. (1981). Control of growth related to pattern specification in chick wing bud mesenchyme. in: Growth and the development of pattern (eds: R. M. Gaze, V. French, M. Snow and D. Summerbell). pp 169-185, Cambridge: Cambridge University Press.

Dekker, E.-J., Pannese, M., Houtzager, E., Timmermans, H., Boncinelli, E. and Durston, A. (1992). Xenopus Hox-2 genes are expressed sequentially after the onset of gastrulation and are differentially inducible by retinoic acid. *Development* **Supplement** 195-202.

Deschamps, J. and Wijgerde, M. (1993). Two phases in the establishment of Hox expression domains. *Dev. Biol.* **156,** 473-480.

Dollé, P., Dierich, A., LeMeur, M., Schimmang, T., Schuhbaur, B., Chambon, P. and Duboule, D. (1993a). Disruption of the Hoxd-13 gene induces localized heterochrony leading to mice with neotenic limbs. *Cell* **75,** 431-441.

Dollé, P., Izpisúa-Belmonte, J.-C., Boncinelli, E., and Duboule, D. (1991b). The Hox-4.8 gene is localised at the 5′ extremity of the HOX-4 complex and is expressed in the most posterior parts of the body during development. *Mech. Dev.* **36,** 3-14.

Dollé, P., Izpisúa-Belmonte, J.-C., Falkenstein, H., Renucci, A., and Duboule, D. (1989). Coordinate expression of the murine Hox-5 complex homeobox-containing genes during limb pattern formation. *Nature* **342,** 767-772.

Dollé, P., Izpisúa-Belmonte, J.-C., Tickle, C., Brown, J. and Duboule, D. (1991a). Hox-4 genes and the morphogenesis of mammalian genitalia. *Genes Dev.* **5,** 1767-1776.

Dollé, P., Lufkin, T., Krumlauf, R., Mark, M., Duboule, D. and Chambon, P. (1993b). Local alterations of Krox-20 and Hox gene expression in the hindbrain of Hox-1.6 null embryos. *Proc. Natl. Acad. Sci. USA* **90,** 7666-7670.

Duboule, D. (1991). Patterning in the vertebrate limb. *Curr. Opin. Genet. Dev.* **1,** 211-216.

Duboule, D. (1992) The vertebrate limb: A model system to study the Hox/HOM gene network during development and evolution. *BioEssays* **14,** 375-384.

Duboule, D., and Dollé, P. (1989). The structural and functional organization of the murine HOX gene family resembles that of Drosophila homeotic genes. *EMBO J.* **8,** 1497-1505.

Edelman, G. M. (1988). *Topobiology: An Introduction to Molecular Embryology*. New York: Basic books.

French, V., Bryant, P. J. and Bryant, S. V. (1976). Pattern regulation in epimorphic fields. *Science* **193,** 969-981.

Gaunt, S. J. (1987). Homeo-box gene Hox-1.5 expression in mouse embryos: earliest detection by in situ hybridization is during gastrulation. *Development* **101,** 51-60.

Gaunt, S. J. (1991). Expression patterns of mouse Hox genes: Clues to an understanding of developmental and evolutionary strategies. *BioEssays* **13,** 234-242.

Gaunt, S. J., Miller, J. R., Powell, D. J. and Duboule, D. (1986). Homeo-box gene expression in mouse embryos varies with position by the primitive streak stage. *Nature* **324,** 662-664.

Gaunt, S. J., Sharpe, P. T. and Duboule, D. (1988). Spatially restricted domains of homeo-gene transcripts in mouse embryos: relation to a segmented body plan. *Development* **104 Supplement** 169-179.

Gendron-Maguire, M., Mallo, M., Zhang, M. and Gridley, T. (1993). Hoxa-2 mutant mice exhibit homeotic transformation of skeletal elements derived from cranial neural crest. *Cell* **75,** 1317-1332.

Gérard, M., Duboule, D. and Zakany, J. (1993). Structure and activity of regulatory elements involved in the activation of the Hoxd-11 gene during late gastrulation. *EMBO J.* **12,** 3539-3551.

Goodrich, E. S. (1913). Metameric Segmentation and Homology. *Quat. J. Micr. Sci.* **59,** 227-248.

Graham, A., Papalopulu, N., and Krumlauf, R. (1989). The murine and Drosophila homeobox gene complexes have common features of organization and expression. *Cell* **57,** 367-378.

Harding, K., Weddeen, C., McGinnis, W., and Levine, M. (1985). Spatially regulated expression of homeotic genes in a Drosophila. *Science* **229,** 1236-1242.

Hinchliffe, J. R. (1991). Developmental approaches to the problem of transformation of limb structure in evolution in: Developmental patterning of the vertebrate limb (ed. Hinchliffe J. R., Huerle J. M. and Summerbell D.), pp. 313-324. New York, London: Plenum publishing corp.

Holder, N. (1981). Pattern formation and growth in the regenerating limbs of urodelean amphibians (ed. R. M. Gaze, V. French, M. Snow and D. Summerbell). pp 19-36, Cambridge: Cambridge University Press.

Horder, T. (1991). Molecular biology and evolution: Two perspectives- a review of concepts. In *Developmental Patterning of the Vertebrate Limb* (eds Hinchliffe J. R., Huerle J. M. and Summerbell D.), pp. 423-438. New York, London. Plenum publishing corp.

Izpisúa-Belmonte, J.-C., Brown, J. M., Duboule D. and Tickle C. (1992). Expression of Hox-4 genes in the chick wing links pattern formation to the epithelial-mesenchymal interactions that mediate growth. *EMBO J.* **11,** 1451-1458.

Izpisúa-Belmonte, J.-C. and Duboule, D. (1992). Homeobox genes and pattern formation in the vertebrate limb. *Dev. Biol.* **152,** 26-36.

Izpisúa-Belmonte, J.-C., Falkenstein, H., Dollé, P., Renucci, A. and Duboule, D. (1991). Murine genes related to the Drosophila AbdB homeotic gene are sequentially expressed during development of the posterior part of the body. *EMBO J.* **10,** 2279-2289.

Jeanotte, L., Lemieux, M., Charron, J., Poirier, F. and Robertson, E. (1993). Specification of axial identity in the mouse: role of the Hoxa-5 gene. *Genes and Dev.* **7,** 2085-2096.

Kaufman, T. C., Seeger, M. A., and Olsen, G. (1990). Molecular and genetics organisation of the Antennapedia gene complex of *Drosophila melanogaster*. In Advances in Genetics: *Genetic regulatory hierarchies in development*, Vol. 27. (ed. J. G. Scandalios and T. R. F. Wright). 309-362. Academic Press.

Kessel, M., Balling, R. and Gruss, P. (1990). Variations of cervical vertebrate after expression of a Hox-1.1 transgene in mice. *Cell* **61,** 301-308.

Kessel, M. and Gruss, P. (1991). Homeotic transformation of murine vertebrae and concommitant alteration of Hox code induced by retinoic acid. *Cell* **67,** 89-104.

Kostic, D. and Capecchi, M. R. (1994). Targeted disruption of the murine Hoxa-4 and Hoxa-6 genes result in homeotic transformations of components of the vertebral column. *Mech. Dev.* (in press).

Lewis, E. B. (1978). A gene complex controlling segmentation in Drosophila. *Nature* **276,** 565-570.

LeMouellic, H., Lallemand, X. and Brûlet, P. (1992). Homeosis in the mouse induced by a null mutation in the Hox-3.1 *Gene. Cell* **69,** 251-264.

Lufkin, T., Dierich, A., LeMeur, M., Mark, M., and Chambon, P. (1991). Disruption of the Hox-1.6 homeobox gene results in defects in a region corresponding to its rostral domain of expression. *Cell* **66,** 1105-1120.

Lufkin, T., Mark, M., Hart, C.P., Dollé, P., LeMeur, M., and Chambon, P. (1992). Homeotic transformation of the occipital bones of the skull by ectopic expression of a homeobox gene. *Nature* **359,** 835-841.

McGinnis, W. and Krumlauf, R. (1992). Homeobox genes and axial patterning. *Cell* **68,** 283-302.

Morata, G., Macias, A., Urquia, N. and Gonzales-Reyes, A. (1990). Homeotic genes. *Seminars in Cell Biol.* **1,** 219-229.

Raff, R. A. (1992). Evolution of developmental decisions and morphogenesis: the view from two camps. *Development* **Supplement** 15-22.

Ramirez-Solis, R., Zheng, H., Whiting, J., Krumlauf, R. and Bradley, A. (1993). Hoxb-4 (Hox-2.6) mutant mice show homeotic transformation of a cervical vertebra and defects in the closure of the stermal rudiments. *Cell* **73,** 279-294.

Riddle, R. D., Johnson, R. L., Laufer, E. and Tabin, C. (1993). Sonic hedgehog mediates the polarizing activity of the ZPA. *Cell* **75,** 1401-1416.

Rijli, F. M., Mark, M., Lakkaraju, S., Dierich, A., Dollé, P. and Chambon, P. (1993). A homeotic transformation is generated in the rostral branchial region of the head by disruption of Hoxa-2, which acts as a selector gene. *Cell* **75,** 1333-1350.

Rose, S. M. (1962). Tissue-arc control of regeneration in the amphibian limb. In *Regeneration* (ed. D. Rudnick). pp 153-176, Ronald Press.

Sander, K. (1983). The evolution of patterning mechanisms: gleanings from insect embryogenesis and spermatogenesis. In *Development and Evolution* (eds. B. C. Goodwin, N. Holder and C. C. Wylie), pp.124-137.

Sham, M., H., Hunt, P., Nonchev, S., Papalopulu, N., Graham, A., Boncinelli, E., and Krumlauf, R. (1992). Analysis of the murine Hox-2.7 gene reveals multiple transcripts with differential distributions in the nervous system. *EMBO J.* **11,** 1825-1836.

Slack, J. M. W. (1980). A serial threshold theory of regeneration. *J. Theor. Biol.* **82,** 105-140.

Slack, J. M. W., Holland, P. W. H. and Graham, C. F. (1993). The zootype and the phylotypic stage. *Nature* **361,** 490-491.

Small, K. M. and Potter, S. S. (1993). Homeotic transformations and a limb defect in Hoxa-11 mutant mice. *Genes Dev.* **7,** 2318-2328.

Summerbell, D. (1981). Evidence for regulation of growth, size and pattern in the developing chick limb bud. In Growth and the Development of Pattern (eds: R. M. Gaze, V. French, M. Snow and D. Summerbell). pp. 129-150, Cambridge University Press.

Summerbell, D., Lewis, J. H., and Wolpert, L. (1973). Positional information in chick limb morphogenesis. *Nature* **244**, 492-496.

Stocum, D. (1981). Distal transformation in regenerating double anterior axolotl limbs. in: Growth and the development of pattern (ed. R. M. Gaze, V. French, M. Snow and D. Summerbell), pp. 3-18. Cambridge University Press.

Whitting J., Marshall, H., Cook, M., Krumlauf, R., Rigby, P., W., J., Stott, D. and Allemann, R., K. (1991). Multiple spatially specific enhancers are required to reconstruct the pattern of Hox-2.6 gene expression. *Genes Dev.* **5,** 2048-2059.

Wolpert, L. (1989). Positional information revisited. *Development* **107, Supplement** 3-12.

Wolpert, L. (1991). *The Triumph of the Embryo.* Oxford University Press.

Development 1994 Supplement, 143-153 (1994)
Printed in Great Britain © The Company of Biologists Limited 1994

Developmental expression of the mouse *Evx-2* gene: relationship with the evolution of the HOM/Hox complex

Pascal Dollé[1], Valérie Fraulob[1] and Denis Duboule[2]

[1]Laboratoire de Génétique Moléculaire des Eucaryotes du CNRS, Unité184 INSERM, Faculté de Médecine, 11, rue Humann, 67085 Strasbourg Cédex, France
[2]Département de Zoologie et Biologie Animale, Université de Genève, Sciences III, Quai Ernest Ansermet 30, 1211 Genève 4, Suisse

SUMMARY

The mouse *Evx-2* gene is located in the immediate vicinity of the *Hoxd-13* gene, the most posteriorly expressed gene of the HOXD complex. While the *Evx-1* gene is also physically linked to the HOXA complex, it is more distantly located from the corresponding *Hoxa-13* gene. We have analysed the expression of *Evx-2* during development and compared it to that of *Evx-1* and *Hoxd-13*. We show that, even though *Evx-2* is expressed in the developing CNS in a pattern resembling that of other *Evx*-related genes, the overall expression profile is similar to that of the neighbouring *Hoxd* genes, in particular with respect to the developing limbs and genitalia. We propose that the acquisition of expression features typical of Hox genes, together with the disappearance of some expression traits common to *Evx* genes, is due to the close physical linkage of *Evx-2* to the HOXD complex, which results in *Evx-2* expression being partly controlled by mechanisms acting in the HOX complex. This transposition of the *Evx-2* gene next to the *Hoxd-13* gene may have occurred soon after the large scale duplications of the HOX complexes. A scheme is proposed to account for the functional evolution of *eve*-related genes in the context of their linkage to the HOM/Hox complexes.

Key words: homeobox, *even-skipped*, development, HOX complex phylogeny

INTRODUCTION

The mouse genome contains 38 Hox genes which are clustered in four genomic loci (the HOXA,B,C and D complexes). These complexes have similar features of structure and organization as they arose by large-scale duplication of an ancestral complex (reviewed in McGinnis and Krumlauf, 1992; Dollé and Duboule, 1993). During development, Hox genes are activated during gastrulation, while the embyo establishes its anteroposterior axis. *Hox* transcripts are found in partially overlapping domains along the embryonic axis, with genes located towards the 5′ extremity of one complex being expressed in progressively more posterior (caudal) areas of the embryo (Gaunt et al., 1988). All these genes are transcribed from the same DNA strand so that each complex can be assigned with a general 5′ (posterior) to 3′ (anterior) orientation. The importance of this gene family to the establishment and realization of the vertebrate body plan has been recently demonstrated by gene inactivation experiments (LeMouellic et al., 1992; Ramirez-Solis et al., 1993; Dollé et al., 1993; Condie and Capecchi, 1993). The results of such inactivations suggest that the function of Hox genes could be mediated through the control of either the local growth or patterning (or both) of structures of neurectodermal and/or mesodermal origin.

In their 5′ regions, the HOXA, C and D complexes contain 4-5 Hox genes that are related to the *Drosophila Abdominal-B* homeotic gene (Izpisúa-Belmonte et al., 1991). The tandem duplication of these Hox sub-groups is thought to have occurred before large scale HOX cluster duplication and may, perhaps, have been linked to evolution of appendicular structures, since such genes are coordinately expressed in the developing limbs and genital tubercle (Dollé et al., 1989, 1991a; Yokouchi et al., 1991) in which they exert an important function (Dollé et al., 1993; Small and Potter, 1993).

The coordinate expression of Hox genes may result from a successive temporal activation of these genes, in a 3′ to 5′ sequence, the *temporal colinearity* (see Dollé et al., 1989; Izpisúa-Belmonte et al., 1991; Duboule, 1992). As a consequence, genes located in 5′ positions display transcript domains that are always more distal (or posterior and distal at earlier stages) than their 3′ neighbour genes (the *spatial colinearity*). In the genital tubercle, quantitative differences are observed rather than true spatial colinear expression domains (Dollé et al., 1991a). The fact that some of these genes are functional in all axial skeleton structures has been confirmed by the experimental knock-out of *Hoxd-13*, which alters both the growth and patterning of skeletal elements arising from the limb buds and genital tubercle (Dollé et al., 1993).

Evx-1 and *Evx-2* are the two genes identified to date in mammals, which contain a homeobox homologous to that of the *even-skipped* (*eve*) gene from *Drosophila* (Bastian and Gruss, 1990). A genomic walk along the upstream (5′) part of the human *HOXD* locus revealed that the human *EVX2* gene is located in the close vicinity of *HOXD13* (D'Esposito et al., 1991) and a similar linkage was reported in mice (Bastian et al., 1992). The human *EVX2* homeobox is found about 13 kilobases (kb) away from that of *HOXD13*. Similarly, the *EVX1* gene is linked to the HOXA complex. In this case, however, its distance from the complex is larger, as it lies approximately 45 kb upstream of the *HOXA13* gene (Faiella et al., 1992). Interestingly, both *EVX1* and *EVX2* are transcribed from the DNA strands that are opposite to those from which all *HOX* sequences are transcribed (Faiella et al., 1992; D'Esposito et al., 1991).

The developmental expression of the murine *Evx-1* gene has been studied in detail; *Evx-1* starts to be expressed at approx. 6.5 days post-coitum (dpc) in part of the embryo epiblast, and then throughout gastrulation (6.5-8.5 dpc) in a graded fashion along the primitive streak and the involuting mesodermal cells, but not anterior to the primitive streak (Dush and Martin, 1992). Later on, *Evx-1* is still expressed in the tail bud, which corresponds to the primitive streak in its histogenic potential to generate various tail structures. Starting at approx. 10.5 dpc, *Evx-1* is also expressed in discrete neural areas along the spinal cord and hindbrain (Bastian and Gruss, 1990). Finally, *Evx-1* transcripts are transiently detected in the limb buds, where they first appear in the posterior distal mesenchyme, and are subsequently maintained only in very distal (subectodermal) cells (Niswander and Martin, 1993). Thus, none of the three components of the *Evx-1* pattern during development (early, neural and limb expression) resembles the coordinate regulation of the neighbouring Hox genes, which are activated at later stages of gastrulation and in progressively more posterior areas (Yokouchi et al., 1991; Haack and Gruss, 1993).

We have analysed further the linkage of *Evx-2* to the mouse HOXD complex and show that it lies in close contact with the *Hoxd-13* gene (approx. 8 kb). Such a distance is in the range of the usual spacing between Hox genes themselves and we investigated whether *Evx-2* gene expression could, in part, be controlled by regulatory mechanisms acting on the neighbouring Hoxd genes, or instead, would be regulated independently from the complex, as seems to be the case for *Evx-1*. We report here that *Evx-2* is regulated as expected for a gene belonging to the HOXD complex, in both the limb buds and the genitalia. Interestingly, *Evx-2* expression follows temporal colinearity (i.e. shortly after the *Hoxd-13* gene), and thus does not share the early expression phase of *Evx-1* during gastrulation. However, an important aspect of *Evx-2* expression (in the CNS) makes it different from other Hoxd genes but similar to *Evx-1*. The relationships between the respective genomic locations of Evx genes and the potential control of their expression by regulatory mechanisms acting upon the HOX complexes are discussed from an evolutionary perspective.

Evx-2 IS NEAR *Hoxd-13* AND IS ACTIVATED LATE IN DEVELOPMENT

The cloning and sequence analysis of DNA fragments located upstream the *Hoxd-13* gene on genomic cosB (Dollé et al., 1991b) confirmed the linkage between the *Evx-2* sequences and the HOXD complex. Sequences were identified that correspond to the previously reported sequence of an *Evx-2* cDNA (Dush and Martin, 1992) and encode the N terminus of the *Evx-2* product. We localized the *Evx-2* ATG initiation codon to 8 kb from the *Hoxd-13* initiation codon, in an inverted orientation (Fig. 1). The spacing between *Evx-2* and *Hoxd-13* coding regions was thus found to be in the range of that observed between all five neighbouring *Abdominal-B* related Hox genes (from 5 to 10 kb; Fig. 1). This genomic organization is distinct from that of the *EVX1* locus, which was mapped, in human, at about 45 kb from *HOXA13* (Faiella et al., 1992; Fig. 1).

Dush and Martin (1992) reported the activation of *Evx-1* shortly before the onset of gastrulation; here we compared *Evx-1* and *Evx-2* expression at corresponding stages by in situ hybridization. No detectable *Evx-2* labeling was observed in 7.5 dpc embryo sections (Fig. 2A). In contrast, a graded dis-

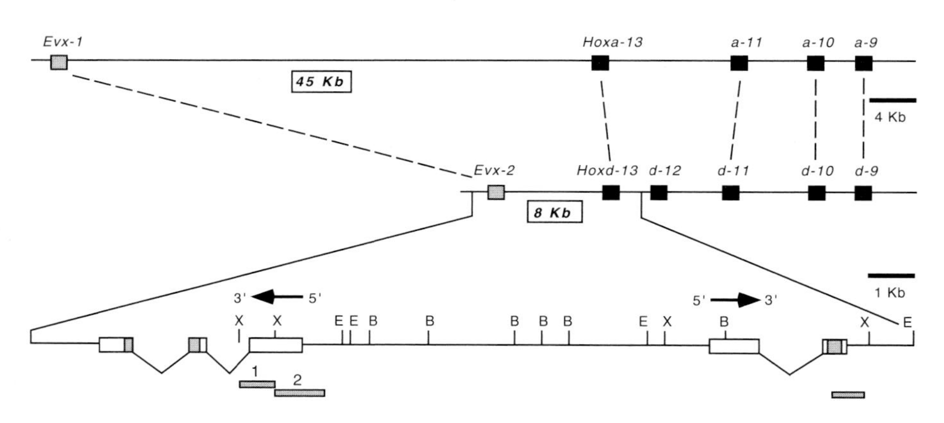

Fig. 1. Structural relationships between the HOXA/Evx-1 and HOXD/Evx-2 complexes. 12 kilobases of genomic DNA extending 5′ from the *Hoxd-13* transcription unit (Dollé et al., 1991b) were subcloned and shown, by Southern blot analysis, to contain *Evx-2* sequences (Bastian et al., 1992). Extensive sequence analysis localized the 5′ extremity of a previously reported *Evx-2* cDNA clone (Dush and Martin,1992) as well as the position of the exon containing the 5′ part of the *Evx-2* homeobox. The *Evx-2* coding sequence was found to lie on the opposite DNA strand to that of *Hoxd-13*. *Evx-2* and *Hoxd-13* ATG initiation codons were mapped and found to be separated by approx. 8 kilobases of genomic DNA. The five *Hox* subgroups related to *Abd-B* (groups 9 to 13) are shown here as black boxes. Dashed lines indicate paralogous genes on the HOXA and HOXD complexes. The HOXA complex lacks a gene member of group 12. The distances between the various genes are drawn to scale to emphasize the important difference in the spacing between *Evx-1/Hoxa-13* and *Evx-2/Hoxd-13*. The bottom line shows an enlargement of the *Evx-2/Hoxd-13* region with arrows indicating the opposite directions of transcription. The transcription units are sketched with, in dark boxes, the positions of the two homeoboxes. Stippled rectangles below the line point to the locations of the various RNA probes used for in situ hybridization experiments. X, *Xho*I; B, *Bam*HI; E, *Eco*RI (data from this work and Bastian et al., 1992; Faiella et al., 1992; D'Esposito et al., 1991; Haack and Gruss, 1993).

tribution of *Evx-1* transcripts was seen in the primitive streak, the embryonic mesoderm and the epiblast (Fig. 2A; and Dush and Martin, 1992). *Evx-2* expression was also not detected in 8.5 dpc embryos, a stage at which the *Evx-1* signal was present in caudal areas (Fig. 2B). At this stage, a restricted *Hoxd-13* signal was seen in the most posterior embryonic mesoderm and hindgut (Fig. 2B, arrow). This probably represents the time of *Hoxd-13* activation, slightly earlier than previously reported (Dollé et al., 1991b). At later stages of development, no *Evx-2* labeling was detected in the tail bud (Fig. 2C), a region equivalent to the primitive streak (Tam, 1984), even though specific signals were observed in other areas of the same embryos (see below). In contrast, both *Evx-1* and *Hoxd-13* transcripts were detected in the tail bud, although with different distributions

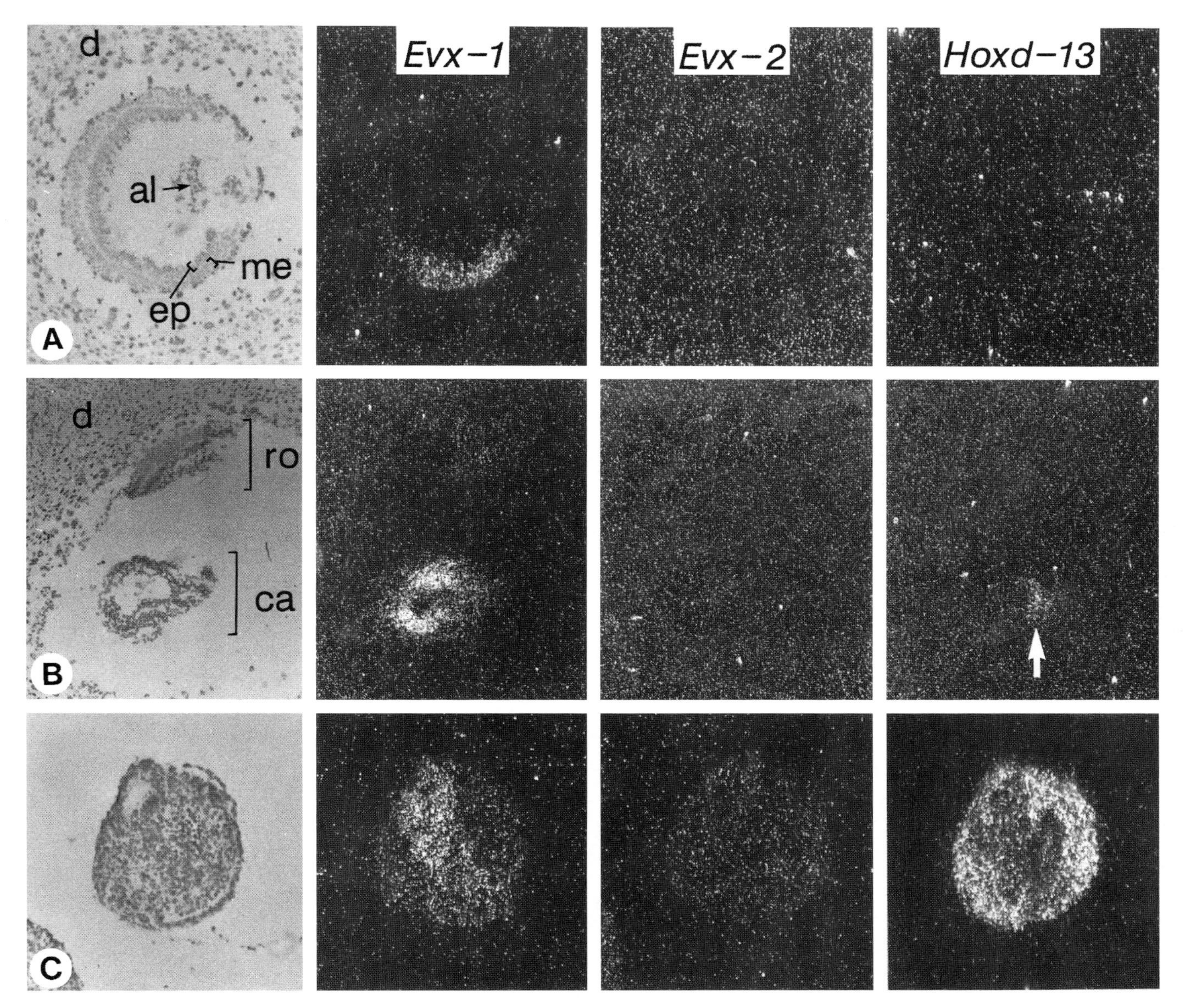

Fig. 2. Activation of *Evx-1, Evx-2* and *Hoxd-13*. (A) Three sections of a 7.5 dpc mouse embryo hybridized to *Evx-1*, *Evx-2* or *Hoxd-13* riboprobes. A corresponding section viewed under bright-field light is shown on the left for histological details. The caudal part of the embryo is towards the bottom of the picture, next to the allantois. (B) Sections through the caudal extremity (ca) of a 8.5 dpc embryo. In addition to the strong *Evx-1* signal, a restricted labeling is detected for *Hoxd-13* (arrow) while no *Evx-2* transcripts are observed yet. The most rostral part of the same embryo (ro) is negative for all three probes. (C) Enlargement of the tail bud area of a 10.5 dpc embryo, which shows distinct distributions of *Evx-1* and *Hoxd-13* transcripts. No *Evx-2* signal is detected even though other areas of the same embryo are positive at this stage (see below). For in situ hybridization, embryos and fetuses were collected at various stages of gestation, and the day of the vaginal plug appearance was considered as day 0.5 post-coitum (dpc). The embryos were either fixed overnight in 4% paraformaldehyde and embedded in paraffin wax, or frozen directly in OCT medium. In situ hybridization was performed both on paraffin-embedded embryo sections and frozen sections; the latter result in higher signal to noise ratio (Décimo et al., 1994) although with a somewhat poorer histology. In both cases, series of adjacent sections were collected on three sets of slides which were subsequently hybridized to *Evx-2*, *Hoxd-13* and *Evx-1* probes. ep, epiblast (embryonic ectoderm); me, mesoderm; al, allantois; d, decidual tissue; ro, rostral; ca, caudal.

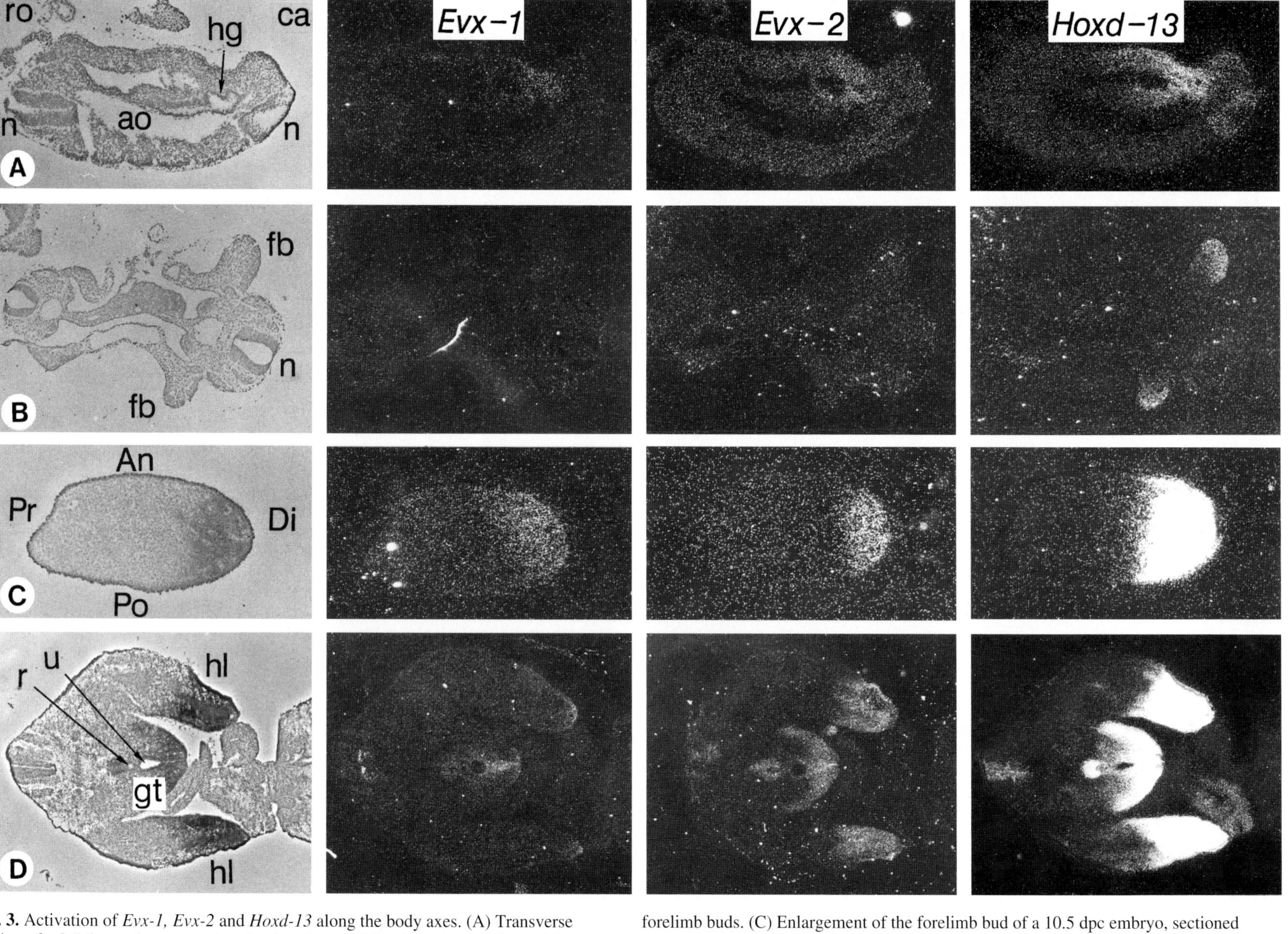

Fig. 3. Activation of *Evx-1, Evx-2* and *Hoxd-13* along the body axes. (A) Transverse section of a 9.5 dpc embryo, crossing the dorsal part of the abdominal cavity and the hindgut (cloaca). Note the similar and limited *Evx-2* and *Evx-1* transcript domains in the vicinity of the hindgut diverticulum. (B) Transverse section of a 10.0 dpc embryo, through both forelimb buds. The specimen was collected as a 10.5 dpc embryo, but was developmentally retarded. At this stage, only *Hoxd-13* transcripts are expressed in the forelimb buds. (C) Enlargement of the forelimb bud of a 10.5 dpc embryo, sectioned sagitally. (D) Transverse section through the hindlimbs and genital tubercle of a 12.5 dpc embryo. At this stage, there is a strong difference between *Evx-2* and *Evx-1* transcripts in both structures. hg, hindgut; ao, aorta; n, neural tube; fb, forelimb buds; hl, hindlimbs; gt, genital tubercle; u, urogenital sinus; r, rectum; ro, rostral; ca, caudal; Pr, proximal; Di, distal; An, anterior; Po, posterior.

(Fig. 2C). *Evx-2* expression was not merely delayed in the primitive streak and tail bud, with respect to *Evx-1*; instead, *Evx-2* was not activated in these structures.

The earliest signs of *Evx-2* transcription were seen in 9.5 dpc embryos, in the cloacal epithelium and surrounding mesoderm (Fig. 3A), from which the genital bud will arise. *Hoxd-13* was also expressed in these structures, but extending over a larger area (Fig. 3A). *Evx-1* was also expressed in this future genital region (Fig. 3A). *Evx-2* transcripts were not visible in 9.5 or 10.0 dpc limb buds while *Hoxd-13* labeling was clearly detected in posterior mesenchyme (Fig. 3B). The first detection of *Evx-2* expression in the forelimb bud was at 10.5 dpc, when it was restricted to a subset of *Hoxd-13*-expressing cells in distal and posterior mesenchyme, within an area that also expressed *Evx-1* (Fig. 3C). The co-expression of *Evx-1* and *Evx-2* in both the genital area and limb buds was transient and the two expression patterns diverged by 11.5 dpc (not shown). In 12.5 dpc fetuses, *Evx-1* was expressed in the very distal extremities of the limb mesenchyme, while *Evx-2* displayed a Hoxd-type of expression pattern, that is, across a wider distal region (Fig. 3D; see below).

EXPRESSION IN LIMB BUDS AND GENITALIA

Hoxd genes are expressed in the limb buds along spatial domains which are colinear with the ordering of the genes along the chromosome (Dollé et al., 1989). In both 11.5 dpc fore- and hindlimbs, the distribution of *Evx-2* transcripts was consistent with this gene being subject to a similar mechanism of regulation. Successive sections of hindlimb buds showed that *Hoxd-13* transcripts were distributed in slightly more extended domains than those of *Evx-2*, both more proximally, along the posterior margin of the bud (Fig. 4A), and more anteriorly, when progressing towards the tip of the bud (Fig. 4B,C). These differences were still visible in 12.5 dpc limbs (Fig. 4D). Both *Evx-2* and *Hoxd-13* transcripts were restricted to the footplates; however, the *Evx-2* domain was not as proximal as the *Hoxd-13* domain along the posterior margin and the anterior margin (Fig. 4D). The weak expression of *Evx-2* in the limbs of 13.5 dpc fetuses indicated that its transcript domain extended into all five digit primordia, up to the base of the most anterior digit I (Fig. 4E), a domain equivalent to that in which *Hoxd-13* is expressed and functional (Dollé et al., 1993). However, while transcripts of 5′-located Hoxd genes are known to accumulate at high levels in the limb extremities, with *Hoxd-13* displaying the strongest expression until at least 17.5 dpc (Dollé et al., 1991b; unpublished data), the *Evx-2* signals remained much weaker at 12.5 and 13.5 dpc (see Fig. 4D-E) and were hardly detectable at 14.5 dpc (not shown).

Hoxd genes are expressed in the genital tubercle where 5′-gene transcripts are more abundant, and maintained for longer times, than 3′-gene transcripts (Dollé et al., 1991a). Serial sections of 11.5 dpc embryos showed that *Evx-2* also had a specific expression domain in the genital bud and urogenital mesenchyme. This domain is similar to that of *Hoxd-13*, both being centred in the genital bud, and around the urogenital sinus and rectum (Fig. 5A). However, *Evx-2* transcripts were not detected in more lateral regions of the genital bud (Fig. 5B, see also Fig. 4B,C), suggesting either that *Evx-2* transcripts may be restricted to a subset of *Hoxd-13*-expressing cells in the genital bud and urogenital mesenchyme or that the transcript level is too low to be detected.

CO-EXPRESSION OF *Evx-2* AND *Evx-1* IN THE CENTRAL NERVOUS SYSTEM

Evx-2 was expressed in discrete cell layers within the embryonic central nervous system (CNS). This was first detected at 10.5 dpc and was clearly apparent until 12.5 dpc. The labeling was observed in a thin and continuous column of cells along the entire spinal cord in near mid-sagittal sections (Fig. 6A) as well as in more scattered cells of the ventral spinal cord, visible in more lateral sections (Fig. 6B). As this labeling was reminiscent of that reported for *Evx-1* (Bastian and Gruss, 1990), we compared the expression of both genes at various levels of the CNS on both transverse and coronal sections. Double-labeling experiments were performed in which the two probes were hybridized, either separately, or together on a third set of sections in order to assess possible co-expression. On transverse sections of 12.5 dpc embryo spinal cord, *Evx-2* labeling was indistinguishable from, although weaker than, that of *Evx-1* (Fig. 6C). In the cervical spinal cord, *Evx-2* signal was found in two symmetrical columns in the medial part of the ventricular zone (Fig. 6C, arrowheads; also in Fig. 6A), as well as in cells located between the ventral horns and the ventricular zone (Fig. 6C, open arrows; also in Fig. 6B) and in the most dorsal layers of the spinal cord (Fig. 6C, filled arrows). Only the former two areas were labeled in more caudal regions of the spinal cord (thoracic and lumbar; data not shown). Sections hybridized to both probes together showed no detectable extension of the signal, which indicated a probable co-expression of the two genes in the same cells (Fig. 6C; panel '1+2'). A similar co-expression was found more rostrally, at the level of the developing hindbrain (Fig. 6D). In the CNS, *Evx-1* and *Evx-2* appeared to be activated at a similar developmental stage, in 10.5 dpc embryo (data not shown).

In contrast to the similarities in the spinal cord, a striking difference was seen in the extent of *Evx-2* and *Evx-1* transcripts domains towards more rostral regions of the developing brain. While *Evx-1* transcripts were expressed up to the rhombencephalic isthmus area (the metencephalon-mesencephalon boundary, see Bastian and Gruss, 1990; Figs 6D, 7), *Evx-2* transcripts extended into the superficial layer of the entire midbrain (Fig. 7A,B). The point of divergence between *Evx-1* and *Evx-2* transcript distributions in the marginal layer was seen on coronal sections of 11.5 dpc brain (Fig. 7A), where it appeared that the *Evx-2* signal was found more rostrally than *Evx-1* in the same layer. Therefore, the expression of *Evx-2* in the superficial layer of the midbrain (Fig. 7B) corresponds to a rostral extension of domains where *Evx-2* and *Evx-1* are coexpressed.

IS *Evx-2* A *Hoxd* GENE?

Members of the vertebrate HOX gene complexes are expressed, during development, according to a set of rules that control the coordination of their functions. An interesting aspect of these rules is that they rely on the genomic organization of this gene family, that is, they rely on the respective

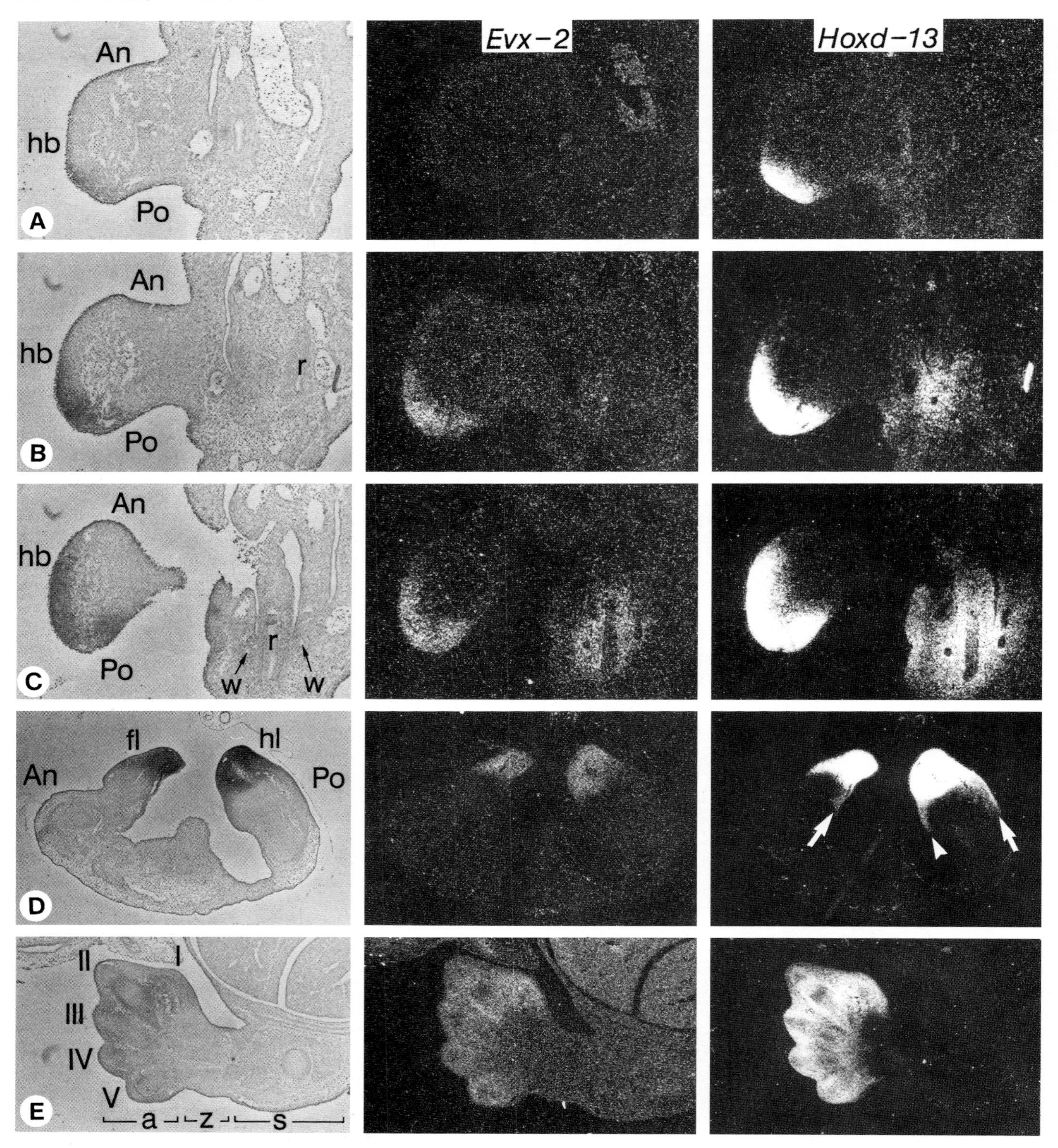

Fig. 4. *Evx-2* expression in developing limbs. (A-C) Three serial sections of an 11.5 dpc hindlimb bud. The bud is sectioned along its A-P axis, and from A to C, the sectioning progresses towards the tip of the bud (i.e. the sections progress from the dorsal side towards the ventral side of the embryo). In addition to the distinct domains of *Evx-2* and *Hoxd-13* transcripts in the hindlimb bud, compare also the labeling patterns in the trunk (urogenital area). (D) Sagittal section of a 12.5 dpc fetus, crossing both the forelimb and hindlimb. (E). Sagittal section of a 13.5 dpc fetus through the footplate of the hindlimb (crossing all five digit anlagen). hb, hindlimb bud; An, anterior; Po, posterior; r, rectum; w, wolffian ducts; fl, forelimb; hl, hindlimb; I, II, III, IV, V, digit primordia; a, autopod (footplate); z, zeugopod; s, stylopod.

position of each gene within its complex. Thus, a particular position will coincide with both the time of activation (temporal colinearity, Dollé et al., 1989; Izpisúa-Belmonte et al., 1991) and the craniocaudal position of the expression domain (spatial colinearity, Gaunt et al., 1988). Though the mechanisms underlying these rules have yet to be discovered, it has been hypothesized that spatial and temporal colinearities were major contributory factors to the conservation of this clustered organization during evolution (Duboule, 1992, 1994).

The two vertebrate genes *Evx-1* and *Evx-2* are related to the *Drosophila even-skipped* gene and are closely linked to the 5′ extremities of the HOXA (*Evx-1*) and HOXD (*Evx-2*) complexes (D'Esposito et al., 1991; Faiella et al., 1992; Bastian et al., 1992). Interestingly, while the distances between paralogous Hox genes have been relatively well conserved among the various HOX complexes (an average of approx. 10 kb), the distances between the Evx genes and the HOXA and D complexes have not been equally conserved; *EVX1* is separated from *HOXA13* by 45 kb, whereas *EVX2* was found in the close vicinity of *HOXD13* (D'Esposito et al., 1991; Faiella et al., 1992). We have further analysed the mouse *Evx-2* locus and found that this gene lies 8 kb from the *Hoxd-13* gene, in an inverted orientation. Thus, while the distance between *EVX1* and *HOXA13* is well above those usually found between vertebrate Hox genes, the spacing between *Evx-2* and *Hoxd-13* is in the range of the distances separating Hoxd genes themselves.

The expression profile of *Evx-1* during development (Bastian and Gruss, 1990, Dush and Martin, 1992, this work) suggests that its physical linkage to the HOXA complex is not associated with a 'Hox type' of regulation. Its early expression, at the onset of gastrulation, does not follow temporal colinearity (it is expressed well before *Hoxa-13*) and, subsequently, no restriction to posterior regions is observed in the spinal cord expression (as is the case for *Hoxa-13*). Even though *Evx-1* transcripts are found in both limb buds and genital tubercle, their distribution in time is distinct from that of *Hox* transcripts in these structures. As the *Evx-2* gene was found to be physically closer to the HOXD complex, we analysed whether its

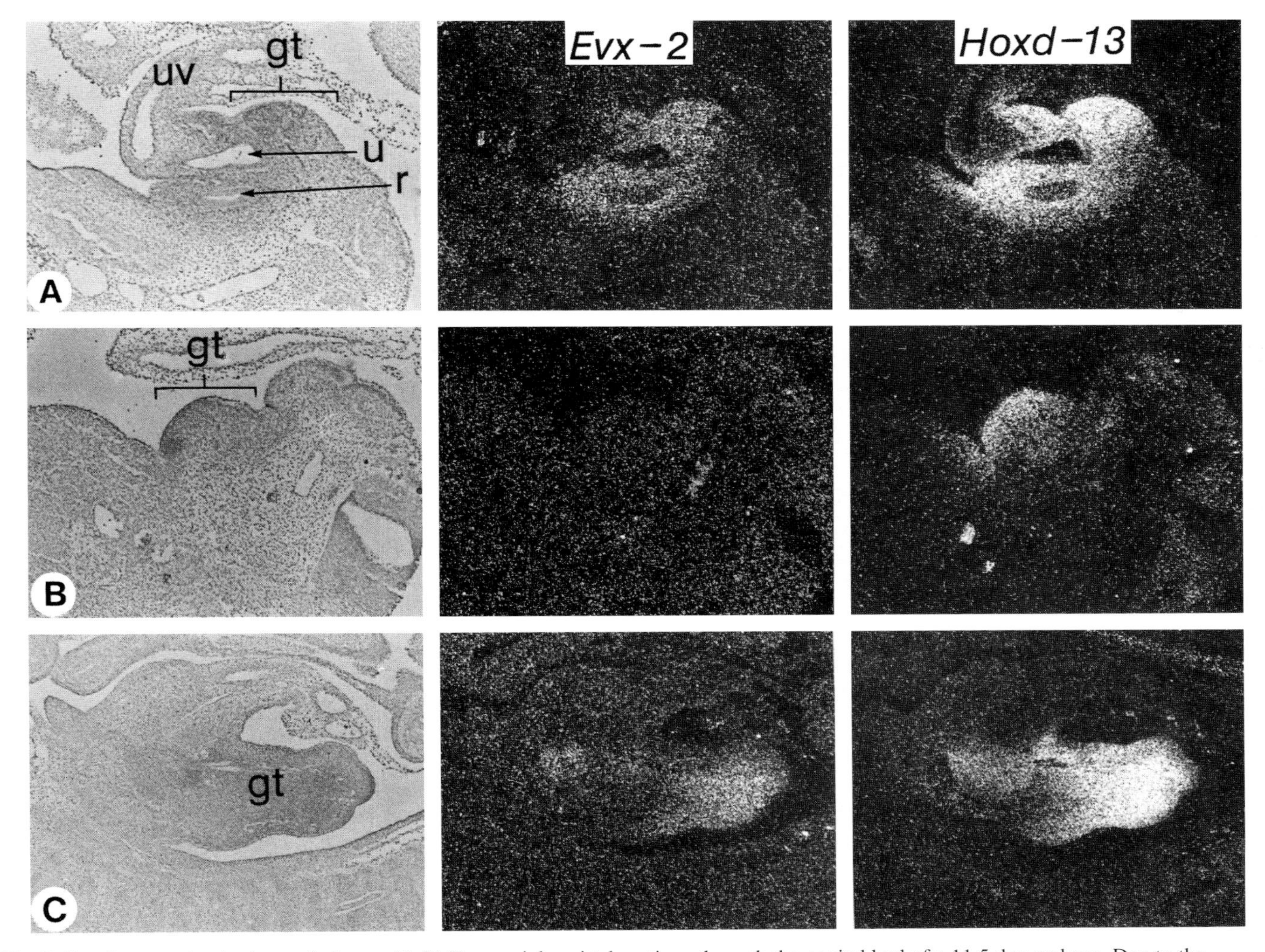

Fig. 5. *Evx-2* expression in the genital area. (A,B) Two serial sagittal sections through the genital bud of a 11.5 dpc embryo. Due to the curvature of the embryo, the sectioning becomes more lateral towards the caudal part of the embryo (right side of the pictures). Section (A) goes through medial parts of the genital region and crosses the umbilical vessels, rectum and urogenital sinus, while (B) covers more lateral regions. (C) Sagittal section through the genital tubercle of a 13.5 dpc fetus. gt, genital tubercle; u, urogenital sinus; r, rectum; uv: umbilical vessels.

expression could be subject to regulatory influences of the nearby HOXD complex. Our results show that the expression of *Evx-2* corresponds, in many respects, to that of a genuine *Hoxd* gene. However, an important aspect of regulation conserved between *Evx-1* and *Evx-2* relates to expression in the central nervous system.

Evx-2 expression seems to comply with temporal colinearity of the HOXD complex (Dollé et al., 1989; Izpisúa-

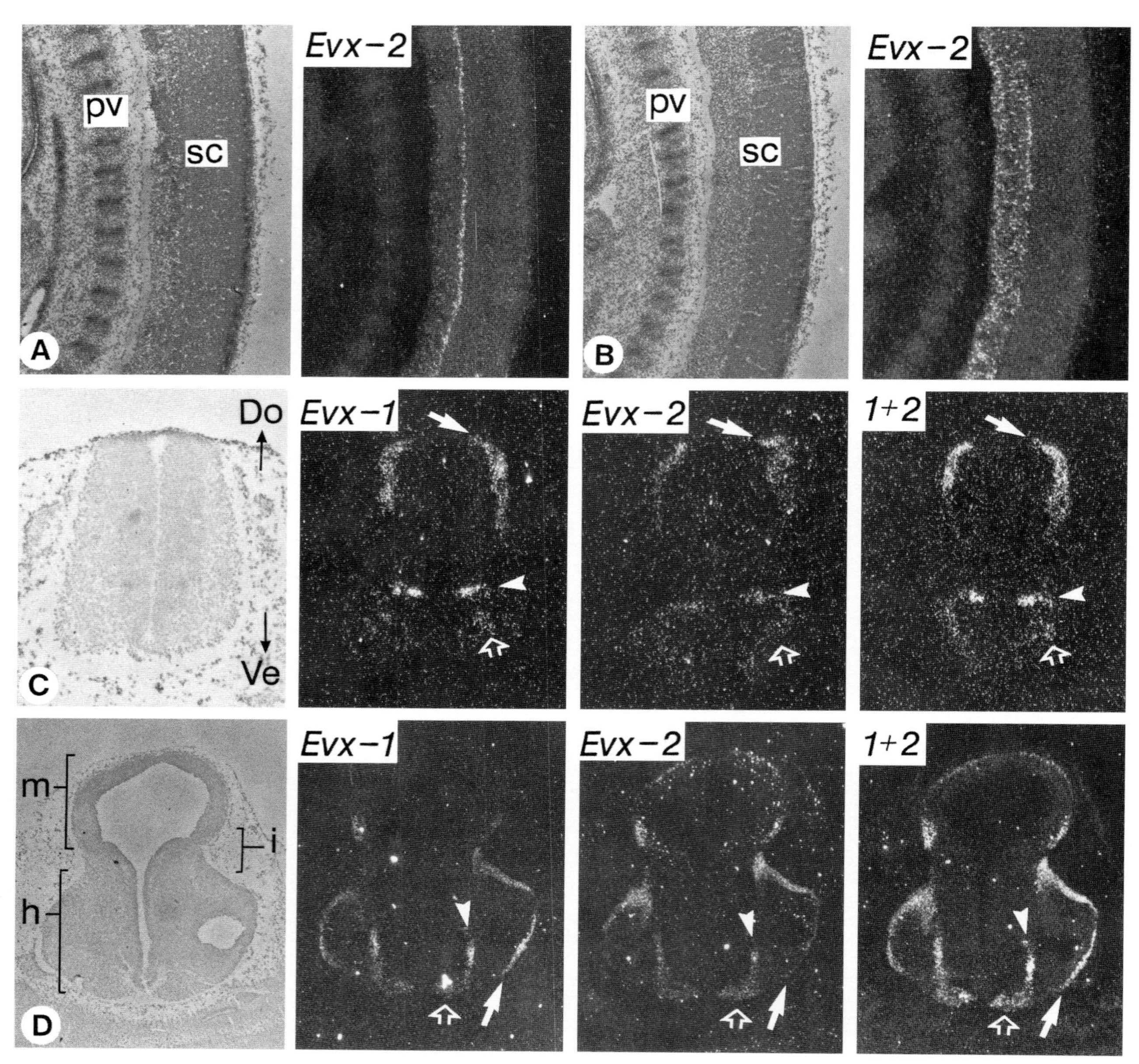

Fig. 6. *Evx-2* expression in the hindbrain and spinal cord. (A,B) Two serial sagittal sections through the thoracic spinal cord of the same 12.5 dpc fetus, showing two distinct components of *Evx-2* expression. Section A, which is rather medial, is essentially through the ventricular zone and shows the thin column of *Evx-2*-expressing cells in this area (also visible in C, arrowhead). Section B is slightly more lateral and crosses the ventral part of the mantle layer. A few scattered cells express *Evx-2* all along this ventral area (visible in C, open arrow). (C). Three adjacent transverse sections of the cervical spinal cord from a 12.5 dpc fetus were hybridized to *Evx-2* (middle panel), *Evx-1* (left panel) and to a mixture of both probes (right panel). This transverse view shows the two populations of labeled cells shown in A and B; the medial column (arrowheads) and the ventral cells (open arrows). In addition, the dorsal-most layers are also labeled at the cervical level (filled arrows). Note the similarity of the labeling patterns for both probes, as well as in the double labeling. (D) Coronal sections through the hindbrain of a 12.5 dpc fetus. The same three components of *Evx-1* and *Evx-2* expression are found along the hindbrain; the thin column of cells (arrowheads), groups of ventral cells (open arrows) and the most superficial layer (filled arrows). The labeling patterns obtained with each probe, and with the mixture of both probes, are indistinguishable in the hindbrain. Note, however, the sharp boundary of *Evx-1* transcripts close to the rhombencephalic isthmus, whereas *Evx-2* transcripts extend in the superficial cells of the midbrain (also observed in the double labeling experiment). sc, spinal cord; pv, prevertebral column; do, dorsal; ve, ventral; h, hindbrain; m, midbrain; i, isthmus.

Belmonte et al., 1991), as there was no area in the embryo where *Evx-2* transcripts could be detected prior to those of *Hoxd-13*. *Evx-2* transcripts were detected earliest in 9.5 dpc embryos, in a region of the genital anlage corresponding to a subset of the *Hoxd-13* domain. This suggests that *Evx-2* is activated slightly after *Hoxd-13* in the cell population that will eventually give rise to the genitalia. This delay was even clearer in limb buds, where *Evx-2* expression was detectable only by day 10.5 in posterior forelimb mesenchyme, a region that expressed *Hoxd-13* at 9.5-10.0 dpc. Possibly as a result of this temporal progression, the *Evx-2* transcript domain in developing limbs appears to be more restricted than that of *Hoxd-13*. This is in good agreement with the spatial colinearity observed amongst Hoxd genes in limbs (Dollé et al., 1989).

Temporal colinearity may be respected in the activation of *Evx-2* in the central nervous system as well, even though this is a structure where the *Evx-2* expression pattern is clearly distinct from that of all the neighbouring Hoxd genes. Therefore, there is a correlation between the close association of *Evx-2* to the HOXD complex and the fact that its developmental expression follows the rules of temporal and spatial colinearities acting upon the adjacent complex. In this respect, and in contrast to *Evx-1*, *Evx-2* could be considered as an additional *Hoxd* gene. Such an assimilation is based on the location and regulation of *Evx-2* and does not necessarily imply similar functions of the gene products. Furthermore, the following reservations are noteworthy; first, the *Evx-2* signal was always much weaker than that of *Hox* genes, making it possible that the distributions of transcripts, in both space and time, were underestimated due to technical problems with detection of gene activity. However, results obtained with other mouse *Evx-2* probes confirm that the gene is expressed at a low level, as is also suggested by comparison of *Evx-2* versus *Hoxd-13* expression in zebrafish (P. Sordino and D. D., unpublished data). Second, the expression of *Evx-2* in the CNS clearly contradicts spatial colinearity as *Hoxd-13* is not expressed in the CNS anterior to the sacral region. In addition, it should be noted that *Evx* and *Hoxd* genes are not expressed in the same cellular subsets in the spinal cord.

IS *Evx-2* AN *Evx* GENE?

Previous analyses of vertebrate and invertebrate genes from the *even-skipped* family revealed two features of expression in common. First, there are similarities in the early expression pattern, which is initiated prior to gastrulation and subsequently restricted to posterior parts of the embryo. This is exemplified by the expression,during gastrulation, of mouse *Evx-1* (Dush and Martin, 1992) or zebrafish *eve-1* (Joly et al., 1993), and by grasshopper and beetle *eve* genes until caudal extension (Patel et al., 1992, 1994). In vertebrates, this early expression is maintained in the tail bud. Second, these genes have specific expression patterns in the CNS, for example, *eve* in *Drosophila* (Doe et al., 1988). *Evx-2* is the first example of an *eve*-related gene that does not show either the early-posterior expression, or its persistence in the tail bud. As such, this gene is rather atypical amongst the *eve* gene family. A posterior expression is however resumed later in development, at a position similar to *Hoxd* gene expression in the genital

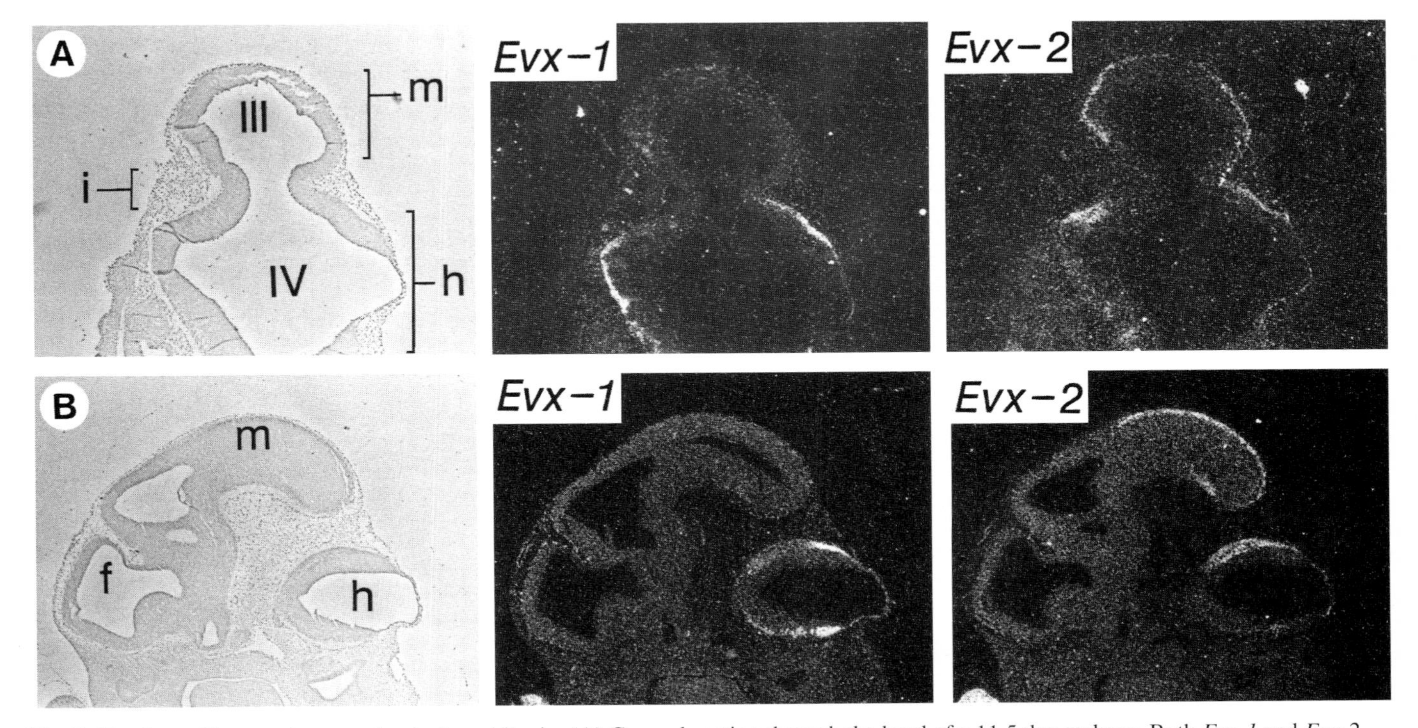

Fig. 7. *Evx-2* specific neural expression in the midbrain. (A) Coronal section through the head of a 11.5 dpc embryo. Both *Evx-1* and *Evx-2* transcripts are expressed in superficial marginal cells of the hindbrain. In addition, *Evx-2* transcripts are found in superficial marginal cells of the midbrain. (B) Sagittal section of the head of a 11.5 dpc fetus. *Evx-2* and *Evx-1* are co-expressed within the hindbrain but only *Evx-2* labelling is seen in superficial cells of the midbrain. m, midbrain; h, hindbrain; f, forebrain; i, isthmus; III, third ventricle; IV, fourth ventricle.

anlage and following the activation of *Hoxd-13*. In limbs and genitalia, *Evx-1* and *Evx-2* expression were similar at first, but soon diverged, essentially through a decrease of *Evx-1* expression. Such differences in the dynamics of the expression domains suggest that the systems of expression maintainance may have differentially evolved between the two genes.

In contrast, *Evx-2* expression in the CNS was almost identical to that of *Evx-1*, both in timing and distribution. Both genes were activated around day 10 of development and *Evx-2* transcripts showed the same cell-type specificity, mostly in post-mitotic cells (early differentiating neurons). These highly restricted expression domains were located at presumptive areas for interneurons (Bastian and Gruss, 1990). It is therefore probable that the function of both *Evx-2* and *Evx-1* in the CNS, is to identify particular types of neural cells; a role with parallels to the *Drosophila eve* gene (Doe et al., 1988). This aspect of *Evx-2* regulation, clearly shared by *Evx-1*, is likely confered by some CNS-specific regulatory sequences which, were conserved after large-scale duplication of an original HOM/Evx complex.

Although both genes are probably co-expressed along the spinal cord and hindbrain, the cranial boundaries of the *Evx-1* and *Evx-2* expression domains in the CNS are different. While the rostral limit of *Evx-1* expression is found at the rhombencephalic isthmus (the hindbrain-midbrain transition), *Evx-2* transcripts extend more rostrally into the midbrain, up to the midbrain-forebrain junction. In the midbrain, the unique expression of *Evx-2* is observed in cell layers similar to those expressing both *Evx* genes at more caudal levels of the CNS. It is thus conceivable that this additional *Evx-2* rostral domain reflects an anterior extension of the original *Evx* function. This functional extension may have followed the duplication of an original HOX/Evx complex, in prechordates or early in the vertebrate lineage (see below), in parallel to the acquisition of a complex anterior nervous system. It is noteworthy that both *Evx-1* and *Evx-2* rostral limits of expression coincide with two important morphological sub-divisions of the embryonic brain, between the mesencephalic-metencephalic and metencephalic-diencepalic vesicles, respectively.

EVOLUTION OF THE *Evx/Hox* FUNCTIONAL RELATIONSHIPS

These results, together with previously published data, suggest the following scheme for the history of the functional relationships between *eve/HOM* related genes. The conservation of an original linkage between an ancestral *eve* gene and an ancestral HOM complex (presumably formed originally by horizontal gene duplication) may have been favoured by a primordial function of *eve* in the early specification of the 'posterior' in development. This assumption is supported by the presence of an *eve*-like gene linked to a potential HOM complex in a Cnidarian (Miller and Miles, 1993). Before the separation between protostomes and deuterostomes, this ancestral gene was secondarily recruited for neuronal specification, probably in association with the evolution of different cell types in an ancient CNS. At this point, temporal colinearity must have been imposed on the HOM complex (if one assumes that this mechanism is used in some present day arthropods or annelids; Duboule, 1992). However, because of its early function, it is probable that the ancestral *eve* gene was not subject to this additional regulation. It nevertheless remained at the vicinity of the complex, perhaps because of the regulatory control determining posterior specificity. In many arthropods, a similar configuration could still exist nowadays, where an *eve*-like gene would be next to the HOM complex (on the 'posterior' side), playing an important function during early posterior development (Patel et al., 1992). In long germ band insects, a novel function (in addition to neural and posterior functions) was developed for *eve,* as reflected by the pair rule expression pattern observed in Diptera *eve* (see Patel, this volume). In fact, a reminiscence of the ancestral posterior pattern may be seen in flies, with the expression of *eve* in the primordial proctodeum at the end of germ band extension (Goto et al., 1989).

In early deuterostomes, a primitive *Evx* gene was thus linked to an ancestral HOX complex and probably had functions during both gastrulation and neuronal specification. The present configuration of the vertebrate HOXA/Evx-1 complex may reflect this ancestral situation, with the *Evx* gene relatively independent from *HOX* regulation. In cephalochordates, a situation quite similar to the ancestral one may be observed, since these close relatives of the vertebrates still retain a single HOX complex (see Holland et al., this volume). In the lineage leading to vertebrates, either in parallel with, or soon after, the large scale complex duplications, a rearrangement might have taken place in the upstream part of the HOXD complex resulting in a reduction in the distance separating *Hoxd-13* from *Evx-2*. (The alternative, that the *Evx-1* gene could have been secondarily removed from the immediate proximity of *Hoxa-13* would imply that an ancestral *Evx* gene, before duplication, had no early posterior function. We consider this unlikely, as discussed above.) Since the large scale HOX complex duplications probably occurred in parrallel to the design of a complex central nervous system, it is possible that at this time one of the duplicated Evx genes (*Evx-2*) was also recruited to extend its original function up to a more anterior region.

The close vicinity of the *Evx-2* gene to the HOXD complex correlates with the two essential differences between *Evx-2* and other members of the *eve* gene family. On the one hand, the aquisition by *Evx-2* of expression features that are characteristic of Hox genes; on the other hand, the suppression of a major trait of the *eve* family, the early phase of expression. We suggest that these two novel components of *Evx-2* expression pattern arose by this gene coming under the influence of the neighbouring HOXD complex. As a consequence, while the *Evx-2* gene became expressed in limbs and genitalia through the control of some of the *Hoxd-13* regulatory (enhancer) elements, its early phase of expression was suppressed by temporal colinearity. Indeed this latter mechanism would not allow the *Evx-2* gene to be transcribed earlier than *Hoxd-13*, that is not before 8.5-9 days of gestation. The molecular mechanisms involved in temporal colinearity are not known. One potential explanation is the sequential opening (accessibility) of the HOX complexes, from 3′ to 5′, to the transcription machinery during gastrulation (Dollé et al., 1989; Duboule, 1992). In this context, the inactivation, in the course of evolution, of the early phase of *Evx-2* expression by its translocation next to the 5′ extremity of the HOXD complex (i.e. near an early 'inactive configuration') may be mechanistically similar to position effect variegation in *Drosophila* (Reuter and Spierer, 1992). If this were the case, however, one might not expect *Evx-2* to be expressed anterior to *Hoxd-13,* even in different cell types. Analysis of the location and structure of the regulatory element(s) conferring CNS specificity on *Evx-2* may, therefore, give insight into the mechanism involved not only in *Evx-2* regulation, but also in temporal colinearity of HOX complexes.

We thank Professor P. Chambon and Drs J.-C. Izpisúa-Belmonte and T. Schimmang for their support and contributions to various phases of this project and J. Zakany and for critical reading of the manuscript. We thank M. Philippe for technical assistance as well as D. Décimo for suggestions in the in situ hybridization procedure. We also thank Drs M. Dush and G. Martin for the *Evx-1* probe and B. Boulay, S. Metz, V. Stengel and C. Werlé for help in preparing the manuscript and illustrations. This work was supported by funds from the Institut National de la Santé et de la Recherche Médicale, the Centre National de la Recherche Scientifique, the Centre Hospitalier Universitaire Régional, the Association pour la Recherche sur le Cancer, the Fondation pour la Recherche Médicale, the Human Science Frontier Program, the EMBL, the Swiss National Science Foundation, the University of Geneva and the Foundation Claraz.

REFERENCES

Bastian, H. and Gruss, P. (1990). A murine even-skipped homologue, Evx-1, is expressed during early embryogenesis and neurogenesis in a biphasic manner. *EMBO J.* **9**, 1839-1852.

Bastian, H., Gruss, P., Duboule, D. and Izpisúa-Belmonte, J.-C. (1992). The murine even-skipped like gene Evx-2 is closely linked to the HOX-4 complex but is transcribed in the opposite direction. *Mammalian Genome* **3**, 241-243.

Condie, B. G. and Capecchi, M. R. (1993). Mice homozygous for a targeted mutation of Hoxd-3 exhibit anterior transformations of the first and second cervical vertebrae, the atlas and the axis. *Development* **119**, 579-591.

D'Esposito, M., Morelli, F., Acampora, D., Migliaccio, E., Simeone, A. and Boncinelli, E. (1991). EVX2, a human homeobox gene homologous to the even-skipped homeotic gene, is localized at the 5′ end of the Hox4 locus on chromosome 2. *Genomics* **10**, 43-50.

Décimo, D. Georges-Labouesse, E. and Dollé, P. (1994). In situ hybridization to cellular RNA. In *Gene Probes, A Practical Approach* vol II (eds. S. Higgins and B. D. Hames). Oxford University Press.

Doe, C. Q., Smouse, D. and Goodmann, C. S. (1988). Control of neuronal fate by the *Drosophila* segmentation gene *even-skipped*. *Nature* **333**, 376-378.

Dollé, P. and Duboule, D. (1993). Structural and functional aspects of mammalian Hox genes. In *Advances in Developmental Biochemistry,* vol. 2 (ed. P. Wassarmann), pp. 57-109. J. A. I. Press.

Dollé, P., Izpisúa-Belmonte, J.-C., Falkenstein, H., Renucci, A. and Duboule, D. (1989). Coordinate expression of the murine Hox-5 complex homeobox-containing genes during limb pattern formation. *Nature* **342**, 767-772.

Dollé, P., Izpisúa-Belmonte, J.-C., Tickle, C., Brown, J. and Duboule, D. (1991a). Hox-4 genes and the morphogenesis of mammalian genitalia. *Genes Dev.* **5**, 1767-1776.

Dollé, P., Izpisúa-Belmonte, J.-C., Boncinelli, E. and Duboule, D. (1991b). The Hox-4. 8 gene is localised at the 5′ extremity of the HOX-4 complex and is expressed in the most posterior parts of the body during development. *Mech. Dev.* **36**, 3-14.

Dollé, P., Dierich, A., LeMeur, M., Schimmang, T., Schuhbaur, B., Chambon, P. and Duboule, D. (1993). Disruption of the *Hoxd-13* gene induces localized heterochrony leading to mice with neotenic limbs. *Cell* **75**, 431-441.

Duboule, D. (1992). The vertebrate limb: A model system to study the HOX/HOM gene network during development and evolution. *BioEssays* **14**, 375-384.

Duboule, D. (1994). Temporal colinearity and the phylotypic progression: a basis for the stability of a vertebrate bauplan and the evolution of morphologies through heterochrony. *Development* **1994 Supplement**.

Dush, M. K. and Martin, G. R. (1992). Analysis of mouse Evx genes: Evx-1 displays graded expression in the primitive streak. *Dev. Biol.* **151**, 273-287.

Faiella, A., D'Esposito, Rambaldi, M., Acampora, D., Balsofiore, S., Stornaiuolo, A., Mallamaci, A., Migliaccio, E., Gulisano, M., Simeone, A. and Boncinelli, E., (1992). Isolation and mapping of EVX1, a human homeobox gene homologue to even-skipped, localized at the 5′ end of the HOX1 locus on chromosome 7. *Nucl. Acid Res.* **19**, 6541-6545.

Gaunt, S. J., Sharpe, P. T. and D. Duboule. (1988). Spatially restricted domains of homeo-gene transcripts in mouse embryos: relation to a segmented body plan. *Development* **104 Supplement**, 71-82.

Goto, T., McDonald, P. and Maniatis, T. (1989). Early and late periodic patterns of eve expression are controlled by distinct regulatory elements that respond to different spatial cues. *Cell* **57**, 413-422.

Haack, H. and Gruss, P. (1993). The establishment of murine Hox-1 expression domains during patterning of the limb. *Dev. Biol.* **157**, 410-422.

Izpisúa-Belmonte, J.-C., Falkenstein, H., Dollé, P., Renucci, A. and Duboule, D. (1991). Murine genes related to the Drosophila *AbdB* homeotic gene are sequentially expressed during development of the posterior part of the body. *EMBO J.* **10**, 2279-2289.

Joly, J. S., Joly, C., Schulte-Merker, S., Boulekbache, H. and Condamine, H. (1993). The ventral and posterior expression of the zebrafish homeobox gene eve1 is perturbed in dorsalized and mutant embryos. *Development* **119**, 1261-1275.

LeMouellic, H., Lallemand, X. and Brûlet, P. (1992). Homeosis in the mouse induced by a null mutation in the Hox-3.1 gene. *Cell* **69**, 251-264.

McGinnis, W. and Krumlauf, R. (1992). Homeobox genes and axial patterning. *Cell* **68**, 283-302.

Miller, D. J. and Miles, A. (1993). Homeobox genes and the zootype. *Nature* **365**, 215-216.

Niswander L. and Martin, G. M. (1993). Fgf-4 regulates expression of Evx-1 in the developing mouse limb. *Development* **119**, 287-294.

Patel, N. H., Ball, E. E. and Goodman, C. S. (1992). Changing role of even-skipped during the evolution of insect pattern formation. *Nature* **357**, 339-341.

Patel, N. H. (1994). The evolution of arthropod segmentation: insights from comparisons of gene expression patterns. *Development* **1994 Supplement**

Patel, N. H., Condron, B. G. and Zinn, K. (1994). Pair rule expression patterns of even-skipped are found in both short- and long-germ beetles. *Nature* **367**, 429-434.

Ramirez-Solis, R., Zheng, H., Whiting, J., Krumlauf, R. and Bradley, A. (1993). Hoxb-4 (Hox-2.6) mutant mice show homeotic transformation of a cervical vertebra and defects in the closure of the sternal rudiments. *Cell* **73**, 279-294.

Reuter, G. and Spierer, P. (1992). Position effect variegation and chromatin proteins. *Bioessays* **14**, 605-612.

Small, K. M. and Potter, S. S. (1993). Homeotic transformations and limb defects in Hoxa-11 mutant mice. *Genes Dev.* **7**, 2318-2328.

Tam, P. P. L., (1984). The histogenetic capacity of tissues in the caudal end of the embryonic axis of the mouse. *J. Embryol. Exp. Morph.* **82**, 253-266.

Yokouchi, Y., Sasaki, H. and Kuroiwa, A. (1991). Homeobox gene expression correlated with the bifurcation process of limb cartilage development. *Nature* **353**, 443-445.

Development 1994 Supplement, 155-161 (1994)
Printed in Great Britain © The Company of Biologists Limited 1994

Gene loss and gain in the evolution of the vertebrates

Frank H. Ruddle[1,4,*], Kevin L. Bentley[1], Michael T. Murtha[2] and Neil Risch[3,4]

[1]Department of Biology, Yale University, New Haven, CT, USA
[2]CuraGen Corporation, Branford CT 06405, USA
[3]Department of Epidemiology and Public Health, Yale University Medical School, New Haven, CT, USA
[4]Department of Genetics, Yale University Medical School, New Haven CT, USA

*Author for correspondence

SUMMARY

Homeobox cluster genes (Hox genes) are highly conserved and can be usefully employed to study phyletic relationships and the process of evolution itself. A phylogenetic survey of Hox genes shows an increase in gene number in some more recently evolved forms, particularly in vertebrates. The gene increase has occurred through a two-step process involving first, gene expansion to form a cluster, and second, cluster duplication to form multiple clusters. We also describe data that suggests that non-Hox genes may be preferrentially associated with the Hox clusters and raise the possibility that this association may have an adaptive biological function. Hox gene loss may also play a role in evolution. Hox gene loss is well substantiated in the vertebrates, and we identify additional possible instances of gene loss in the echinoderms and urochordates based on PCR surveys. We point out the possible adaptive role of gene loss in evolution, and urge the extension of gene mapping studies to relevant species as a means of its substantiation.

Key words: homeobox, echinoderms, ascidians, gene families, genome duplication

INTRODUCTION

The homeobox system of genes is generally recognized as having useful attributes for the study of evolution (Shashikant et al., 1991; Kappen et al., 1993). Most prominent among these is its high level of conservation exemplified in the homeobox sequences and the structural organization of the Hox gene clusters. These properties make it possible to identify sequence motifs, genes, and gene clusters that are homologous, and thus can be compared both quantitatively and qualitatively with confidence over a broad spectrum of metazoans. That the homeobox genes play a fundamental role in metazoan development also suggests that they may themselves be important to the evolutionary process.

In a recent review (Ruddle et al., 1994), we have shown that homeobox genes have been reported for all the major phyla, this is also true for the clustered Hox genes with the exception of the sponges (Seimiya et al., 1992). Hox gene clusters have been directly demonstrated in *Caenorhabditis elegans* (Bürglin et al., 1991; Bürglin and Ruvkun, 1993), *Tribolium castaneum* (Beeman, 1987), *Drosophila pseudoobscura* (Randazzo et al., 1993), *D. melanogaster* (Lewis, 1978; Kaufman et al., 1990), *Branchiostoma floridae* (Holland et al., this volume; Pendleton et al., 1993), *Petromyzon marinus* (Pendleton et al., 1993), and all jawed vertebrates so far examined. Moreover, good evidence has been presented for Hox cluster duplication giving rise to four clusters, each on a different chromosome, in all amniotes (Kappen et al., 1989; Schughart et al., 1989).

In addition to the clustered Hox genes there exist a number of related homeobox genes which are divergent with respect to the homeobox and other features (Kappen et al., 1993). These fall into a number of groups based on similarity, as for example the *Paired*, *Caudal*, and *Distal-less* type genes. We will refer to these genes as non-clustered or diverged homeobox genes. The homeobox genes have been shown to undergo duplication by both *cis* (laterally within a chromosome) or *trans* (chromosome duplication within a genome) processes (Kappen and Ruddle, 1993). *Cis* duplication can arise by unequal crossing over and *trans* duplication by polyploidization, although other mechanisms are also possible.

Ohno has suggested that gene duplication by polyploidy serves an important role in evolution (Ohno, 1970). It is argued that developmentally relevant genes become integrated into developmental pathways that are hierarchical and highly interdependent, and thus they cannot readily mutate or take on new functions without disrupting the overall developmental plan. Gene duplication provides a way around this impasse by the retention of old developmental interrelations and the incorporation of newly duplicated genes into new pathways and relationships. The genomic and functional conservation of the homeobox system is consistent with this view (Ruddle et al., 1994). It is interesting that gene duplication is often cited as being adaptive, because it introduces genetic redundancy into developmental systems. However, as viewed here, redundancy might simply be a consequence of developmental conservatism. Developmental genes and particularly homeotic genes are interactive transcriptionally and have been shown to have the properties of a combinatorial system (Wagner, 1994). The insertion of new genes into a network of genes can be expected to introduce new degrees of freedom, and thereby multiple possible avenues of evolutionary divergence. In this respect, the increase in Hox clusters and the duplication of many non-clustered homeobox genes in the forerunners of the vertebrates can be postulated to have had a profound effect on their evolution and possibly to have contributed to vertebrate radiation (Gould and Eldredge, 1993).

Previous studies have provided support for the idea that the vertebrate Hox cluster gene family has arisen by means of a two step process: firstly, the expansion of the cluster by lateral gene duplication, and secondly the duplication of the clusters by a large domain duplication events, as for example, chromosome or genome duplication (Ruddle, 1989; Kappen et al., 1989; Schughart et al., 1989). Sequence comparisons of the homeobox domain indicate a relatedness between paralogy groups 1-3, 4-8, and 9-13*. These groups have been termed anterior, medial, and posterior, respectively, based on their patterns of expression along the anterior/posterior axis (Kappen et al., 1989; Schubert et al., 1993; Ruddle et al., 1994) . These relationships have suggested that these three groups of genes have arisen from an ancestral cluster of three genes (Schubert et al., 1993).

In this report, we will discuss two systems that relate to chordate evolution and gene duplication. One deals with the amplification of non-homeobox genes which are in linkage with the Hox clusters. The second deals with the possible loss of Hox genes in echinoderms and tunicates.

PARALOGOUS GENES IN LINKAGE WITH THE Hox GENE CLUSTERS

An examination of genes in the vicinity of the Hox clusters shows that many are paralogous and map to two or more of the four Hox gene cluster chromosomes (Rabin, et al., 1986; Ferguson-Smith and Ruddle, 1988; Schughart, et al., 1988; Schughart, et al., 1989; Ruddle, 1989; Craig and Craig, 1991, 1992; Hart et al., 1992; Lundin, 1993; Bentley et al., 1993). In order to study these genes more systematically, we have tabulated all the mouse genes that are members of gene families and of which at least one member maps to the chromosomes bearing Hox gene clusters, namely chromosomes 2, 6, 11 and 15 (Silver, 1993; Siracusa and Abbot, 1993; Moore and Elliott, 1993; Buchberg and Camper, 1993; Mock et al., 1993). The Hox gene clusters themselves are not included in the sample to avoid bias. The identification of gene families is based on sequence similarity. Seventy-four families with a total of 323 genes were identified using these criteria. A representative sample of these genes including 30 families and 203 genes is shown in Fig. 1. A statistical analysis was used to test whether there was excess clustering of members of the gene families on these four chromosomes†. The hypothesis of no excess clustering could be rejected at the 0.01 level of confidence. Four other chromosomes selected on the basis of size similarity to the Hox cluster-bearing chromosomes and number of mapped loci involved, namely mouse chromosomes 4, 9, 12, and 16, were subjected to the same analysis. In this instance the null hypothesis could be not rejected ($P = 0.11$).

The Hox gene clusters are estimated to have undergone duplication minimally 350 million years ago (Kappen et al., 1989). This figure may represent a gross underestimate since Forey and Janvier (1993) have determined the divergence date between lamprey and gnathostomes to be 435 million years. Sufficient time has elapsed since the duplication event to randomize genes throughout the genome. We base this supposition on the randomization of linkage relationships (demonstrated) between the mouse and human genomes over a period of approximately 100 million years (Nadeau, 1989, 1991). However, our analysis indicates a proclivity of genes linked to the clusters to retain their primordial linkage relationships. This relationship is all the more striking when one limits consideration to genes closely linked to the Hox gene clusters. In an extension of this study, we confined our analysis to genes mapping within approximately 30 centiMorgans (cM) centered on the Hox clusters. This distance represents approximately one half of chromosome 15, the smallest of the four chromosomes bearing Hox clusters. Paralogous genes linked within 15 cM to each side of the clusters showed a highly significant association with the clusters ($P = 2\times10^{-5}$). The probability score for paralogous genes outside this region on the same chromosomes was not significant ($P = 0.18$). Hence, the excess clustering of gene families initially observed for chromosomes 2, 6, 11, and 15 is due to specific clustering around the Hox gene complexes.

A possible explanation for these findings is that the linkage of genes to the Hox clusters is a simple structural remnant of the ancestral linkage pattern prior to cluster duplication. A second explanation is that the linkage of paralogous genes is conserved, because it serves an important biological function. This adaptive point of view is strengthened by the fact that many of the genes in linkage to the Hox cluster genes also serve a developmental function, such as growth factors, receptors, members of signaling pathways, and structural proteins having a developmental role such as the cytokeratins and collagens (see Fig. 1). It is also of interest and of possible significance that genes bearing a sequence or functional similarity to the mammalian genes in linkage to the Hox clusters

*Throughout this article paralogy is defined generally as homologies within a genome, but also more specifically to *trans*-homologies between chromosomes.

†First, we calculated the percentage of total mapped loci that fall on each chromosome. For example, the percentages for chromosomes 2, 6, 11, and 15 are 7.4%, 4.6%, 6.3% and 3.8%, respectively. For a gene family, a hit on a particular chromosome is defined as at least one member of the family mapping to that chromosome. For each gene family, there can be a total of 1 to 4 hits, depending on the number of chromosomes containing hits. There are four single hit possibilities (one for each chromosome), six possible two-hit combinations, four possible three-hit combinations, and one four-hit combination. The probability of each of these outcomes can be calculated directly from the percentages give above. We note that this probability needs to be corrected for the fact that the family must have contained at least one hit to be ascertained; hence, each multi-hit outcome probability is divided by the probability of at least one hit. This correction is similar, in spirit, to that used in segregation analysis for human recessive diseases, where families are ascertained through at least one affected child (Elandt-Johnson, 1971).

From these probabilities, we then calculate the expected number of single, double, triple, and quadruple hits, and compare these with the observed numbers. From the probability distribution for number of hits, we calculate the exact probability of obtaining the actual observed number of hits or greater using simulation; it is these P values that we report. A significant excess of hits over expected indicates clustering of gene families.

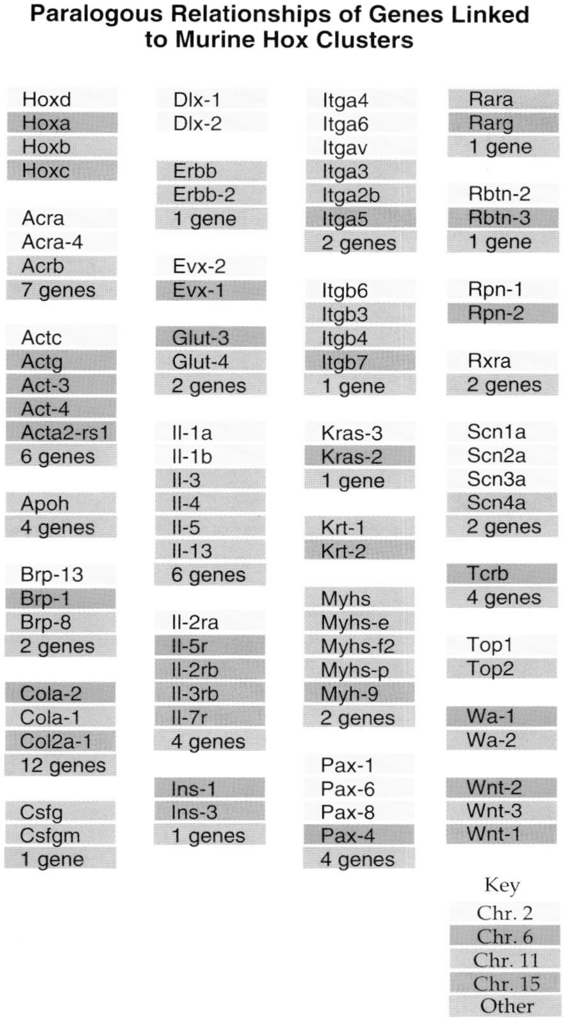

Paralogous Relationships of Genes Linked to Murine Hox Clusters

Hoxd	Dlx-1	Itga4	Rara
Hoxa	Dlx-2	Itga6	Rarg
Hoxb		Itgav	1 gene
Hoxc	Erbb	Itga3	
	Erbb-2	Itga2b	Rbtn-2
Acra	1 gene	Itga5	Rbtn-3
Acra-4		2 genes	1 gene
Acrb	Evx-2		
7 genes	Evx-1	Itgb6	Rpn-1
		Itgb3	Rpn-2
Actc	Glut-3	Itgb4	
Actg	Glut-4	Itgb7	Rxra
Act-3	2 genes	1 gene	2 genes
Act-4			
Acta2-rs1	Il-1a	Kras-3	Scn1a
6 genes	Il-1b	Kras-2	Scn2a
	Il-3	1 gene	Scn3a
Apoh	Il-4		Scn4a
4 genes	Il-5	Krt-1	2 genes
	Il-13	Krt-2	
Brp-13	6 genes		Tcrb
Brp-1		Myhs	4 genes
Brp-8	Il-2ra	Myhs-e	
2 genes	Il-5r	Myhs-f2	Top1
	Il-2rb	Myhs-p	Top2
Cola-2	Il-3rb	Myh-9	
Cola-1	Il-7r	2 genes	Wa-1
Col2a-1	4 genes		Wa-2
12 genes		Pax-1	
	Ins-1	Pax-6	Wnt-2
Csfg	Ins-3	Pax-8	Wnt-3
Csfgm	1 genes	Pax-4	Wnt-1
1 gene		4 genes	

Key
Chr. 2
Chr. 6
Chr. 11
Chr. 15
Other

Fig. 1. Paralogous relationships of genes linked to murine Hox clusters. Data used in preparing this figure were derived from The Encyclopedia of the Mouse Genome accessed at Jackson Laboratory and sources listed in the text. This table is a representative sampling of the multi-gene families selected and analyzed as described in the text. The gene families are named as follows, vertically in each column. Column 1: Homeobox, Acetylcholine receptor, Actin, Apolipoprotein, Brain protein, Collagen, Colony stimulating factor; Column 2: Distal-less, Epidermal growth factor receptor, Even-skipped, Glucose transport, Interleukin, Interleukin receptor, Insulin; Column 3: a-Integrin, b-Integrin, Kirsten rat sarcoma, Keratin, Myosin heavy chain, Paired box; Column 4: Retinoic acid receptors, Rhombotin, Ribophorin, Retinoic acid X receptor, Sodium channel, T-cell receptor, Topoisomerase, Waved, Wingless.

Table 1. Possible orthologous genes linked to the *HOM* or Hox complexes in *C. elegans*, mouse, and human

C. elegans (Chromosome III)	Chromosome	
	Mouse	Human
ceh11 (homeobox group 9)	2,6,11,15	2,7,12,17
mab 5 (homeobox group 6)	6,11,15	7,12,17
Distal-less	2	2,7
Collagen	6, 11, 15	2,7,12,17
Ras family	2, 6, 11	2,7,12
Integrin, α subunit	2, 11, 15	2,12,17
Myosin heavy chain 3	11,15	7,17
Acetylcholine receptor	2,11	2,17
Protein kinase,ser/threo	11	17
Erbb	11	7,12,17
α-β-crystallin	11	2,17
Calcium channel, α-2b subunit	?	2,17
Synaptobrevin	6,11	12,17
GTP binding protein	6,11	7,12
ATP synthase, subunit (two)	?	2,12,17
Topoisomerase 2	2,11	17
Acyl CoA dehydrogenase	?	2,12
Engrailed	–	2,7
Vitronectin receptor, α subunit	2	2
Proton pump (Vpp-1)	11	2,17
Cyclic GMP phosphodiesterase	11	17
β-Spectrin	11	2
α-Spectrin	2	–
Calcium binding protein	11	–
Glutamate receptor	11	–
Notch repeat	2	–
Gelsolin	2	–

Data used in preparing this table were derived from the GDB, Human Genome Data Base accessed at Johns Hopkins University, and The Encyclopedia of the Mouse Genome accessed at Jackson Laboratory. *C. elegans* data were taken from Wilson et al. (1994). No attempt has been made to compare the *C. elegans* sequences with the possible orthologs located on any of the mouse or human chromosomes. The *HOM* complex in *C. elegans* is located on chromosome III. All *C. elegans* genes listed above are located within 2.2 megabases of the *HOM* complex. HOX clusters (*A*, *B*, *C*, *D*) are located on human chromosomes 7,17, 12, and 2, respectively. The locations of the mouse Hox clusters (*a*, *b*, *c*, *d*) are chromosomes 6, 11, 15, and 2, respectively. More than one mouse or human chromosome listed indicates the location of possible paralogs of these genes. – indicates that the gene is not located on a Hox cluster-bearing chromosome. ? indicates map location is not known.

are also located in proximity to the Hox gene clusters in *Caenorhabditis* (Table 1) and *Drosophila* (Table 2).

Assuming that the linkage pattern described above is indeed adaptive, one might speculate on its biological basis. One possibility is that the expression of the linked genes is regulated by *cis* regulatory effects that extend over large distances throughout the region bounding the Hox gene cluster. This is consistent with findings that enhancers can modulate gene activity over distances in the range of a hundred kilobases (Forrester et al., 1987; Qian et al., 1991). A second possibility is that products of genes in the Hox cluster domains interact functionally, and thus are co-adaptive and require coordinated evolutionary modification in the sense of positive epistatic interactions. One can postulate that this can be most efficiently accomplished if the genes are in linkage and tend to assort together in populations. At present these and other notions must be regarded as highly speculative, but can serve as the bases of hypotheses to be tested by experimentation.

Hox CLUSTERS IN ECHINODERMS AND LOWER CHORDATES

The structure of the four Hox gene clusters in vertebrates implies that some Hox genes were lost following duplication (Kappen and Ruddle, 1993; Ruddle et al., 1994). Several possibilities exist concerning the history of cluster duplication (Kappen and Ruddle, 1993), and one of the simplest models involves the two-fold duplication of a single Hox gene cluster

Table 2. Possible orthologous genes linked to the *HOM* and *Hox* complexes in *D. melanogaster*, *C. elegans*, human, and mouse.

Drosophila	*C. elegans*	Mouse	Human
lab (homeobox group 1)	*C. elegans*	2,6,11	2,7,17
pb (homeobox group 2)		6,11	7,17
Dfd (homeobox group 4)	*C. elegans*	2,6,11,15	2,7,12,17
Scr (homeobox group 5)		6,11,15	7,12,17
Antp (homeobox group 6)	*C. elegans*	6,11,15	7,12,17
Ubx (homeobox group 7)		6,11	7,17
abd-A (homeobox group 8)		2,11,15	2,12,17
Abd-B (homeobox group 9-13)	*C. elegans*	2,6,11,15	2,7,12,17
Protein kinase, ser/thr	*C. elegans*	11	17
Actin		2,6	2,7,17
Tubulin	*C. elegans*		2,12
paired box-mesoderm		2,6	7
Ras proto-oncogene	*C. elegans*	2,6,11	2,7,12
Delta (Egf receptor-like)	*C. elegans*	11	7,12,17
Na^+, K^+ ATPase, α subunit		11	17
c-abl oncogene		2	–
Rhodopsin 2, 3, 4		6	–
Catalase		2	–
Arylsulfatase		15	–
U3, small nuclear RNA		11	–
U1, small nuclear RNA		11	–
Protein kinase, cAMP dep.	*C. elegans*	?	7,17
Casein kinase II, a subunit	*C. elegans*	?	–
Ribosomal protein L21	*C. elegans*	?	–
Calmodulin	*C. elegans*	–	7
RNA polymerase II, subunits		11	17
elongation factor 2a	*C. elegans*	?	?
empty spiracles		2	?
Nk homeobox		2	?

Data used in preparing this table were compiled from *Flybase* (on-line *Drosophila* database accessed by Gopher) and Lindsley and Zimm (1992). All of the *Drosophila* genes are located on chromosome 3 from 3-43 to 3-72. The split HOM-C complex is located at 3-47 and 3-59. Sources for mouse, human, and *C. elegans* data are listed in the legend of Table 1. *C. elegans* genes are all on chromosome III, and only include those genes located within 2.2 megabases of the HOM complex (Wilson, et al., 1994). Locations listed for human, mouse, and *C. elegans* indicate possible orthologs of the genes given for *Drosophila*, but not necessarily the exact ortholog of the *Drosophila* gene. – indicates that the gene is not located on a Hox cluster-bearing chromosome. ? indicates map location is not known.

containing ancestral representatives of all 13 paralogy groups, followed by relatively rapid loss of individual genes or cluster segments. Sequence conservation within the homeodomain allows us to estimate the number of Hox cluster genes (paralogy groups 1-10) in a species by the polymerase chain reaction (PCR; Frohman et al., 1990; Murtha et al., 1991; Pendleton et al., 1993). In cases where the number of Hox cluster genes is known, such PCR surveys have shown a recovery rate of more than 85% in a single sampling (Pendleton et al., 1993; and unpublished data).

Surveys of Hox cluster gene number in primitive chordates and echinoderms have provided some curious insights concerning the relationship between vertebrates and other deuterostomes. The hemichordate *Saccoglossus kowaleskii* most likely contains a single Hox gene cluster, with representatives from each of paralogy groups 1-9 (Pendleton et al., 1993). The Hox cluster number in the cephalochordate *Branchiostoma floridae* is more difficult to assess. Data from a PCR survey revealed the presence of 11 Hox cluster genes in amphioxus (Pendleton et al., 1993). On the basis of amino acid sequence similarity, the data predicted that these Hox genes would be distributed over two clusters. Due to the short sequence (82 bp) amplified by PCR and the high similarity between some paralogy groups, paralogy assignments based on homology are necessarily speculative. Detailed linkage analysis of *B. floridae* Hox genes (Holland et al., this volume; Garcia-Fernàndez and Holland, personal communication) reveals a single Hox cluster containing representatives from each of paralogy groups 1-10. Comparison of the sequences from both data sets would be quite informative. Our data could be consistent with a single Hox cluster in *B. floridae,* although an argument could also be made for the existence of two genes in each of paralogy groups 1 and 3. Nineteen different Hox cluster genes were sampled in the agnathan *Petromyzon marinus*, suggesting that the closest extant relatives of the true vertebrates, the jawless fish, have at least two and most likely three or four Hox clusters (Pendleton et al., 1993).

Considering the important regulatory role that Hox cluster genes play during animal development (Shashikant et al., 1991; Ruddle et al., 1994), it is appropriate to examine Hox cluster structure in phyla that exhibit unique developmental qualities. The echinoderms comprise a deuterostome phylum that shows early developmental affinities with the the hemichordates, a phylum suspected to have close affinities with the chordates, but they have a radically different adult body plan that includes secondarily derived radial symmetry. Four Hox cluster genes have been previously isolated from the Hawaiian sea urchin *Tripneustes gratilla* using hybridization techniques (Dolecki et al., 1986, 1989; Wang et al., 1990). Three of these are related to 'medial' paralogy groups (Table 3), and one is most likely a member of 'posterior' paralogy group 9. Our own PCR survey of the sea urchin species *Strongylocentrotus purpuratus* and *Lytechinus variegatus* identified homologs of these *T. gratilla* genes, as well as three other Hox cluster genes (Table 3). One of these, *Hbox9*, is most likely a fourth homolog of 'medial' group genes. Two others, *Hbox7* and *Hbox10*, are most highly homologous to 'posterior' group 9, and do not appear to be related to 'posterior' paralogy groups 10-13. Interestingly, *Hbox7* and *Hbox10* are more closely related to one another than either are to posterior paralogy group 9. Paralogy groups 10-13 in amniotes are thought to have arisen by serial tandem duplication events beginning from paralogy group 9 (Kappen et al., 1993; Schubert et al., 1993). This mechanism might also explain the origin of *Hbox7* and *Hbox10* in echinoderms, but the lack of homology between these sequences and paralogy groups 10-13 in amniotes suggests that such duplications took place independently after the divergence of echinoderms and other deuterostomes.

Another unique distinction of sea urchin Hox clusters appears to be the curious absence of genes from 'anterior' paralogy groups 1-3. 'Anterior' Hox cluster genes have been reported for all other metazoan species examined (Except sponges, where no Hox cluster genes have yet been identified; Ruddle et al., 1994). One possible reason for failing to detect Hox cluster sequences in genomic DNA by PCR could be the presence of introns in the homeodomain. Introns that disrupt the homeodomain are rare, but the *Drosophila* 'anterior' Hox cluster genes *labial* and *proboscipedia* do contain homeodomain introns. None of the known vertebrate Hox cluster genes have introns in the homeodomain. The absence of 'anterior' Hox cluster genes in the echinoderms is also supported by the isolation of only four non-'anterior' genes by

Table 3. Sea urchin Hox cluster sequences

Gene name	Species			Derived sequence (aa 21-47)	Paralogy assignment
	T	S	L		
Hbox1	x	x		HFNRYLTRRRRIELSHLLGLTERQIKI	'medial' groups 4-8
Hbox3	x	x	x	HFSRYVTRRRRFEIAQSLGLSERQIKI	'medial' groups 4-8
Hbox4	x	x	x	LFNMYLTRDRRLEIARLLSLTERQVKI	'posterior' group 9
Hbox6	x		x	HYNRYLTRKRRIEIAQAVCLSERQIKI	'medial' groups 4-8
Hbox7	x		x	QANMYLTRDRRSKLSQALDLTERQVKI	derived from 'posterior' group 9
Hbox9		x	x	HFNRYLTRRRRIEIAHALGLTERQIKI	'medial' groups 4-8
Hbox10		x	x	LYNMYLTRDRRSHISRALSLTERQVKI	derived from 'posterior' group 9

PCR surveys for Hox cluster sequences were carried out as described (Murtha et al., 1991; Pendleton et al., 1993) using genomic DNA from *S. purpuratus* (=S) and *L. variegatus* (=L). The resulting sequences were conceptually transcribed, and are listed in comparison to previously reported sequences from *T. gratilla* (=T) Hox cluster genes (Dolecki et al., 1986, 1989; Wang et al., 1990). The nomenclature used here is based on that used for *T. gratilla*. An 'x' in the species columns indicates those Hox cluster genes found for each species. Paralogy assignments are based on the best homologies within that sequence compared with other known Hox cluster genes from paralogy groups 1-13 (see text).

Table 4. Ascidian Hox cluster sequences

Gene name	Species	Derived sequence (aa 21-47)	Paralogy assignment
CI-1	*Ciona intestinalis*	HYNRYLTRRRRIEVAHTLCLTERQIKI	'medial' groups 4-8
SC-6	*Styela clava*	HFNRYLTRRRRIEIAHSLCLSERQIKI	'medial' groups 4-8
MO-4	*Molgula oculata*	HFNQYLTRERRLEVAKSVNLSDRQVKI	'posterior' group 10

PCR surveys for Hox cluster sequences were carried out as described by Murtha et al. (1991) and Pendleton et al. (1993), using genomic DNA. The listed sequences are conceptual translations of the PCR products. Paralogy assignments are based on the best homologies within that sequence compared with other known Hox cluster genes from paralogy groups 1-13 (see text).

hybridization in *T. gratilla* (see above). Also compelling is the failure to identify 'anterior' Hox cluster genes in sea urchin RNA (P. Martinez, personnal communication and unpublished data).

The urochordates are a large and varied subphylum, but share distinct developmental affinities with the chordates. The ascidians, for example, have a tadpole larval stage where they quite literally resemble what could be described as primitive chordates. The larvae are free-swimming and contain structures such as a notochord, dorsal nervous system, a primitive brain with paired sensory organs, mesenchyme cells, etc. (Jeffery and Swalla, 1992). After the larvae attach to a substrate, metamorphosis occurs generating a sessile, filter-feeding adult form rather unique among coelomates. A previously reported search for Hox cluster genes by hybridization in the ascidian *Halocynthia roretzi* (Saiga et al., 1991) failed to isolate any Hox cluster genes, as only one very diverged non-Hox cluster gene, *AHox1*, was described.

We have surveyed four different ascidian species for Hox cluster sequences by PCR. In each case, only one Hox cluster gene sequence could be identified (Table 4). Even more intriguing is the fact that the single Hox gene sequence detected in one ascidian species is entirely different from the ones present in other species. *Ciona intestinalis* surveys show one Hox cluster sequence related to 'medial' (Hox paralogs 4-8) class paralogy groups (Table 4), while *Styela clava* shows a clearly different 'medial' class Hox cluster sequence. Survey data from the more distantly related genus *Molgula* relate an even more curious tale. *Molgula oculata* represents a urodele ascidian species with a tailed larva from which a single Hox cluster gene related to 'posterior' paralogy group 10 was identified. The closely related species *Molgula occulta* represents an anural ascidian which displays a tailless larval form. No Hox cluster sequences were detected in *M. occulta*, while the same non-Hox cluster genes were found that were present in the survey of *M. oculata* (unpublished data). Hybrids can be formed between these two *Molgula* species resulting in a hatched larva with a short tail (Swalla and Jeffery, 1990). In a review article on ascidian development, Jeffery and Swalla (1992) comment that anural development has probably evolved more than once in ascidians, suggesting that it may be the consequence of a relatively small number of loss-of-function mutations.

The urochordate Hox cluster data is indeed puzzling. A PCR survey of the pelagic (non-sessile) tunicate *Oikopleura dioica* (Holland et al., this volume) again is compatible with the presence of a single Hox cluster sequence, and in this case it is most closely related to 'anterior' paralogy group 1. It therefore appears that Hox cluster genes from 'anterior', 'medial' and 'posterior' paralogy groups are represented in this phylum. Have the bulk of urochordate Hox cluster gene sequences diverged beyond detection? Have wholesale changes in genomic organization (e.g. intron insertions) occurred? These scenarios seem unlikely since a different Hox cluster gene is present for each genus examined. One possible model considers that the forerunner of the urochordates had a single, complete Hox gene cluster. By any of a number of mechanisms of adaptation, the requirement for Hox cluster function in this species was lost. The single Hox cluster genes that remain in the different urochordate species may have been co-opted for different roles in the newly evolved developmental mechanisms. An example of the kind of role taken by these single Hox cluster genes may have already been alluded to, that is, could the single *Molgula* Hox cluster gene described above be necessary for formation of the larval tail, with its loss resulting in anural development? Do any extant species exist that contain more than one Hox cluster gene? More rigorous examination of Hox cluster structure in the urochordates will be necessary to provide answers to these speculations.

DISCUSSION

In surveys for Hox cluster genes, representatives have been recovered for all of the major phyla with the sole exception of sponges. Moreover, a general correlation can be made between Hox gene number and the more recently evolved phyla (Ruddle et al., 1994). This result is consistent with the postulated conservative notion of development espoused by Ohno (1970). We refer to this process as the 'gene freeze' hypothesis. This hypothesis states that evolutionary innovations are facilitated by gene duplication of developmentally significant genes which allows the retention of old developmental functions and the introduction of new. The duplicated gene(s) then becomes integrated into the developmental plan of the organism and likewise becomes constrained with respect to change and is 'frozen'. This view of the developmental-evolutionary process has additional interesting properties. If we assume that developmental control is combinatorial and that developmental genes (eg., Hox genes) are interactive in the form of gene networks then the addition of genes by duplication may increase developmental possibilities geometrically. In other words, the introduction of a single gene may introduce a broad range of developmental possibilities and corresponding evolutionary options. Increasingly, evidence supports the view that evolution progresses in spurts (Gould and Eldredge, 1993). We submit that a discontinuous rate of evolutionary change is consistent with the gene freeze model. In this respect, the duplication of clusters of Hox genes in the antecedents of the vertebrates may have had important immediate consequences with respect to the vertebrate radiation.

Gene loss may also play an important role in the evolution of developmental mechanisms. In vertebrates, duplication events that created the four Hox gene clusters were presumably quickly followed by gene loss until extant cluster structures became 'frozen' (Kappen and Ruddle, 1993). As yet no remnants of the lost Hox cluster genes (e.g. pseudogenes) have been detected. It will be of great interest to examine the vertebrate classes in a detailed fashion with respect to the presence and absence of Hox cluster genes, since such data may possibly reveal the detailed patterns of gene loss, which in turn can give insight into class affinities. PCR surveys have provided compelling evidence that Hox cluster gene loss may also have occurred in other deuterostomes, particularly in the echinoderms and urochordates. Sea urchins appear to lack 'anterior' Hox cluster genes when they would be predicted to contain them on the basis of their phylogenetic position; they share a common ancestor with arthropods and chordates in which 'anterior' Hox cluster genes have been identified. It will be of interest to determine the extent to which 'anterior' Hox cluster gene loss, if true, has contributed to the unique anatomical characteristics shared by echinoderms. The ascidian Hox cluster gene data inspire curious speculation considering the central position accorded them with respect to vertebrate origins (Berrill, 1955). One historically prominent theory posits that the ascidian larva, by means of a neotenic process, represents the ancestral form for vertebrate evolution (Garstang, 1894). The apparent loss of most Hox cluster genes and associated functions, if true, suggests that the urochordates are a derived group, diverging rather early from the stem lineage leading to the vertebrates. In addition, Hox cluster gene loss in urochordates, as speculated for the echinoderms, may have had profound consequences regarding developmental pathways and adaptive adult morphology.

The Hox clusters provide an exceptional system for the study of developmental processes, especially with regard to morphology. We have demonstrated the great potential of various deuterostome phyla as model systems to study the role of Hox cluster genes in the evolution of morphology, and the advancement to more complex forms. The detailed study of Hox cluster structure and its evolution will provide new and exciting insights into the origins of the vertebrates and the mechanism of their development.

We thank several individuals for supplying materials used in these studies: specifically, Joe Minor for *Strongylocentrotus* DNA, Bill Klein for *Lytechinus* DNA, Tom Meedel for *Ciona* sperm, Billie Swalla for *Styela* and *Molgula* DNA, and Pedro Martinez for personal communications and insightful discussion. Thanks to S. Pafka for graphics and photography. This work was supported in part by NIH-grant GM09966 to FHR.

REFERENCES

Beeman, R. W. (1987). A homoeotic gene cluster in the red flour beetle. *Nature (London)* **327**, 247-249.

Bentley, K. L, Bradshaw, M. S. and Ruddle, F. H. (1993). Physical linkage of the murine *Hoxb* cluster and nerve growth factor receptor on yeast artificial chromosomes. *Genomics* **18**, 43-53.

Berrill, N. J. (1955). *The Origin of the Vertebrates*. Oxford: Clarendon Press.

Buchberg, A. M. and Camper, S. A. (1993). Mouse Chromosome 11. *Mammalian Genome* **4 (Special Issue)**, S164-S175.

Bürglin, T. R. and Ruvkun, G. (1993). The *Caenorhabditis elegans* homeobox gene cluster. *Curr. Opin. Genet. Develop.* **3**, 615-620.

Bürglin, T. R., Ruvkun, G., Coulson, A., Hawkins, N. C., McGhee, J. D., Schaller, D., Wittmann, C., Müller, F. and Waterston, R. H. (1991). Nematode homeobox cluster. *Nature* **351**, 703.

Craig, I. and Craig, S. (1991). Genomic distribution of coding sequences homologous to those on chromosome 12. *Cytogenet. Cell Genet.* **58**, 1851-1852.

Craig, I. and Craig, S. (1992). Genomic distribution of coding sequences homologous to those on chromosome 12 paralogy map II. *Cytogenet. Cell Genet.* **61**, 255.

Dolecki, G. J., Wang, G. and Humphreys, T. (1989). Stage- and tissue-specific expression of two homeo box genes in sea urchin embryos and adults. *Nucl. Acids. Res.* **16**, 11543-11558.

Dolecki, G. J., Wannakrairoj, S., Lum, R., Wang, G., Riley, H. D., Carlos, R., Wang, A. and Humphreys, T. (1986). Stage-specific expression of a homeo box-containing gene in the non-segmented sea urchin embryo. *EMBO J.* **5**, 925-930.

Elandt-Johnson, R. C. (1971). *Probability Models and Statistical Methods in Genetics*. pp. 458-496. New York: John Wiley and Sons.

Ferguson-Smith, A. C. and Ruddle, F. H. (1988). The genomics of human homeobox-containing loci. *Pathol. Immunopathol. Res.* **7**, 119-126.

Forey, P. and Janvier, P. (1993). Agnathans and the origin of jawed vertebrates. *Nature* **361**, 129-134.

Forrester, W. C., Takegawa, S., Papayannopoulou, T., Stamatoyannopoulos, G. and Groudine, M. (1987). Evidence for a locus activation region: the formation of developmentally stable hypersensitive sites in globin-expressing hybrids. *Nucl. Acids Res.* **15**, 10159-10177.

Frohman, M. A., Boyle, M. and Martin, G. R. (1990). Isolation of the mouse *Hox-2.9* gene; analysis of embryonic expression suggests that positional

information along the anterior-posterior axis is specified by mesoderm. *Development* **110**, 589-607.
Garstang, W. (1894). Preliminary note on a new theory of the phylogeny of the chordata. *Zool. Anz.* **17**, 119.
Gould, S. J. and Eldredge, N. (1993). Punctuated equilibrium comes of age. *Nature* **366**, 233-227.
Hart, C. P., Compton, J. G., Langely, S. H., Hunihan, L., LeClair, K. P., Zelent, A., Chambon, P., Roderick, T. H. and Ruddle, F. H. (1992). Genetic linkage analysis of the murine developmental mutant velvet coat (*Ve*) and the distal chromosome 15 developmental genes *Hox-3.1*, *Wnt-1*, and *Krt-2*. *J. Exp. Zool.* **263**, 83-95.
Holland, P. W. H., Garcia-Fernàndez, J., Holland, L. Z., Williams, N. A. and Holland, N. D. (1993). The molecular control of spatial patterning in amphioxus. *J. Mar. Biol. Assoc. U.K.* **74**, 49-60.
Holland, P. W. H., Holland, L. Z., Williams, N. A. and Holland, N. D. (1992). An amphioxus homeobox gene: sequence conservation, spatial expression during development and insights into vertebrate evolution. *Development* **116**, 653-661.
Holland, P. W. H., Garcia-Fernàndez, J., Williams, N. A. and Sidow, A. (1994). Gene duplications and the origins of vertebrate development. *Development* **Supplement**, 125-133.
Jeffery, W. R. and Swalla, B. J. (1992). Evolution of alternate modes of development in ascidians. *BioEssays* **14**, 219-226.
Kappen, C. and Ruddle, F. H. (1993). Evolution of a regulatory gene family: *HOM/HOX* genes. *Curr. Opin. Genet. Dev.* **3**, 931-938.
Kappen, C., Schughart, K. and Ruddle, F. H. (1989). Two steps in the evolution of *Antennapedia*-class vertebrate homeobox genes. *Proc. Natl. Acad. Sci., USA* **86**, 5459-5463.
Kappen, C., Schughart, K. and Ruddle, F. H. (1993). Early evolutionary origin of major homeodomain sequence classes. *Genomics* **18**, 54-70.
Kaufman, T. C., Seeger, M. A. and Olsen, G. (1990). Molecular and genetic organization of the *Antennapedia* gene complex of *Drosophila melanogaster*. *Adv. Genet.* **27**, 309-362.
Lewis, E. B. (1978). A gene complex controlling segmentation in *Drosophila*. *Nature* **276**, 565.
Lindsley, D. L. and Zimm, G. G. (1992). *The Genome of Drosophila melanogaster*. San Diego: Academic Press.
Lundin, L. (1993). Evolution of the vertebrate genome as reflected in paralogous chromosomal regions in man and house mouse. *Genomics* **16**, 1-19.
Mock, B. A., Neumann, P. E., Eppig, J. T., Duncan, R. C. and Huppi, K. E. (1993). Mouse Chromosome 15. *Mammalian Genome* **4 (Special Issue)**, S211-S222.
Moore, K. J. and Elliott, R. W. (1993). Mouse Chromosome 6. *Mammalian Genome* **4 (Special Issue)**, S88-S109.
Murtha, M. T., Leckman, J. F. and Ruddle, F. H. (1991). Detection of homeobox genes in development and evolution. *Proc. Natl. Acad. Sci., USA* **88**, 10711-10715.
Nadeau, J. H. (1989). Maps of linkage and synteny homologies between mouse and man. *Trends in Genetics* **5**, 82-86.
Nadeau, J. H. (1991). Genome duplication and comparative gene mapping. In *Advanced Techniques in Chromosome Research* (ed. K. Adolph), pp. 269-296. New York: Dekker.
Ohno, S. (1970). *Evolution by Gene Duplication*. Heidelberg: Springer-Verlag.
Pendleton, J. W., Nagai, B. K., Murtha, M. T. and Ruddle, F. H. (1993). Expansion of the *Hox* gene family and the evolution of chordates. *Proc. Natl. Acad. Sci., USA* **90**, 6300-6304.
Qian, S., Capovilla, M. and Pirrotta, V. (1991). The *bx* region enhancer, a distant *cis*-control element of the *Drosophila Ubx* gene and its regulation by *hunchback* and other segmentation genes. *EMBO J.* **10**, 1415-1425.
Rabin, M., Ferguson-Smith, A., Hart, C. P. and Ruddle, F. H. (1986). Cognate homeobox loci mapped on homologous human and mouse chromosomes. *Proc. Natl. Acad. Sci., USA* **83**, 9104-9108.
Randazzo, F. M., Seeger, M. A., Huss, C. A., Sweeney, M. A., Cecil, J. K. and Kaufman, T. C. (1993). Structural changes in the *Antennapedia* complex of *Drosophila pseudoobscura*. *Genetics* **134**, 319-330.
Ruddle, F. H. (1989). Genomics and evolution of murine homeobox genes. In *The Physiology of Human Growth* (ed. J. M. Tanner and M. A. Preece), pp 47-66. Cambridge: Cambridge University Press.
Ruddle, F. H., Bartels, J. L., Bentley, K. L., Kappen, C., Murtha, M. T. and Pendleton, J. W. (1994). Evolution of *Hox* genes. *Ann. Rev. Genet.* **28**, In press.
Saiga, H., Mizokami, A., Makabe, K. W., Satoh, N. and Mita, T. (1991). Molecular cloning and expression of a novel homeobox gene *AHox1* of the ascidian, *Halocynthia roretzi*. *Development* **111**, 821-828.
Schubert, F. R., Nieselt-Struwe, K. and Gruss, P. (1993). The *Antennapedia*-type homeobox genes have evolved from three precursors separated early in metazoan evolution. *Proc. Natl. Acad. Sci., USA* **90**, 143-147.
Schughart, K., Kappen, C. and Ruddle, F. H. (1989). Duplication of large genomic regions during the evolution of vertebrate homeobox genes. *Proc. Natl. Acad. Sci., USA* **86**, 7067-7071.
Schughart, K., Kappen, C., and Ruddle, F. H. (1988). Mammalian homeobox-containing genes: Genome organization, structure, expression, and evolution. *Br. J. Cancer* **58** (Suppl. IX) 9-13.
Seimiya, M., Watanabe, Y. and Kurosawa, Y. (1992). Identification of homeobox-containing genes in the most primitive metazoa, sponge. *Zool. Sci.* **9**, 1213 (Abstr.).
Shashikant, C. S., Utset, M. F., Violette, S. M., Wise, T. L., Einat, P., Einat, M., Pendleton, J. W., Schugart, K. and Ruddle, F. H. (1991). Homeobox genes in mouse development. *Crit. Rev. Eukary. Gene Expres.* **1**, pp. 207-245.
Silver, L. M. (1993). Master locus list. *Mammal. Genome* **4 (Special Issue)**, S2-S9.
Siracusa, L. D. and C. M. Abbot, C. M. (1993). Mouse Chromosome 2. *Mammalian Genome* **4 (Special Issue)**, S31-S46.
Swalla, B. J. and Jeffery, W. R. (1990). Interspecific hybridization between an anural and urodele ascidian: Differential expression of urodele features suggests multiple mechanisms control anural development. *Dev. Biol.* **142**, 319-334.
Wagner, A. (1994). Evolution of gene networks by gene duplications: A mathematical model and its implications on genome organization. *Proc. Natl. Acad. Sci., USA* **91**, 4387-4391.
Wang, G. V., Dolecki, G. J., Carlos, R. and Humphreys, T. (1990). Characterization and expression of two sea urchin homeobox sequences. *Dev. Genet.* **11**, 77-87.
Wilson, R., Ainscough, R., Anderson et al. (1994). 2.2 Mb of contiguous nucleotide sequence from chromosome III of *C. elegans*. *Nature* **368**, 32-38.

Development 1994 Supplement, 163-168 (1994)
Printed in Great Britain © The Company of Biologists Limited 1994

Evolutionary developmental biology of the tetrapod limb

J. Richard Hinchliffe

Institute of Biological Sciences, University of Wales, Aberystwyth, Wales, UK

SUMMARY

New insights into the origin of the tetrapod limb, and its early development and patterning, are emerging from a variety of fields. A wide diversity of approaches was reported at the BSDB Spring Symposium on 'The Evolution of Developmental Mechanisms' (Edinburgh, 1994); here I review the contributions these various approaches have made to understanding the evolutionary developmental biology of the tetrapod limb. The fields covered included palaeontology, descriptive embryology, experimental embryological analysis of interactions within developing limbs plus description and manipulation of homeobox gene expression in early limb buds. Concepts are equally varied, sometimes conflicting, sometimes overlapping. Some concern the limb 'archetype' (can the palaeontologists and morphologists still define this with precision? how far is there a limb developmental bauplan?); others are based on identification of epigenetic factors (eg secondary inductions), as generating pattern; while yet others assume a direct gene-morphology relationship. But all the contributors ask the same compelling question: can we explain both the similarity (homology) and variety of tetrapod limbs (and the fins of the Crossopterygians) in terms of developmental mechanisms?

Key words: limb development, palaeontogeny, Hox gene, tetrapod

INTRODUCTION

The developmental basis of macroevolution has long been ignored by the neo-Darwinian synthesis, with its emphasis on the population genetics approach to understanding evolutionary change. In recent years, however, the relationship between genes, development and macroevolution has aroused much greater interest (Gould, 1977; Hall, 1992; Duboule, 1992; Holland, 1992; Alberch, 1989). This represents a return to Charles Darwin's own view of embryology as second to none in importance, with his repeated citing of the embryological work of von Baer as the best evidence for the 'unity of structural plan' amongst vertebrates, and as a source of information about their phylogeny. In Darwin's time, of course, only descriptive embryological observations were available, but the most recent attempts to 'explain' phylogenetic changes integrate information from experimental embryology, and from cell and molecular biology. A recent and striking example is Carl Gans's citation (Gans and Northcutt, 1983) of the neural crest as a critical chordate invention, and a necessary precursor to the whole of skull and dermal axial skeleton evolution in vertebrates. In the field of limb development and evolution, current discussion has been enlivened by recent discoveries ranging from palaeontology to molecular biology. Palaeontologists investigating the Devonian period (360 Myr) have recently described (Coates and Clack, 1991) the polydactylous limbs of the first tetrapods - at a stroke undermining the classic pentadactyl archetype – as well as Panderichthyid lobe fin fish representing a sister group much closer to the first tetrapods (Vorobyeva and Schultze, 1991), than the longer known forms (such as *Eusthenopteron*) traditionally allocated in textbooks of vertebrate evolution the role of the first conquerors of the land, and of tetrapod ancestors. In the earliest tetrapods, these first limbs with digit numbers ranging from 6-8 appear more experimental than later more strictly pentadactyl versions. Thus polydactylous tetrapods present novel patterns which developmental biologists need to take into account in their explanations. Recent attempts incorporate evidence from developmental processes and include models based on (i) generation processes elaborating the endoskeleton and attenuating the exoskeleton (e.g. the fin rays) (Thorogood, 1991), (ii) a 'developmental bauplan' of segmenting and asymmetrical branching of the skeletogenic mesenchyme of the limb bud (Shubin, 1991; Shubin and Alberch, 1986) and (iii) evidence that the developmental system in the limb bud is biased towards generating stability proximally and variability distally (Hinchliffe, 1991).

From the very different perspective of molecular biology, analysis of genetic regulation of limb development has provoked reassessment of the evolution of the vertebrate limb. In the chick and mouse limbs, homeobox expression domains along both proximodistal and anterior-posterior axes give the mesenchyme cells unique 'positional addresses' in terms of their patterns of Hox gene expression (Duboule, 1992). Evidence from mutants and from experimental manipulation of the domains (Morgan et al., 1992) suggests that the particular combinations of expression of homeobox genes may relate to the type of digit (or other structure) which forms, regulating its pattern and determining whether the structure is posterior or anterior, proximal or distal.

PALAEONTOLOGICAL FINDINGS AND THE DECONSTRUCTION OF THE ARCHETYPE

To the developmental biologist, the palaeontological record of fins and limbs and, more specifically, the fin/limb transition, sets out the full range of adult fin/limb skeletal patterning that morphogenetic mechanisms must generate (or have generated). To rely on extant forms – lungfish, the coelacanth, the highly derived urodeles – would produce an impoverished set of fin and limb skeletons. Coates (1994 – this volume) sets out the rich diversity of vertebrate appendages (both fins and limbs) with forms very unfamiliar to developmental biologists who may just about remember the distinction between cartilaginous and bony fish. From his survey, Coates suggests that paired fins may have evolved twice and that pectoral fins preceded pelvic fins. Developmental biologists will be most interested in his account of the origin and diversification of the tetrapod limb. Classically, the origin has been sought among the Sarcopterygians, the fleshy or lobe-finned fish (forms in which a substantial endoskeleton enters the fin base and which include the present day lungfish (dipnoans) and the coelacanth, as well as extinct forms). The classification of sarcopterygians is controversial (Coates, 1991): Schultze (1991) identifies a 'rhipidistian crossopterygian' subgroup which includes most of the fossil forms which are divided into osteolepiforms, such as *Eusthenopteron*, (and their probably close relative, Sauripterus), the panderichthyids, and strictly also the tetrapods. The homology of the proximal part (eg humerus, radius, ulna of the forelimb) of the tetrapod limb and osteolepiform paired fin endoskeleton (see Fig. 1) has long been recognised (Jarvik, 1980; Gregory and Raven, 1941) and used as an argument to support the hypothesis of an osteolepiform origin of tetrapods. At a more distal level it has been difficult to detect homologies; for example, the digits do not correspond with elements of the simple branching patterns of the 'rhipidistian crossopterygian' paired fin skeleton. Thus the digits have generally been regarded as evolutionary novelties. The textbook scenario has been of osteolepiform fish dragging themselves with their fleshy lobed fins over the land from one drying out pool to a still-surviving water source. Later, a strengthened and elaborated fin skeleton could become permanently weight bearing on the land as envisaged in the traditional reconstruction of the early tetrapod, *Ichthyostega*, (eg Jarvik, 1980) as a thoroughly terrestrial form.

There are now many problems with this interpretation (Ahlberg and Milner, 1994). The osteolepid *Eusthenopteron* frequently selected for the role of terrestial conquest, appears to have been a fast free-swimming form unlikely to move at all convincingly in shallow water or on land. A more likely sister group for the first tetrapods is the order Panderichthyida (Vorobyeva and Schultze, 1991) named after *Panderichthys*, a species with several features shared with early tetrapods, and whose life-style was probably shore- and shallow water-based, which used the relatively massive proximal skeletal parts of the fore fin (humerus, radius, ulna) for hard substrate locomotion on the shore (Vorobyeva, 1991, 1992a,b; Vorobyeva and Kuznecov, 1992). *Panderichthys* is itself unlikely to be a tetrapod ancestor, on account of the early termination of the branching of the fin endoskeleton (only two branching events – see Fig. 1) which is composed of only five elements and is distally quite unlike that of the tetrapods.

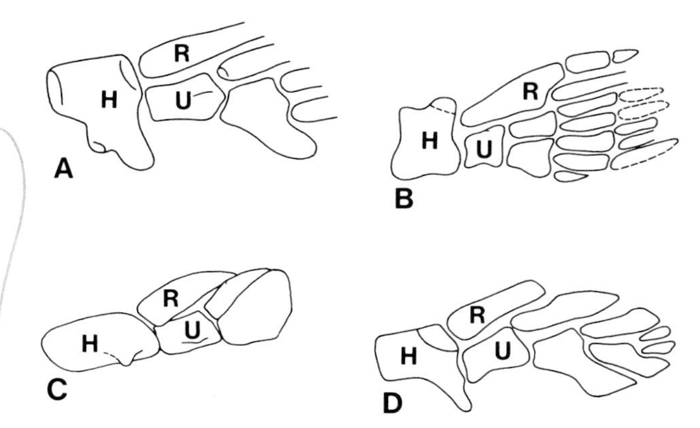

Fig. 1. Distal variation and proximal stability in the pectoral fin endoskeletons from the fossil 'rhipidistian crossopterygians', close relatives of the first tetrapods. ('Rhipidistian crossopterygians' form a subgroup of the sarcopterygians, Schultze, 1991). Proximally the structure is homologous with that of tetrapods (H, humerus; R, radius; U, ulna). Distally, the number of branchings shows variation. (A) *Sterropterygion* (an osteolepid); (B) *Sauripterus* (a rhizodont); (C) *Panderichthys*; (D) *Eusthenopteron* (an osteolepid).

The newly discovered (or reinterpreted) polydactylous tetrapods (Coates and Clack, 1990; Coates, 1994 – this volume) also challenge the traditional interpretation. There are 3 species of these from the Devonian period whose limbs have been preserved: from Greenland, *Ichthyostega* (Jarvik, 1980) and *Acanthostega* (see Coates, Fig. 4) at the base of the amphibian radiation, and *Tulerpeton* (Lebedev, 1985), a reptiliomorph, from Russia. All have lost the fin rays of their rhipidistian crossopterygian relatives, and all have digits, being polydactylous with digit numbers ranging from 8 (*Acanthostega*, fore and hindlimbs), 7 (*Ichthyostega*, hindlimb), to 6 (*Tulerpeton*, fore and possibly hindlimb). Also the low number of carpus or tarsus elements, and their position seems quite different from that typical for the tetrapod limb. These limbs appear likely to have been used horizontally as props and/or aquatic paddles exploiting a shore and shallow water niche, rather than as terrestrial weight-bearing structures. Thus the tetrapod limb may well have evolved initially as a shallow water rather than a terrestrial adaptation (Coates and Clack, 1991; Vorobyeva, 1992b). Later forms of the tetrapod limb are more terrestrially adapted, with the digit number reduced to 5 as a maximum (later amphibians have a maximum of 4 in the forelimb) and with carpus and tarsus elements closer in both number and spatial arrangement to the traditional tetrapod archetype. It appears that pentadactyly – having 5 digits – has been independently evolved along both the reptile and amphibian lines of evolution (Coates, 1991).

One casualty of this discovery of digit variation in early tetrapod limbs is the pentadactyl archetype (see Fig. 2B). This represents the attempt to define a canonical formula or a primitive, numerically fixed pattern of skeletal elements. But with variation in both digit number, and in tarsal pattern, no archetype can be defined. In fact in the interpretation of this subject, the archetype has been very persistent, generating attempts based on recapitulation (ontogeny repeating phylogeny) to read into development at the prechondrogenic stage an ancestral pattern based on the adult limb skeleton of the first tetrapods. This pattern itself supposedly corresponds with that of the hypothesised dipnoan or osteolepiform tetrapod ancestors (see Hinchliffe, 1977, 1989 for criticism of archetype-based analyses by Holmgren, 1933; Montagna, 1945; Jarvik, 1980). The absence of a precisely defined archetype must affect the credibility of such ancestral archetype 'explanations' of the development of tetrapod limb structure, which attempt to account for both the general common features of fin/limbs (eg the proximal parts) as well as the specialised and variable features. Absence of an archetype gives added weight to process-based explanations.

DEVELOPMENTAL PROCESSES

One common feature of recent attempts to explain limb pattern generation in developmental terms is that they are all process based. Developmental biology has revealed a hierarchy of such processes ranging, for example, from apical ridge induction of limb mesenchyme outgrowth, through prechondrogenic condensation, to polarising zone role in control of the anteroposterior axis (Fig. 2A). According to one scheme, the mesenchyme will generate branching precartilaginous condensations whose number relates to the area of available mesenchyme (Pautou, 1973). Positional identity is then imprinted on the condensations (the 'prepattern') by the polarising zone. Variants of this scheme are given by Wolpert, 1989; Ede, 1982, 1991; Hinchliffe, 1991. The most detailed of such schemes (Fig. 2B) emphasising generation of the limb skeleton through branching has been put forward by Alberch and his co-workers (both at the conference and in Shubin and Alberch, 1986; Shubin, 1991), and aims to explain both general and specific features of developing limbs. Condensing mesenchyme is considered to branch asymmetrically (eg the ulna branches distally but not the radius) in generating precartilaginous pattern in a proximodistal direction. This pattern is common to anuran amphibians and amniotes (though urodeles may do it rather differently). The branches segment thus forming individual condensations. Specific variation may be produced by differences in the position of segmentation divisions. From this model, Alberch has evolved the concept of 'developmental constraints' – in particular the last-forming elements being the most easily lost in conditions of mesenchymal cell shortage – which fits evidence provided by digit reduction examples occurring either naturally or experimentally. The model also fits rather well digit reduction and other pattern modifications in birds as compared with chelonians (turtles), and archosaurs (eg crocodiles) in Muller's scheme (1991) of heterochronic modification through paedomorphic truncation of the skeletogenic branching and segmenting patttern.

One characteristic of the evolution of tetrapod limbs and sarcopterygian fins (especially osteolepiforms, *Sauripterus* and *Panderichthys*) (Fig. 1) is proximal stability and distal variability in structure. Unlike the proximal elements, digits are frequently lost in tetrapod limb evolution. This probably relates to the developmental properties of the limb bud, since in experimental manipulation such as ZPA grafting, and in mutants, it is relatively easy to add supernumerary digits, but difficult to affect zeugopod and stylopod. Hinchliffe (1991) suggests this structural outcome is related to relatively minor changes in limb bud boundaries arising through alteration in mesenchyme

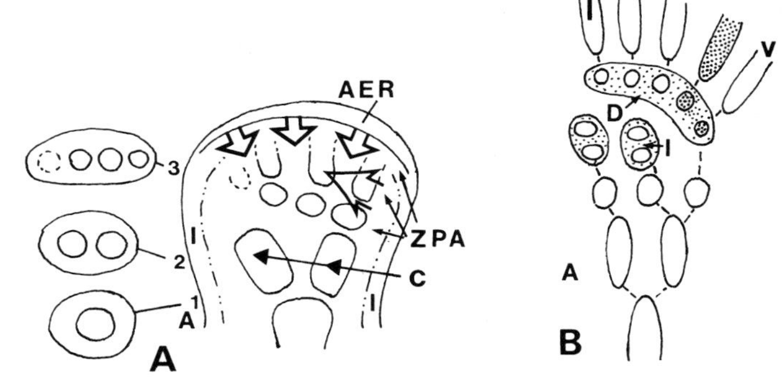

Fig. 2. (A) Schematic diagram of main processes in chick limb development. These processes may represent the common developmental basis to the homology of the tetrapod limb (possibly urodeles are an exception). Cross-sectional areas represented (left) at limb levels 1, 2, 3. Abbreviations. A, anterior; AER, apical ectodermal ridge, mesoderm outgrowth inducer (and site of activity of FGF, fibroblast growth factor); C, branching and segmenting condensations; I, ectoderm-induced zone of inhibition of chondrogenesis; ZPA, zone of polarising activity controlling posterior to anterior axis (also site of *HoxD13* (stage 22) and '*hedgehog*' gene expression). (B) Patterns of branching and segmentation of pre-cartilaginous condensations in Shubin's (1991) scheme for amniotes. Dotted lines represent embryonic connections between condensations. While this partly resembles Holmgren's eutetrapod archetype (1933), this pattern is not a fixed archetype since the number of elements within each stippled area may vary in a species-specific way. Heavy shading represents earliest forming elements in digital arch or among digits. Abbreviations. A, anterior; D, digital arch; I, intermedium/centrale series; I-V, digit ray condensations.

cell number or in the period of AER activity. Quantitative changes in cell number would readily result in distal qualitative changes eg in digit number. Such distal channelling of structural change may represent a developmental constraint very similar to that proposed by Alberch.

A further process-based scheme has been put forward by Thorogood (1991) in relation to one of the most striking features of the fin/limb evolutionary transition: the loss of the distal fin ray skeleton. Most analyses of the transition focus on the endoskeleton, but limb evolution can be seen as a process of elaboration of this, and elimination of the exoskeleton, paralleling a similar loss or reduction of fin rays from the unpaired fins of early tetrapods and their sister groups. Separate mechanisms govern endoskeletal and exoskeletal (fin ray) formation in teleosts. The endoskeleton forms from mesoderm generated by an apical ectodermal ridge, but this is succeeded by a fin fold within which the actinotrich precursors of the fin rays develop. Different cell lineages are likely to be involved: the endoskeleton being mesoderm derived and the exoskeleton probably neural crest derived. Neural crest participation in the exoskeleton is likely by analogy with unpaired fins where there is now good evidence in a teleost, the zebrafish, that the dermal skeleton depends for its formation on a neural-crest-derived mesenchyme (Smith et al., 1994). Thorogood (1991) proposes a heterochronic shift that would explain the distinction between ray-finned and lobe-finned fish: in the latter, the endoskeleton generation process would be extended, and the exoskeletal generation process abbreviated. In the transition to tetrapods, the latter process would be entirely eliminated. The model cannot be tested directly, but its predictions can be tested by examining fin development in extant fishes. Preliminary findings (Thorogood, personal communication) suggest that, in the dipnoan *Neoceratodus,* there is such a late transition from apical ridge to fin fold in the development of their paired fins.

Generic properties of developmental processes were shown by Muller in his paper (see also Muller, 1991; Muller and Wagner, 1991) to play an active role in the generation of evolutionary novelty, in contrast to the concept of developmental constraint emphasised by Alberch. Muller has attempted to analyse some of the secondary interactions that generate adult structure. In general, these have been more difficult to analyse than the more familiar overall limb field interactions (for example, of ridge induced outgrowth and ZPA patterning control in early limb buds). This is partly due to the difficulty of clean surgical intervention and grafting in the post-limb-bud stages.

An example of a morphological novelty resulting from this property of vertebrate tissues is the sesamoid osseous fibular crest on the tibia, which links the reduced fibula to the tibia in birds. Sesamoid bones appear in tendons, usually initiated by pressure. The fibular crest is a neomorphic structure appearing first in the theropod dinosaurs, the presumed ancestors of birds. Developmentally the crest forms initially as a separate cartilaginous sesamoid which ossifies and becomes incorporated first into the tibia. As Hall (1986) has demonstrated, embryonic movement acts as an inducer and is necessary for the formation of sesamoids and secondary cartilages. The dependence of fibular crest on embryonic movement has been demonstrated in paralyzed embryos, which lose their crest. The necessity for the presence of the fibula for the developmental interactions has been shown also by deficiency experiments on chick leg buds which eliminate the fibula: in consequence the crest is also missing (Muller and Streicher, 1989). In the evolution of dinosaurs, reduction of the archosaur fibula could have increased mechanical instability during embryonic movement, thus creating pressure stress on the connective tissue and eliciting formation of the sesamoid cartilage precursor of the crest. Muller argues that the appearance of this novelty is based on specific conditions: skeletal proportions, biomechanical changes and the potential of connective tissues to react to mechanical stimulation. The initial trigger may well have been quantitative – an alteration in mesenchyme cell number in the zeugopod region of the limb bud generating a qualitative change through its effects on a critical threshold.

GENE-MORPHOLOGY RELATIONSHIP

Interesting as is the evidence from palaeontology and from analysis of developmental processes, the dynamic for the current 'buzz' of animated discussion is new molecular evidence for genetic regulation of limb pattern. This falls into a number of categories. One is the comparison of Hox gene complexes in different chordates (and invertebrates) enabling phylogenetic relations to be analysed on the basis of molecular homology, but without necessarily an understanding of the precise developmental role of the genes. Thus amphioxus has a single cluster of Hox genes, contrasting with mammal (mouse and man) which have four Hox clusters apparently resulting from duplication of the initial cluster (Holland, 1994 – this volume). Holland (1992) has suggested that Hox gene duplication is necessary to support the increasing complexity of vertebrate development in evolution: duplication of the complexes permitting related genes to diversify their function. A second category is the expression pattern of the Hox genes, not only along the main body axis of the embryo, but also in the limb buds of mouse and chick embryos. A third category is the experimental modification of the expression domains of the Hox genes, following grafting experiments (e.g. preaxial grafting of ZPAs or of retinoic-acid-containing beads) or transfection experiments. The aim here is investigation of the relation between the domains and the structures they regulate.

One hypothesis is that in the evolution of the tetrapod limb, the Hox genes originally expressed co-linearly along the main body axis of the embryo, and presumed to act as positional information markers in regulating its patterning, have later been co-opted into regulation of limb patterning (Duboule, 1992). Two Hox gene complexes have been shown to be expressed in mouse and chick limb buds. HoxD component genes are expressed as a series of nested domains along an axis from posterior distal to anterior proximal in mouse limb buds. In chick limb buds, HoxA domains are arranged more or less proximally to distally. Both fore and hind mouse and chick limb buds have rather similar HoxA and HoxD domains. All have the same *HoxD13* domain found at the posterior end of the main embryonic axis and present in limb buds along the posterior distal margin. Since this corresponds in chick limb buds with the ZPA position (and with '*hedgehog*' gene expression in both chick and mouse as reported by Riddle et al., 1993; Laufer et al., 1994, Fietz et al., 1994 – this volume), this domain may represent the molecular basis of posterior positional value.

In chick limb buds, experimental embryology may be combined with domain mappings. In the classic experiment on pattern control, grafting the ZPA into a preaxial position provokes the formation of supernumerary digits in tissue from the adjacent host limb bud. The most posterior digit is always the closest to the graft. Normally this anterior limb bud mesenchyme expresses 'anterior' HoxD domains, but following the graft it is switched to a posterior mode of HoxD expression (Izpisua-Belmonte et al., 1991). Retinoic-acid-saturated beads provoke the same response: 'posterior' HoxD expression in an anterior region preceding supernumerary posterior digit formation. Manipulation of the domains was also achieved by transfection experiments in which HoxD genes appropriate to a more posterior position are expressed in the region prospective for the anterior leg digit. In some 30% of cases, a more posterior than normal digit ('2' rather than '1') formed from this region (Morgan et al., 1992). The hypothesis of correlation of HoxD gene expression with specific digits is supported by evidence from a mutant chick embryo, talpid, whose polydactylous digits lack anteroposterior specificity, and in whose limb buds HoxD genes are expressed throughout, rather than demonstrating the normal posterior to anterior sequence of domains (Izpisua-Belmonte et al., 1992).

These experiments and observations are interpreted as evidence that the HoxD code specifies the anterior posterior axis, allocating an identity to individual digits. According to the theory, proximodistal specification is through HoxA activity: mesenchyme cells' unique combination of HoxA and

HoxD expression will determine their differentiation appropriate to their position along the two axes. A more specific variation of this identity allocation hypothesis is given by Tabin (1992) in response to the question "Why do we have 5 digits?". His answer is that the 5 nested HoxD domains of the distal part of the mouse limb bud (Duboule, 1992) code for the 5 digits of the pentadactyl limb. In the chick wing and limb buds, Morgan et al. (1992), describe 5 corresponding HoxD domains (the 'loss' of the supposed ancestral number 1 bird digit is attributed to the intervention of cell death anteriorly in the limb bud). Tabin attempts to homologise digits of both fossil and extant forms on the basis of conserved genes and expression domains, applying the 5 domains to the 8 digit limb of *Acanthostega* in which 5 different digit types are supposedly specified. Perhaps because of its simplicity, such a direct explanation for the (later) pentadactyl nature of tetrapod limbs has proved immediately attractive, with accounts in recent textbooks (Alberts et al., 1994; Gilbert, 1994).

Such a reductionist 'one domain, one digit type' hypothesis has had to be modified however, (Morgan and Tabin, 1994 – this volume), since it now appears that, in the chick limb bud, there are not 5 distinct regional nested HoxD domains and that these expression areas in the distal part shift with time in development. For example, *HoxD13* begins its expression posteriorly at stage 22, but then extends its area anteriorly until it is expressed in most of the distal mesenchyme by stages 23 and 25 (Morgan and Tabin; Fig. 2). In these later stages, almost all the distal mesenchyme is expressing all the *HoxD9-HoxD13* genes. As development continues Hox D expression fades proximally with the exception of the perichondrial tissue, surrounding the cartilage skeletal elements. Morgan and Tabin suggest now that early Hox expression is concerned in growth regulation (rather than pattern specification) of the limb bud while later expression may be regulating bone growth.

FUTURE PROSPECTS

A note of caution should be sounded about the ability of the new homeobox expression data, even when combined with experimental techniques, to solve the problem of the evolutionary developmental biology of the limb. We have information from two species, chick and mouse, and probably we can expect data soon from Zebrafish (a teleost) and from *Xenopus*. Developmental biologists sometimes give the appearance of believing evolution is from Zebrafish via axolotl and *Xenopus* to either chick or mouse and man (Hanken, 1993) and of forgetting that teleosts and urodeles should not be regarded as 'primitive' ancestors, for example of the amniotes. Coates account is a useful antidote to this belief. But in principle at least we can obtain data about gene expression in fins and limbs additionally from lungfish, coelacanth and urodeles. This at least will provide data relevant to Holland's hypothesis of gene duplication as a necessary basis for vertebrate complexity and add molecular to structural data on homology, potentially, for example, clarifying the controversial relation between urodele and non-urodele (anuran, amniote) limbs.

Implicit in much of the recent emphasis on homeobox expression domains encoding positional address is the belief in a simple relation between genes and morphology, exemplified by the 5 domains-5 digit hypothesis. But, in my view, the development of structure cannot be fragmented in this way. Parts of the skeleton are formed by co-related cellular processes and inductions organised in both time and space. A consequence of the domination of analysis of limb development by single factor one-step mechanisms (whether domains or ZPA) is the diversion of attention from experiments demonstrating sequential local interactions, including epithelial/mesenchymal ones (Hinchliffe and Horder, 1993). The extradigit phenomenon, in which an incision in interdigital mesenchyme in the developing chick leg triggers the formation of an additional digit, with joints and associated tendons, is difficult to explain in terms of ZPA or gene domain control, emphasising instead the importance of multiple local interactions (Hinchliffe and Horder, 1993). For example, one way of looking at generation of differences between the chick leg digits is to see them in terms of apical ridge stimulating the supply of mesenchyme distally to a forming digit condensation ray which is being divided into phalangeal elements by a segmentation process. Early, or late, termination of ridge activity will then control the number of phalangeal elements, respectively by reducing or increasing their number. Looked at in this way, differences in digit structure can be attributed to modification of interacting hierarchical processes (rather than to a single factor causation) and considered in Waddington's terminology (1962) as 'condition generated': a purely genetic theory of the development and evolution of structure is deficient because, as Alberch (1989) argues, genes and developmental processes cannot be dissociated as different levels of interaction.

On the positive side in our struggle to explain pattern generation within individual species, and in evolution as a whole, important evidence is provided by the new homeobox expression analysis. It is clear that at many different levels developmental mechanisms are conservative; at the molecular level, for example, in Hox D gene duplication and diversification in vertebrate evolution, and in the similarities in anteroposterior expression domains of Hox D and '*hedgehog*' genes in the tetrapod limb buds of the two species (mouse and chick) thus far examined. But mechanisms are conservative also at other levels analysed; in the iterative asymmetric branching of prechondrogenic condensations, and in ectoderm/mesoderm inductive mechanisms shown to be common to reptile, bird and mammal limb buds (since the tissue of one class understands instructive messages issued by the tissue of another class – Hinchliffe, 1991). We can surely hope to assemble in the next few years a model of limb development, with both regulatory gene and developmental process components, which, given some heterochronic variation and some novelty generation, will go far to provide a developmental basis to the 'variation on a theme' which Richard Owen identified (1849) in his pentadactyl limb paradigm model of homology. If so, this would bring development back to its true – though neglected – role as a major actor in our interpretation of the drama of vertebrate evolution.

REFERENCES

Ahlberg, P. and Milner, A. R. (1994). The origin and early diversification of tetrapods. *Nature* **368**, 507-514.

Alberch, P. (1989). The logic of monsters: evidence for internal constraint in development and evolution. In *Ontogenese et Evolution*. Colloque International du CNRS. Geobios, Supple 12. pp. 21-57.

Alberts, B. et al. (1994). *Molecular Biology of the Cell*. 3rd edn. New York: Garland.

Coates, M. I. (1991). New palaeontological approaches to limb ontogeny and phylogeny. In *Developmental Patterning of the Vertebrate Limb* (eds J. R. Hinchliffe, J. M Hurle and D. Summerbell), pp. 325-338, New York: Plenum Press.

Coates, M. I. and Clack, J. (1991). Polydactyly in the earliest known tetrapod limbs. *Nature* **347**, 66-69.

Coates, M. I. (1994). The origin of vertebrate limbs. *Development* **1994 Supplement.**

Duboule, D. (1992). The vertebrate limb: a model system to study the Hox/Hom gene network during development and evolution. *BioEssays* **14**, 375-384.

Duboule, D. (1994). Temporal colinearity and the phylotypic progression: a basis for the stability of a vertebrate bauplan and the evolution of morphologies through heterochrony. *Development* **1994 Supplement.**

Ede, D. A. (1991). Mutation and limb evolution. In *Developmental Patterning of the Vertebrate Limb* (eds J. R. Hinchliffe, J. M. Hurle and D. Summerbell), pp. 365-371. New York: Plenum Press.

Ede, D. A. (1982). Levels of complexity in limb-mesoderm cell culture systems. In *Differentiation in Vitro* (eds M. M. Yeoman and D. E. S. Truman), pp. 207-229. Cambridge: University Press.

Fietz, M., Concordet, J. P., Barbosa, R., Johnson, R.,Krauss,S., McMahon, A. P.,Tabin, C. and Ingham, P. W. (1994). The *hedgehog* gene family in *Drosophila* and vertebrate development. *Development* **1994 Supplement.**

Gans, C. and Northcutt, R. G. (1983). Neural crest and the origin of vertebrates: a new head. *Science* **220**, 268-274. Washington.

Gilbert, S. F. (1994). *Developmental Biology,* 4th edn. Massachusetts: Sinauer.

Gould, S. J. (1977). *Ontogeny and Phylogeny*. Cambridge, Mass: Belknap Press.

Gregory, W. K. and Raven, H. C. (1941). Studies on the origin and early evolution of paired fins and limbs. *Annals New York Acad. Sci.* **42,** 273-360.

Hall, B. K. (1986). The role of movement and tissue interactions in the development and growth of bone and secondary cartilage in the clavicle of the embryonic chick. *J. Embryol Exp. Morph* **93**, 133-152.

Hall, B. K. (1992). *Evolutionary Developmental Biology*. London: Chapman and Hall.

Hanken, J. (1993). Model systems versus outgroups: alternative approaches to the study of head development and evolution. *Am. Zool.* **33**, 448-456.

Hinchliffe, J. R. (1977). The chondrogenic pattern in chick limb morphogenesis: a problem of development and evolution. In *Vertebrate Limb and Somite Morphogenesis* (eds D. A. Ede, J. R. Hinchliffe and M. Balls), pp 293-309, Cambridge: University Press.

Hinchliffe, R. (1989). Reconstructing the archetype: innovation and conservatism in the evolution and development of the pentadactyl limb. In *Complex Organismal Function: Integration and Evolution in Vertebrates*, (eds D. B. Wake and G. Roth), pp. 171-189. Chichester: J. Wiley.

Hinchliffe, J. R. (1991). Developmental approaches to the problems of transformation of limb structure in evolution. In *Developmental patterning of the vertebrate limb* (eds J. R. Hinchliffe, J. R. Hurle and D. Summerbell), pp. 313-323. New York: Plenum Press.

Hinchliffe, J. R. and Horder, T. J. (1993). Testing the theoretical models for limb patterning. In *Experimental and Theoretical Advances in Biological Pattern Formation* (eds H. G. Othmer, P. K. Maini and J. D. Murray), pp. 105-119, New York: Plenum Press.

Holland, P. (1992). Homeobox genes in vertebrate evolution. *Bioessays* **14**, 267-273.

Holland, P. (1994). Gene duplications and the origins of vertebrate development. *Development* **120, Supplement**, 125-133.

Holmgren, N. (1933). On the origin of the tetrapod limb. *Acta Zoologica* **14**, 185-295.

Izpisua-Belmonte, J. C., Tickle, C., Dolle, P., Wolpert, L. and Duboule, D. (1991). Expression of the homeobox Hox-4 genes and the specification of position in chick wing development. *Nature* **350**, 585-589.

Izpisua-Belmonte, J. C., Ede, D. A., Tickle, C. and Duboule, D. (1992). The misexpression of posterior Hox-4 genes in talpid (ta3) mutant wings correlates with the absence of anteroposterior polarity. *Development* **114**, 959-963.

Jarvik, E. (1980). *Basic Structure and Evolution of Vertebrates.* Vols 1, 2, London, New York: Academic Press.

Laufer, E. M., Riddle, R., Johnson, R. L. and Tabin, C. J. (1994). Sonic hedgehog mediates the polarising activity of the ZPA. *J. Morphol.* **220**, 365.

Lebedev, O. A. (1985). First discovery of a Devonian tetrapod vertebrate in USSR. *Dokl. Akad. Nauk. SSR* **278**, 1470-1473.

Montagna, W. (1945). A re-investigation of the development of the wing of the fowl. *J. Morphol* **76**, 87-113.

Morgan, B. A., Izpisua-Belmonte, J. C., Duboule, D. and Tabin, C. J. (1992). Targeted misexpression of Hox 4.6 in the avian limb bud causes apparent homeotic transformations. *Nature* **358**, 236-7.

Morgan, B. A. and Tabin, C. (1994). Hox genes and growth: early and late roles in limb bud morphogenesis**.** *Development* **1994 Supplement.**

Muller, G. B. (1991). Evolutionary transformation of limb pattern: heterochrony and secondary fusion. In *Developmental Patterning of the Vertebrate Limb* (eds J. R. Hinchliffe et al). pp. 395-405, New York: Plenum Press.

Muller, G. B. and Streicher, J. (1989). Ontogeny of the syndesmosis tibiafibularis and the evolution of the bird hindlimb: a caenogenetic feature triggers phenotypic novelty. *Anat. Embryol* **179**, 327-339.

Muller, G. B. and Wagner, G. P. (1991). Novelty in evolution: restructuring the concept. *Annu. Rev. Ecol. Syst.* **22**, 229-256.

Owen, R. (1849). *On the Nature of Limbs*. London: J Van Voor.

Pautou, M. P. (1973). Analyse de la morphogenese du pied des Oiseaux a l'aide de melanges cellulaires interspecifiques. 1 Etude morphologique. *J. Embryol. Exp. Morph.* **29**, 175-196.

Riddle, R. D., Johnson, R. L., Laufer, E. and Tabin, C. (1993). Sonic hedgehog mediates the polarizing activity of the ZPA. *Cell* **75**, 1401-1416.

Schultze, H-P. (1991). Controversial hypotheses on the origin of tetrapods. In *Origins of the Higher Groups of Tetrapods* (eds H-P. Schultze and L. Trueb) pp. 29-67, Ithaca: Comstock.

Shubin, N. H. (1991). Implications of the 'Bauplan' for development and evolution of the tetrapod limb. In *Developmental Patterning of the Vertebrate Limb* (eds J. R. Hinchliffe, J. M. Hurle and D. Summerbell), pp 411-422. New York: Plenum Press.

Shubin, N. H. and Alberch, P. (1986). A morphogenetic approach to the origin and basic organisation of the tetrapod limb. *Evolutionary Biology* **20**, 318-390.

Smith, M., Hickman, A., Amanze, D., Lumbsden, A. and Thorogood, P. (1994). Trunk neural crest origin of caudal fin mesenchyme in the Zebrafish Brachydanio rerio. *Proc. R. Soc. Lond. B* **256**, 137-145.

Tabin, C. J. (1992). Why we have (only) five fingers per hand: Hox genes and the evolution of paired limbs. *Development* **116**, 289-296.

Thorogood, P. V. (1991). The development of the teleost fin and implications for our understanding of tetrapod limb evolution.. In *Developmental Patterning of the Vertebrate Limb* (eds J. R. Hinchliffe, J. M. Hurle and D. Summerbell), pp. 347-354. New York: Plenum Press.

Vorobyeva, E. I. (1991). The fin-limb transformation: palaeontological and embryological evidence. In *Developmental Patterning of the Vertebrate Limb* (eds J. R. Hinchliffe et al), pp. 339-345. New York: Plenum Press.

Vorobyeva, E. I. (1992a). The role of development and function in formation of 'tetrapod-like' pectoral fins. *J. Common Biology* 53, 149-158.

Vorobyeva, E. I. (1992b). *The Problem of Tetrapod Origin*. Moscow: Nauka (In Russian, English summary).

Vorobyeva, E. I. and Kuznetsov, A. (1992). The locomotor apparatus of *Panderichthys rhombolepis* (Gross), a supplement to the problem of fish-tetrapod transition. In *Fossil Fishes as Living Animals*. (ed. E. Mark-Kurik), pp. 131-140, Inst of Geology, Acad. Sci. of Estonia, Tallin.

Vorobyeva, E. I. and Schultze, H-P. (1991). Description and systematics of panderichthyid fishes with comments on their relationship to tetrapods. In *Origin of the Higher Groups of Tetrapods* (eds H-P. Schultze and L. Trueb), pp. 68-109, Ithaca: Comstock.

Waddington, C. H. (1962). *New Patterns in Genetics and Development.* Columbia Univ. Press.

Wolpert, L. (1989). Positional information revisited. *Development* **107**, **Supplement**, 3-12.

Development 1994 Supplement, 169-180 (1994)
Printed in Great Britain © The Company of Biologists Limited 1994

The origin of vertebrate limbs

Michael I. Coates

University Museum of Zoology, Downing Street, Cambridge CB2 3EJ, UK

SUMMARY

The earliest tetrapod limbs are polydactylous, morphologically varied and do not conform to an archetypal pattern. These discoveries, combined with the unravelling of limb developmental morphogenetic and regulatory mechanisms, have prompted a re-examination of vertebrate limb evolution. The rich fossil record of vertebrate fins/limbs, although restricted to skeletal tissues, exceeds the morphological diversity of the extant biota, and a systematic approach to limb evolution produces an informative picture of evolutionary change. A composite framework of several phylogenetic hypotheses is presented incorporating living and fossil taxa, including the first report of an acanthodian metapterygium and a new reconstruction of the axial skeleton and caudal fin of *Acanthostega gunnari*. Although significant nodes in vertebrate phylogeny remain poorly resolved, clear patterns of morphogenetic evolution emerge: median fin origination and elaboration initially precedes that of paired fins; pectoral fins initially precede pelvic fin development; evolving patterns of fin distribution, skeletal tissue diversity and structural complexity become decoupled with increased taxonomic divergence. Transformational sequences apparent from the fish-tetrapod transition are reiterated among extant lungfishes, indicating further directions for comparative experimental research. The evolutionary diversification of vertebrate fin and limb patterns challenges a simple linkage between Hox gene conservation, expression and morphology. A phylogenetic framework is necessary in order to distinguish shared from derived characters in experimental model regulatory systems. Hox and related genomic evolution may include convergent patterns underlying functional and morphological diversification. *Brachydanio* is suggested as an example where tail-driven patterning demands may have converged with the regulation of highly differentiated limbs in tetrapods.

Key words: vertebrate, limb, fin, evolution, phylogeny, Hox gene, digits

INTRODUCTION

The history of vertebrate limbs has long been a popular subject for comparative anatomical research, combining fossil and recent morphologies with speculations about developmental evolution. More recently, advances in the investigation of developmental regulation have prompted experimental researchers to speculate about evolutionary morphology (cf. Coates, 1993a; Tabin and Laufer, 1993). This article continues in the tradition of morphology and speculation, but is also intended to provide an informative database emphasising the diversity of fin/limb patterns within an explicit phylogenetic framework. And an attempt has been made to include testable developmental-evolutionary speculations.

A glossary at the end of the text includes short explanations of selected phylogenetic and anatomical jargon (first use of included terms marked).*

FIN ARCHETYPES AND HYPOTHETICAL ANCESTORS

The striking similarities of tetrapod limb skeletons, or those of actinopterygian pectoral fins and girdles, have had a sustained influence upon the development of theoretical biology. Explanations of the apparent shared, underlying patterns include Geoffroy's special analogies and (attempted) unified theory of form, Owen's formulation of homology*, and Darwin's recognition of evolutionary descent from remote, common ancestry (Russell, 1916; Appel, 1987). Owen's archetype was subsequently reified as an actual ancestor (discussed in Goodwin and Trainor, 1983), and pre-Darwinian typological baggage was thereby incorporated into the new evolutionary paradigm. Gegenbaur (1878: translation and summary), probably the first evolutionist to consider tetrapod limbs in detail, promptly abstracted a theoretical ground-plan of fin skeletons, which he derived in turn from an ancestral branchial arch (Fig. 1A; note resemblence to the endoskeletal lungfish fin, Fig. 1C). But this surprisingly resilient tree-like 'archipterygium' (eg. Horder, 1989) was soon challenged by the lateral fin-fold theory (Thacher, 1877; Mivart, 1879; Balfour, 1881), in which paired appendages evolved from an ancestral, continuous lateral fin (Fig. 1B), resembling the embryonic fold precursing median (fish) fins. Corroborative evidence from the lateral inter-fin ridge of *Torpedo* embryos, plus support from Haeckel's influential biogenetic law (Gould, 1977), soon established this as the preferred, evolutionary scenario. Instead of Gegenbaur's archipterygium (1878), paired fin skeletons were thought to have evolved from parallel radials*. After an unspecified evolutionary period, pectorals then differentiated from pelvic fins, and basal constriction produced a metapterygium* (Fig. 1D). Corresponding anterior and central sets of apparently fused radials were called the pro*- and mesopterygium* (as in the tri-basal lateral fin pattern characterising recent batoids* and sharks). No gill arch contribution was ever observed and, with increasing evidence for the exclusively mesodermal origin of endoskeletal pectoral girdles (Burke, 1991), little support remains for Gegenbaur's original theory (but see Tabin, 1992, noting the expression of otherwise gill patterning genes in the anteroventral corner of an amniote* pectoral girdle).

Although the continuous lateral fin-fold theory remains widely accepted (eg. Jarvik, 1980; Tabin and Laufer, 1993), it, too, has long been criticised for its assumption of an idealised, hypothetical ancestor (eg. Westoll, 1958; Romer, 1962) (Fig. 1B), which, as an *average*, vitiates the interpretation of the fossil record (Ghiselin, 1988). The embryological support is similarly questionable: Gegenbaur regarded the inter-fin ridge as recapitulating the pelvic fin's posterior, evolutionary migration (suggesting that pelvic fins predated pectorals, cf. Tabin, 1992), although Kerr (1899; deriving limbs from external gills) interpreted it as a specialisation of batoid fins. Goodrich's (1906) study of fin development in '*Scyllium*' (= the shark *Scyliorhinus*) remains probably the most authoritative embryological support for the fin fold theory, wherein abortive inter-fin muscle buds were interpreted as vestiges of ancestral, continuous folds. If these embryonic morphologies are interpreted within a phylogenetic framework (eg. Fig. 2A), then their significance is less convincing. Although these abortive muscle buds resemble the continuous, inter-limb bud amniote Wolffian ridges, in anamniotes* and teleosts such mesodermal outgrowths are clearly disconnected (Balinsky, 1975). Furthermore, Ekman (1941) actually illustrates their discontinuity during the development of '*Acanthius* '(= the shark *Squalus*). These patchy, phylogenetic distributions of ontogenetic data are inconclusive as indicators of evolutionary polarity (Mabee, 1993). Phylogenetic trees are therefore fundamentally important to any further discussion of evolutionary patterns and processes. A detailed discussion of phylogenetic reconstruction methods is beyond the scope of this article, but the following references are suggested as clear, concise accounts of the application of phylogenetic systematics and the logic of evolutionary theory: Harvey and Pagel (1991) and Panchen (1992).

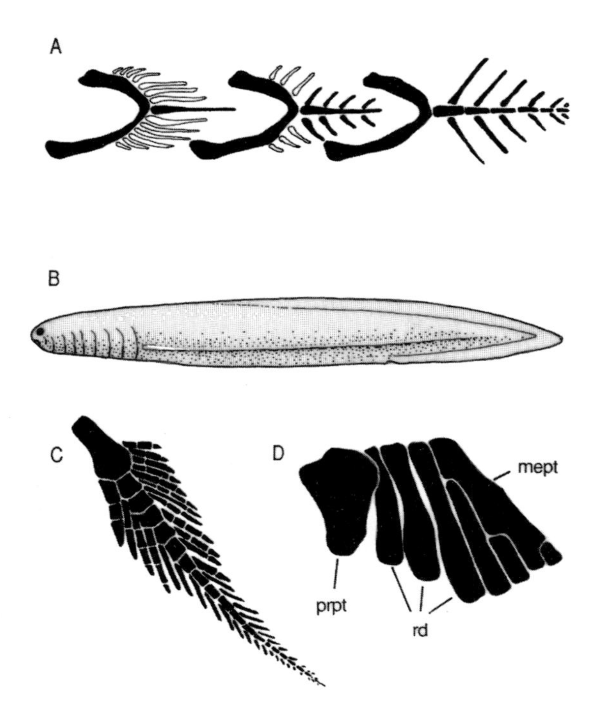

Fig. 1. Vertebrate fin archetypes. (A) An idealised gill arch, at left of figure, transformed into an archipterygial fin and girdle endoskeleton at right of figure (after Gegenbaur, in Jarvik, 1980). Conserved structures shown in black: the gill arch becomes a pectoral girdle resembling a large chondrichthyan scapulocoracoid; the most prominent gill-ray extends and trifurcates serially to produce a fin endoskeleton; and outlined gill-rays are those supposed to have been lost during evolution. Anterior to right of figure. (B) Balfour and other's (see text for refs) continuous lateral fin fold manifest in Jarvik's (1980) hypothetical ancestral vertebrate. (C) The archipterygium-like, but in fact exclusively metapterygial pectoral endoskeleton of an australian lungfish (*Neoceratodus*). Each axial segment articulates with a single preaxial radial (left side); there is no simple, equivalent relation with the number of postaxial radials (after Haswell, 1882; includes figures of other fins showing significant variation in endoskeletal patterns). (D) The pectoral endoskeleton of a sturgeon *Acipenser sturio*, including a metapterygium (mept) supporting one postaxial and two preaxial radials, three median proximal radials (rd) and an anterior propterygium (prpt) (only proximal radials illustrated: after Grande and Bemis, 1991). Anterior of all structures to left of figure.

PHYLOGENETIC AND DEVELOPMENTAL HYPOTHESES

Theories of vertebrate phylogeny have not reached a stable consensus; certain nodes are relatively well resolved, but the origins of other major radiations remain contentious (eg. agnathan* interrelationships surrounding the base of the lampreys; the origin of gnathostomes*: Fig. 2A). This uncertainty affects directly theories of morphogenetic change, because altered phylogenetic topologies transform the way in which taxonomic ground-plans disintegrate (as all summaries of essential characters must, within an evolutionary paradigm: Simpson, 1961; Ghiselin, 1988).

The vertebrate phylogeny (considered in this work to be a monophyletic* subgroup of the Chordata) shown in Fig. 2, is a composite of several analyses incorporating morphological and molecular data: Forey and Janvier (1993) on agnathans and stem*-gnathostomes; Maisey on gnathostomes (1984) and chondrichthyans (1986); Young (1986) on placoderms; Long (1986) on acanthodians; and Meyer and Dolven (1992), Hedges et al., (1993) and Ahlberg (1989) on osteichthyans. The combined results do not consititute a highly corroborated set of interrelationships, and neither does the apparently high degree of resolution signify that all of the nodes are equally robust (Lanyon, 1993). For example, thelodonts and anaspids, although included within the agnathans (generally agreed to be paraphyletic*) are regarded by Wilson and Caldwell (1991) as stem-gnathostomes.

Morphological character distribution among stem-taxa is critically important for the detection of primitive conditions and convergent evolution (Panchen and Smithson, 1987; Gauthier et al., 1988). Although molecular phylogenies, constructed independently of morphological data, can supply

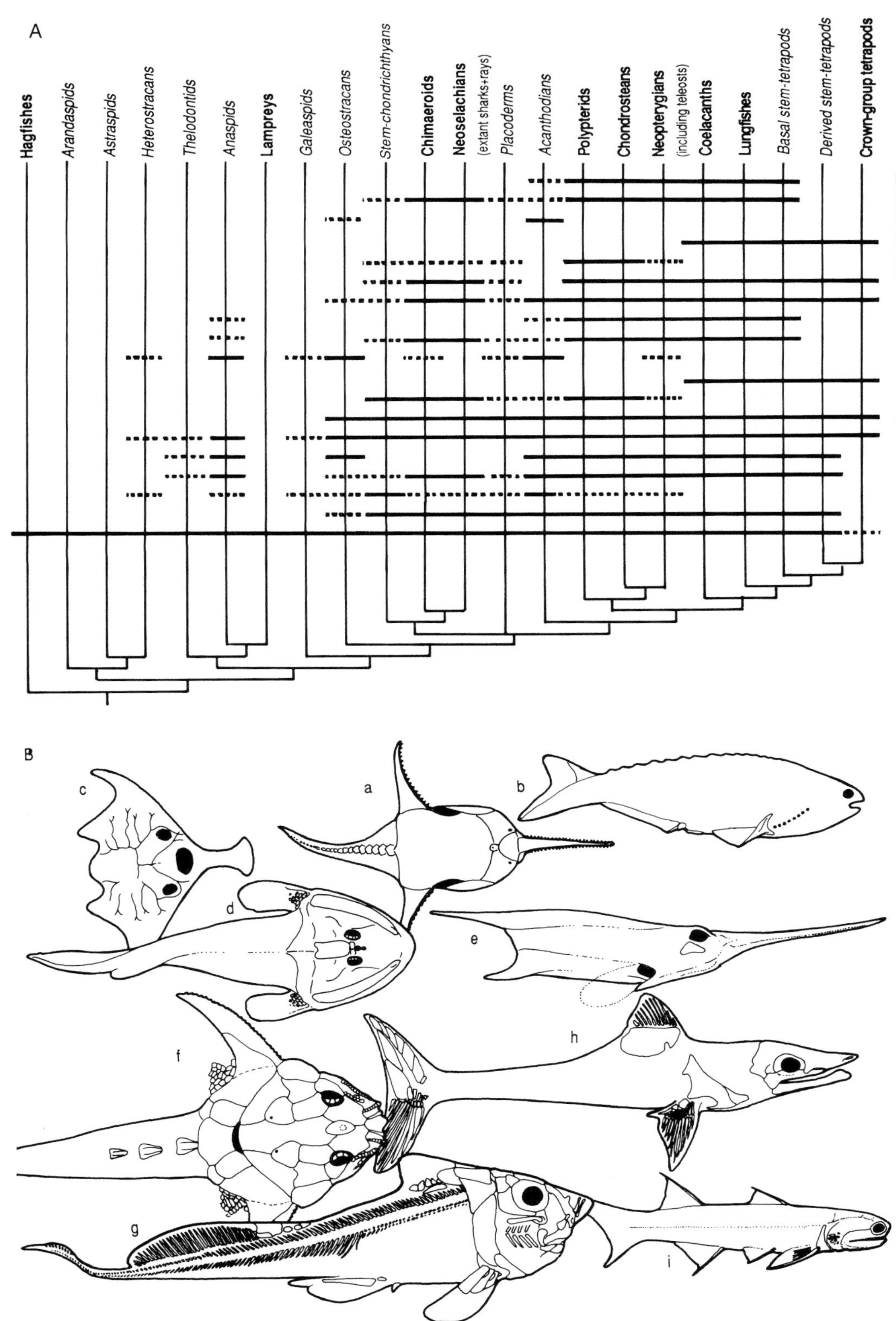

Fig. 2. (A) The phylogenetic distribution of vertebrate fin morphologies and principal skeletal structures. Living taxa are listed in bold type; fossil taxa are listed in italics. The branching sequence of vertebrate interrelationships is presented at the base (sources for the phylogeny are given in the text). Stem-taxa refer to an extinct taxon's closest relationships to a monophyletic group of living taxa (the crown-group). Thus polypterids, chondrosteans, neopterygians, coelacanths, lungfish and tetrapods constitute crown-group osteichthyans (bony fishes). Acanthodians in this scheme emerged after the chondrichthyan: osteichthyan evolutionary split, and are therefore stem-osteichthyans; similarly, galeaspids and osteostracans are stem-gnathostomes. Such stem-groups often supply unique data on the primitive conditions of living taxa. Horizontal bars crossing vertical branches indicate distribution of fin characteristics listed on vertical axis: solid bars = general or primitive condition; broken bars = uncertain data or significantly incomplete character distribution in particular taxon. (B) Assorted extinct vertebrates, selected to exemplify morphological diversity lost from the extant biota (compare these with hypothetical ancestors such as that in Fig. 1B): (a) *Doryaspis*, dorsal view, a heterostracan (Carroll, 1988); (b) *Rhyncholepis*, lateral view, an anaspid (Ritchie, 1980); (c) *Sanchaspis*, head shield, dorsal view, a galeaspid (Janvier, 1984); (d) *Hemiclaspis*, dorsal view, an osteostracan (Janvier, 1984); (e) *Pituriaspis*, head shield, lateral view, an osteostracan (Young, 1991); (f) *Lunaspis*, dorsal view, a petalichthyid placoderm (Denison, 1978); (g) *Ctenurella*, lateral view, a ptyctodontid placoderm (Denison, 1978); (h) *Caseodus*, lateral view, a stem-group chondrichthyan (Zangerl, 1981); (i) *Ischnacanthus*, lateral view, an acanthodian (Denison, 1979).

equally important corroborative results (eg. Meyer and Dolven, 1992; Hedges et al., 1993), trees including recent and fossil taxa have a much greater information content, especially along internodal branches. Fossils may reveal unanticipated primitive morphologies (Coates, 1991), refute conjectured homologies, challenge recapitulatory inferences derived from ontogenetic sequences, overturn tenuous phylogenetic hypotheses (eg. Fig. 2A), and record minimum dates of taxonomic divergence (Patterson, 1981). Fossils therefore supply a unique source of data lost from the extant biota, and record the earliest examples of morphologies resulting from developmental processes known in living taxa.

The relationship between developmental research and phylogenetic reconstuction is reciprocal. Developmental research underpins a large body of work on the ontogeny of phylogenetically early, dermal skeletogenic and odontogenic tissues (eg. Smith and Hall, 1993, and references therein). Homologies based upon ontogenetic criteria are clearly dependent upon developmental research, which may therefore challenge phylogenetic hypotheses. The value of the amniote astragalus as a key character of established tetrapod phylogenies has been questioned recently in precisely this way (Rieppel, 1993a,b). Conversely, phylogenetic hypotheses frequently underpin interpretations of experimental research (eg. Tabin's, 1992, developmental-genetic explanation of the prevalence of pentadactyly assumes tetrapod monophyly). And comparative developmental theories are highly sensitive to phylogenetic changes. In Fig. 2A, the independent origins of paired fins suggest that whereas the lobate pectorals of stem-gnathostomes are scaled paddles, those of stem-lampreys have differentiated dermal tissues, implying earlier deployment of ectodermal derivatives. Furthermore, if a fin-related morphogenetic or regulatory system was identified in lampreys, it could imply retention from their finned ancestry, consistent with a gradualistic model of developmental evolution. Alternatively, if early finned taxa are unrelated to lampreys (cf. Wilson and Caldwell, 1993), then the resultant model may be saltatory (ie. aquisition of complete regulatory systems before functional diversification). These kinds of scenarios need to be assessed relative to the robustness of the underlying phylogeny.

VERTEBRATE LIMB STRUCTURES: PATTERNS, COMPOSITIONS AND RECONSTRUCTED EVOLUTIONARY PATHWAYS

The assumed absence of paired fins in most agnathans (including lampreys, hagfish, arandaspids, astraspids, heterostracans and galeaspids) may be an oversimplification (Fig. 2). Lateral fin-like processes, such as dermal branchial plates or postbranchial spines or cornuae* (Moy-Thomas and Miles, 1971; Janvier, 1993) (Fig. 2Ba,c,e,f) are widespread. These structures precede (complex) fins as they are usually recognised, perhaps recording the accumulation of basic fin components.

The first differentiated fin-ray like structures are found in median fins. Fork-tailed thelodonts (Wilson and Caldwell, 1991) and all anaspids (Fig. 2Bb) (Moy-Thomas and Miles, 1971; Ritchie, 1980) bear rows of lepidotrichia*-like, aligned dermal denticles. Certain thelodonts, like heterostracans, have suprabranchial lateral fins, while others, and all anaspids, have pectoral fin-folds immediately behind the branchial openings. All thelodont paired fins are short-based (restricted to the pectoral region), whereas anaspids have both short and long ventrolateral fins. Long-based anaspid lateral fins are specialised and not primitive continuous folds. Thelodont pectoral fin scales resemble those covering the flank, whereas paired anaspid fins, supported anteriorly by spines, when long, may have lepidotrichia. None of these forms preserve direct evidence of dermotrichia*, which may be necessary precursors for lepidotrichial outgrowth (Schaeffer, 1987; Thorogood, 1991). Fossil endoskeletons are similarly absent, but in living cyclostomes (hagfish and lampreys: a paraphyletic* group) the cartilagenous fins supports are neural and haemal arch prolongations, with no dermal contribution (Goodrich, 1904, 1930). Lampreys sometimes have branching, unsegmented radials intercalating between the neural spines (personal observation).

Stem-group gnathostomes repeat the sequence of median preceding paired fin elaboration. Some galeaspid (Fig. 2Bc) head shields bear near-pectoral spines/cornuae (cf. heterostracans) above the branchial openings, but median fin details are uncertain. In contrast, osteostracan (sister-group* to the crown*- gnathostome radiation: Fig. 2Bd) head shields have canalised pectoral fin insertions on the rear of the cornuae, and median fins with lepidotrichia and radials (Moy-Thomas and Miles, 1971). Pectoral insertions similarly extend to a point above the second or third gill openings (Westoll, 1958). Ventrolateral ridges extending from the head shield have been interpreted as further fossil evidence for ancestral continuous lateral fin-folds, but their overlapping conjunction (Patterson, 1981) with pectoral fins challenges an ancestor-descendent relationship (Forey, 1984). Alternatively, pituriaspids have pelvic fin-like flanges (Young, 1991: 'subanal laminae'; Fig. 2e) extending from the rear of a craniothoracic shield, but their phylogenetic affinities are unclear.

The major living lineages of crown-group gnathostomes, chondrichthyans and osteichthyans, are accompanied by two major extinct groups, placoderms and acanthodians, the relationships of which are uncertain (Fig. 2). Young (1986) inserts placoderms in a trichotomy between osteichthyans and chondrichthyans, whereas Forey and Gardiner (1986) interpret placoderms as stem-osteichthyans. Acanthodians may be stem-gnathostomes (Rosen et al., 1981), or stem-osteichthyans (Maisey, 1986; alternative hypotheses of basal gnathostome interrelationships are reviewed in these four references).

Placoderms (Fig. 2Bf,g) are relative late-comers to the gnathostome fossil record, appearing to have radiated, diversified and become extinct mostly within the Devonian period; their subgroup interrelationships are uncertain. Most placoderms have pectoral and pelvic fins (absence of pelvic fins does not appear to be primitive in current phylogenies, summarised in Carroll, 1988). Radials are rarely well preserved; certain taxa have metapterygia, some have dermotrichia, while others' fins are near-plesodic* (Denison, 1978; Jarvik, 1980).

Stem-chondrichthyans (Fig. 2Bh) may be the earliest crown-gnathostomes (Lower Silurian: Karatajute-Talimaa, 1992); their fins may also be plesodic primitively. Some of the earliest (Devonian) articulated specimens, eg. *Cladoselache* (Zangerl, 1981), have branched, unsegmented radials interspersed distally with intercalaries, resembling lamprey (median) fin supports, and experimentally induced interdigitals, (cf. Hurle et al., 1991). Dermotrichia (ceratotrichia) may have been present primitively (Zangerl, 1981) as in living taxa where their growth is persistant and extensive. If edestids and cladoselachians represent the primitive chondrichthyan condition (Maisey, 1984), then pectoral fin growth and elaboration greatly exceeded that of the pelvics (when present). In taxa with two dorsal fins, anterior fin development similarly exceeds that of the posterior. Chondrichthyan pelvic fins may not have been primitively metapterygial (Zangerl, 1981). Sexual dimorphism obscures the issue, because males have branched, metapterygium-like claspers*, while corresponding female pelvics are anaxial.

Acanthodians (Fig. 2Bi) are considered to be the sister-group of crown-group osteichthyans (actinopterygians and sarcopterygians) (cf. Maisey, 1986). Fins preceded by spines (the inter-fin spines of certain taxa are probably not primitive, Long, 1986), bear the first evidence of gnathostome lepidotrichially aligned scales, sometimes underlain with proximally mineralised dermotrichia (Fig. 3A). Perichondrally ossified pectoral girdles are present occasionally (Denison, 1979). Endoskeletal pectoral fin supports, thought to consist of "a few small nodules arranged irregularly" (Rosen et al., 1981; Miles, 1973), are metapterygial (refuting a principal argument used to exclude acanthodians from the gnathostome crown-group: Rosen et al., 1981; sustained in Forey, 1984; Forey and Gardiner, 1986). Furthermore, acanthodian-like fin spines are found in the earliest actinopterygian-like osteichthyan remains (Gross, 1969), although they are absent from the earliest specimens of complete fish (Pearson and Westoll, 1979; Gardiner, 1984). It may also be significant that the pelvic fins of acanthodians and primitive actinopterygians tend to be less constricted proximally than pectorals. Actinopterygian pelvic fins usually enclose a series of simple radials, although pelvic metapterygia are present in certain chondrosteans (sturgeons and paddlefishes; Fig. 3B, Sewertzoff, 1924; Grande and Bemis, 1991). Pectoral radials primitively include a propterygium and a metapterygium (Fig. 1D) but the latter is secondarily absent in living (Goodrich, 1930) and many stem-teleosts. Adult actinopterygian dermotrichia are restricted to an apical fringe (cf. acanthodians), with the exception of certain teleosts' adipose fins which are supported exclusively by elongate dermotrichia.

Sarcopterygians (living groups include tetrapods, lungfish,

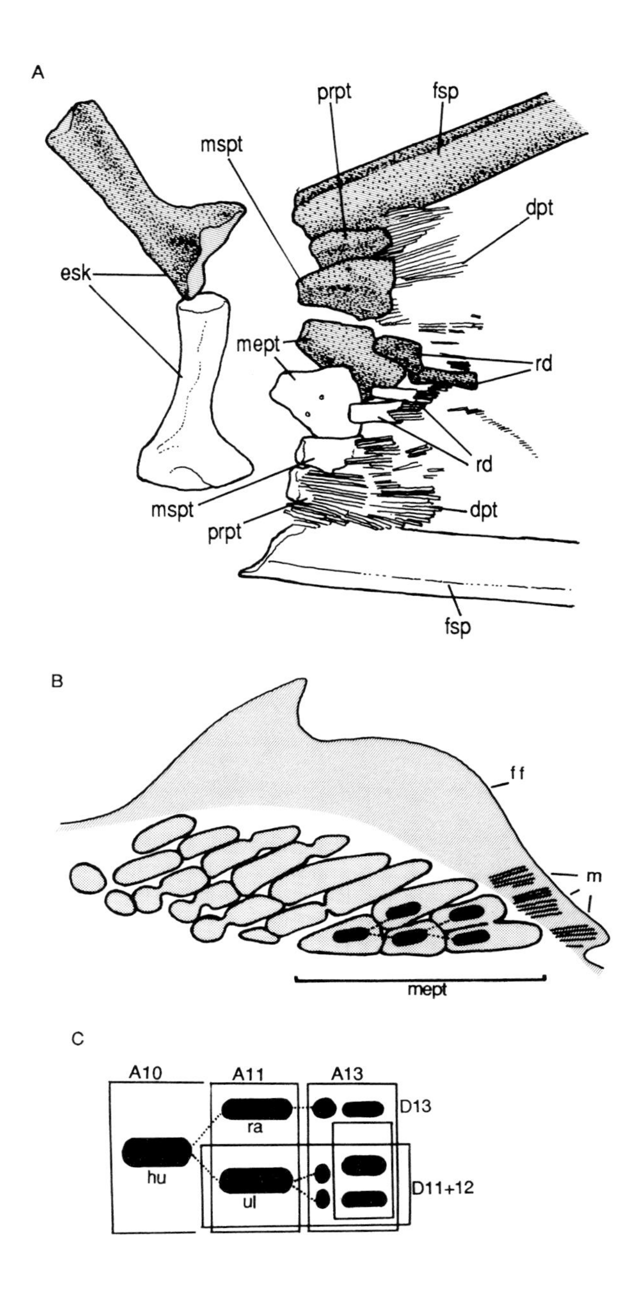

Fig. 3. (A) New data on the pectoral fin skeleton of *Acanthodes bronni*, (Permian: Lebach, Germany) specimen UMZC (University Museum of Zoology, Cambridge) GN15. Abbreviations: dpt, dermotrichia; esk, endoskeletal pectoral girdle; fsp, fin spine; mept, metapterygium; mspt, mesopterygium; prpt, propterygium; rd, radials; stippled structures = left fin; all viewed from ventral aspect. Acanthodian pectoral fin endoskeletons are known only from the perichondrally ossified radials of *Acanthodes*. The metapterygium is perforated by 2+ foramina and supports two radials (displaying a characteristic 1:2 ratio of proximal to distal elements), and the propterygium, unlike those of actinopterans, is imperforate. The proximally mineralised dermotrichia conform to Miles' (1973) description. The accepted reconstruction of an acanthodian pectoral fin endoskeleton consists of three cylindrical rods, as reinterpreted by Miles (1973) after Watson's (1937) earlier attempt (also based upon UMCZ GN15), including extensive proximal and distal radial series. But Miles based his reconstruction on casts of lost, original German specimens, which now appear to be incomplete; UMCZ GN15 was not re-examined. The new data presented here was obtained from high definition silicone rubber peels of the original material after further, detailed, negative preparation. (B) Pelvic fin development in *Acipenser ruthens*, after Sewertzoff (1924: Figs 1 and 3), showing prechondrogenic cell clusters (stippled) segmenting to form radials distally and basals proximally (secondarily fused basals form the rudimentary pelvic girdle). Posterior-most condensations form 1:2 ratio branching pattern characteristic of metapterygium, here superimposed with a segmentation and bifurcation diagram, resembling those devised by Shubin and Alberch (1986). Differentiation and chondrification proceeds from posterior to anterior of fin base (right to left across page). Abbreviations: ff, fin fold (linear shading); m, muscular bundles associated with developing radials, *not* dermotrichia; mept, metapterygium. (C) Schematic diagram of endoskeletal pattern development in a st. 28 chick forelimb (after Yokouchi et al., 1991, Fig. 4f). Shubin and Alberch-style branching diagram shows 1:2 proximodistal ratio inferred as conserved from metapterygial fin axis; boxes outline Hox expression domains, with boundaries correlating with skeletal pattern branching events. Abbreviations: hu, humerus; r, radius; ul, ulna.

and coelacanths) have exclusively metapterygial paired fin endoskeletons. Although biserial fins (eg. Fig. 1C) resemble Gegenbaur's archipterygium, out-group analysis suggests that short uniserial metapterygia are primitive (Ahlberg, 1989). Pelvic metapterygia usually consist of fewer segments than pectoral metapterygia. Dermotrichial distribution is uncertain: they occur in the second dorsal fin of the extant coelacanth, *Latimeria* (Geraudie and Meunier, 1980), but are reported as absent from late juvenile and adult lungfish (Goodrich, 1904; Geraudie and Meunier, 1984). However, an untraced paper by Geraudie (1985, perhaps recorded incorrectly in Musick et al., 1991), entitled "...actinotrichia...in developing teleost and Dipnoi fish fins" implies their transient ontogenetic appearence. Lungfish lepidotrichial equivalents, camptotrichia (Goodrich, 1904; Geraudie and Meunier, 1984) are primitively well ossified, segmented and branched distally. In *Neoceratodus* and *Protopterus* the camptotrichia remain branched and segmented although demineralised distally but, in *Lepidosiren,* the median (caudal) camptotrichia are shorter, very soft, unbranched and scarcely jointed, and paired fin camptotrichia are entirely absent.

Differences between the fins of *Lepidosiren* and those of other lungfish resemble the evolutionary transformations between finned and increasingly finless, digited stem-tetrapods (Fig. 4A). Basal stem-tetrapods (eg. *Eusthenopteron*) have several median fins and paired fins with short, uniserial metapterygia (Jarvik, 1980). The lepidotrichia are ossified, segmented and branched distally, but dermotrichia are unknown. A more derived, finned (rather than digited) stem-tetrapod, *Panderichthys*, retains only a single, median, caudal fin, but still has the primitive pattern of paired fins with short, uniserial metapterygia and ossified, segmented and branched lepidotrichia (Vorobyeva and Schultze, 1991). In comparison

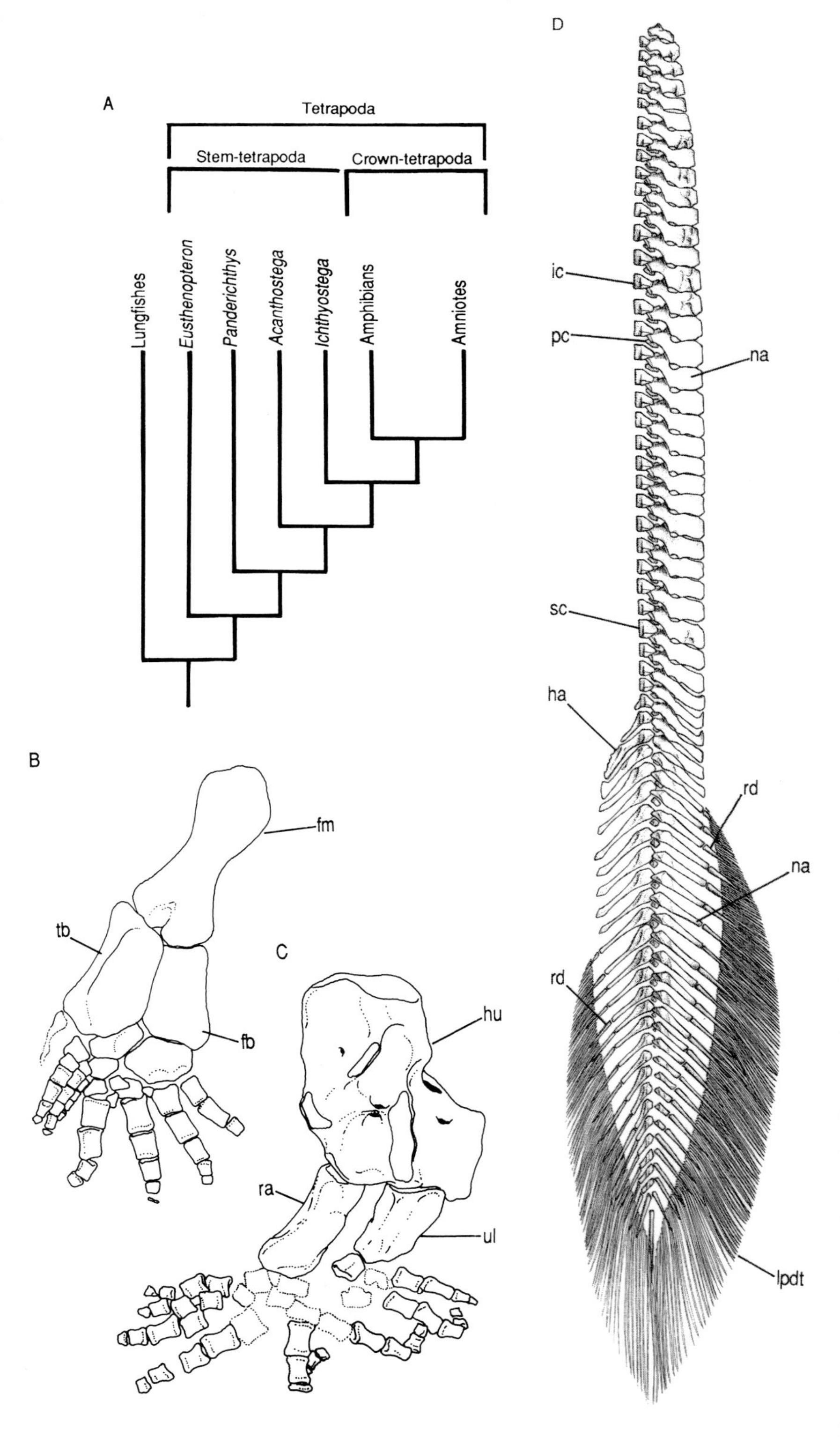

Fig. 4. (A) Cladogram illustrating an Adams consensus tree of early crown- and stem-group tetrapods, and other selected sarcopterygian fish (work in progress; summarised in Coates, 1994). Note that *Acanthostega* and *Ichthyostega* fall outside of the tetrapod crown-group, and, using a node-based definition (Gauthier et al., 1988; Rowe and Gauthier, 1992), must be considered stem-tetrapods along with other extinct taxa such as *Eusthenopteron* and *Panderichthys*, which are usually described as fish (a taxonomically imprecise grade-group). Taxonomic definitions based upon the presence of key characters/ evolutionary novelties (eg. digits) are fraught with typological difficulties and post hoc explanatory scenarios of evolutionary success (see Cracraft, 1990, and previous refs). (B) *Ichthyostega*, hind limb, specimen MGUH 1349 (Geological Museum of University Copenhagen). Abbreviations: fb, fibula; fm, femur; tb, tibia. (C) *Acanthostega* pectoral limb, specimen MGUH 1227. Abbreviations: hu, humerus; ra, radius; ul, ulna. (D) *Acanthostega*, new reconstruction of the axial skeleton showing tail fin structure, based upon specimens UMCZ T1300, MGUH 1227, 1258 and 1324. Vertebrae bipartite with small pleurocentra and large intercentra; notochordal canal unconstricted; weakly developed regionalization: cervical neural arch height gradually reduced anteriorly, anterior caudal vertebrae with specialised neural and haemal arches, presacral count = 30. Abbreviations: ha, haemal arch; ic, intercentrum; lpdt, lepidotrichia; na, neural arch; pc, pleurocentrum; rd, radial; sc, sacral intercentrum.

with these, *Acanthostega* (a digited stem-tetrapod), like *Lepidosiren*, has lost paired fin lepidotrichia, but retains a well-developed caudal fin with unsegmented and unbranched lepidotrichia supported by radials (Fig. 4D). Furthermore, like *Lepidosiren*, the paired fin metapterygia are extended distally, although these are uniserial (preaxial radials proximally and postaxials, ie. digits, distally: Fig. 4C) and the entire axis appears to have been twisted anteriorly within the limb-bud (Shubin and Alberch, 1986).

Digited limbs originated after tetrapods diverged from their shared ancestry with lungfish (tetrapod's living sister-group: Panchen and Smithson, 1987; Meyer and Dolven, 1992), but before the evolutionary radiation of living forms. Digited limbs are therefore neither coincident with tetrapods in the restricted (exclusively crown-group) or the broad/total (crown- plus stem-group) definitions of the group (Fig. 4A). In effect, digited limbs initially constitute another derivation of a plesodic fin pattern. The general points which emerged from a previous review of early limb morphologies (Coates, 1991), are supported by a recently completed analysis of early tetrapod interrelationships. Early reptiliomorph (eg. *Tulerpeton*, Lebedev, 1984; Coates, 1994), crown-group tetrapods, and digited stem-tetrapods, are polydactylous (Coates and Clack, 1990). Furthermore, polydactylous limbs accompany retained, lungfish-like tails in which unsegmented and unbranched lepidotrichia are supported by radials articulating with neural and haemal spines (Fig. 4D). Thus pentadactyly in batrachomorph (anamniote) hindlimbs, and reptiliomorph fore- and hindlimbs, probably originated and stabilised independently. Digits (segmented postaxial radials, supporting no lepidotrichia or dermotrichia) appear before the elaboration of ossified wrist and ankle joints; carpal remains tend to be rarer than tarsal, perhaps reflecting a trend of delayed ossification. Early tarsal patterns (eg. Fig. 4C) include relatively few large elements which cannot be related simply to those of more recent examples, and early carpal patterns appear to be similarly abbreviated. In fact, a striking feature of these early digited limbs is their failure to exhibit fixed primitive patterns or canonical formulae.

DISCUSSION

Morphological change

A fairly robust series of parallel or convergent morphogenetic trends emerges from this overview of vertebrate limb evolution (Fig. 2A). The phylogenetic development and elaboration of median fins precedes that of paired fins, and pectoral fins initially precede pelvic fin development. The first fin-like outgrowths are usually associated with dermal cornuae, spines, and/or specialised scales, but these are impersistent; their phylogenetic distribution is patchy. Dermotrichial and lepidotrichial development is consistently secondary (assuming, as earlier, that lepidotrichial growth is dependent upon dermotrichial precursors), although these are the most widely distributed dermal components of fin skeletons. Early fin endoskeletons are much rarer, perhaps reflecting the apparent independence of dermal from endoskeletal skeletogenic systems (Patterson, 1977). Alternatively, the absence of early radials may be an artifact of the poor preservational potential of cartilage (chondrichthyan prismatic cartilage being a notable exception). The occurrence of these dermal and endoskeletal features in median before paired fins is a striking feature of the earlier taxa in this phylogeny, although with further branching events the sequence is obscured by changes in tissue diversity and structural repatterning.

Endoskeletal changes preceding tetrapod limb evolution mostly concern metapterygial transformations, ie. evolutionarily conserved, specialised posterior basal radials and their associated distal structures (Figs 1D, 3). Metapterygia are unique to crown-group gnathostomes, and their characteristic asymmetry may record the evolutionary imposition of a regulatory mechanism, such as a ZPA, upon the phylogenetically more primitive branching properties of limb/fin bud mesenchyme (Pautou, 1973; Hinchliffe, 1989). Only sarcopterygian osteichthyans (including tetrapods) have consistently similar pectoral and pelvic fin patterns. Actinopterygian pectorals and pelvics differ: clear pelvic metapterygia are restricted to chondrosteans. Such strongly divergent osteichthyan fin evolution has been explained in part by the suggestion that most actinopterygian pelvic metapterygia are incorporated into the pelvic girdle (Sewertzoff, 1924; Rosen et al., 1981). Most actinopterygians emphasise the anterior of the fin skeleton; pectoral metapterygia are lost in the vast majority (teleosts), whereas propterygia are enlarged and elaborated. Sarcopterygians, in contrast, emphasise the posterior of the fin skeleton, losing all proximal radials anterior to the metapterygia (contrast Fig. 1C,D). Thus the simple pectoral endoskeleton of *Brachydanio rerio* has a complex evolutionary history of secondary reduction.

There appear to be similarly significant differences between non-sarcopterygian and sarcopterygian patterns of metapterygial development. Shubin and Alberch (1986) described phylogenetically conserved, dynamic sequences of prechondrogenic focal condensation, segmentation and bifurcation producing the branching endoskeletal patterns of tetrapod limbs. These patterns include a single continuous path of bifurcation and segmentation which they inferred to be a synapomorphy* of tetrapod limbs and sarcopterygian fins, conserving the developmental pattern of a metapterygial axis. In limbs these bifurcation nodes have been correlated closely with Hox gene expression boundaries (Yokouchi et al., 1991; Blanco et al., 1994; Fig. 3C) associated with pattern regulation. However, chondrostean pelvic metapterygia also include secondarily fused, parallel prechondrogenic cell-clusters; branching events are restricted to a small region next to the posterior fin-bud boundary (Fig. 3B, Sewertzoff, 1924). Chondrichthyan metapterygia appear to be formed similarly. Thus primitive metapterygia probably also included these numerous, segmenting proximal focal condensations, characteristically absent in sarcopterygians (including tetrapods). The transformation to a sarcopterygian pattern could have resulted from changed tissue domain dimensions (ie. fin-base constriction as invoked by Goodrich, 1930, or Jarvik, 1980), and segmentation versus branching events may be related to structural packing properties (Shubin and Alberch, 1986; Oster et al., 1988) rather than direct genetic regulation (cf. Yokouchi et al., 1991). However, these kinds of explanations appear to be uninformative about the maintenance of limb- or fin-bud dimensions, or the regulation of meta- versus meso- or propterygial domains.

Dermal skeletal loss is an equally significant event in tetrapod limb evolution. Fin-bud apical ectodermal folds enclosing lepidotrichial development must have been transformed into the short, apical ectodermal ridges of tetrapod limb buds (Thorogood, 1991). Fin-ray* loss also suggests changes affecting neural crest cellular migration, although there is no direct evidence of skeletogenic neural crest tissue in any paired fin bud (Smith and Hall, 1993; neural crest now confirmed in median fins, Smith et al., 1994). Thorogood and Ferretti (1993), however, have already suggested that paired limb buds with exclusively mesodermal mesenchyme may be a tetrapod specialisation (contra Thomson, 1987), as corroborated by the entry of melanophores into the mesenchyme of teleost pectoral fins (Trinkaus, 1988a,b). Devonian tetrapods illustrate the phylogenetic sequence of fin loss, with ray-less (ie. digited) limbs accompanying tails retaining radials and unbranched, unsegmented lepidotrichia (Jarvik, 1980; Coates, 1991; Fig. 4D). As the living sister-group of tetrapods (Hedges et al., 1991; Meyer and Dolven, 1992), the parallel patterns of lungfish fin reduction (described earlier; Fig. 2) probably provide the best subjects for investigating these developmental changes which affected our own fin-limb transition. This "last hired, first fired" (Gould, 1991) pattern of fin loss (assuming that pelvics are novel relative to pectorals, Fig. 2) suggests that pelvic fins may have lost their rays before pectorals. Perhaps the more fish-like pectoral than pelvic skeletons in *Acanthostega* record this evolutionary sequence (Coates, 1991; Ahlberg, 1991). Paired fin origins, probably pectoral-led, may therefore have been superimposed by limb patterning, perhaps pelvic-led.

Developmental regulation, phylogeny and convergence

The use of tetrapod limbs as model systems within which to explore the developmental-regulatory role of homeobox genes has stimulated much of the renewed interest in vertebrate limb evolution. Homeobox genes are distributed and apparently conserved across a spectacularly diverse taxonomic range (Slack et al., 1993). In particular, expression patterns of members of the HoxA (Yokouchi et al., 1991) and HoxD clusters (Dolle et al., 1989) have been mapped during amniote limb bud development (Fig. 3C), and targeted misexpression of the HoxD complement appears to transform digit identities (Morgan et al., 1992). Consequently, the nested, pentate HoxD expression domains in limb buds have been interpreted as specifying five identities (via combinatorial codes) underlying tetrapod pentadactyly (Tabin, 1992). These five are even considered present within the Devonian (360+ million years BP) octodactylous array of *Acanthostega* (Tabin, 1992; Fig. 4C), and therefore retained from an unspecified pre-tetrapod, and perhaps pre-gnathostome condition. Influential biological theoreticians such as Goodwin (1993) have already used Tabin's interpretation of *Acanthostega*'s forelimb as evidence for an historically invarient genetic constraint.

This inference appears to be consistent with the theory that amniote Hox clusters are conserved from an episode of amplification and four-fold duplication early in vertebrate evolution (relative to the homologous *Antennapedia*-class genes of *Drosophila melanogaster*, and a hypothetical common ancestral complement: Krumlauf, 1992; Holland, 1992; Fig. 5A). Functionally, gene duplication and differentiation is thought to have provided regulatory systems for morphological differentiation (Lewis, 1978; Akam et al., 1988) in response to the increased demands of vertebrate embryogenesis (Holland, 1992; Holland et al., 1994). So the relation of morphological evolution to gene duplication, differentiation and functional diversification may be represented as a correlation between the phylogenetic distribution of morphological characters and the spatiotemporal distribution of regulatory gene expression. Similarities between cognate genes' (ie. putative homologues) expression domains should be associated with symplesiomorphies*, while differences may be associated with autapomorphies*. This suggests that more attention should be paid to the structural and expression diversity of vertebrate Hox networks, such as those of *Xenopus* and *Brachydanio* which have yet to be published in detail equivalent to those of mice and humans (eg. Scott, 1992) instead of focusing upon their conserved features. Such an approach would test Tabin's model, or at least provide information about the evolution and/or phylogenetic insertion of the proposed genetic constraint.

In fact, the results of *HoxD11*(=*HoxD4.6*, Scott, 1992) overexpression (Morgan et al., 1992) are ambiguous, and may alternatively suggest a scenario of evolutionary change rather than invariance. The apparently reduced digit diversity (digit I transformed to resemble digit II) in an avian limb bud appears to result from a less diverse combination of overlapping Hox domains. And if applied to the less highly differentiated digits of *Acanthostega* (Fig. 4C), then, contra Tabin (1992), it may imply regulation by a similarly less elaborate Hox cluster (Coates, 1993b). Shubin and Alberch's (1986) analysis of limb development demonstrates clearly the serial, iterative generation of digits. If acanthostegid digits therefore represent a primitively undifferentiated sequence of serial homologues, then anatomical distinctions such as digit length and number could result simply from structural properties (eg. breadth of enclosing limb bud tissue domain: Oster et al., 1988). And the distinctive identities of first and last digits could result from the properties of 'endedness' (Bateson, 1913, discussed in Roth, 1984) superimposed upon an array preceding digit 'individualization' (Wagner, 1989) as exemplified by the nested genetic identities proposed for amniote limbs. Therefore, when attempting to draw homologies between our own five digits and those of *Acanthostega*, the biological content of such decisions covers a variety of inferences. The dynamic sequence of digit production (cf. Shubin and Alberch, 1986), indicates that the five furthest from the leading edge of the limb are topographic equivalents to those of pentadactyl amniotes. But if statements about taxic homology are conjectural inferences of conserved ontogenetic potential (discussed further in Coates, 1993b), then these early digits, like fin radials, may not share with us a nested sequence of elaborate, combinatorial Hox codes supplying an address for each of five digit-types: radials and early digits are evolutionarily indistinct. Simpler, more iteratively patterned, endoskeletal paired fin or limb morphologies (cf. *Neoceratodus* and *Acanthostega*, Figs 1C, 4D) may be regulated by either less elaborately expressed or differentiated Hox clusters. Furthermore, correspondingly simpler morphologies may be found in other structures linked pleiotropically by the same regulatory genes.

Phylogenetic hypotheses of morphological change can also suggest potential examples of evolutionary convergence. The HoxA and HoxD genes expressed in amniote limbs are members of the 5′-located and caudally expressed Abd-B subfamily (Fig. 5A) (Yokouchi et al., 1991; Dolle et al., 1989).

This subfamily includes members of all four clusters aligned as five subgroups related to the single Abd-B gene in *Drosophila* (Krumlauf, 1992). While this resemblence suggests a period of evolutionary 5′-cluster expansion, it is not clear how this relates to cluster duplication. Insufficient current evidence supports the inference of a single HoxB Abd-B gene as primitive relative to the multiple members of A, C, and D clusters (Izpisua Belmonte et al., 1989), although there is some suggestion that four-fold Hox clusters may have appeared within the gnathostome stem-group (lampreys appearing to have three clusters and cephalochordates two: Pendleton et al., 1993). The equivalent Hox complements of *Brachydanio* (a cypriniform teleost) will therefore provide a valuable comparison with those of tetrapods. If Hox elaboration is linked closely to the regulatory demands of morphological complexity, then perhaps the teleost equivalents of groups 11-13 (Scott, 1992) should be correspondly less diverse, given the simplicity of the pectoral endoskeleton.

However, without out-group data distinctions of conserved from convergent features in Hox networks may be difficult to support and, in *Brachydanio*, the situation is probably confounded by the derived morphological complexity of the caudal skeleton. Teleost tails are highly specialised, asymmetric and taxon specific (Fig. 5B; Patterson and Rosen, 1977). The primary expression domains of the Abd-B subfamily are caudal (mapped in mice: Kessel and Gruss, 1991). This suggests that the phylogenetically accumulated complexities of teleost tails may well have required similar, but independent, instances of Hox gene duplication or functional diversification. These could resemble those driven by the similar regulatory requirements of secondary expression in tetrapod limbs, especially if Hox elaboration is constrained structurally to the thirteen paralogous sites proposed by Scott (1992) and others. Therefore, the apparently simple teleost pectoral endoskeleton may be over-written by secondarily expressed, redundant but highly differentiated tail-driven Abd-B genes. This reverses the equivalent expression pattern in mice, where elaborated D

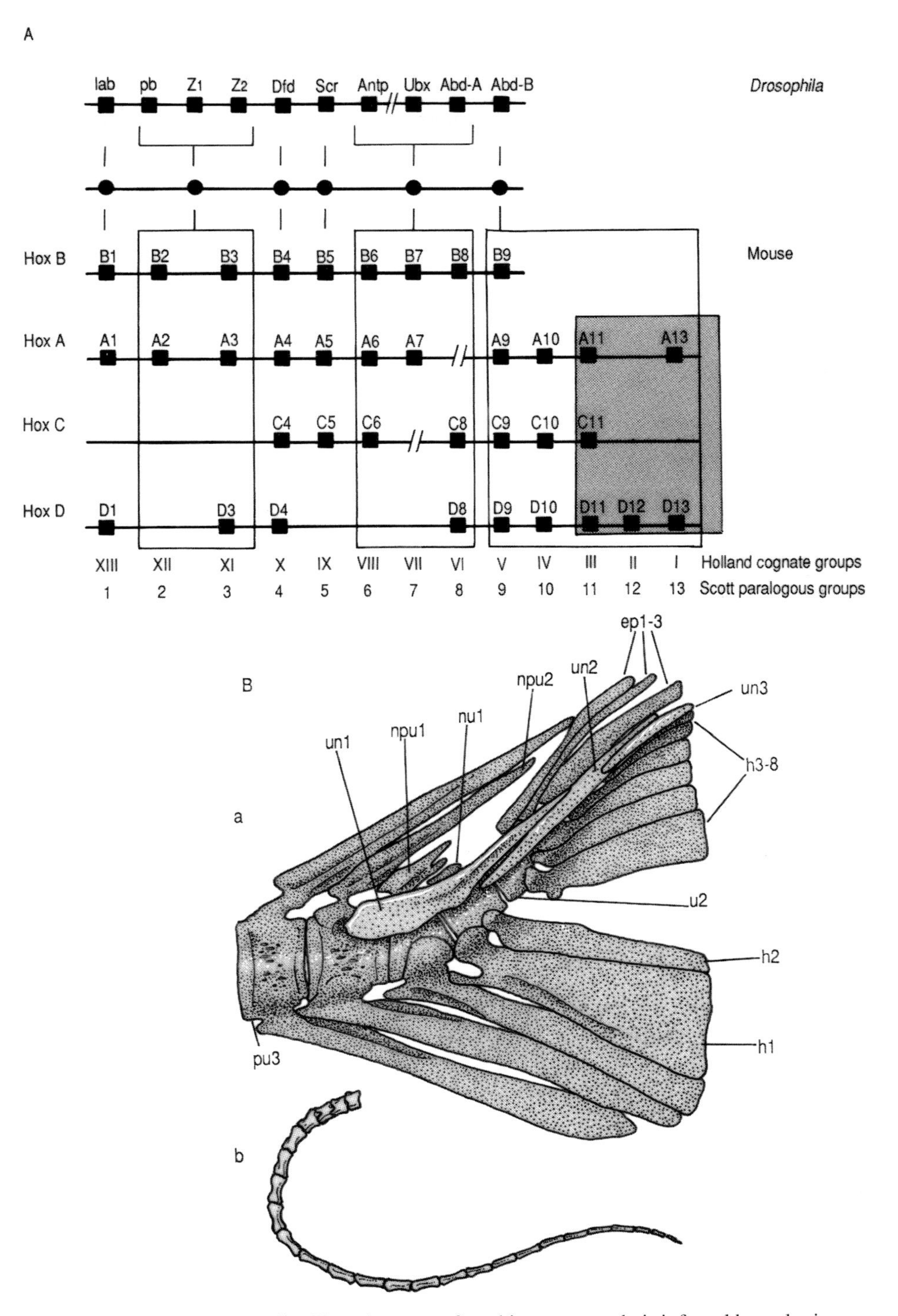

Fig. 5. (A) The four mammalian Hox clusters as found in a mouse; their inferred homologies with the *Drosophila* Hom-C complex, and a hypothetical most recent common ancestor of insects and vertebrates (Based on Krumlauf, 1992; Holland, 1992; Scott, 1992). The brackets/boxes define where homological relationships are established with greater confidence. Shaded box indicates paralogous groups associated with fin/limb patterning expected to differ most in non-tetrapod vertebrates. (B) A pair of vertebrate, osteichthyan, tail endoskeletons, illustrating the taxon specific complexity of (a) a teleost, *Anaethalion* (after Patterson and Rosen, 1977), contrasted with the simple, iterative pattern of (b) a rodent tail, *Paramys* (Carroll, 1988), both of which are assumed to be regulated by members of the Abd-B subfamily (Kessel and Gruss, 1991). Abbreviations: ep, epurals; h, hypurals; npu, neural arch/spine of numbered preural centrum; nu1, neural arch of first ural centrum; pu3, third preural centrum; u2, second ural centrum; un, uroneural bones.

Abd-B genes, apparently required for limb patterning (Morgan et al., 1992), are expressed primarily and redundantly in a simple, iteratively patterned rodent tail (Kessel and Gruss, 1991) (Fig. 5B). A phylogenetic framework is therefore needed to inform the selection of appropriate further out-group comparisons. In this example, a dipnoan, chondrostean or chondrichthyan would provide more clearly informative results, because the phylogenetic history of each reveals none of the extreme caudal axial patterning of teleost evolution.

Finally, and somewhat speculatively, expression distributions with redundancy between primary, secondary and even tertiary fields suggest a plausible mechanism for Muller's (1990) 'side-effect' hypothesis for the evolution of morphological novelties. In this teleost example, redundant expression in the secondary (fin) field could deploy a sequence of pattern regulators available for co-option in an independently evolved suite of highly derived appendages. Antenneriid teleosts, including several pseudo-limbed taxa with complex appendicular skeletons and musculature, may exhibit this phenomenon. Examples such as this and other mechanisms of functional diversification (Holland, 1992) are probably related to the failure to obtain a close correlation between genetic and morphological evolution (John and Miklos, 1988).

CONCLUSION

The early evolution of vertebrate appendages consists of parallel trends of increasing morphological complexity spreading from the median to anterior paired fins. But as phylogeny diversified, this simple pattern of fin elaboration became complicated by tissue loss and repatterning of the remaining structures. Pelvic fins may have originated as a reiteration of pectorals, but as with the discussion about digit evolution, pectoral and pelvic individualisation diverged within the different gnathostome lineages. Thus, in agreement with Hall (1991), elements of several past theories fit the morphological, ontogenetic and phylogenetic data, but none is individually sufficient. Hox gene expression and manipulation in limb development is beginning to provide new ways of thinking about morphological conservation, change and perhaps individualisation or homology. But this approach requires caution. It has long been argued that homology cannot be defined in terms of invarient gene action (de Beer, 1971; Goodwin, 1993), but perhaps more success will be achieved with reference to hypotheses of gene evolution. The relation of Hox genes to morphology may well have become obscured by functional diversification and convergence in experimental subjects' individual evolutionary backgrounds. This re-emphasises the importance of a phylogenetic framework within which to identify such convergent events. Subsequent comparisons between carefully selected out-groups' regulatory systems, and induced misexpressions in the primary experimental subjects, should enable the dissection of developmental regulation networks into individually specialised and shared general characteristics. I think that this kind of approach will become increasingly necessary in order to assess the relevence of choices such as "the fish as a simple model for our own development" (Kimmel, 1989).

I thank Drs Adrian Friday, Neil Shubin, Pete Holland and an anonymous referee for constructive criticism. Thanks also to Sarah Finney for exceptional preparation of acanthodian and *Acanthostega* specimens. The research was funded by awards from NERC, GR3/7215'A', and BBSRC.

GLOSSARY

Agnathan: paraphyletic group of jawless vertebrates.
Amniote: a monophyletic group including turtles, lizards, crocodiles, birds, mammals, and their fossil relatives.
Anamniotes: used here to define a monophyletic group including anurans, urodeles, apodans, and their fossil relatives.
Autapomorphy: a character unique to a monophyletic group.
Batoid: skates and rays.
Claspers: male intromittent organ extending from rear of pelvic fin, supported by metapterygium-like endoskeleton.
Cornuae: horn-like projections from the craniothoracic dermal armour.
Crown-group: a monophyletic group defined by living taxa and their most recent common ancestor, including collateral fossil descendents.
Dermotrichia: protein fin rays produced within developing apical ectodermal fold of fin bud.
Fin ray: dermal fin supports: includes dermotrichia and lepidotrichia; not to be confused with radials.
Gnathostomes: the jawed vertebrates, a monophyletic group.
Homology: in this article used in the sense of 'taxic homology' (Patterson, 1982), equating in practise with synapomorphy (see below), ie. a conjectural inference of shared derived ontogenetic potential relative to a phylogenetic hypothesis. This definition applies equally to hypotheses of genomic evolution (paralogy) and the developmental concept of identity; for current debates about the meaning and utility of homology see Hall, 1994.
Lepidotrichia: specialised scales forming dermal fin rays: aligned ontogenetically with subjacent dermotrichia.
Mesopterygium: middle compound radial.
Metapterygium: posterior compound radial, usually branching to support distal radials.
Monophyletic: a group including a common ancestor and all of its descendents.
Paraphyletic: a group descended from a common ancestor but with incomplete membership.
Plesodic: a fin endoskeleton where the radials extend to the fin perimeter.
Propterygium: anterior compound radial, usually penetrated by appendicular nerve and blood supply.
Radials: rod-like endoskeletal fin supports; not dermal.
Sister-group: a pair of most closely related terminal taxa in a dichotomously branching phylogeny.
Stem-group: a series of fossil taxa most closely related to, but not included within, a crown group.
Symplesiomorphy: characters defining a group at a higher rank than that under consideration.
Synapomorphy: uniquely shared derived characters, uniting a group with another to which it is most closely related.

REFERENCES

Ahlberg, P. E. (1989). Paired fin skeletons and relationships of the fossil group Porolepiformes (Osteichthyes: Sarcopterygii). *Zoo. J. Linn. Soc.* **96**, 119-166.

Ahlberg, P. E. (1991). Tetrapod or near-tetrapod fossils from the Upper Devonian of Scotland. *Nature* **354**, 298-301.

Akam, M., Dawson, I. and Tear, G. (1988). Homeotic genes and the control of segment diversity. *Development* **104** (Supplement), 123-133.

Appel, T. A. (1987). *The Cuvier-Geoffroy Debate*. Oxford: Oxford University Press.

Balfour, F. M. (1881). On the development of the skeleton of the paired fins of Elasmobranchii, considered in relation to its bearings on the nature of the limbs of the vertebrata. *Proc. Zool. Soc. Lond.* **1881**, 656-671.

Balinsky, B. I. (1975). *An introduction to embryology*. Philadelphia: Saunders.

Blanco, M., Misof, B. Y., and Wagner, G. P. (1994). Generation of morphological change: a case study on Xenopus limbs. *Journal of Morphology*. **220**, 326-327.

Burke, A. C. (1991). Proximal elements in the vertebrate limb: evolutionary and developmental origin of the pectoral girdle. In *Developmental Patterning of the Vertebrate Limb* (ed. J. R. Hinchliffe, J. M. Hurle and D. Summerbell), NATO ASI, Series A, **205** pp 385-394. New York: Plenum.

Carroll, R. (1988). *Vertebrate Paleontology and Evolution*. San Francisco: W. H. Freeman and Co.

Coates, M. I. (1991). New palaeontological contributions to limb ontogeny and phylogeny. In *Developmental Patterning of the Vertebrate Limb* (ed. J. R. Hinchliffe, J. M. Hurle and D. Summerbell), NATO ASI, Series A, **205** pp 325-337. New York: Plenum.

Coates, M. I. (1993a). *Hox* genes, fin folds and symmetry. *Nature* **364**, 195-196.

Coates, M. I. (1993b). Ancestors and homology. The origin of the tetrapod limb. In *Evolution, Development, and behaviour. In search of generative orders of biological form and process* (ed. J. L. Dubbeldam, B. C. Goodwin and K. Kortmulder). *Acta Biotheoretica* **41**, 411-424.

Coates, M. I. (1994). Evolutionary patterns and early tetrapod limbs. *Journal of Morphology* **220**, 334.

Coates, M. I. and Clack, J. A. (1990). Polydactyly in the earliest tetrapod limbs. *Nature* **347**, 66-69.

Cracraft, J. (1990). The origin of evolutionary novelties: patterns and process at different heirarchical levelss. In *Evolutionary Innovations* (ed. M. H. Nitecki), pp 21-44. Chicago: University of Chicago Press.

de Beer, G. R. (1971). Homology, an unsolved problem. *Oxford Biol. Readers*, **11**. Oxford Univ. Press, London.

Denison, R. H. (1978). *Placodermi. Handbook of Paleoichthyology.* (ed. H-P Schultze). Stuttgart: Gustav Fischer Verlag.

Denison, R. H. (1979). *Acanthodii. Handbook of Paleoichthyology.* (ed. H-P Schultze). Stuttgart: Gustav Fischer Verlag.

Dolle, P., Izpisua-Belmonte, J. C., Falkenstein, H., Renucci, A. and Dubuole, D. (1989). Coordinate expression of the murine Hox-5 complex-containing genes during limb pattern formation. *Nature* **342**, 762-772.

Ekman, S. (1941). Ein laterales Flossensaumrudiment bei Haiembryonen. *Nova Acta R. Soc. Scient. upsal. IV.* **12 (7)**, 5-44.

Forey, P. L. (1984). Yet more reflections on agnathan-gnathostome relationships. *J. Vert. Paleo.* **4**, 330-343.

Forey, P. L. and Gardiner, B. G. (1986). Observations on *Ctenurella* (Ptyctodontida) and the classification of placoderm fishes. *Zoo. J. Linn. Soc.* **86**, 43-74.

Forey, P. and Janvier, P. (1993). Agnathans and the origin of jawed vertebrates. *Nature* **361**, 129-134.

Gardiner, B. G. (1984). The relationships of placoderms. *J. Vert. Paleo.* **4**, 379-395.

Gauthier, J. A., Kluge, A. G., and Rowe, T. (1988). Amniote phylogeny and the importance of fossils. *Cladistics* **4**, 105-209.

Gegenbaur, C. (1878). *Elements of Comparative Anatomy*. London: Macmillan & C°.

Geraudie, J. and Meunier, F-J. (1980). Elastoidin actinotrichia in Coelacanth fins: a comparison with teleosts. *Tissue and Cell* **12**, 637-645.

Geraudie, J. and Meunier, F-J. (1984). Structure and comparative morphology of camptotrichia of lungfish fins. *Tissue and Cell* **16**, 217-236.

Ghiselin, M. T. (1988). The origin of molluscs in the light of molecular evidence. In *Oxford Surveys in Evolutionary Biology* (ed. P. H. Harvey and L. A. Partridge) **5**, pp 66-95. Oxford: Oxford University Press.

Goodrich, E. S. (1904). On the dermal fin-rays of fishes, living and extinct. Q. Jl. microsp. Sci. **47**, 465-522.

Goodrich, E. S. (1906). Notes on the development, structure, and origin of the median and paired fins of fish. *Q. Jl. microsc. Sci.* **50**, 333-376.

Goodrich, E. S. (1930). *Studies on the structure and development of vertebrates*. London: Macmillan.

Goodwin, B. (1993). Homology and a generative theory of biological form. In *Evolution, Development, and behaviour. In search of generative orders of biological form and process* (ed. J. L. Dubbeldam, B. C. Goodwin and K. Kortmulder). *Acta Biotheoretica* **41**, 305-314.

Goodwin, B. and Trainor, L. (1983). The ontogeny and phylogeny of the pentadactyl limb. In *Development and Evolution* (ed. N. Holder and C. Wylie) pp 75-98. Cambridge: Cambridge University Press.

Gould, S. J. (1977). *Ontogeny and phylogeny*. Cambridge Mass. : Belknap, Harvard University Press.

Gould, S. J. (1991). Eight (or fewer) little piggies. *Natural History* **1991**(1), 22-29.

Grande, L. and Bemis, W. E. (1991). Osteology and phylogenetic relationships of fossil and Recent paddlefishes (Polyodontidae) with comments on the interrelationships of Acipenseriformes. *J. Vert. Paleo.* **11** (supplement), 1-121.

Gross, W. (1969). *Lophosteus superbus* Pander, ein Teleostome aus dem Silur Gotlands. *Lethaia* **2**, 15-47.

Hall, B. K. (1991). Evolution of connective and skeletal tissues. In *Developmental Patterning of the Vertebrate Limb* (ed. J. R. Hinchliffe, J. M. Hurle and D. Summerbell), NATO ASI, Series A, **205** pp 303-312. New York: Plenum.

Hall, B. K. (1994). *Homology, the hierarchical basis of comparative biology*. London: Academic Press.

Harvey, P. H. and Pagel, M. D. (1991). *The comparative method in evolutionary biology*. Oxford: Oxford University Press.

Haswell, W. A. (1882). On the structure of the paired fins of *Ceratodus*, with remarks on the general theory of the vertebrate limb. *Proceedings of the Linnean Society of New South Wales* **7**, 2-11.

Hedges, S. B., Hass, C. A. and Maxson, L. R. (1993). Relations of fish and tetrapods. *Nature* **363**, 501-502.

Hinchliffe, J. R. (1989). Reconstructing the archetype: innovation and conservatism in the evolution and development of the pentadactyl limb. In *Complex organismal finctions: integration and evolution in vertebrates* (ed. D. B. Wake and G. Roth), pp 171-189. Chichester: J. Wiley.

Holland, P. (1992). Homeobox genes in vertebrate evolution. *BioEssays* **14**, 267-273.

Holland, P. W. H., Garcia-Fernandez, J., Holland, L. Z., Williams, N. A. and Holland, N. D. (1994). The molecular control of spatial patterning in amphioxus. *J. mar. biol. Ass. U. K.* **74**, 49-60.

Horder, T. J. (1989). Syllabus for an embryological synthesis. In *Complex organismal functions: integration and evolution in vertebrates.* (ed. D. B. Wake and G. Roth), pp 315-348. Chichester: J. Wiley.

Hurle, J. M., Macias, D., Ganan, Y., Ros, M. A., and Fernandez-Teran, M. A. (1991). The interdigital spaces of the chick leg bud as a model for analysing limb morphogenesis and cell differentiation. In *Developmental Patterning of the Vertebrate Limb* (ed. J. R. Hinchliffe, J. M. Hurle and D. Summerbell), NATO ASI, Series A, **205** pp 249-260. New York: Plenum.

Izpisua-Belmonte, J., Falkenstein, H., Dolle, P., Renucci, A. and Duboule, D. (1991). Murine genes related to the *Drosophila AbdB* homeotic gene are sequentially expressed during development of he posterior part of the body. *EMBO. J.* **10**, 2279-2289.

Janvier, P. (1993). Patterns of diversity in the skull of jawless fishes. In *The Skull* (ed. J. Hanken and B. K. Hall) 2, pp 131-188. Chicago: University of Chicago Press.

Janvier, P. (1984). The relationships of the Osteostraci and Galeaspida. *J. Vert. Paleont.* **4**, 344-358.

Jarvik, E. (1980). *Basic structure and evolution of vertebrates,* 2 Vols. London: Academic Press.

John, B. and Miklos, G. (1988). *The Eukaryotic Genome in Development and Evolution*. London: Allen and Unwin.

Karatajute-Talimaa, V. (1992). The early stages of the dermal skeleton formation in chondrichthyans. In *Fossil fishes as living animals* (ed. E. Mark-Kurik) pp 223-242. Tallin: Academy of Sciences of Estonia.

Kessel, M. and Gruss, P. (1991). Homeotic transformations of murine vertebrae and concommitant alteration of Hox codes induced by retinoic acid. *Cell* **67**, 89-104.

Kerr, J. G. (1899). Note on hypotheses as to the origin of the paired limbs of vertebrates. *Proc. Camb. Phil. Soc.* **10**, 227-235.

Kimmel, C. B. (1989). Genetics and early development of zebrafish. *Trends Genet.* **5**, 283-288.

Krumlauf, R. (1992). Evolution of the vertebrate *Hox* homeobox genes. *BioEssays* **14**, 345-252.

Lanyon, S. M. (1993). Phylogenetic frameworks: towards a firmer foundation for the comparative approach. *Biol. J. Linn. Soc.* **49**, 45-61.

Lebedev, O. A. (1984). The first find of a Devonian tetrapod in USSR. *Doklady Academii Nauk SSSR.* **278**,1470-1473 (in Russian).

Lewis, E. B. (1978). A gene complex controlling segmentation in *Drosophila. Nature* **276**, 565-570.

Long, J. A. (1986). New ischnacanthid acanthodians from the Early Devonian of Australia, with comments on acanthodian interrelationships. *Zoo. J. Linn. Soc.* **87,** 321-339.

Mabee, P. M. (1993). Phylogenetic interpretation of ontogenetic change: sorting out the actual and artefactual in an empirical case study of centrarchid fishes. *Zoo. J. Linn. Soc.* **107**, 175-291.

Maisey, J. G. (1984). Chondrichthyan phylogeny: a look at the evidence. *J. Vert. Paleo.* **4**,359-371.

Maisey, J. G. (1986). Heads and Tails: a chordate phylogeny. *Cladistics* **2**, 201-256.

Meyer, A. and Dolven, S. I. (1992). Molecules, fossils, and the origin of tetrapods. *J. Mol. Evol.* **35**, 93-101.

Miles, R. S. (1973). Articulated acanthodian fishes from the Old Red Sandstone of England, with a review of the structure and evolution of the acanthodian shoulder-girdle. *Bull. Brit. Mus. (Natur. Hist.), Geol.* **24**, 113-213.

Mivart, St G. (1879). On the fins of elasmobranchii. *Trans. Zoo. Soc. Lond.* **10**, 439-484.

Morgan, B. A., Izpisua-Belmonte, J. C., Duboule, D. and Tabin, C. J. (1992). Targeted misexpression of Hox-4. 6 in the avian limb bud causes apparent homeotic transformations. *Nature* **358**, 236-237.

Moy-Thomas, J. A. and Miles, R. S. (1971). *Palaeozoic Fishes.* London: Chapman and Hall.

Muller, G. B. (1990). Developmental mechanisms at the origin of morphological novelty: a side effect hypothesis. In *Evolutionary Innovations* (ed. M. H. Nitecki), pp 99-130. Chicago: University of Chicago Press.

Musick, J. A., Bruton, M. N. and Balon, E. (eds) (1991). *The biology of Latimeria chalumnae and the evolution of coelacanths.* Boston: Klewer Academic Publishers.

Oster, G. F., Shubin, N., Murray, J. D. and Alberch, P. (1988). Evolution and morphogenetic rules: the shape of the vertebrate limb in ontogeny and phylogeny. *Evolution* **42**, 864-884.

Panchen, A. L. (1992). *Classification, Evolution, and the Nature of Biology.* Cambridge: Cambridge University Press.

Panchen, A. L. and Smithson, T. R. (1987). Character diagnosis, fossils and the origin of tetrapods. *Biol. Rev.* **62**, 341-438.

Patterson, C. (1977). Cartilage bones, dermal bones and membrane bones, or the exoskeleton versus the endoskeleton. In *Problems in Vertebrate Evolution* (ed. S. M. Andrews, R. S. Miles and A. D. Walker) pp 77-122. London: Academic Press.

Patterson, C. (1981). Significance of fossils in determining evolutionary relationships. *Ann. Rev. Ecol. Syst.* **12**, 195-223.

Patterson, C. (1982). Morphological characters and homology. In *Problems of Phylogenetic Reconstruction* (ed. K. A. Joysey and A. E. Friday) pp 21-74. Systematics Association Special Volume **21**. London: Academic Press.

Patterson, C. and Rosen, D. E. (1977). Review of ichthyodectiform and other Mesozoic teleost fishes and the theory and practice of classifying fossils. *Bull. Amer. Mus. Nat. Hist.* **158**, 81-172.

Pautou, M.-P. (1973). Analyse de la morphogenese du pied des oiseaux a morphologique. *J. Embryo. Exp. Morph.* **29**,175-196.

Pendleton, J. W., Nagai, B. K., Murtha, M. T., and Ruddle, F. H. (1993). Expansion of the Hox gene family and the evolution of chordates. *P. N. A. S.* **90**, 6300-6304.

Pearson, D. M. and Westoll, T. S. (1979). The Devonian actinopterygian *Cheirolepis* Agassiz. *Trans. R. Soc. Edinb.* **70**, 337-399.

Rieppel, O. (1993a). Studies on skeleton formation in reptiles. v. Patterns of ossification in the skeleton of *Alligator mississippiensis* DAUDIN (Reptilia, Crocodylia). *Zoo. J. Linn. Soc.* **109**, 301-325.

Rieppel, O. (1993b). Studies on skeleton formation in Reptiles. II. The homology of the reptilian (amniote) astragalus revisited. *J. Vert. Paleo.* **13**, 31-47.

Ritchie, A. (1980). The Late Silurian Anaspid Genus Rhyncholepis from Oesal, Estonia, and Ringerike, Norway. *Amer. Mus. Novitates* **2699**, 1-18.

Romer, A. S. (1962). *The vertebrate body.* Philadelphia: W. B. Saunders.

Rosen, D. E., Forey, P. L., Gardiner, B. G. and Patterson, C. (1981). Lungfishes, tetrapods, paleontology, and plesiomorphy. *Bull. Amer. Mus. Nat. Hist.* **167**, 159-276.

Roth, V. L. (1984). On Homology. *Biol. J. Linn. Soc.* **22**, 13-29.

Rowe, T. and Gauthier, J. (1992). Ancestry, paleontology and the definition of the name Mammalia. *Systematic Biology* **41**, 371-378.

Russell, E. S. (1916). *Form and Function.* London: John Murray.

Schaeffer, B. (1987). Deuterostome monophyly and phylogeny. In *Evolutionary Biology* (ed. M. K. Hecht, B. Wallace and G. I. Prance), **21**, 179-235. New York: Plenum Press.

Scott, M. P. (1992). Vertebrate homeobox gene nomenclature. *Cell* **71**, 551-553.

Sewertzoff, A. N. (1924). Development of the pelvic fins of *Acipenser ruthenus.* New data for the theory of the paired fins of fishes. *J. Morph. Physiol.* **41**, 547-580.

Shubin, N. H. and Alberch, P. (1986). A morphogenetic approach to the origin and basic organization of the tetrapod limb. In *Evolutionary Biology* (ed. M. K. Hecht, B. Wallace and G. I. Prance), **20**, pp 319-387. New York: Plenum Press.

Simpson, G. G. (1961). *Principles of Animal Taxonomy.* London. Oxford University Press.

Slack, J. M. W., Holland, P. W. H. and Graham, C. F. (1993). The zootype and the phylotypic stage. *Nature* **361**, 490-492.

Smith, M. M. and Hall, B. K. (1993). A developmental model for evolution of the vertebrate exoskeleton and teeth. The role of cranial and trunk neural crest. In *Evolutionary Biology* (ed. M. K. Hecht, B. Wallace and G. I. Prance), **27**, pp 387-448. New York: Plenum Press.

Smith, M., Hickman, A., Amanze, D., Lumsden, A., and Thorogood, P. (1994). Trunk neural crest origin of caudal mesenchyme in the zebrafish *Brachydanio rerio. Proc. R. Soc. Lond. B* **256**, 137-145.

Tabin, C. J. (1992). Why we have (only) five fingers per hand: Hox genes and the evolution of paired limbs. *Development* **116**, 289-296.

Tabin, C. J. and Laufer, E. (1993). *Hox* genes and serial homology. *Nature* **361**, 692-693.

Thacher, J. K. (1877). Median and paired fins, a contribution to the history of vertebrate limbs. *Transactions of the Connecticut Academy*, **3**, 281-310.

Thomson, K. S. (1987). The neural crest and the morphogenesis and evoluton of the dermal skeleton in vertebrates. In *Developmental and Evolutionary Aspects of the Neural Crest* (ed. P. F. A. Maderson) pp 301-338. New York: J. Wiley.

Thorogood, P. (1991). The development of the Teleost fin and implications for our understanding of tetrapod limb evolution. In *Developmental Patterning of the Vertebrate Limb* (ed. J. R. Hinchliffe, J. M. Hurle and D. Summerbell), NATO ASI, Series A, **205** pp 347-354. New York: Plenum.

Thorogood, P. and Ferretti, P. (1993). *Hox* genes, fin folds and symmetry. *Nature* **364**, 196.

Trinkaus, J. P. (1988a). Directional cell movement during early development of the Teleost *Blennius pholis*: I. Formation of epithelial cell clusters and their pattern and mechanism of movement. *J. Exp. Zool.* **245**, 157-186.

Trinkaus, J. P. (1988b) Directional cell movement during early development of the Teleost *Blennius pholis*: II. Transformation of the cells of epithelial clusters into dendritic melanocytes, their dissociation from each other, and their migration to and invasion of the pectoral fin buds. *J. Exp. Zool.* **248**, 55-72.

Vorobyeva, E. and Schultze, H-P. (1991). Description and systematics of Panderichthyid fishes with comments on their relationship to tetrapods. In *Origins of the Higher Groups of Tetrapods* (ed. H-P. Schultze and L. Treub) pp 68-109. Ithaca: Comstock, Cornell University Press.

Wagner, G. (1989). The origin of morphological characters and the biological basis of homology. *Evolution* **43**, 1157-1171.

Watson, D. M. S. (1937). The acanthodian fishes. *Phil. Trans. R. Soc. Lond. (B)* **228**, 49-146.

Westoll, T. S. (1958). The lateral fin-fold theory and the pectoral fins of ostracoderms and early fishes. In *Studies on Fossil Vertebrates* (ed. T. S. Westoll) pp 181-211. London: Athlone Press.

Wilson, M. V. H. and Caldwell, M. W. (1993). New Silurian and Devonian forktailed 'thelodonts' are jawless vertebrates with stomachs and deep bodies. *Nature* **361**, 442-444.

Yokouchi, Y., Sasaki, H. and Kuroiwa, A. (1991). Homeobox gene expression correlated with the bifurcation process of limb cartilage development. *Nature* **353**, 443-445.

Young, G. C. (1986). The relationships of placoderm fishes. *Zoo. J. Linn. Soc.* **88**, 1-57.

Young, G. C. (1991). The first armoured agnathan vertebrates from the Devonian of Australia. In *Early Vertebrates and Related Problems of Evolutionary Biology* (ed. M. M. Chang, Y. H. Liu and G. R. Zhang), pp 67-85. Beijing: Science Press.

Zangerl, R. (1981). *Chondrichthyes I. Handbook of Paleoichthyology* (ed. H-P Schultze). Stuttgart: Gustav Fischer Verlag.

Development 1994 Supplement, 181-186 (1994)
Printed in Great Britain © The Company of Biologists Limited 1994

Hox genes and growth: early and late roles in limb bud morphogenesis

Bruce A. Morgan[1,2] **and Cliff Tabin**[3]

[1]Cutaneous Biology Research Center, MGH East, Charlestown MA 02129, USA
[2]Department of Dermatology, Harvard Medical School, Boston MA 02115, USA
[3]Department of Genetics, Harvard Medical School, Boston MA 02115, USA

SUMMARY

In recent years, molecular analysis has led to the identification of some of the key genes that control the morphogenesis of the developing embryo. Detailed functional analysis of these genes is rapidly leading to a new level of understanding of how embryonic form is regulated. Understanding the roles that these genes play in development can additionally provide insights into the evolution of morphology.

The 5′ genes of the vertebrate Hox clusters are expressed in complex patterns during limb morphogenesis. Various models suggest that the Hoxd genes specify positional identity along the anteroposterior (A-P) axis of the limb. Close examination of the pattern of Hoxd gene expression in the limb suggests that a distinct combination of Hoxd gene expressed in different digit primordia is unlikely to specify each digit independently. The effects of altering the pattern of expression of the *Hoxd-11* gene at different times during limb development indicate that the Hoxd genes have separable early and late roles in limb morphogenesis. In their early role, the Hoxd genes are involved in regulating the growth of the undifferentiated limb mesenchyme. Restriction of the expression of successive 5′ Hoxd genes to progressively more posterior regions of the bud results in the asymmetric outgrowth of the limb mesenchyme. Later in limb development, Hoxd genes also regulate the maturation of the nascent skeletal elements. The degree of overlap in function between different Hoxd genes may be different in these early and late roles. The combined action of many Hox genes on distinct developmental processes contribute to pattern asymmetry along the A-P axis.

Key words: Hox gene, limb bud morphogenesis, axis specification, anteroposterior determination

Comparative molecular analysis suggests that the genes that gave rise to the modern Hox clusters have been specifying regional differences in the animal body for possibly more than one billion years (Kappen and Ruddle, 1993; Shubert et al., 1993). Sequence comparisons suggest that, at the time of the origin of the chordate, nematode and arthropod lineages, there were between 5 and 7 members of the Hox complex (Shubert et al., 1993). Since that time, these complexes have undergone expansion and duplication independently in different lineages ultimately generating, for example, 38 members in the four clusters of Hox genes observed in gnathostome vertebrates. During the course of this expansion, these genes retained their original functions in patterning the anterior-posterior (A-P) body axis and also acquired new functions in regulating other aspects of morphogenesis. When novel developmental innovations arise in evolution, modifying preexisting body plans, the derived embryonic steps tend to occur late in ontogeny ("the general precedes the specialized" in von Baer's formulation; Gould, 1977). In performing patterning roles that are phylogenetically old, and therefore occur developmentally early, there may be a high degree of overlap in function between paralogous genes in different Hox clusters, and even between sequentially arranged genes within a cluster (see Fig. 1 for nomenclature). Continued function in these early roles constrains their divergence (as discussed by Holland et al., 1992). However, in roles that arose later in evolution and are frequently observed later in development of the embryo, the differences in function between paralogue groups and between different genes in a paralogue group are more pronounced, taking advantage of divergence in regions not required for primitive function. This concept is generally useful in attempting to decipher the role of the Hox genes in morphogenesis and is illustrated in our consideration of the role of the Hox genes in limb development.

The limb arises from an accumulation of cells in the lateral plate mesenchyme. These cells induce a specialized structure in the overlying ectoderm, the apical ectodermal ridge (AER). Subsequent proliferation and outgrowth of the limb mesenchyme is dependent on signals from the AER; in turn the AER is dependent on the underlying mesenchyme for its maintenance. Mesenchymal cells in the region subjacent to the AER (the so-called progress zone) remain in a highly proliferative and undifferentiated state. As the limb bud grows, cells are continuously displaced from this zone. Displaced cells decrease their rate of proliferation and subsequently begin to differentiate, leading to a proximal-to-distal wave of differentiation. Cells that will participate in the formation of distal structures are still being generated when cells participating in the formation of proximal structures are beginning to differentiate.

The Hox genes are expressed in intriguing patterns during morphogenesis of the wing and leg. These patterns evolve in complex ways as development proceeds. In earlier work in the mouse limb, these changing patterns of expression were described as the continuing evolution of a single expression domain (Dolle et al., 1989). Examination of Hox gene expression in the chick limb bud reveals that the evolving pattern of gene expression described for each gene actually represents temporal and spatial overlap of several distinctly regulated expression domains.

Members of the Hoxd cluster are sequentially expressed during the development of the chick limb bud (Fig. 2). In the early phases of limb bud development, transcripts of the *Hoxd-9* and *Hoxd-10* genes are expressed across the A-P extent of the nascent wing and leg bud. Activation of the nine paralogues is synchronous with the initial outgrowth of the bud. It is during this phase of development (stages 17-19) that the presumptive stylopod is displaced from the progress zone (Saunders, 1948).

The next phase of limb outgrowth involves the activation of the *Hoxd-11* and *Hoxd-12* genes in progressively restricted domains in the posterior half of the bud. This occurs at a stage when the presumptive zeugopod is being displaced from the progress zone (stage 19 through 21/22). Finally, *Hoxd-13* is activated in the posterior distal region of the limb bud at stage 20. As development proceeds, expression of *Hoxd-10, Hoxd-11, Hoxd-12* and *Hoxd-13* come to occupy very similar domains in the distal segment (autopod) and remain strongly expressed in this region while the nested domains of expression in the proximal segments of the limb are fading. For most of these genes, a clear separation of proximal and distal expression domains is observed. *Hoxd-13* is only weakly expressed in the posterior zeugopod at this stage, but is strongly expressed in the autopod in a domain that extends slightly anterior to that of the other Hoxd genes.

The apparent separation of Hoxd gene expression into several phases is confirmed by recent identification of *Sonic hedgehog* as a gene involved in the regulation of the Hoxd gene expression in distal mesenchyme. Early expression of Hoxd genes is independent of *Sonic* regulation. *Hoxd-9* and *Hoxd-10* appear in proximal regions before *Sonic* is expressed (Laufer et al., unpublished data). The posterior distal expression of *Hoxd-11, Hoxd-12* and *Hoxd-13* in the limb appears to be regulated by *Sonic*; ectopic expression of *Sonic* in the anterior region of the bud is sufficient to elicit ectopic expression of these genes in the anterior distal region (Riddle et al., 1993; Laufer et al., unpublished data).

Analysis of the sequences required to drive expression of the mouse *Hoxd-11* gene suggests that Hox gene expression is also independently regulated in proximal and distal regions of the mouse limb bud. The degree of overlap between domains prevents their observation as distinct entities in the wild-type animal. However, transgenic mice bearing different segments of the *Hoxd-11* fused to a β-galactosidase reporter construct demonstrate that expression of *Hoxd-11* may be independently regulated in the proximal (stylopod), central (zeugopod), and distal (autopod) segments of the limb (Gerard et al., 1993). In particular, expression in the distal segment of the limb is separable from expression in the central segment.

As development proceeds, Hoxd expression fades in the mesenchyme with the exception of the perichondrial regions. The maintenance of Hoxd gene expression in the perichondrial regions correlates with the expression of a second vertebrate hedgehog homologue, *Indian hedgehog*. This late pattern of expression may represent an independently regulated phase of Hox gene transcription indicative of a separable role in limb morphogenesis.

The early pattern of Hoxd gene expression in the chick limb correlates with the outgrowth of the bud. The onset of expression of *Hoxd-9* and *Hoxd-10* throughout the bud is synchronous with initial limb outgrowth and at this stage bud outgrowth is symmetric along the A-P axis. Growth of the bud becomes markedly asymmetric along the A-P axis with a distinct posterior bias when the *Hoxd-11* and *Hoxd-12* genes are activated in the posterior half of the bud. This bias in outgrowth is observed during stages 19 to 23, a period when the presumptive sylopod and zeugopod constituents are being displaced from the progress zone. Subsequent growth of the bud becomes less asymmetric as the distal domains of Hoxd gene expression spread to the anterior regions of the bud.

This correlation between the timing and position of early

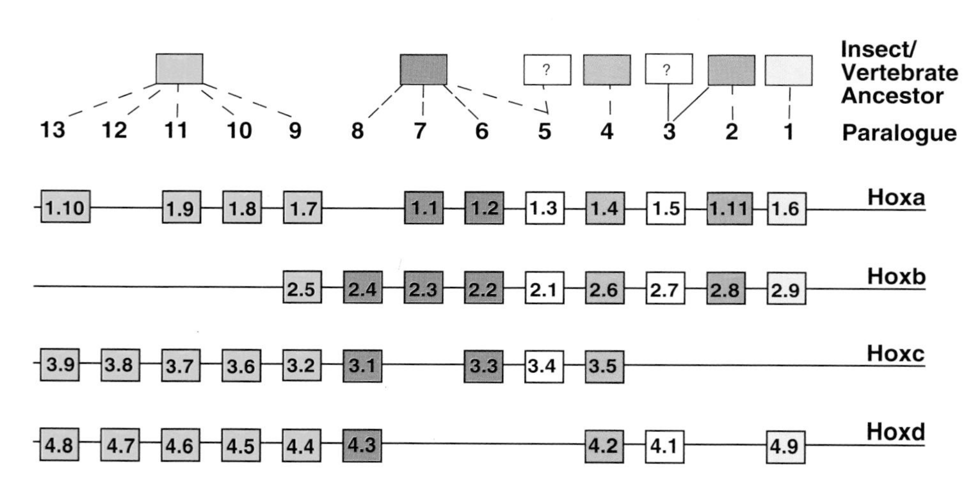

Fig. 1. Vertebrate Hox genes. There are four clusters of vertebrate Hox genes. Numbers in the boxes represent previous nomenclature. Currently, genes are referred to by their cluster letter (right) and paralogue number (1-13 listed above). At the time of insect/vertebrate divergence, there were between 5 and 7 Hox genes (diagrammed above the vertebrate clusters). The four modern vertebrate clusters apparently arose by serial duplication of a 13 member cluster with subsequent deletions. As a result, the homologous genes in different clusters (referred to as paralogues) are more similar to each other than they are to adjacent genes in the same cluster. Paralogous genes differ by a few amino acids in the homeobox region and are therefore expected to have similar DNA-binding characteristics. Most divergence between paralogues is observed in the N-terminal half of the protein. The N-terminal region is presumed to interact with other proteins or otherwise modulate activity of the protein, but direct evidence of its function in vertebrates has not yet been obtained.

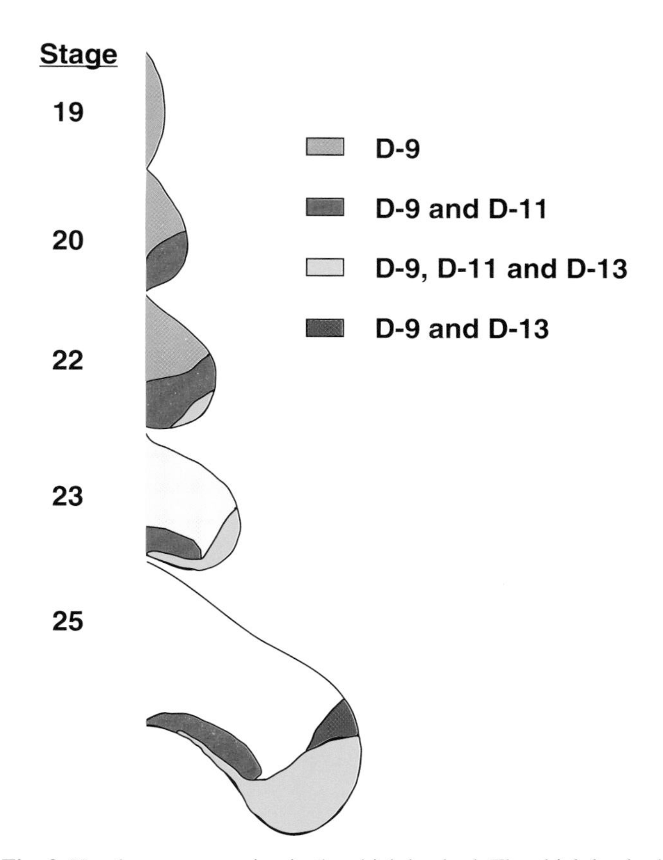

Fig. 2. Hoxd gene expression in the chick leg bud. The chick leg bud is diagrammed at various stages of development (Hamburger and Hamilton, 1951, stages listed at left). These sketches represent a view of the dorsal surface of the right limb bud (anterior is towards the top of the page). The expression of the *Hoxd-9, Hoxd-11* and *Hoxd-13* genes are diagrammed. To simplify the diagram, *Hoxd-10* and *Hoxd-12* are not shown. At early stages (19-22) *Hoxd-10* expression resembles that of *Hoxd-9.* During stages 22 and 23, *Hoxd-10* expression fades in the anterior regions and its expression domain approximates that shown in yellow for *Hoxd-9, Hoxd-11* plus *Hoxd-13*. By stage 25 *Hoxd-9* expression is negligible and the color codes shown represent the designated gene combinations without *Hoxd-9*. The *Hoxd-10* expression domain at later stages resembles that of *Hoxd-11*. The expression of *Hoxd-12* is very similar to that of *Hoxd-11* throughout development, although it is not found in the anterior most regions of the *Hoxd-11* expression domain through stage 23.

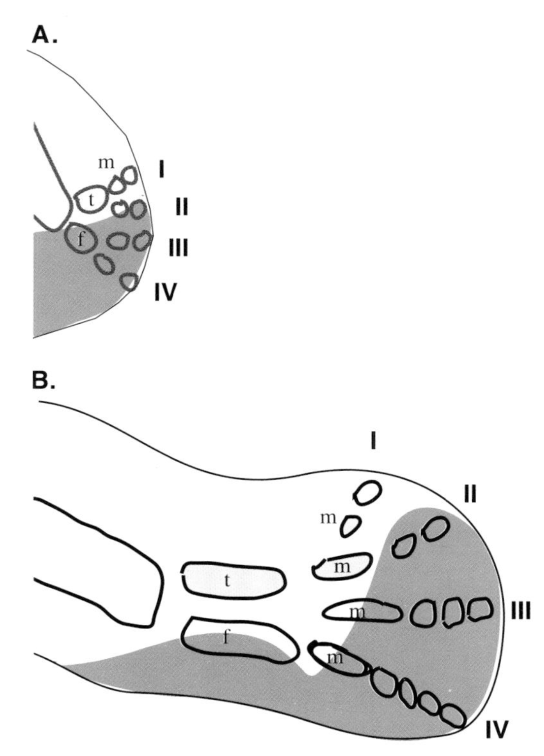

Fig. 3. *Hoxd-11* expression compared to a fate map of the leg bud. At either stage 21 or stage 26, domains of *Hoxd-11* expression in the leg bud (blue) are restricted to the regions posterior to the primordia of digit I. At early stages, *Hoxd-11* expression encompasses the primordia of the fibula (f), the posterior metatarsals (m) and digits II through IV. The position of the primordia of distal skeletal element can only be approximated at this stage. As development proceeds, expression of *Hoxd-11* remains strong in digits II, III and IV as well as the posterior metatarsals and the perichondrium of the fibula. Skeletal elements whose growth is inhibited by ectopic expression of *Hoxd-11* are shown in violet.

Hoxd gene expression and the growth of the undifferentiated limb mesenchyme could reflect the response of both Hox gene expression and cell proliferation to a common inducer, or even a requirement for additional cell divisions to achieve the sequential activation of increasingly 5′ Hox genes. Alternatively, one role of the 5′ Hoxd genes early in limb bud development may be to mediate the asymmetric growth of the bud. In such a case, the discrete regulation of these genes in the three segments of the limb may reflect a fundamental role of these genes in the evolution of the appendage; the sequential appearance of elements capable of activating these genes would correspond to the addition of segments along the proximal distal axis.

The effects of ectopic expression of *Hoxd-11* suggest that regulation of the growth of undifferentiated mesenchyme is an early role of the Hoxd genes in the limb. These ectopic expression experiments were performed using a replication competent retroviral vector to express the mouse *Hoxd-11* protein specifically in the developing chick limb bud (Morgan et al., 1992) *Hoxd-11* is normally expressed in the posterior regions of the bud in a domain that encompasses the primordia of the fibula, posterior metatarsals and digits II, III and IV (see Fig. 3). The retrovirus was used to infect the entire limb bud and therefore expand that domain of expression into anterior regions. These infections led to complex and variable phenotypes which reflects the mechanistic constraint on virus-mediated ectopic expression (see Fig. 4 legend).

Perhaps the most striking phenotype associated with the mis-expression of *Hoxd-11* in regions anterior to its normal domain of expression in the leg bud was the appearance of an additional phalange in digit I leading to a morphology similar to that of digit II (Fig. 4A,B). Anterior digits containing either an additional phalange or a single elongated phalange were observed, while no effect was observed on the bones of the posterior digits II, III and IV, which arise from the region in which *Hoxd-11* is normally expressed.

A conceptually similar phenotype is observed in *Hoxd-11*-

injected wings. Cells in the anterior region of the chick wing which do not express *Hoxd-11* do not normally give rise to skeletal elements. Ectopic expression of *Hoxd-11* in the anterior region of the wing results in the formation of an additional digit at the anterior edge of the wing which resembles digit II in structure.

The apparent homeotic change was observed in roughly 30% of the infected legs showing any phenotype. This level of penetrance indicates that the digit-affecting action of *Hoxd-11* occurs early in limb development, when only 30% of injected limb buds are already fully infected by the spreading virus.

Both the appearance of an additional phalange in digit I in the leg and an additional anterior digit in the wing can be ascribed to the proposed effect of the Hoxd genes on outgrowth or survival of undifferentiated limb mesenchyme. We propose that procedures that increase the proliferation or survival of limb bud mesenchyme lead to the formation of additional cartilaginous condensations. We suggest that, when these condensations are sufficiently separated, they give rise to additional bones. Insufficiently separated condensations fuse to form a single enlarged bone. Hence the effects on distal elements of ectopic *Hoxd-11* expression in the anterior region of the wing bud resemble those observed when elevated levels of bFGF are achieved either by implantation of an FGF-soaked bead or by implantation of FGF-secreting cells in the anterior wing bud (Riley et al., 1993). In both cases, increased proliferation or survival of undifferentiated limb mesenchyme leads to the formation of additional bones. Indeed, earlier activation of *Hoxd-11* in the anterior region of the hind limb bud can have a more pronounced effect on limb growth leading to the formation of an additional digit anterior to the normal digit I.

A role for the Hox genes in stimulating growth of limb precursors is sufficient to explain the appearance of an additional bone in the anterior digit after ectopic expression of *Hoxd-11* in the anterior region. However, this postulated function is not sufficient to explain other skeletal phenotypes observed in response to ectopic *Hoxd-11*. Most of these reflect a failure of specific bones to mature properly after apparently normal initial development (Fig. 4). Ectopic *Hoxd-11* expression in the anterior region leads to an abnormally short second tarsometatarsal, while the third and fourth tarsometatarsals are comparatively normal. In a similar fashion, the tibia is reduced

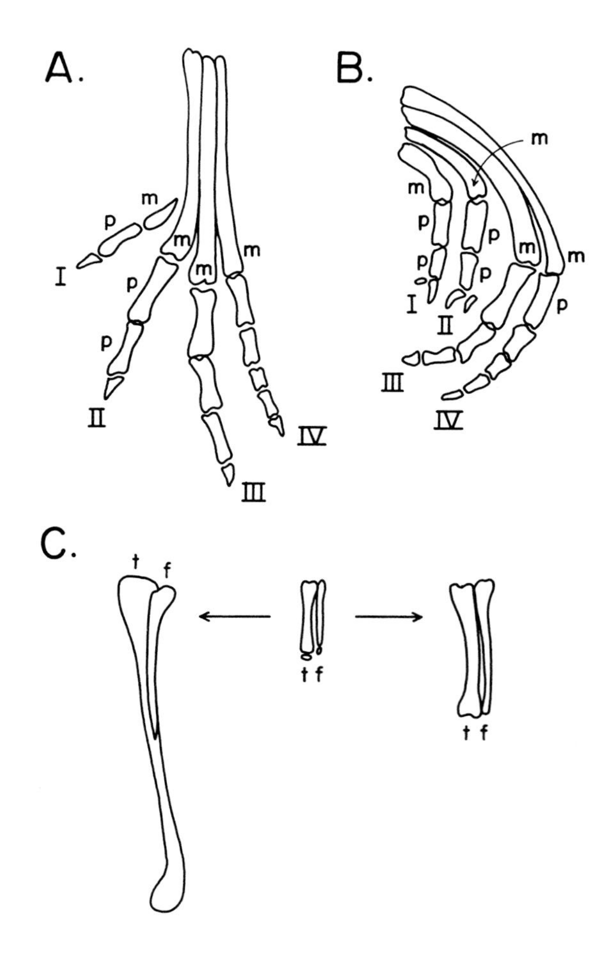

Fig. 4. Phenotypic consequences of ectopic *Hoxd-11* expression. A retrovirus encoding the mouse *Hoxd-11* cDNA was used to express ectopically this protein in the chick leg bud (Morgan et al., 1992). This wild-type chick foot is shown in A. Note that the first metatarsal (m) is a deltoid bone arising distally. Excluding the terminal claw, digit I has one phalange (p) while digits II, III and IV have 2, 3 and 4 phalanges respectively. In a foot infected with the *Hoxd-11* virus (B), the anterior two digits have a similar structure which includes the two phalanges normally found in the second digit. The anterior metatarsal now arises proximally and has a structure similar to that of the posterior tarsometatarsals. The first and second metatarsals are approximately half the length of a normal second metatarsal. Digits III, IV and the posterior metatarsals are relatively normal. The curvature of the posterior metatarsals and digits may be caused by the lack of growth of the anterior metatarsals and failure of interdigital cell death. (C) In a similar fashion, the tibia (t) and fibula (f) of a wild-type leg at day 11 of incubation (left) or day 5 of incubation (center) are shown. At day 5 of incubation, the primordia of the tibia and fibula are approximately equal in length. The tibia and fibula of an infected embryo are very similar to that of an uninfected embryo at this stage. However, as development proceeds the tibia of an infected embryo fails to elongate normally and the tibia and fibula remain the same length at day 11 of incubation (right). To achieve a domain of contiguous infected cells roughly encompassing the entire limb bud, several focal infections are induced by microinjection of virus in the lateral plate mesenchyme early in development. As development proceeds these infections spread to adjacent cells and coalesce to encompass the entire limb bud. However, the precise position of the initial infections is somewhat variable, as is the time when infection has spread sufficiently to emcompass the entire bud. For some buds, this process is complete by stage 21, while for others it may be as late as stage 24 or 25. Effects on distal skeletal elements cannot be reliably assayed before day 11, making it impossible to directly relate the degree of infection at an early stage with a particular phenotype in an affected embryo. Therefore, population approaches must be employed, correlating the degree of infection of early harvested specimens with the range of phenotypes later in development. The continuous spread of the virus during the course of incubation increases the penetrance of phenotypes which result from Hox gene activity later in development. Because there is a proximal to distal progression of differentiation in the limb, this will be observed in two ways. The effects of altered Hox expression on a given developmental process (e.g. cartilage condensation) will be evident more frequently in distal structures where this process occurs later, allowing more time for viral spread. Furthermore, at a given level along the proximal distal axis, phenotypes resulting from influences on later developmental events will also be observed more frequently than those reflecting earlier activity.

to half its normal length and is now very similar in length to the fibula. This phenotype was observed in more than half of the affected specimens as compared to the one third that showed an additional phalange in digit I. The higher penetrance suggests that these phenotypes result from a later effect of altered *Hoxd-4* expression on the limb bud. (When the injected virus has had time to spread completely in a higher percentage of limbs.)

If the digital phenotypic effect were more prevalent than the proximal effect, one might ascribe the difference in penetrance to the fact that there is a proximal-to-distal wave of differentiation in the limb bud. Thus the Hox genes could act at a single developmental stage (e.g. cartilage condensation); and when that stage occurred proximally (early) there would naturally be less frequent complete infection than when that same stage occurred in distal regions (late). However this is not the case: the proximal phenotype is more frequent, and hence occurs later in time. This strongly implies that phenotypes in regions proximal to the digits are not due to the effect on early proliferation (that gives rise to the digit phenotypes), but rather that they result from a distinct later effect of altered Hoxd expression on the developing limb. Consistent with this hypothesis is the observation that at early stages of development, the initial condensations that will become the tibia and fibula appear normal in infected limb buds. Through stage 26, there is no obvious difference between the zeugopod of the infected and uninfected contra-lateral limbs. At this stage, the cartilaginous condensations that will contribute to the tibia and fibula are approximately equal in length. The normal disparity between the size of these bones at later stages arises as a consequence of greater subsequent growth of the tibia compared to the fibula. *Hoxd-11* expression in anterior regions prevents this preferential growth leading to a phenotype in which both bones are similar lengths. In a similar fashion, the second matatarsal also fails to undergo normal elongation and is reduced to roughly half the size of the normal bone. Even when the anterior metatarsal shows a complete phenotypic conversion to resemble the posterior metatarsal, both are half the length of a normal metatarsal. (Fig. 4B).

These effects could be due to persistent expression of the *Hoxd-11* gene in differentiating cells where it is normally turned off. However, such an explanation would predict that all bones in the limb would be affected equally since Hoxd genes are normally down regulated in all but the perichondrial regions within the diagrammed domain of expression. The fact that this inhibitory effect on bone elongation is observed only in bones from the regions where *Hoxd-11* is not normally expressed suggests that it is more specific. Some other aspect of differential gene expression such as the expression of the endogenous gene or *Hoxd-12* in the normal expression domain of *Hoxd-11* prevents the inhibitory action of the exogenous gene in this region. Alternatively the exogenous *Hoxd-11* interferes with the activity of factors that are only responsible for stimulating the preferential growth of the tibia and anterior metatarsal. Other Hox genes are likely targets for this interaction; protein-protein interactions between different Hoxd gene products which inhibit transcriptional activity have been described (Zappavigna et al., 1994). This late effect on bone growth has also been observed in gene ablation experiments with the mouse *Hoxd-13* gene. Absence of the *Hoxd-13* gene product leads to delayed and incomplete ossification of the proximal phalanges in the anterior and posterior digits (Dolle et al., 1993).

It has been proposed that the combination of Hoxd genes expressed in a digit primordium specifies the unique identity of each digit in a combinatorial code and mediates its characteristic morphogenesis (Tabin, 1992). The apparent homeotic nature of the morphological changes in the anterior digit and metatarsal resulting from ectopic expression of *Hoxd-11* can be interpreted to support such a model. The separable early and late roles of Hox genes that emerge from the analysis presented here do not in themselves contradict this view. Rather they provide a two-phase mechanistic basis for the effect of Hox genes on limb (including digit) morphology.

However, the simple Hox-code model is excluded when one also takes into account the extremely dynamic expression patterns of the multiple domains of each Hox gene within the limb (Nelson, Morgan and Tabin, unpublished data). The Hoxd genes do appear to act early when there are nested expression patterns of the Hoxd genes along the anterior-posterior limb axis. However, they also clearly function later as well, when their relative domains are very different, and not at all aligned with unique digit primordia. Indeed, at this later time all of the Hoxd genes are expressed across the entire distal limb bud and may play partially redundant roles in this domain. This is consistent with the finding that deletion of the *Hoxd-13* gene affects all the digits (not just the most posterior one), but results in a fairly subtle effect in mice whose other Hoxd genes are intact (Dolle et al., 1993). Thus, while the expression of Hox genes in the limb bud appear to regulate digit morphology, they do not encode digit identity by a simple combinatoral code.

An attempt was recently made to apply the Hox code model to the problem of understanding the origin of pentadactyly (five digits) in modern tetrapods (Tabin, 1992). A two part argument was proposed.

(1) It is know that the regulation of the number of digits produced in a limb field is independent of the regulation of the morphology of the digits. If there were a constraint on the number of distinct morphological types of digits that could be specified in a given limb field such that only five 'different' digit types were permissible, then more than five digits might never be able to evolve with a selective advantage in a population. Evidence presented in support of this hypothesis included the fact that polydactylous individuals arising spontaneously in many species do not typically have a novel extra digit but rather have a morphological duplicate of either the most preaxial or postaxial digit. Similarly, polydactylous primitive tetrapods such as *Acanthostega* (8 total digits) have only five or fewer morphological types of digits. Finally, when extra distal 'digits' arise in tetrapod evolution with unique morphologies, they are invariably produced by modification of a wrist bone and are not true extra digits (e.g. Panda Bears, some moles, some frogs). Thus, there does indeed appear to be some constraint, limiting number of potential digit 'types' to five.

(2) It was further argued that the Hox code could directly produce such a constraint since only five combinatorial codes are possible from the Hoxd genes expressed in the limb. However, as we have seen, the Hoxd genes do not act in a simple combinatorial code for 'digit identity'. While they do contribute to the regulation of digit morphology, our current understanding of their action does not provide an indication of a constraint on potential morphologies. Either such a constraint

remains to be discovered in the subtler aspects of Hox gene action or else one will have to look elsewhere for it.

In summary, rather than an alteration in Hoxd code producing the apparent homeotic effect of *Hoxd-11* misexpression, we suggest that the effect on the anterior digit results from an early stimulatory effect on limb outgrowth which generates sufficient additional mesenchyme to generate an additional chondrification center. At this stage of development, all of the 5′ Hoxd genes may have qualitatively similar roles. Appropriately timed ectopic expression of another 5′ Hoxd gene might well achieve a similar result. Later effects on skeletal morphogenesis may reflect paralogue specific roles in regulating bone growth. The fact that these genes are having an effect at a stage when the expression domains of the *Hoxd-10, Hoxd-11* and *Hoxd-12* genes are very similar in the autopod renders the hypothesis that different combinations of Hoxd genes specify digit identity improbable. Rather, the combined early action on the accumulation of undifferentiated mesenchyme and later action on skeletal growth lead to the characteristic pattern of the chick limb skeleton.

REFERENCES

Dolle, P., Dierich, A., LeMeur, M., Schimmang, T., Schubaur, B., Chambon, P. and Duboule, D. (1993). Disruption of the Hoxd-13 gene induces localized heterochrony leading to mice with neotenic limbs. *Cell* **75**, 431-41.

Dolle, P., Izpisua-Belmonte, J.-C., Falkenstein, H., Renucci, A. and Duboule, D. (1989). Coordinate expression of the murine Hox-5 complex-containing genes during limb pattern formation. *Nature* **342**, 767-772.

Gerard, M., Duboule, D. and Zakany, J. (1993). Structure and activity of regulatory elements involved in the activation of the Hoxd-11 gene during late gastrulation. *EMBO J.* **9**, 3539-3550.

Gould, S. J. (1977). *Ontogeny and Phylogeny.* Cambridge, MA:Harvard University Press.

Hamburger, V. and Hamilton, H. (1951). A series of normal stages in the development of the chick embryo. *J. Exp. Morph.* **88**, 49-92.

Holland, P. W. H., Holland, L. Z., Williams, N. A. and Holland, N. D. (1992). An amphioxus homeobox gene: sequence conservation, spatial expression during development, and insights into vertebrate evolution. *Development* **116**, 653-661.

Kappen, C. and Ruddle, F. H. (1993). Evolution of a regulatory gene family: HOM/HOX genes. *Current Opin. Gen. Devel.* **3**, 931-938.

Morgan, B., Izpisua-Belmonte, J., Duboule, D. and Tabin, C. (1992). Targeted mis-expression of Hox-4. 6 in the avian limb bud causes apparent homeotic transformations. *Nature* **358**, 236-239.

Riddle, R., Johnson, R., Laufer, E. and Tabin, C. (1993). Sonic hedgehog mediates the polarizing activity of the ZPA. *Cell* **75**, 1401-1416.

Riley, B. B., Savage, M. P., Simandl, B. K., Olwin, B. B. and Fallon, J. F. (1993) Retroviral expression of FGF-2 (bFGF) affects patterning in chick limb bud. *Development* **118**, 95-104.

Saunders, J. (1948). The proximo-distal sequence of origin of the parts of the chick wing and the role of the ectoderm. *J. Exp. Zool.* **108**, 363-402.

Shubert, F., Nieslelt-Struwe, K. and Gruss, P. (1993). The Antennapedia-type homeobox genes have evolved from three precursors separated early in metazoan evolution. *Proc. Natn. Acad. Sci. USA* **90**, 143-147.

Tabin, C. (1992) Why we have (only) five fingers per hand: Hox genes and the evolution of paired limbs. *Development* **116**, 289-296.

Zappavinga, V., Sartori, D. and Mavilio, F. (1994). Specificity of Hox protein function depends on DNA-protein and protein-protein interactions, both mediated by the homeo domain. *Genes Dev.* **8**, 732-744.

Development 1994 Supplement, 187-191 (1994)
Printed in Great Britain © The Company of Biologists Limited 1994

The evolution of insect patterning mechanisms: a survey of progress and problems in comparative molecular embryology

Klaus Sander

Institut für Biologie I (Zoologie) der Universität, Albertstrasse 21a, D-79104 Freiburg i.Br., Germany

SUMMARY

This report surveys data and interpretations presented by speakers in the Arthropod Session of the 1994 BSDB Spring Symposium. After a short review of phylogenetical aspects in premolecular insect embryology, the following topics are discussed: the ancestral germ type of pterygote insects, correlations between oogenesis and embryonic pattern formation, the universality or otherwise of *bicoid* as the anterior morphogen, novel functions in the insect Hox complex, the formal asymmetry between evolution and decay of complex gene networks, novel regulatory interactions as the main cause of evolutive changes, the repeated activity of conserved gene networks in successive steps of ontogenesis and strategies for future research. Interspersed are some unpublished data on oogenesis and pattern formation in lower dipterans, and their possible evolutionary implications.

Key words: insect, patterning, germ type, *bicoid*, lower dipterans, oogenesis, pattern formation, Hox

INTRODUCTION

Research into embryogenesis was tremendously stimulated by Charles Darwin's 'Origin of Species', and especially by his view that the study of ontogenesis might offer the most convincing testimony of evolution. His expectations were based on the morphology of embryonic stages, but many articles in the present volume show that the study of ontogenesis with molecular methods holds even more promise. Not only can it yield a wealth of novel information on the probable course of evolution but it can also reveal molecular and developmental changes that have led – and may still lead – to overt morphological change.

The molecular approach to evolution has advanced farther (as yet) in the insects than in any other group, because insects combine a sufficient degree of complexity and variation with many technical advantages – a combination that has made *Drosophila* the organism of choice for the pioneers of classical as well as developmental genetics, and now qualifies this species to serve as the reference system for comparative molecular embryology.

Below I shall give a brief outline of past thoughts on the evolutionary aspect of comparative insect embryology, and then try and highlight some of the progress and the problems that have become evident during the session on arthropod development in this BSDB symposium. The individual contributions published in the present volume will be quoted here by authors' names only, without the year, except when they are first mentioned.

THE PHYLOGENETICAL ASPECT IN PREMOLECULAR INSECT EMBRYOLOGY

Even before Darwin's 'Origin', the (very few) authors writing on insect embryogenesis had considered problems akin to phylogeny, in that they searched for unifying concepts under which to subsume the course of embryogenesis in both insects and vertebrates (reviewed in Nübler-Jung and Arendt, 1994). As in comparative anatomy, their method was to search for homologies. While the origins of these homologies were usually left open, some interspersed statements have a ring of evolutionary thinking. Albert Kölliker for instance wrote in 1842 that "as we all know, in higher animals an organ at the onset and during its development reflects the form with which it is endowed in lower animals". Such statements probably betray a widespread feeling in the years between Lamarck and Darwin that life (contrary to Lamarck's concept) was created only once and that therefore "only the external form of the appearances of life is subject to continuous change, tied to the external conditions which either favour or prevent its development..." (J. C. Pander 1821, as quoted by Bäumer-Schleinkofer, 1993; translation by the present author).

Almost a century later, the comparison of embryogenesis between different insect species – now of course firmly embedded in concepts of phylogenesis – reached its first climax in the doctoral dissertation of Friedrich Seidel (1924). At that time, Seidel relied exclusively on morphological data, but his interpretations expressly implied profound differences between the patterning mechanisms involved; indeed he classified insect embryos in a series ranging from the 'non-determinative' to the strictly 'determinative' developmental type. In subsequent experimental work (reviewed e.g. in Counce and Waddington, 1972; Sander, 1976), Seidel and his students demonstrated that corresponding functional differences exist. They thereby abolished the view that the insect egg cell embodies a complex mosaic of localized maternal determinants (an accepted wisdom hailing largely from Hegner's work on germ cell determination, reviewed in Sander, 1984). Their assembled descriptive and experimental data enabled Krause (1939a,b) to propose a graded series of 'insect egg types' which were primarily intended to provide a conceptual frame for physiological investigations. But this series, according to Krause (1939a) "will also please the taxonomist... in that the eggs as types can be arranged in correspondence with the adults" – and thus, implicitly, with the phylogenetic tree.

Krause's terms for the main types or modes of early insect embryogenesis – short, half-long (now called intermediate) and long germ – are still with us and pervade the relevant contributions in this volume. It may therefore be worthwhile to define them in his own words (translated from Krause, 1939b by the present author). "The 'short germ' mainly represents the head region (*Tachycines*), the 'long germ' maintains the natural proportions of the body regions of the larva (*Apis*). Therefore short germ, half-long germ and long germ differ by the number of presumptive segments within the segment formation zone", the posterior blastema which would generate these segments successively by proliferation.

Some 30 years later, at the eve of the 'molecular revolution' of insect embryology to be rung in by the herculean genetic studies of Lewis (1978) and Nüsslein-Volhard and Wieschaus (1980), embryogenesis in various groups of the Articulata was exploited by Anderson as a guide to establishing phylogenetic relationships (Anderson, 1973). Soon afterwards, the evolution of patterning mechanisms was briefly invoked by the present writer when the increasing role of anterior polar determinants during insect phylogenesis was recognized (Sander, 1976), and was treated more extensively at a subsquent BSDB meeting (Sander, 1983).

THE ANCESTRAL MODE OF EARLY EMBRYOGENESIS IN PTERYGOTE INSECTS – SHORT-GERM OR INTERMEDIATE TYPE?

On this question, opinion was and still is divided. Patel (1994) implies the possibility that short-germ development is ancestral. In favour of this view speaks the sequential budding of segments that is reminiscent of development in marine annelids. Its shortcoming is that orthopterans do not start segmenting (and expressing *engrailed*, see Patel) right behind the head lobes but rather in the thorax; however, this might be a secondary specialization, to be viewed as a corollary of the fact that the largest limb buds form in that region (Fleig, 1990). Tautz et al. (1994), in contrast, like Anderson (1972, 1973) favour the intermediate mode as ancestral, drawing on the fact that this mode apparently prevails in the Odonata (damsel- and dragonflies), considered by many to be the most primitive living pterygotes.

The short-germ hypothesis implies that the pair-rule level of patterning (Nüsslein-Volhard and Wieschaus, 1980) evolved within the Pterygota – that is unless pair-rule patterning should yet be found in orthopterans. The concept of Anderson (1972) and Tautz et al., however, would mean that patterning in the anterior body half might from the beginning have employed a pair-rule mechanism (perhaps of diplopod origin? cf. Sander, 1988), which was subsequently lost in the lineages leading to short-germ insects. This view might prevail if pair-rule patterning were indeed demonstrated in the anterior regions of odonate embryos, and if everyone agreed that the Odonata are really closer to the ancestral pterygote than the orthopterans or other primitive forms, among them the ephemerids and stone flies, of which the latter clearly follow the short-germ mode (Miller, 1939). In favour of an odonate-like ancestor is the fact that only in the Odonata do the yolk cells yield part of the midgut wall (Ando, 1962; see also Anderson, 1972) whereas midgut development in all other pterygote insects seems highly derived.

GERM TYPE CLASSIFICATION AT THE MOLECULAR LEVEL OF RESOLUTION?

The controversy over the ancestral germ type may resolve in yet another and hitherto unsuspected way, namely by a reclassification of germ types using molecular in addition to morphological criteria. The need for a more sophisticated classification is apparent from Patel's demonstration that the short-germ mode may exhibit considerable molecular differences, and from the observations of Kraft and Jäckle (quoted by Tautz et al.) that the seemingly short-germ tobacco hawkmoth expresses in its blastoderm the full number of stripes of some pair-rule and segment polarity genes. Conversely, the honeybee, which morphologically represents the prototype of long-germ development (see quotation from Krause above), was shown to generate its abdominal *engrailed* stripes in an anteroposterior sequence reminiscent of intermediate- or even short-germ development (Fleig, 1990). These and other discrepancies, some of which were noted already by Krause (1939b), call for a revision, and Patel's approach of exploiting the temporal relationship between gastrulation and molecular segment specification may be a first step towards a more satisfactory classification.

OOGENESIS IN RELATION TO GERM TYPES AND PATTERN FORMATION

The first generalized correlations between modes of oogenesis and types of early embryogenesis were established by Bier (1970) who linked short-germ development with panoistic, and long-germ development with meroistic-polytroph oogenesis (reviewed in Sander, 1976). However, comparing this generalization with the examples listed by Krause (1939b) for the different germ types will reveal exceptions, for instance the apparent long-germ development in cockroaches (where oogenesis is panoistic) or the apparent short-germ development in some – meroistic – beetles. May be these inconsistencies will disappear with a better classification of germ types (see the previous section) but it is worthwhile to ask, with Tautz et al., whether long-germ development really requires molecular and cellular innovation in oogenesis, specifically the addition of nurse cells. Perhaps the cytoskeletal mechanisms known from panoistic development (and general cell biology) might suffice for determinant localization, as suggested by Tautz et al., but the fact is that in *Drosophila* the nurse cells enforce an anterior course of development on the oocyte pole(s) to which they are attached by ring canals (Bohrmann and Sander, 1987). The related question whether *bcd* is a general determinant of 'anteriorness' throughout the pterygota will be discussed below – after a look at another maternal patterning component, namely the terminal class genes which Tautz et al. suggest to be ancestral.

The expression of the maternal terminal gene *torso-like* occurs in the polar cells of the follicular epithelium (Martin et al., 1994). Of these, the anterior group are known in *Drosophila* as the border cells once they have migrated through the nurse cell cluster to reach the anterior oocyte pole. The border cells produce, during late oogenesis, the micropylar canal and part of the surrounding egg envelopes (reviewed in Spradling, 1993). Surprisingly, in lower dipterans, these cells do not migrate but stay anterior to the nurse cell(s) and assemble the micropyle in this ectopic position. The micropyle then comes to touch the anterior oocyte pole only when the nurse cell(s) have shrivelled away. This holds not only for *Bradysia* (syn. *Sciara*) (Wenzel et al., 1990), where the single nurse cell might leave no space for migration, but also for psychodids (moth midges) which, like *Drosophila,* have 15 nurse cells. These findings (our unpublished results) might gain wider evolutionary interest when viewed together with two other pieces of information: (1) the *torso-like* signal is believed to reside temporarily in the vitelline envelope (see St Johnston and Nüsslein-Volhard, 1992), and (2) in many insect species (e.g. the stick insects reviewed in Sander, 1983) the micropyle is situated in the posterior egg half and the head lobes form next to it while the more anterior egg parts are probably dispensable for pattern formation. It might be worthwhile testing, once suitable molecular probes become available, whether in such species the micropyle-forming follicle cells also emit a 'polar' signal which in turn might define the anterior limits, and maybe axial polarity, of the future germ band.

IS BICOID UNIVERSAL?

The *bicoid* protein has provided the first and as yet best-analyzed molecular example of a maternally specified morphogenetic gradient (reviewed in St Johnston and Nüsslein-Volhard, 1992). Notwithstanding these epochal merits, it may be legitimate to question its universality. Akam et al. (1994) point out that *bcd* is one of a group of rapidly evolving non-homeotic genes within the Antennapedia complex of insects (but not found in any other taxon), and that *bcd* homologs have been recognized so far only in fruit-, house- and blowflies. Even within these groups, some differences are evident, especially among the blowflies (Schröder and Sander, 1993) where *Calliphora* differs from the other species in both mRNA localization and the fact that its anterior ooplasm has so far failed to rescue *Drosophila bicoid* embryos.

Akam et al. suggest that a Hox class 3 gene recently isolated from the locust *Schistocerca* may share a 'common ancestor' with both *zen* and *bicoid*, which would imply that the *bicoid* function arose late in insect evolution. Assuming that lower dipterans reflect the ancestral dipteran stock, they might provide some relevant information. As mentioned earlier (Sander, 1988), centrifugation can easily and quantitatively induce the eggs of lower dipterans to form mirror-image patterns of the double cephalon or double abdomen type, whereas this is very hard to achieve in wild-type *Drosopila* embryos. However, centrifuged eggs from *bcd* mutant flies are quite prone to double abdomen formation; the yield is inversely correlated to the strength of the *bcd* allele(s) used, and with the strongest alleles (e.g. E1) can approach 100% (Schröder, 1992). There may be many explanations for this, but in our context the most tempting would be that lower dipterans develop without *bicoid*, perhaps using *hunchback* as the anterior determinant as envisioned by St Johnston and Nüsslein-Volhard (1992). In line with this, attempts to isolate from the lower dipteran *Psychoda* a *bcd* homolog have failed so far (R. Schröder, unpublished result). Another potentially relevant difference between lower and higher dipterans was observed in mirror-image duplication patterns. In *Drosophila*, symmetrical double abdomens comprise less segments than in lower diperans (Percy et al., 1986, and our unpublished data), while *Drosophila* double cephalons contain significantly more segments than their counterparts in lower dipterans (our unpublished results). These findings signal that the longitudinal patterning mechanisms of lower dipterans differ considerably from those of *Drosophila* – perhaps owing to the absence of *bcd* and a concomitant shift of gap gene expression domains?

CONSERVED GENES AND NOVEL FUNCTIONS IN THE HOX COMPLEX

The extreme conservation of the homeotic genes in the Hox complex(es) enabled Carroll (1994) and Akam et al. to test a famous proposal concerning a gene of this class. Lewis (1978) had suggested that the *Ubx* gene is an evolutionary novelty of the dipterans which evolved with (or rather for) the suppression of wings in the metathorax. However, Akam et al. show that *Ubx* and all other homeotic *Drosophila* Hox genes have their homologs in crustaceans, even in species with almost uniform trunk segments. Moreover, butterflies according to Carroll express *Ubx* in their winged metathorax. Both findings mean that the *Ubx* protein as such must be much older than its apparent function in *Drosophila*; what has changed during evolution is obviously the network of target genes regulated by the *Ubx* protein (see below). Interestingly, the regulation of this gene itself seems to differ even within the holometabolous insects. For instance, Carroll has shown that in the abdominal segments of the silkmoth both *Ubx* and *abd A* are locally down-regulated in the abdominal cell patches that subsequently give rise to the proleg buds.

ASYMMETRY BETWEEN EVOLUTION AND DECAY OF COMPLEX SYSTEMS

The extreme conservation of the homeotic genes certainly has to do with their fundamental regulatory role in establishing the phylotypic body plan, which should require very complex networks. Akam et al. by their comparative analysis of the non-homeotic Hox gene *ftz*, now provide a quantitative measure for this conservation: *ftz* evolves about ten times faster than the homeotics in the same complex. The *ftz* gene may be less subject to stabilizing selection because, as indicated by the homeobox sequences, the interaction of its protein with other regulatory proteins may be less complex.

The new data on regulatory networks are highlighting an old but sometimes forgotten insight concerning acquisition and loss, respectively, of complex characters. Regulatory networks can evolve by successive accumulation of many changes in the target genes. However, abolition of characters generated by this tedious process may require just the mutative loss of function in a single regulatory gene. This asymmetry must be borne in mind when it comes to deciding whether a given, seemingly ancestral trait, for instance homonomous segmentation (Akam et al.) or wings in a dipteran metathorax (Carroll), is really primitive or not.

MORPHOLOGICAL INNOVATION FOLLOWS FROM NOVEL REGULATORY INTERACTIONS RATHER THAN NEW PROTEINS

To quote Carroll, "the chemical evolution of of animals has not been nearly as great as their morphological evolution". This insight, too, is not quite new, but it can now be supplemented with a wealth of hard molecular data. Instances are the sophisticated regulation mechanisms for and by *Ubx*, evolved perhaps by changes at the enhancer level, which enable a single protein to specify the characters of several abdominal segments (Akam et al.), or the recruiting of genes that serve to specify cell fates in the central nervous system (e.g. *ftz* and *eve*) for pair-rule functions in the segmentation cascade (Patel). The most striking evidence, however, comes from the repeated 'deployment' of certain genes and regulatory networks in the course of a single ontogenesis.

REPEATED ACTIVITY OF CONSERVED REGULATORY NETWORKS AND GENES

The classical example for this is the expression of certain vertebrate Hox complexes that specify cell fates first along the body axis and thereafter in the appendages (see Duboule, 1994). Among insects, wing development in butterflies (Carroll; Nijhout, 1994) has now provided another instance. The genes which in *Drosophila* play key roles in specifying the spatial organization of the wing disc have homologs with similar functions in the butterfly *Precis coenia*. This reflects a common overall organization of wing morphogenesis in both species. However, in the butterfly, these genes later on are transcribed again, this time in each of the 'wing cells' (demarcated by the wing veins) where they apparently are involved in specifying the beautiful patterns of coloured scales. The most striking pattern element, the eye spot, is organized from its center by yet another 're-deployed' gene, namely *Distalless*, which earlier on is expressed (and required) in the prospective distal parts of embryonic appendage buds and in the imaginal discs of late larval stages.

STRATEGIES FOR FUTURE RESEARCH IN COMPARATIVE MOLECULAR EMBRYOLOGY

The strategies for future research mentioned implicitly or expressly in the relevant contributions to this symposium fall in two classes: strategies that are confined to the concepts and probes provided by molecular research on *Drosophila* development, and others that propose to evade the limitations that the prevalent "*Drosophila*-centric view" (Carroll) imposes on our prospects. As Patel points out, the highly successful exercise of identifying homologs of *Drosophila* segmentation genes in other insects will fail to identify any patterning mechanisms that *Drosophila* might be lacking. He therefore proposes to seek out additional organisms amenable to genetic analysis. But remembering the long history of *Drosophila* research and the concurrent input of both intellect and money, this may not be easily achieved. Those eager to try might be well advised to study the list of requirements that Carroll has drawn up for his rather less ambitious approach. As a counterpart to *Drosophila* for genetical analysis, this writer would recommend saprophagous lower dipterans, e.g. the psychodids and scatopsids (see Schmidt-Ott et al., 1994). They are easily mass-reared with short generation times, have beautiful embryos and, last but not least, differ surprisingly from *Drosophila* both developmentally (see above) and, if the first relevant data (Sommer et al., 1992) are representative, also at the molecular level.

I am indebted to the organizers of this BSDB meeting, particularly Michael Akam, for inviting me to chair the session on the evolution of insect development; to Diethard Tautz and to my collaborators Dieter Zissler, Karl-Heinz Fecht, Katrin Serries, Reinhard Schröder and Klaus Rohr for cooperation in research on lower dipterans; and to the Deutsche Forschungsgemeinschaft for funding some of this research.

REFERENCES

Akam, M., Averof, M., Castelli-Gair, J., Dawes, R., Falciani, F. and Ferrier, D. (1994). The evolving role of Hox genes in arthropods. *Development* **1994 Supplement** (in press).
Anderson, D. T. (1972). The development of hemimetabolous insects. In *Developmental Systems: Insects.* vol. 1 (eds J. Counce and C. H. Waddington), pp. 95-163. New York: Academic Press.
Anderson, D. T. (1973). *Embryology and Phylogeny in Annelids and Arthropods.* Oxford: Pergamon Press.
Ando, H. (1962). The comparative embryology of Odonata with special reference to a relic dragonfly *Epiophlebia superstes* Selys. Tokyo: The Japan Society for the Promotion of Science.
Bäumer-Schleinkofer, Ä. (1993). *Die Geschichte der beobachtenden Embryologie.* Frankfurt am Main: Peter Lang Verlag
Bier, K. H. (1970). Oogenesetypen bei Insekten und Vertebraten, ihre Bedeutung für die Embryogenese und Phylogenese. *Zool. Anz. Suppl.* **33**, 7-29.
Bohrmann, J. and Sander, K. (1987). Aberrant oogenesis in the patterning mutant *dicephalic* of *Drosophila melanogaster*: time lapse recordings and volumetry in vitro. *Roux's Arch Dev Biol.* **196**, 279-285.
Carroll, S. B. (1994). Developmental regulatory mechanisms in the evolution of insect diversity. *Development* **1994 Supplement** (in press).
Counce, S. J. and Waddington, C. (eds.) (1972). *Developmental Systems: Insects.* 2 vols. New York: Academic Press.
Duboule, D. (1994). Temporal colinearity and the phylotypic progression: a basis for the stability of a vertebrate bauplan and the evolution of morphologies through heterochrony *Development* **1994 Supplement**, 135-142.
Fleig, R. (1990). *Engrailed* expression and body segmentation in the honeybee *Apis mellifera. Roux's Arch. Dev. Biol.* **198**, 467-473.
King, R. C. (1970). *Ovarian Development in Drosophila melanogaster.* New York: Academic Press.
Kölliker, A. (1842) Observationes de prima insectorum genesi, adjecta articulorum evolutionis cum vertebratorum comparatione. Zürich: Meyer and Zeller.
Krause, G. (1939a). Neue Erkenntnisse über die verschiedenen Eitypen der Insekten und ihre Bedeutung für Entwicklungsphysiologie und Systematik. *VII. Internat. Congr. Entomol. Berlin.* pp. 772-779.
Krause, G. (1939b). Die Eitypen der Insekten. *Biol. Zbl.* **59**, 495-536.

Lewis, E. B. (1978). A gene complex controlling segmentation in *Drosophila*. *Nature* **276**, 761-769.

Martin, J.-R., Ralbaud, A. and Ollo, R. (1994). Terminal pattern elements in *Drosophila* embryo induced by the *torso-like* protein. *Nature* **367**, 741-745

Miller, A. (1939). The egg and early development of the stonefly *Pteronarcys proteus*. *J. Morphol.* **64**, 555-609.

Nijhout, H. F. (1994). Symmetry systems and compartments in lepidopteran wings: the evolution of a patterning mechanism. *Development* **1994 Supplement** (in press).

Nübler-Jung, K. and Arendt, D. (1994). Is ventral in insects dorsal in vertebrates? - A history of embryological arguments favouring phylogenetic axis inversion. Centennial Essay, *Roux's Arch. Dev. Biol.* (in press).

Nüsslein-Volhard, C. and Wieschaus, E. (1980). Mutations affecting segment number and polarity in *Drosophila*. *Nature* **287**, 795-801.

Patel, N. H. (1994). The evolution of arthropod segmentation: insights from comparisons of gene expression patterns. *Development* **1994 Supplement** (in press).

Percy, J., Kuhn, K. L. and Kalthoff, K. (1986). Scanning electron microscopic analysis of spontaneous and UV-induced abnormal segment patterns in *Chironomus samoensis* (Diptera, Chironomidae). *Roux's Arch. Dev. Biol.* **195**, 92-102.

Sander, K. (1976). Specification of the basic body pattern in insect embryogenesis. *Adv. Insect Physiol.* **12**, 125-238.

Sander, K. (1983). The evolution of patterning mechanisms: gleanings from insect embryogenesis and spermatogenesis. In *Development and Evolution* (eds. B. C. Goodwin, N. Holder and C. G. Wylie), pp. 137-159. Cambridge: University Press.

Sander, K. (1984). Embryonic pattern formation in insects: Basic concepts and their experimental foundations. In *Pattern formation. A primer in developmental biology* (ed. G. M. Malacinski), pp. 245-268. New York: Macmillan

Sander, K. (1988). Studies in insect segmentation: from teratology to phenogenetics. *Development* **104 Supplement**, 112-121.

Schmidt-Ott, U., Sander, K., Technau, G. H. (1994). Expression of *engrailed* in embryos of a beetle and five dipteran species with special reference to the terminal regions. *Roux's Arch. Dev. Biol.* **203**, 298-303

Schröder, R. (1992). Mechanismen der frühembryonalen Musterbildung bei verschiedenen Insektenarten. PhD thesis, Fakultät für Biologie, Universität Freiburg i. Br.

Schröder, R. and Sander, K. (1993). A comparison of transplantable *bicoid* activity and partial *bicoid* homeobox sequences in several *Drosophila* and blowfly species (Calliphoridae). *Roux's Arch. Dev. Biol.* **203**, 34-43.

Seidel, F. (1924). Die Geschlechtsorgane in der embryonalen Entwicklung von *Pyrrhocoris apterus*. *Zeitschr. Morph. Ökol. Tiere* **1**, 429-506.

Sommer, R. J., Retzlaff, M., Goerlich, K., Sander, K. and Tautz, D. (1992). Evolutionary conservation pattern of zinc-finger domains of *Drosophila* segmentation genes. *Proc. Natl. Acad. Sci. USA* **89**, 10782-10786.

Spradling, A. (1993) Developmental genetics of oogenesis. In *The Development of Drosophila melanogaster* (eds M. Bate and A. Martinez-Arias). Cold Spring Harbour: CSH Laboratory Press

St Johnston, D. and Nüsslein-Volhard, C. (1992). The origin of pattern and polarity in the *Drosophila* embryo. *Cell* **68**, 201-219.

Tautz, D., Friedrich, M. and Schröder, R. (1994). Insect embryogenesis - what is ancestral and what is derived? *Development* **1994 Supplement** (in press)

Wenzel, F., Gutzeit, H. O. and Zissler, D. (1990). Morphogenesis of the micropylar apparatus in ovarian follicles of the fungus gnat *Bradysia tritici* (syn. *Sciara ocellaris*). *Roux's Arch. Dev. Biol.* **199**, 146-155.

Development 1994 Supplement, 193-199 (1994)
Printed in Great Britain © The Company of Biologists Limited 1994

Insect embryogenesis – what is ancestral and what is derived?

Diethard Tautz, Markus Friedrich and Reinhard Schröder

Zoologisches Institut der Universität München, Luisenstrasse 14, 80333 München, Germany

SUMMARY

The systematic genetic analysis of *Drosophila* development has provided us with a deep insight into the molecular pathways of early embryogenesis. The question arises now whether these insights can serve as a more general paradigm of early development, or whether they apply only to advanced insect orders. Though it is too early to give a definitive answer to this question, we suggest that there is currently no firm reason to believe that the molecular mechanisms that were elucidated in *Drosophila* may not also apply to other forms of insect embryogenesis. Thus, many of the *Drosophila* genes involved in early pattern formation may have comparable functions in other insects and possibly throughout the arthropods.

Key words: evolution, insect embryogenesis, oogenesis, segmentation

INTRODUCTION

There is a long tradition of research in comparative insect embryology. Representative taxa of almost all insect orders were studied in detail and inferences were made on ancestral and derived traits of embryogenesis. Among these, *Drosophila* clearly represents a derived mode of insect embryogenesis. However, the choice of *Drosophila* as an embryological system was entirely governed by the unique suitability of this organism for genetic analysis. This genetic approach to embryogenesis (Nüsslein-Volhard and Wieschaus, 1980) turned out to be very successful. There is now an almost complete understanding of the principles of early *Drosophila* development at the molecular level (reviewed in Bate and Martinez Arias, 1993). It is therefore time to ask which of these processes may also be utilized in other insect orders and which may be special to the *Drosophila* mode of development.

The key to such studies is the possibility of using the *Drosophila* genes as molecular tools to isolate homologs of segmentation genes from other insects and to study their expression pattern in these species. The patterns can then be related to the patterns known from *Drosophila*. Thus, comparative insect embryology can now be done at the molecular level. A number of genes have already been studied in this way and the results have recently been summarized (Tautz and Sommer, 1994). Here we want to review the literature on previous work of comparative insect embryogenesis and to reassess it in the light of the new molecular results. We feel that this may serve as a basis for developing new ideas and new experimental directions in the future.

PHYLOGENY

To discriminate ancestral from derived traits, one needs a reliable phylogeny in order to carry out the necessary outgroup comparisons (Kitching, 1992). The phylogeny of major insect groups, as it is supported by morphological analysis (Kristensen, 1991) is depicted in Fig. 1. We found that the same tree is also supported by comparisons of ribosomal RNA sequences from representative taxa (Friedrich and Tautz, unpublished data). Thus, there is little doubt about the correct grouping of these insect orders. The picture is however less clear for the more primitive entognathan hexapods. These relationships are therefore left unresolved in Fig. 1. The following discussions will deal mainly with the ectognathan insects, since the relationships among these are most clearly resolved.

LONG, INTERMEDIATE AND SHORT GERM EMBRYOS

The most obvious difference among the embryos of different insect taxa is the way in which the early germ band is formed. Krause (1939) introduced a classification according to the length of the early germ band, whereby he has used two different descriptive terms, namely *Kleinkeim* versus *Grosskeim* (small germ versus large germ) and *Kurzkeim* versus *Langkeim* (short germ versus long germ). The former terms describe merely how large the germ anlage is with respect to the size of the egg. In contrast, the terms short or long germ were meant to imply functional differences, namely a germ anlage which, among other criteria, does, or does not show a secondary growth process after blastoderm stage (Fig. 2). Krause also used the term *halblang* (semi-long) which is now more frequently called intermediate germ. Short and intermediate germ embryos are found in the more primitive, hemimetabolous insect orders, such as Orthoptera and Ephemeroptera, while long germ embryos are restricted to the more advanced holometabolous orders, such as Hymenoptera and Diptera. In long germ embryos, all segments are already defined at blastoderm stage, while in intermediate and short germ embryos, the more posterior segments are produced during a secondary segmental growth process. In extreme cases, for example, in *Schistocerca* (Orthoptera), the early germ anlage at blastoderm stage shows only the head lobes and a growth zone from which the remainder of the segments will be generated. These extremely short germ embryos thus share a superficial similarity with Trochophora larvae that are characteristic for taxa with spiral embryogenesis, such as annelids and molluscs. Therefore, it seemed reasonable to conclude that these extremely short-germ type embryos represent the ancestral mode of insect embryogenesis (Krause, 1939; Sander, 1983). This interpretation is, however, not unequivocal for two reasons. First, not all Orthopteran species show the extreme short germ mode seen in *Schistocerca* (Kanelis, 1952, and see below), and second, more primitive insect groups such as the Odonata are of the intermediate germ type (see below). Similarly, in the bristle tail *Petrobius* (Archaeognatha), the head lobes and at least the mandibular segments are found in the early germ anlage (Larink, 1969). Thus, it seems possible that the extremely short germ embryos are not ancestral, but represent a secondary specialization (Anderson, 1973).

The long germ mode exemplified by *Drosophila* becomes particularly clear when one looks at the expression pattern of early segmentation genes. Crucial for the following discussion is the class of pair rule genes that are responsible for a transient double segmental organization of the developing embryo (Nüsslein-Volhard and Wieschaus, 1980). In *Drosophila*, the pair rule genes are usually expressed in seven stripes at blastoderm stage (Ingham, 1988), corresponding to the three mandibular, three thoracic and eight abdominal segments that will eventually be formed. The expression pattern of the pair rule genes can thus be taken as a direct molecular marker for the blastoderm fate map.

Such a molecular fate map has also been created for the moth *Manduca sexta* (Lepidoptera). *Manduca* is phylogenetically close to the dipterans (Fig. 1), but morphologically, *Manduca* looks more like a short or intermediate germ type. Only the headlobes and a region that looks superficially like a growth zone become visible at the end of blastoderm stage. However, rather than undergoing a secondary growth process, the originally rather broad germ anlage elongates via tissue reorganization (Broadie et al., 1991). Homologues of *Drosophila* gap, pair rule and segment polarity genes were cloned and their early expression pattern was studied (Kraft and Jäckle, 1994). It was found that the expression of these genes was very similar to that in *Drosophila*. In particular, the pair rule gene *runt* was seen to be expressed in eight stripes and the segment polarity gene *wingless* in sixteen stripes in the early embryo. This suggests that all sixteen segments (three mandibular, three thoracic and ten abdominal) may already be specified at blastoderm stage, even though they become morphologically visible only later and form in the progressive manner typical of a short germ embryo. Thus, from the molecular point of view, *Manduca* shows clearly a long germ embryogenesis (Kraft and Jäckle, 1994). Interestingly, this conclusion is also supported by UV irradiation experiments in the related lepidopteran *Tineola*. By destroying certain cell groups at blastoderm stage with UV-light, Lüscher (1944) found that specific larval structures were affected at later stages. Most importantly, he found that practically all larval pattern elements could be destroyed in this way, depending on which region of the egg was treated with the UV-light. This allowed him to conclude that all segments were already specified at blastoderm stage. These types of destructive fate mapping experiments were also applied to *Drosophila* where basically the same results were obtained (Lohs-Schardin et al., 1979). However, there are some conceptual caveats about the use of destructive methods for fate mapping. On the one hand, it is possible that cells that had already become specified may be replaced by other cells after they were destroyed. In this way, an early specification would be missed. On the other hand, cells that are not yet committed may not be destroyed completely, but only loose their capability to respond to the signals that would normally specify them at later stages. In this way, an early determination would be incorrectly assumed. However, even though these possibilities may exist, at least in *Drosophila*, but apparently also in lepidopteran embryos, the destructive fate maps seem to conform very closely with the molecular results that were obtained later. Thus, fate maps constructed in this way appear to be reliable indicators for the underlying molecular principles, at least in insects. Therefore, we are going to use arguments that are based on such fate maps in the following discussion.

While the dipterans and the lepidopterans have apparently a very similar molecular fate map, the situation is entirely

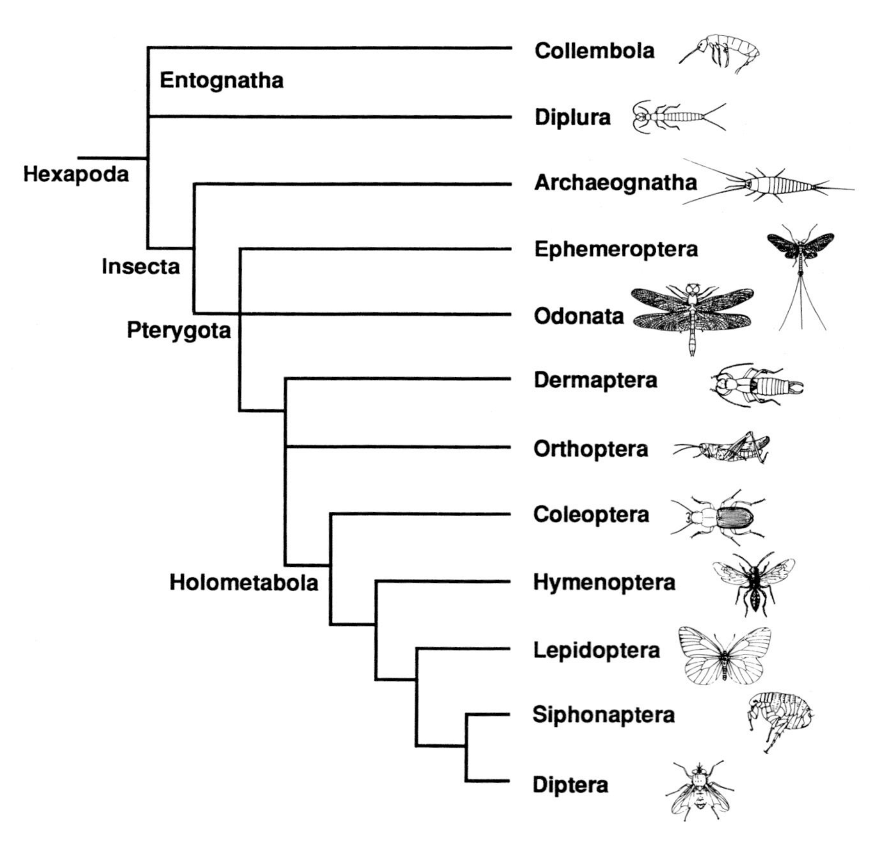

Fig. 1. Phylogeny of the insect orders discussed in the text (after Kristensen, 1991).

different in beetles (Coleoptera). Again, it is the expression pattern of the pair rule genes that shows this most clearly. In the flour beetle *Tribolium*, only three stripes of the pair rule genes *hairy* and *even skipped* are formed at blastoderm stage (Sommer and Tautz, 1993; Patel et al., 1994). According to the definitions of Krause (1939), *Tribolium* belongs to the intermediate germ type of embryo, where the three gnathal and the three thoracic segments become specified at blastoderm stage. Thus, six segments need to be defined, which is in line with the three pair rule stripes observed. Most interestingly, a striped pair rule gene expression is also evident after blastoderm stage, during the secondary segmental growth process (Sommer and Tautz, 1993; Patel et al., 1994). This suggests that these genes are involved in defining the segments even at these later stages.

Only one pair rule gene, *even skipped*, has so far been analysed in the Orthopteran *Schistocerca*, which shows the extremely short germ mode of embryogenesis. Pair rule stripes were not found in these embryos, either at blastoderm stage, or during the secondary segmental growth process, though *even skipped* is transiently expressed in the growth zone (Patel et al., 1992; Patel et al., 1994). It was therefore suggested that pair rule gene activity may not be present in the more primitive hemimetabolous insect orders (Patel et al., 1992). However, it is still possible that another one of the several pair rule genes known from *Drosophila* has substituted for the function of *even skipped* in *Schistocerca*. An alternative interpretation could be that *Schistocerca* represents a secondary reduction of the intermediate germ type (see above). In this case one would not necessarily have to postulate that pair rule genes play a role in these types of embryos, since they could have become secondarily lost. This assumption is supported from a comparison of the mode of embryogenesis in *Gryllus domesticus* (Orthoptera). This species is relatively closely related to *Schistocerca*, yet shows an intermediate germ as can be inferred from a blastoderm fate map (Kanelis, 1952). Moreover, staining with an antibody against *even skipped* (Patel et al., 1994) shows that stripes are formed in the growth zone of these embryos (unpublished results). However, it is not yet clear whether these are organized in double segmental units and might thus represent a pair rule activity, or whether they are segmentally reiterated and might thus have a different function. Still, one can conlude that the situation in *Schistocerca* may be derived and that the answer to how segmentation is achieved in this species may be less relevant for assessing which of the molecular mechanisms are more ancestral.

To find an answer to the question of the ancestral or derived status of the pair rule genes, it will be necessary to look in the oldest insect orders that show the intermediate type of embryogenesis. Particularly well studied in this respect is the damselfly *Platycnemis* (Odonata) (Seidel, 1935). The blastoderm fate map of this organism, as it was derived from experimental embryology, is depicted in Fig. 3A. On the basis of the arguments given above, namely that experimentally produced fate maps concur with molecular fate maps in those cases where this has been analysed, we should like to use the *Platycnemis* fate map to make the equivalent inferences. It appears that at least six segments are laid down at very early stages (Fig. 3A), very similar to *Tribolium*. It seems therefore reasonable to predict that this pattern is generated by similar molecular mechanisms, i.e. that three pair rule stripes should appear at blastoderm stage.

The *Platycnemis* fate map can in fact be taken as representative of a more general fate map of intermediate germ insects (Anderson, 1973). Thus, an archetypic fate map for insects might look like the one depicted in Fig. 3B. This intermediate germ type is characterized by two separate phases of segmentation, namely one at blastoderm stage and one during the germ band extension phase. Since the *Tribolium* results show that the same genes are utilized during both of these phases, it is easy to see how the two derived forms of embryogenesis, the long and the extremely short germ mode, may have evolved. The extremely short germ forms would have discarded the blastoderm stage phase of segmentation and accordingly, some gene functions required at this stage, such as the pair rule genes, may have become lost, or their function may have become modified. The long germ embryos on the other hand, would have replaced the secondary growth phase by an extension of the blastoderm stage phase of segmentation without the need to recruit additional gene functions, at least at the level of the pair rule genes. Thus, in this interpretation, the *Drosophila* mode of segmentation is likely to have retained ancestral gene functions.

OOGENESIS AND MATERNAL GENES

Not only the mode of segmentation, but also the mode of oogenesis subdivides the insects. Egg production can be either meroistic or panoistic, i.e. with or without nurse cells. In a first view, this subdivision appears to be correlated with the phylogenetic relationships of the insect orders. The more ancestral, hemimetabolous orders show panoistic oogenesis, suggesting

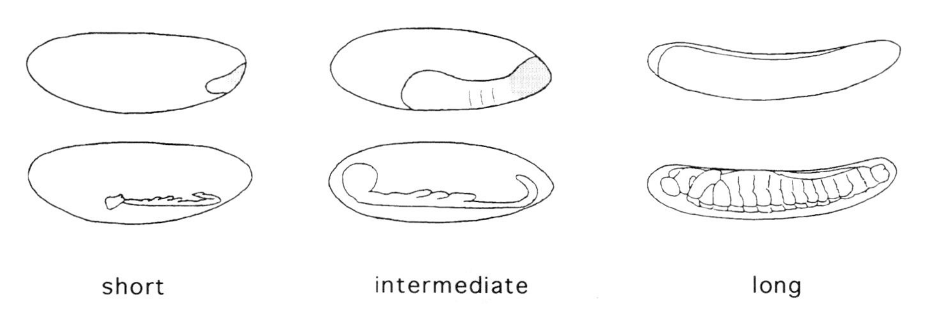

Fig. 2. Germ types in insects. The top row represents blastoderm stages, the bottom row stages during which the first segments become visible. Anterior is to the left and ventral is up. The areas that represent the early germ band are indicated in the top row. The stippled areas represent the regions of the growth zone in the short germ and intermediate germ embryos. The stippled lines in the intermediate germ embryo indicate the fact that some segments are already specified at this stage. This is in contrast to the short germ embryo, where only the headlobes and the growth zone are formed. Note that the total length of the germ band in relation to the egg is not the decisive criterium for classifying short and long germ embryos, but is rather the basis for a second classification system, namely small and large germ embryos (see text; modified after Krause and Sander, 1962).

that this is the ancestral mode of oogenesis (King and Büning, 1985). However, the picture is not so clear. Meroistic modes can be found in phylogenetically basal orders, such as the Ephemeroptera (Gottanka and Büning, 1993) and Dermaptera (King and Büning, 1985) and panoistic modes are seen in rather advanced orders such as the Siphonaptera (King and Büning, 1985). While the latter ones have been classified as secondarily panoistic (Büning and Sohst, 1988), this is less clear for the situation in the older orders. Meroistic oogenesis can also be found among the entognathan hexapod orders, the Collembola and the Diplura (Stys and Bilinski, 1990). These observations suggest that the mode of oogenesis is not a stable evolutionary character and may have changed independently several times.

Why does the mode of oogenesis matter for embryonic development? In *Drosophila* it was found that both, the antero-posterior and the dorsoventral axis become originally specified by maternally localized factors. The means by which this localization is achieved is however rather different for the different factors. Four sets of genes are involved, the anterior group (key gene *bicoid*), the posterior group (key gene *nanos*), the terminal group (key gene *torso*) and the dorsoventral group (key gene *dorsal*) (St. Johnston and Nüsslein-Volhard, 1992).

Among these, only the localization of the anterior factor *bicoid* seems to depend on the mode of oogenesis. *bicoid* RNA is synthesized in the nurse cells and delivered to the growing oocyte where it becomes trapped at the anterior pole by gene products that are homogeneously distributed in the oocyte (St. Johnston et al., 1989). Thus, the fact that the *bicoid* RNA becomes anteriorly localized would be due to the anterior-posterior polarity information caused by the asymmetric localization of the nurse cells. In fact when the nurse cell clusters are aberrantly located on both sides of the oocytes, as occurs in the *Drosophila* mutant *dicephalic*, one can observe a duplication of head structures (Lohs-Schardin, 1982; Bohrmann and Sander, 1987)

The localization of the posterior determinant *nanos* relies on a different mechanism. *nanos* RNA is also produced in the nurse cells and delivered into the oocyte. However, it then becomes trapped by a specific receptor at the posterior pole (Wang and Lehmann, 1991). The localization signal for this receptor might in turn be determined by a signal from the follicle cells that surround the oocyte and would therefore be nurse cell independent. A similar mechanism applies to the realization of the terminal information, as well as to the dorso-ventral axis formation where again the follicle cells are involved in providing the respective primary signal (St. Johnston and Nüsslein-Volhard, 1992). Intriguingly, in the latter case it was shown that the follicle cells themselves receive the signal from a gene product that is produced by the oocyte nucleus (Neumann-Silberberg and Schüpbach, 1993).

Thus, it appears that only the localization of *bicoid* would be a conceptual problem in the panoistic mode of oogenesis. However, a reassessment of the way in which *bicoid* becomes localized in the *Drosophila* embryo suggests that even this process could be nurse cell independent. The protein involved in anchoring the *bicoid* mRNA appears to be capable of actively transporting the *bicoid* RNA along a microtubule scaffold (Pokrywka and Stephenson, 1991; St. Johnston and Nüsslein-Volhard, 1992). Such a localization mechanism could of course also work in a panoistic oocyte. Asymmetric localization of an RNA within a single cell is in fact also seen for other genes (Ding and Lipshitz, 1993) and is obviously no particular problem for a cell.

Thus, none of the *Drosophila* results imply directly that embryonic axis formation has to occur differently in species with meroistic or panoistic oogenesis. However, unequivocal homologs of the genes involved in primary axis determination in *Drosophila* have not yet been recovered from more distantly related insects. On the other hand, there is at least indirect evidence for a similar mode of dorso-ventral axis formation (Sommer and Tautz, 1994), a terminal activity (Nagy and Carroll, 1994; and see below) as well as for a *nanos*-like activity in *Tribolium* (Wolff et al., unpublished data).

CONSERVED PATTERNING DECISIONS IN INSECTS

The differences seen in the mode of oogenesis and segmentation among the insect taxa conceal, somewhat, the fact that other modes of patterning decisions are much more conserved.

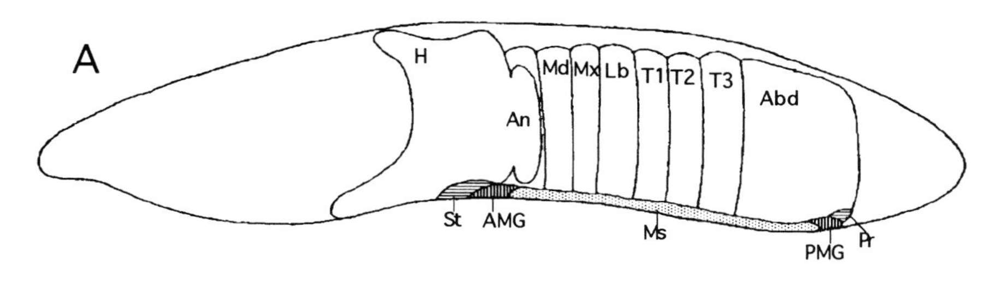

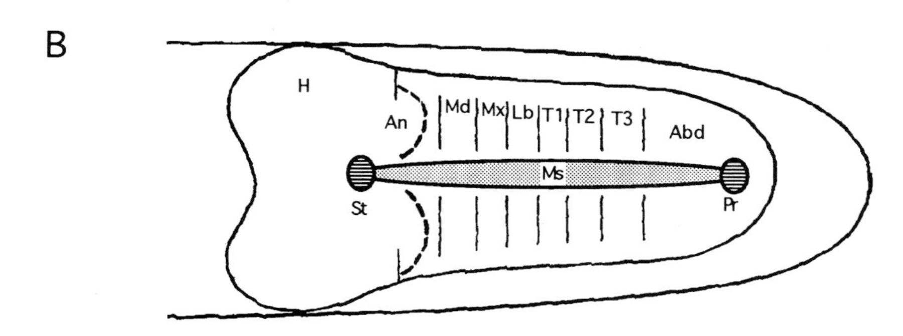

Fig. 3. Blastoderm fate maps of (A) *Plactynemis* (after Seidel, 1935) and (B) a generalized fate map for the intermediate germ type (see text). Anterior is left, the embryo in A is viewed from the lateral side, the one in B from the ventral side. The anlagen of the mesoderm and the gut are indicates by the stippled and hatched areas respectively. H, head; An, antennae; Md, mandible; Mx, maxilla; Lb, labium; T1-T3, thorax 1-3; Abd, abdominal growthzone; St, stomodaeum; Pr, proctodaeum; Ms, mesoderm; AMG, anterior midgut; PMG, posterior midgut.

These concern head segment formation, mesoderm formation, formation of the gut and the formation of the nervous system. Two of these processes, namely mesoderm formation and gut formation, are particularly pertinent to the question of determination by maternal positional cues and shall therefore be discussed in more detail here.

Mesoderm formation

The mesoderm in *Drosophila* is formed in response to maternal dorso-ventral cues and involves the specific expression of the genes *twist* and *snail* at the ventral side at blastoderm stage (St. Johnston and Nüsslein-Volhard 1992; Leptin, 1991). The expression domains of these two genes are directly regulated by the maternal *dorsal* gene product, which forms a gradient of nuclear localization in the early embryo. *twist* and *snail* determine the mesodermal fate of cells in which they are expressed and these cells invaginate during gastrulation. The expression pattern of *twist* and *snail* was also studied in *Tribolium* and found to be basically the same as in *Drosophila*. Both genes are expressed at the ventral side and the cells expressing them are the ones that will invaginate (Sommer and Tautz, 1994). Thus, this establishes a link between morphology and gene expression pattern and, it suggests also that the underlying maternal mode of dorso-ventral specification may be conserved.

What happens in the other insects? The mesoderm invagination (called the "gastral groove") occurs always in a very similar manner along the whole germ anlage, independent of the length of the germ (Anderson, 1972a,b). In the short germ embryos, the invagination continues until all segments are formed and in *Tribolium* it was in fact found that *snail* and *twist* expression persist until the process is completed (Sommer and Tautz, 1994). Thus, the morphological and genetic mode of mesoderm invagination appear to be very similar among insects. Most importantly however, the mesodermal anlagen can be defined on the blastoderm fate map (Fig. 3B) even in the most primitive insects, including the Archaeognathans (Jura, 1972). It seems possible therefore that the maternal mode of the definition of the mesoderm via the regulation of the zygotic genes *twist* and *snail* is the ancestral one, at least in insects. It should be noted, however, that there are some observations that would argue against this assumption. In several insect species it is possible to produce "parallel twins", i.e. twinned embryos along the longitudinal axis, by different types of experimental manipulation (reviewed by Sander, 1976). Clearly, the dorso-ventral axis for these additional embryos must have become specified in a way that is difficult to reconcile with a maternal specification. However, these experiments do not exclude the possibility that at least the zygotic pathway, as it is reflected in the expression of *twist* and *snail*, is conserved. Interestingly, homologues of *twist* and *snail* are also expressed in vertebrates in the developing mesoderm (Hopwood et al., 1989; Smith et al., 1992), though the mode of mesoderm formation is apparently rather different and nothing is known about maternal influences.

Formation of the gut

The gut is formed from several primordia in insects. The stomodaeum at the anterior and the proctodaeum at the posterior are derived from ectodermal tissues and the midgut is of endodermal origin. The genes defining the posterior anlagen of the gut in *Drosophila* are regulated by the maternal terminal system (Weigel et al., 1990). Though homologues of these genes have not yet been analysed in more primitive insects, it is nonetheless evident, that similar blastoderm anlagen of the gut can be defined (Fig. 3B), even in the most basic orders (Anderson, 1973). Thus, this suggests, indirectly, that a maternal system equivalent to the terminal class of genes is also ancestral, though this speculation has to be verified by data. It is, however, interesting to note that even the further development of the gut occurs in a fairly stereotypic manner in most insects. In *Drosophila*, the stomodaeum and the proctodaeum begin development by forming epithelial tubes during gastrulation, after the completion of the segmentation process. The midgut does not form a tube, but consists at first of two lateral strands of cells that migrate from the ends of the ectodermal parts of the gut towards the middle of the embryo. Once they have met, they spread out ventrally and dorsally and eventually engulf the remaining yolk. The same course of events, sometimes with modifications, is basically found in all insects. The details of this process are beginning to be studied at the genetic level in *Drosophila* (Reuter et al., 1993) and it will be interesting to see which of the genes involved in the process can also be found in other insects.

ARTHROPOD EMBRYOGENESIS

Comparison with representatives from the other arthropod classes (myriapoda, chilopoda, crustacea and chelicerata) may also be helpful in identifying ancestral features of hexapod embryogenesis. An extensive effort in this direction was made by Anderson (1973). Unfortunately, his studies were strongly influenced by Manton's (1973) theory of a polyphyletic origin of the arthropods and some of his inferences have to be treated with caution. Today there is a consensus, based on morphological (Lauterbach, 1973) and molecular data (Turbeville et al., 1991; Ballard et al., 1992), that arthropods do share a common ancestor, although, due to the long time of evolutionary separation, most extant arthropod taxa are presumably very derived with respect to embryological features. Nonetheless, at least some inferences can be made, since there are general similarities among the different forms of arthropod embryogenesis. One concerns the mode of blastoderm formation. Though some taxa begin their development with a total cleavage of the egg, almost all (exception: lower crustacean orders) also form a syncytial blastoderm stage later on (Anderson, 1973), as is characteristic for insects. Given the importance of the syncytial blastoderm for the early patterning decisions in insects, this suggests strongly that similar decisions may be necessary in the embryos of all arthropods. Another intriguing parallel is the fact that a large part of the anterior segment pattern may become specified at, or shortly after blastoderm stage, while the remainder of the segments are generated in a secondary growth process (Anderson, 1973). We have seen above that a similar mode of development may be the ancestral form of embryogenesis in insects. However, direct comparisons of embryogenesis between the classes are difficult, since the morphological details can look rather different. Nonetheless, this does not preclude that similar molecular mechanisms could be at work, since it has also become clear from the work in insects that morphologically different embryonic forms may be generated by the very similar molecular mechanisms (see above). Moreover, we think that it has become clear that the blastoderm fate map is a reliable indicator of the underlying molecular processes and it is in fact the blastoderm fate map that suggests the parallels among arthropods (Anderson, 1973). At least some homologs of early segmentation genes have already been recovered from representatives of the other arthropod classes (Sommer et al., 1992) and it will, therefore, be highly interesting to study their expression pattern in the future.

THE CONCEPT OF THE PHYLOTYPIC STAGE

The concept of the phylotypic stage was proposed because of a seeming paradox of embryogenesis. The very earliest stages of development, namely egg production, blastoderm formation, gastrulation and secondary growth processes seem to be fairly dissimilar between different taxa and, as dicussed above, may not even be related to the phylogeny of the respective taxa. Yet these early events all seem to channel into a highly stereotypic stage at which the full segmental pattern becomes morphologically visible. Intriguingly, this stage looks very similar between different taxa, not only among insects, but even among the arthropods as a whole and thus represents the general bauplan of the phylum. This stage was therefore called the "Körpergrundgestalt" or the phylotypic stage (Seidel, 1960; Sander, 1983). This phenomenon is not merely restricted to the arthropods, but is also seen in the other animal phyla, for example among the vertebrates. Furthermore, comparative analysis of the expression pattern of homologs of the homeotic genes known from *Drosophila* at this phylotypic stage in different taxa has shown that they are expressed in a comparable spatial and temporal order as in *Drosophila* (McGinnis and Krumlauf, 1992). This provides a strong argument for the universality of the animal bauplan (Slack et al., 1993).

However, homeotic genes are not themselves involved in generating the segment pattern. Rather, they depend on the information from the preceding levels of the gene hierarchy for delimiting their own segmentally organized expression domains. Thus, it seems reasonable to propose that the gene network regulating the expression pattern of the homeotic genes should also be more or less conserved between organisms (for an alternative view see Sander, 1983). This way of reasoning, together with the molecular results mentioned above, suggest that there may be no paradox after all with respect to the phylotypic stage. Though it remains true that the morphological pathways towards the phylotypic stage may look rather diverse, there is currently no reason to believe that the regulatory genetic pathways may not be conserved to some degree.

The way to analyse this experimentally would be to take the regulatory regions from homeotic genes and to test them in different animals to see whether they are regulated in a similar manner. So far, one such experiment has been successfully performed. It has been shown that a particular regulatory element of a homeobox gene expressed in the head (*Deformed*) can be functionally interchanged between vertebrates and *Drosophila* (Awgulewitsch and Jacobs, 1992; Malicki et al, 1992). However, the element analysed in these studies is essentially only necessary for autoregulation of the gene. Thus, the conservation of the underlying regulatory network has not been finally proved with this experiment. However, these are clearly the kinds of experiments that have to be done to obtain an insight into the general degree of conservation of the regulatory hierachy and thus eventually into the evolution of the general animal bauplan.

We should like to thank Klaus Sander for his comments on the manuscript and the members of the laboratory for fruitful discussions. The work in our laboratory is supported by grants from the Deutsche Forschungsgemeinschaft and by the Human Frontier of Science Program.

REFERENCES

Anderson, D. T. (1972a). The development of hemimetabolous insects. In *Developmental Sytems: Insects* (ed. S. J. Counce and C. H. Waddington), pp. 96-163. London: Academic Press.

Anderson, D. T. (1972b). The development of holometabolous insects. In *Developmental Sytems: Insects* (ed. S. J. Counce and C. H. Waddington), pp 166-242. London: Academic Press.

Anderson D. T. (1973). *Embryology and Phylogeny in Annelids and Arthropods*. Oxford: Pergamon Press.

Awgulewitsch, A. and Jacobs, D. (1992). *Deformed* autoregulatory element from *Drosophila* functions in a conserved manner in transgenic mice. *Nature* **358**, 341-344.

Ballard, J. W. O., Olsen, G. J., Faith, D. P., Odgers, W. A., Rowell, D. M. and Atkinson, P. W. (1992). Evidence from 12S ribosomal RNA sequences that onychophorans are modified arthropods. *Science* **258**, 1345-1348.

Bate, M. and Martinez Arias, A. (1993). *The Development of Drosophila melanogaster*. New York: Cold Spring Harbour Laboratory Press.

Bohrmann, J. and Sander, K. (1987). Aberrant oogenesis in the patterning mutant *dicephalic* of *Drosophila melanogaster*: time-lapse recordings and volumetry in vitro. *Roux's Arch. Dev. Biol.* **196**, 279-285.

Büning, J. and Sohst, S. (1988). The flea ovary: ultrastructure and analysis of cell clusters. *Tissue Cell* **20**, 783-795.

Broadie, K. S., Bate, M. and Tublitz, N. J. (1991). Quantitative staging of embryonic development of the tobacco hawkmoth, *Manduca sexta*. *Roux's Arch. Dev. Biol.* **199**, 327-334.

Ding, D. and Lipshitz, H. D. (1993). Localized RNAs and their functions. *BioEssays* **15**, 651-658.

Gottanka, J. and Büning, J. (1993). Mayflies (Ephemeroptera) the most "primitive" winged insects, have teloptrophic meroistic ovaries. *Roux's Arch. Dev. Biol.* **203**, 18-27.

Hopwood, N. D., Pluck, A. and Gurdon, J. B. (1989). A *Xenopus* mRNA related to *Drosophila twist* is expressed in response to induction in the mesoderm and the neural crest. *Cell* **59**, 893-903.

Ingham, P. (1988). The molecular genetics of embryonic pattern formation in *Drosophila*. *Nature* **335**, 25-34.

Jura, C. (1972). Development of apterygote insects. In *Developmental Systems: Insects* (ed. S. J. Counce and C. H. Waddington), pp. 49-94. London: Academic Press.

Kanelis, A. (1952). Anlagenplan und Regulationserscheinungen in der Keimanlage des Eies von *Gryllus domesticus*. *Wilhelm Roux Arch. EntwMech. Org.* **145**, 417-461.

King, R. C. and Büning, J. (1985). The origin and function of insect oocytes and nurse cells. In *Comprehensive Insect Phsiology, Biochemistry and Pharmacology* vol. 1. (ed. G. A. Kerkut and L. I. Gilbert), pp. 37-82. Oxford: Pergamon Press.

Kitching, I. J. (1992). The determination of character polarity. In *Cladistics: a practical course in systematics* (ed. P. L. Forey et al.), pp. 22-42. Oxford: Clarendon Press.

Kraft, R. and Jäckle, H. (1994) *Drosophila* mode of metamerization in the embryogenesis of the intermediate germband insect *Manduca sexta* (Lepidoptera). *Proc. Natl. Acad. Sci. USA* **91**, 6634-6638.

Krause, G. (1939). Die Eitypen der Insekten. *Biol. Zentralblatt* **59**, 495-536.

Krause, G. and Sander, K. (1962). Ooplasmic reaction systems in insect embryogenesis. *Adv. Morphogen.* **2**, 259-303.

Kristensen, N. P. (1991). Phylogeny of extant hexapods. In *The insects of Australia* 2nd ed. (ed. I. D. Naumann et al.) pp. 125- 140, CSIRO, Melbourne University Press, Melbourne.

Larink, O. (1969). Zur Entwicklungsgeschichte von *Petrobius brevistylus* (Thysanura, Insecta). *Helgoländer wiss. Meeresunters.* **19**, 111-155.

Lauterbach, K. E. (1973). Schlüsselereignisse in der Evolution der Stammgruppe der Euarthropoda. *Zool. Beitr. N.F.* **19**, 251-299.

Leptin, M. (1991). *twist* and *snail* as positive and negative regulators during Drosophila development. *Genes Dev.* **5**, 1568-1576.

Lohs-Schardin, M., Cremer, C. and Nüsslein-Volhard, C. (1979). A fate map for the larval epidermis of *Drosophila melanogaster*: localized cuticle defects following irradiation of the blastoderm with an ultraviolet laser microbeam. *Dev. Biol.* **73**, 239-255.

Lohs-Schardin, M. (1982). *dicephalic* - a *Drosophila* mutant affecting polarity in follicle organization and embryonic patterning. *Roux's Arch. Dev. Biol.* **191**, 28-36.

Lüscher, M. (1944). Experimentelle Untersuchungen über die larvale und die imaginale Determination im Ei der Kleidermotte (*Tineola biselliella* Hum.) *Revue suisse Zool.* **51**, 531-627.

Malicki, J., Cianetti, C., Peschle, C. and McGinnis, W. (1992). A human HOX4B regulatory element provides head-specific expression in *Drosophila* embryos. *Nature* **358**, 345-347.

Manton, S. M. (1973). Arthropod phylogeny - a modern synthesis. *J. Zool. Lond.* **171**, 111-130.

McGinnis, W. and Krumlauf, R. (1992). Homeobox genes and axial patterning. *Cell* **68**, 283-302.

Nagy, L. M. and Carroll, S. (1994). Conservation of *wingless* patterning functions in the short-germ embryos of *Tribolium castaneum*. *Nature* **367**, 460-463.

Neumann-Silberberg, F. S. and Schüpbach, T. (1993). The *Drosophila* dorsoventral patterning gene *gurken* produces a dorsally localized RNA and encodes a TGFα-like protein. *Cell* **75**, 165-174.

Nüsslein-Volhard, C. and Wieschaus, E. (1980). Mutations affecting segment number and polarity in *Drosophila*. *Nature* **287**, 795-801.

Patel, N. H., Ball, E. E. and Goodman, C. S. (1992). Changing role of *even skipped* during the evolution of insect pattern formation. *Nature* **357**, 339-342.

Patel, N. P., Condron, B. G. and Zinn, K. (1994) Pair rule expression patterns of *even-skipped* are found in both short- and long-germ beetles. *Nature* **367**, 429-434.

Pokrywka, N. J. and Stephenson, E. C. (1991). Microtubules mediate the localization of *bicoid* RNA during *Drosophila* oogenesis. *Development* **113**, 55-66.

Reuter, R., Grunewald, B. and Leptin, M. (1993). A role for the mesoderm in endodermal migration and morphogenesis in Drosophila. *Development* **119** 1135-1145.

Sander, K. (1976). Specification of the basic body pattern in insect embryogenesis. *Adv. Insect Physiol.* **12**, 125-238.

Sander, K. (1983). The evolution of patterning mechanisms: gleanings from insect embryogenesis and spermatogenesis. In *Development and Evolution: The Sixth Symposium of the British Society for Developmental Biology.* (ed. B. C. Goodwin, N. Holder and C. C. Wylie), pp. 137-159. Cambridge: Cambridge University Press.

Seidel, F. (1935) Der Anlagenplan im Libellenkeim, zugleich eine Untersuchung über die allgemeinen Bedingungen für defekte Entwicklung und Regulation bei dotterreichen Eiern. *Wilhelm, Roux Arch. EntwMech. Org.* **132**, 671-751.

Seidel, F. (1960). Körpergrundgestalt und Keimstruktur. Eine Erörterung über die Grundlagen der vergleichenden experimentellen Embryologie und deren Gültigkeit bei phylogenetischen Überlegungen. *Zool. Anz.* **164**, 245-305.

Slack, J. M. W., Holland, P. W. H. and Graham, C. F. (1993). The zootype and the phylotypic stage. *Nature* **361**, 490-492.

Smith, D. E., Franco Del Amo, F. and Gridley, T. (1992). Isolation of SNA, a mouse gene homologous to the *Drosophila* genes *snail* and *escargot*: its expression pattern suggests multiple roles during postimplantation development. *Development* **116**, 1033-1039.

Sommer, R. J., Retzlaff, M., Görlich, K., Sander, K. and Tautz, D. (1992). Evolutionary conservation pattern of zinc-finger domains of *Drosophila* segmentation genes. *Proc. Natl. Acad. Sci. USA* **89**, 10782-10786.

Sommer, R. J. and Tautz, D. (1993). Involvement of an orthologue of the *Drosophila* pair rule gene *hairy* in segment formation of the short germ band embryo of *Tribolium* (Coleoptera). *Nature* **361**, 448-450.

Sommer, R. J. and Tautz, D. (1994). The expression patterns of *twist* and *snail* in *Tribolium* (Coleoptera) suggest a homologous formation of mesoderm in long and short germ band insects. *Dev. Genet.* **15**, 32-37.

St. Johnston, D. and Nüsslein-Volhard, C. (1992). The origin of pattern and polarity in the *Drosophila* embryo. *Cell* **68**, 201-219.

St. Johnston, D., Driever, W., Berleth, T., Richstein, S. and Nüsslein-Volhard, C. (1989). Multiple steps in the localization of *bicoid* RNA to the anterior pole of the *Drosophila* oocyte. *Development Supplement* **107**, 13-19.

Stys, P. and Bilinski, S. (1990). Ovariole types and the phylogeny of hexapods. *Biol. Rev.* **65**, 401-429.

Tautz, D. and Sommer, R. J. (1994). Evolution of segmentation genes in insects. *Trends Genet.* (in press).

Turbeville, J. M., Pfeiffer, D. M., Field, K. G. and Raff, R. A. (1991). The phylogenetic status of arthropods as inferred from 18S rRNA sequences. *Mol. Biol. Evol.* **8**, 669.

Wang, C. and Lehmann, R. (1991). *nanos* is the localized posterior determinant in *Drosophila*. *Cell* **66**, 637-647.

Weigel, D., Jürgens, G., Klingler, M. and Jäckle, H. (1990). Two gap genes mediate maternal terminal pattern information in *Drosophila*. *Science* **248**, 495-498.

Development 1994 Supplement, 201-207 (1994)
Printed in Great Britain © The Company of Biologists Limited 1994

The evolution of arthropod segmentation: insights from comparisons of gene expression patterns

Nipam H. Patel

Department of Embryology, Carnegie Institution of Washington, 115 West University Parkway, Baltimore, Maryland 21210, USA

SUMMARY

The comparison of gene expression patterns in a number of insect and crustacean species has led to some insight into the evolution of arthropod patterning mechanisms. These studies have revealed the fundamental nature of the parasegment in a number of organisms, shown that segments can be generated sequentially at the molecular level, and suggested that pair-rule pre-patterning might not be shared by all insects.

Key words: evolution, segmentation, *Drosophila*, *Manduca*, *Tribolium*, *Schistocerca*, Acheta, Crustacea, *engrailed*, *even-skipped*

INTRODUCTION

The 1988 Development Supplement entitled 'Mechanisms of Segmentation' contained twenty-three articles dealing with the establishment of metameric pattern in animals as diverse as *Drosophila*, mice and leeches. At that time – and at present, the situation remains largely unchanged – the process generating anterior-posterior segmental pattern during development was best understood in *Drosophila*, and this detailed knowledge was built on a foundation of intensive study of *Drosophila* development at the genetic level. At least six articles in that Development Supplement went on to pose specific questions regarding the extent to which the *Drosophila* paradigm could be applied to understanding segmentation in other arthropods. Several of these articles also suggested that studies of pattern formation in additional insects could help us understand the evolution of the developmental system seen in *Drosophila*.

Few other insects, however, are as amenable to genetic analysis as *Drosophila*, and several authors outlined an alternative method that might provide some initial information about segmentation in other insects and arthropods. This approach involved the isolation of homologs of *Drosophila* segmentation genes from additional arthropods and the subsequent comparison of the expression patterns of these genes in various arthropod embryos. Six years later, many of the questions raised in that 1988 Development Supplement are still with us, but we have obtained a number of answers and some significant insights by pursuing the comparative molecular approach. I will summarize a portion of the progress that has been made and describe answers to several of the questions that had been posed.

(1) ARE PARASEGMENTS UNIVERSAL IN INSECTS?

While segment boundaries are morphologically obvious and segments are historically the units used to describe the metameric properties of insects, genetic and molecular studies of *Drosophila* point to a more developmentally relevant unit - the parasegment. An individual parasegment includes the posterior portion of one segment plus the anterior portion of the next more posterior segment. Thus, parasegments span the same length as segments, but their boundaries lie between the segment boundaries. Molecularly, the parasegment boundaries lie at the interface of the *engrailed* and *wingless* expression domains (reviewed by Martinez Arias, 1993). In the 1988 Development Supplement, Lawrence (1988) provided a concise summary of the data supporting the hypothesis that parasegments are the fundamental units of design in the *Drosophila* embryo (Martinez Arias and Lawrence, 1985): (1) parasegments are the first metameres to be defined during development, (2) parasegment boundaries provide important lineage boundaries during development, and (3) parasegments are the domains in which key genes (such as homeotic genes) are expressed during development. Lawrence ended his discussion on parasegments by stating that ' ..... it would be astounding if other insects, and even annelids, were made of fundamentally different units.' A few pages later, Sander (1988) indicated that comparisons of gene expression patterns between *Drosophila* and other insects might be used to determine whether parasegment organization was a common feature of insect development and that this approach was being actively pursued by a number of laboratories.

Ultimately, these sorts of comparative molecular studies showed that parasegment metamery is found throughout the insects and crustaceans. Homologs of two *Drosophila* genes in particular, *engrailed* (a segment polarity gene) and *abdominal-A* (a homeotic gene), have been studied in a number of species, and an analysis of their expression patterns reveals the evolutionary conservation of parasegmental domains. In the case of *engrailed*, expression is seen in the posterior portion of each segment in all insects and crustaceans examined so far (Fig. 1; for additional examples see Patel et al., 1989a, 1989b; Fleig, 1990, Sommer and Tautz, 1993; Scholtz et al., 1993; Manzanares et al., 1993; Brown et al., 1994). In these species, as in *Drosophila*, the anterior (parasegmental) border of each *engrailed* stripe rapidly resolves into a sharp boundary, whereas the posterior (segmental) border is not clearly demarcated until much later in development, thus illustrating that the parasegments are the first metameres to be resolved (Patel et al., 1989b). Furthermore, *wingless* expression has been characterized in the beetle, *Tribolium castaneum*, and is found to abut the anterior margin of the *engrailed* stripes, just as in *Drosophila* (Nagy and Carroll, 1994). Expression of *abdominal-A* has been studied in *Manduca* (tobacco hawkmoth), *Tribolium*, and *Schistocerca* (grasshopper), and in all three of these insects, as in *Drosophila*, the anterior boundary of expression coincides with the parasegment boundary within the first abdominal segment (Nagy et al., 1991; Staurt et al., 1993, Tear et al., 1990). In addition, mutations in the *Tribolium abdominal-A* homolog lead to defects that transform parasegmental domains (Stuart et al., 1993). Thus, at least some homeotic genes obey parasegmental boundaries in a variety of insects.

In *Drosophila*, careful lineage analysis reveals that the parasegmental boundary, as marked by the anterior margin of each *engrailed* stripe, defines a stable lineage boundary throughout much of development (Vincent and O'Farrell, 1992; Martinez Arias, 1993). Studies of the development of a number of crustacean species reveal even more striking ectodermal lineage units, which have been termed 'genealogical units' (see for example Dohle, 1976; Dohle and Scholtz, 1988; Scholtz, 1992). Each genealogical unit starts as a single row of cells (roman numeral row), which then undergoes two rounds of division to yield four orderly rows of cells (Fig. 2D). Dohle showed that these genealogical units do not correspond to segments since a segment groove lies within each genealogical unit and not between each genealogical unit. It was speculated that these geneological units might bear some relationship to *Drosophila* parasegments (Martinez Arias and Lawrence, 1985; Dohle and Scholtz, 1988). Indeed, the analysis of *engrailed* expression in the crayfish, *Procambarus clarki*, suggests that these lineage units show striking similarities to *Drosophila* parasegments as the anterior boundary of each of the crustacean geneological units is demarcated by the stable anterior margin of each *engrailed* stripe (Patel et al., 1989b), and similar results have been obtained in additional crustacean species (Fig. 2; Scholtz et al., 1993; N. Patel, unpublished data). Thus, parasegmental units are easily visualized in a wide range of insects and crustaceans. More importantly, these parasegment units appear to satisfy the criteria for fundamental units of design as stipulated by Lawrence (1988).

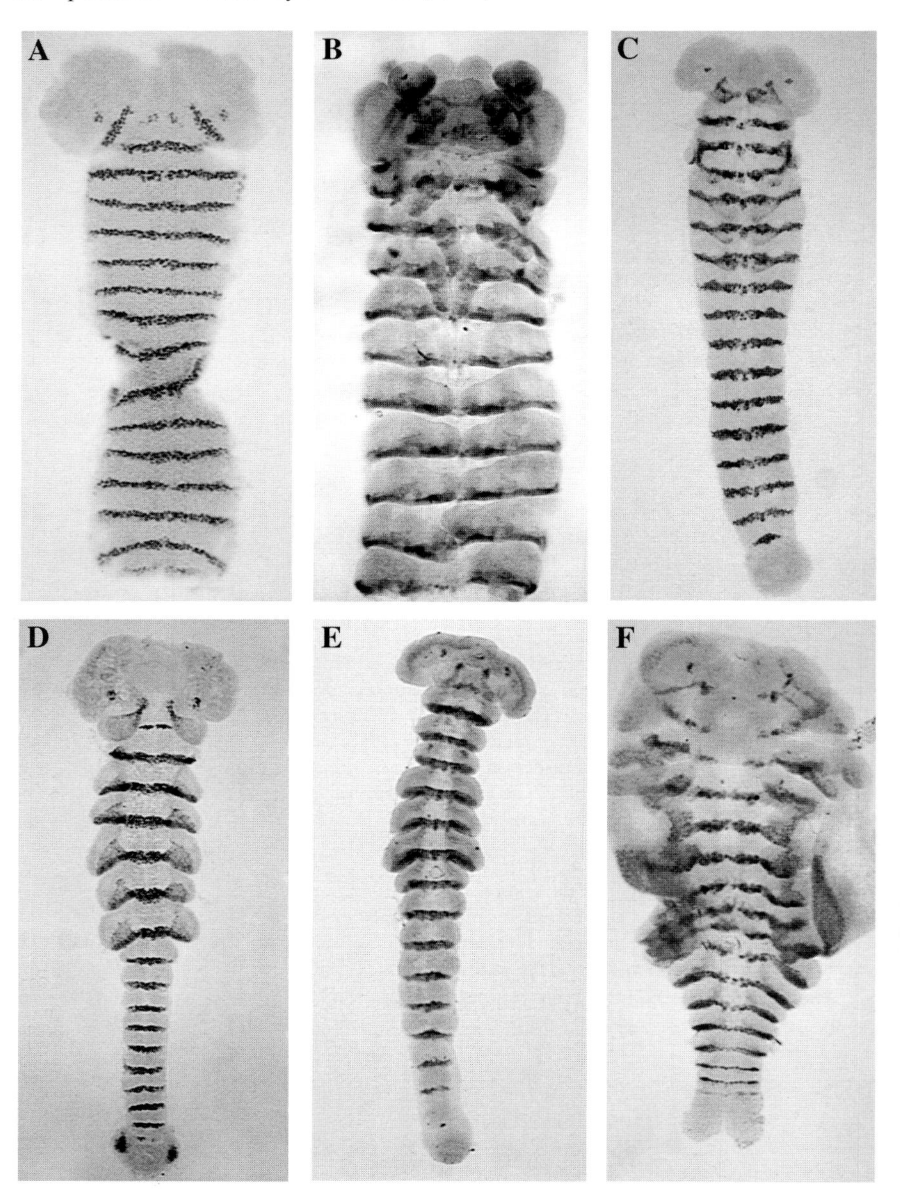

Fig. 1. Expression of *engrailed* in a variety of insects and a crustacean. (A) *Drosophila melanogaster* (fruit fly), (B) *Manduca sexta* (tobacco hawkmoth), (C) *Tribolium castaneum* (red flour beetle), (D) *Acheta domestica* (cricket), (E) *Schistocerca americana* (grasshopper), (F) *Procambarus clarki* (crayfish). *engrailed* is expressed in the posterior portion of each body segment. The stage 9 *Drosophila* embryo in A has been dissected flat and the proctodeum and posterior midgut have been removed (see Fig. 3A,B for undissected specimens). The faint *engrailed* stripe at the posterior end of this *Drosophila* embryo is part of the ninth abdominal segment. In B, all segments posterior to A7 have been removed. Embryos in C–F have not completed the formation of some of the most posterior *engrailed* stripes. All embryos are oriented anterior up and are viewed from the ventral side. mAb 4D9 was used to detect *engrailed* in all embryos except *Manduca*, where mAb 4F11 was used instead (Patel et al., 1989a).

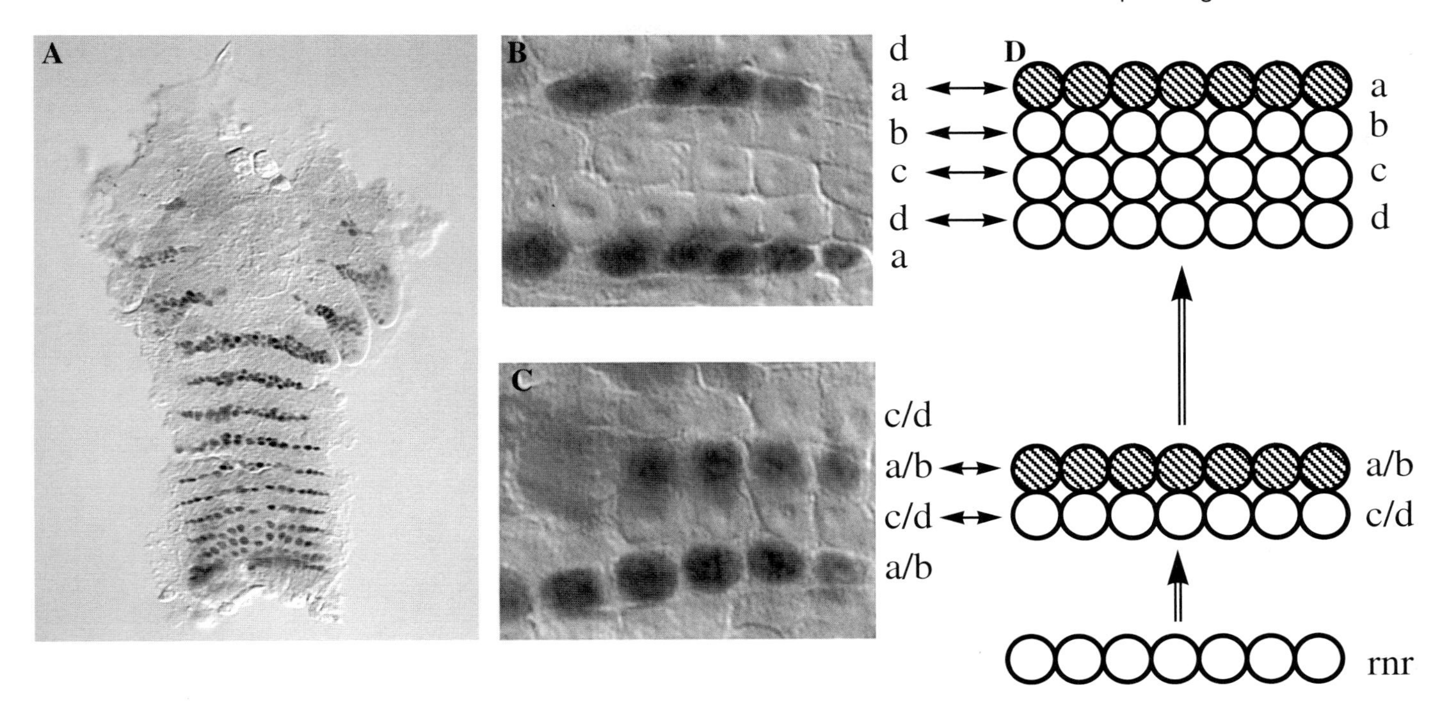

Fig. 2. Expression of *engrailed* during the formation of genealogical units in the crustacean, *Mysidium columbiae*. (A) *Mysidium* embryo stained with mAb 4D9 to visualize the expression of *engrailed*. (B,C) Higher magnification views of some of the rows of *engrailed*-expressing cells. (D) Schematic illustration of the generation of 'genealogical' units in *Mysidium*. In *Mysidium*, segments are generated sequentially from a posterior growth zone. The *Mysidium* growth zone consists of an organized row of cells called ectoteloblasts. The ectoteloblasts undergo a series of asymmetric divisions, each division generating a row of cells known as a 'roman numeral row' (rnr; Dohle, 1976; Dohle and Scholtz, 1988). Each roman numeral row divides symmetrically to generate a/b and c/d rows. Each of these two rows then divides symmetrically to yield a, b, c, and d rows. Since this division pattern provides an inherent temporal and spatial gradient, all steps are visible in a single embryo of the appropriate stage; ectoteloblasts at the very posterior, newly formed roman numeral rows just anterior to the ectoteloblasts, a/b and c/d rows slightly more anterior, and a, b, c, and d rows even further anterior. *engrailed* expression is not seen in the ectoteloblasts nor in roman numeral rows. C shows a level of the *Mysidium* embryo in which a/b and c/d rows have just formed and at this time *engrailed* is expressed in the a/b row (shaded in the corresponding section of D). After the a/b row divides, *engrailed* expression is lost from the b row cells and is maintained in the a row cells. B shows a region of the *Mysidium* embryo in which a, b, c, and d rows of cells are visible and *engrailed* protein is only in the a row cells (shaded in the corresponding section of D). Later, some b row cells will regain *engrailed* expression, and after the next round of division, *engrailed* will be maintained in all a row progeny, but in only the more anterior b row progeny. When the segmental groove forms, it will traverse between the progeny of the b row. For further details, see the description of *engrailed* expression in *Procambarus* (Patel et al., 1989b) and *Cherax* (Scholtz et al., 1993).

(2) ARE SEGMENTS ESTABLISHED SEQUENTIALLY AT THE MOLECULAR LEVEL IN SHORT GERM INSECTS?

The articles by French (1988), Tear et al. (1988), and Sander (1988) discussed an important classification system that has been used to divide insect embryos into three large developmental categories (reviewed by Sander, 1976). This classification scheme uses data from a variety of embryonic manipulations to determine the extent to which the body plan is established in the initial germ anlage. Embryos such as those of *Drosophila*, which have established a complete body plan by the onset of gastrulation, are termed long germ embryos. *Schistocerca* embryos, which belong to the short germ category, appear to consist of only a head region and a subterminal proliferative zone at the end of the blastoderm stage. All the body segments appear to be generated subsequently as the embryo elongates by cell proliferation (Mee and French, 1986). Finally, embryos whose segments are established as far posterior as the thorax or anterior abdomen at the blastoderm stage and that specify the remaining, more posterior segments after gastrulation are termed intermediate germ embryos.

In long germ *Drosophila*, the establishment of the entire segmental pattern of the body is revealed at the molecular level by the patterns of segmentation and homeotic genes during the blastoderm stage (reviewed byAkam, 1987; Ingham, 1988). For example, all fourteen *engrailed* stripes of the body are visible in the *Drosophila* embryo by the time gastrulation begins, and these stripes are more or less evenly spaced out over the body region of the germ anlage (Fig. 3A,B). A priori, there were two possibilities for molecular specification of segments in short germ embryos: all segments could be established at the molecular level within the prospective proliferative zone during the blastoderm stage and simply expand and differentiate sequentially during the growth phase, or the proliferative zone could generate a sheet of cells that would be sequentially segmented at the molecular level after the growth phase. Studies with both *engrailed* and a number of homeotic genes suggests that the latter is true. In short germ embryos of *Schistocerca*, the first *engrailed* stripes appear in the thorax and at this time there is no compressed pattern of *engrailed* stripes in the region that will proliferate to provide the cells of the abdomen (Fig. 3E). Instead, abdominal *engrailed* stripes appear sequentially, later in development, in a region that has been newly generated by cell proliferation (Fig. 3F; Patel et al., 1989b). In *Tribolium*, the first *engrailed* stripe (mandibular segment) appears at the onset of gastrulation, and the remaining *engrailed* stripes of the body appear sequentially during development (Fig. 3C,D; Brown et al., 1994). Similar sequential appearance of body stripes has also been found for the pattern of *Tribolium wingless* (Nagy and Carroll, 1994). The analysis of homeotic gene expression patterns also supports the notion that segments are not specified simultaneously during short germ development – *Schistocerca Antp*, *Ubx*, *abd-A*, and *Abd-B* appear in sequence as the embryo develops (Tear et al., 1990; Kelsh et al., 1994, 1993; E. Ball, N. Patel, D. Hayward, and C. Goodman, unpublished results).

(3) WHAT ARE THE EVOLUTIONARY ORIGINS OF PAIR-RULE ORGANIZATION?

In *Drosophila*, the expression patterns of segment polarity genes are established by the actions of the pair-rule class of

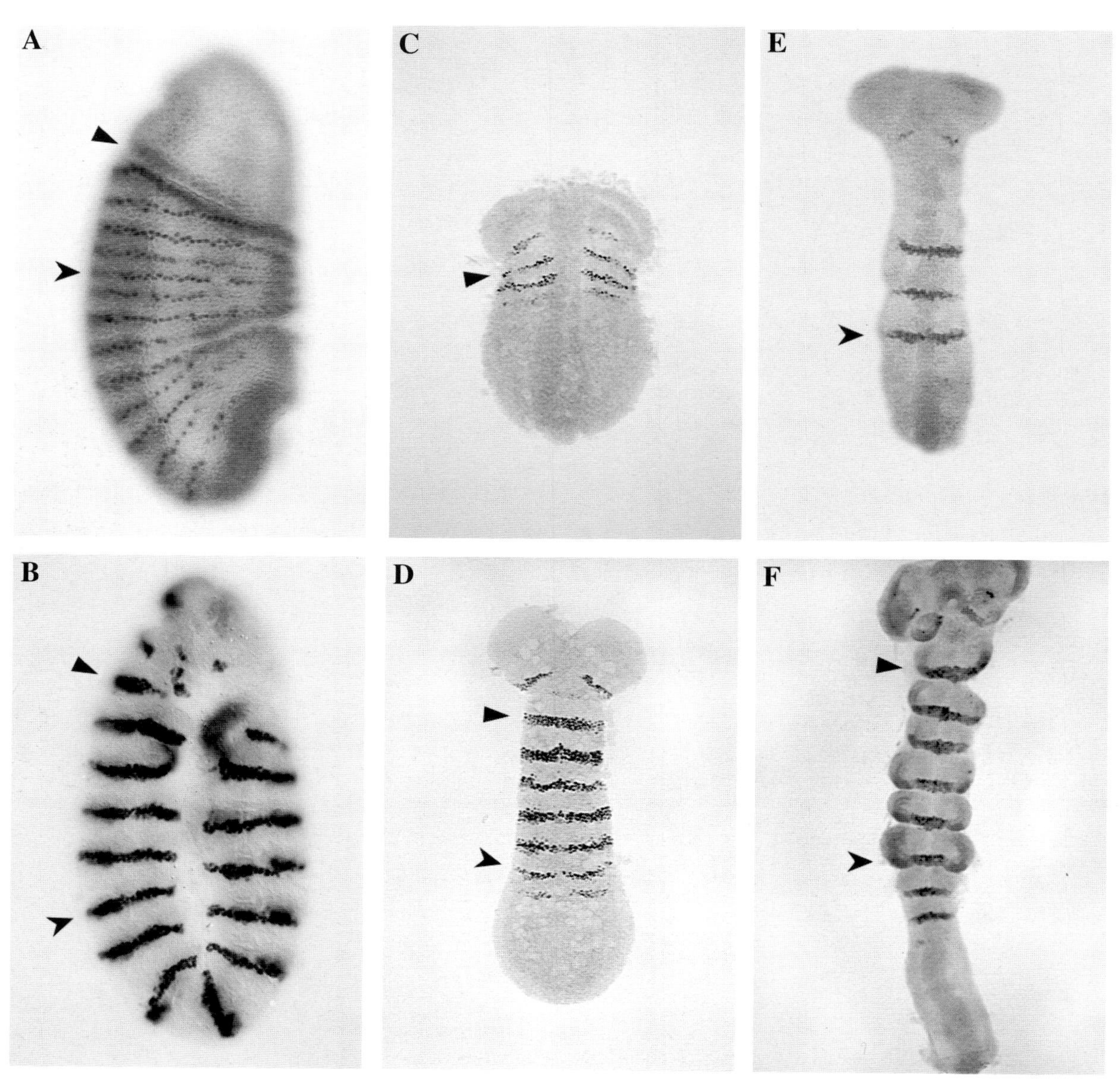

Fig. 3. Establishment of *engrailed* patterns in *Drosophila*, *Tribolium*, and *Schistocerca*. In all panels, the triangle marks the *engrailed* stripe of the mandibular segment and the arrowhead indicates the position of the *engrailed* stripe of the third thoracic segment. (A) At stage 6, the *Drosophila* embryo has just started gastrulation and germband extension and all fourteen *engrailed* stripes of the body are already visible. The mandibular stripe is slightly hidden as it is folded into the cephalic furrow. (B) By stage 11, germband formation is completed and additional patches of *engrailed* expression are now present in the terminal regions. In *Tribolium*, only the most anterior *engrailed* stripes are present shortly after the start of gastrulation and germband extension (C) and the remaining stripes appear as the germband elongates (D). In *Schistocerca*, *engrailed* stripes first appear in the thorax (E) and more posterior *engrailed* stripes appear only after the abdominal region has expanded by cell proliferation (F).

genes (Fig. 4A,B; reviewed by Martinez Arias, 1993). The maintenance and refinement of the segment polarity expression patterns, however, involves a system of cell-cell interactions mediated by the products of the segment polarity genes themselves (reviewed by Martinez Arias, 1993). Because *Schistocerca* pattern formation appears to occur in a cellular (as opposed to syncytial) environment, it was proposed that the generation of segment polarity expression patterns in short germ *Schistocerca* embryos might not involve a pair-rule pre-patterning system, but instead might rely on a system of cell-cell interactions like those used to maintain segment polarity patterns in *Drosophila* (Tear et al., 1988; Patel et al., 1989b). Furthermore, whereas most *Drosophila* segment polarity genes, including *engrailed*, show an initial 'pair-rule' pattern of intensity as their expression first begins, no such 'pair-rule' patterns are seen during the generation of *engrailed* stripes in *Schistocerca* (Patel et al., 1989b). Sander (1988), however, argued that the pair-rule pre-patterning system is of ancient

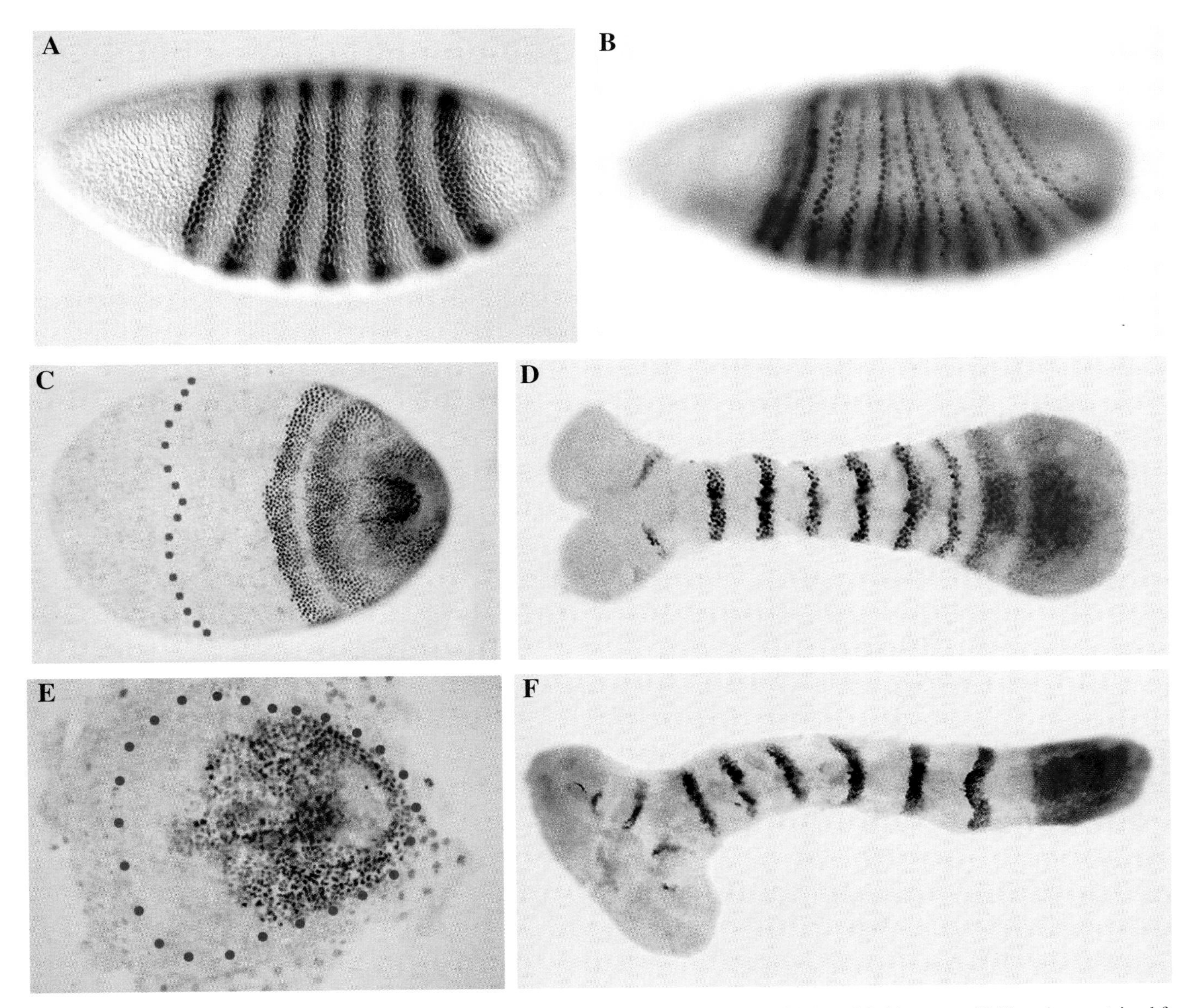

Fig. 4. Relationship of *even-skipped* and *engrailed* expression. *Drosophila* (A,B), *Tribolium* (C,D), and *Schistocerca* (E,F) embryos stained for *even-skipped* alone (A,C,E) or *even-skipped* and *engrailed* (B,D,F; *even-skipped* in brown, *engrailed* in black). In *Drosophila*, all seven *even-skipped* pair-rule stripes are present at the blastoderm stage (A). At the onset of gastrulation, odd numbered *engrailed* stripes appear at the anterior margin of each *even-skipped* pair-rule stripe (B). In *Tribolium*, *even-skipped* also displays a pair-rule expression pattern, but only the first two pair-rule stripes are present at the onset of gastrulation (C). The remaining *even-skipped* stripes appear as the embryo elongates (D). As in *Drosophila*, the anterior border of each *Tribolium even-skipped* pair-rule stripe marks the position of odd-numbered *engrailed* stripes (see Patel et al., 1994 for a more detailed discussion of *even-skipped* expression in *Tribolium*, particulary for details concerning the formation of segmental secondary stripes). In *Schistocerca*, *even-skipped* is expressed in a posterior domain at the onset of gastrulation with a crescent of unstained cells at the posterior end of the gastral furrow (E), and expression remains in a posterior domain as the embryo elongates (F). No pair-rule patterns have been observed for *Schistocerca even-skipped*. The blue dots in C and E indicate the boundary between embryonic and extra-embryonic cells. *Even-skipped* was detected using mAb 2B8 (Patel et al., 1994) and *engrailed* with mAb 4D9 (Patel et al., 1989a).

origin based on the morphologically visible patterns of two segment periodicity in the myriapods, although it should be noted that *Drosophila* does not itself display these sorts of morphological two segment periodicities.

Schistocerca homologs of the *Drosophila* pair-rule genes *fushi tarazu* and *even-skipped* have been characterized (Dawes et al, 1994; Patel et al., 1992, 1994). Neither is expressed in any discernable pair-rule pattern, although both show conserved expression patterns within the developing nervous system. Curiously, both *Schistocerca even-skipped* and *fushi tarazu* are expressed in the posterior region of the embryo during early development, but the function of this expression is unknown (Fig. 4E,F).

In contrast to *Schistocerca* embryos, however, short germ *Tribolium* embryos do display pair-rule pre-patterning, as revealed by the expression of *Tribolium* homologs of *hairy*, *fushi tarazu*, and *even-skipped* (Fig. 4C,D; Sommer and Tautz, 1993; Patel et al., 1994; S. Brown, J. Parrish, and R. Denell, personal communication). In addition, *even-skipped* expression patterns reveal pair-rule pre-patterning in the long germ beetle, *Callosobruchus maculatus*, and the intermediate germ beetle, *Dermestes frischi*, but consistent with their germ type designations, these beetles differ from *Tribolium* and each other in the relative number of *even-skipped* stripes that have been formed by the onset of gastrulation (Patel et al., 1994).

At first, the results from *Schistocerca* and *Tribolium* might seem contradictory: both have embryos of the short germ type, but only *Tribolium* seems to use pair-rule pre-patterning. A resolution to this apparent contradiction comes from a more precise consideration of germ type designations. Germ type designations simply reflect the timing of segment specification - both *Tribolium* and *Schistocerca* embryos establish most body segments after gastrulation. Intermediate germ *Dermestes* is patterned through the thorax by the start of gastrulation. Long germ *Drosophila* and *Callosobruchus* have established virtually all segments by the onset of gastrulation. On the other hand, germ type classification, with its simple three category system, is insufficient to describe the mechanistic details and evolutionary origins of insect pattern formation (Sander et al., 1985; Patel et al., 1994). Thus, embryos with shared germ type designations, such a *Schistocerca* and *Tribolium*, do not necessarily share identical pattern formation mechanisms.

The data obtained by examining *even-skipped* expression patterns in *Schistocerca* and the various beetles is easier to understand when viewed in a phylogenetic framework. Both beetles and *Drosophila* belong to phylogenetically advanced insect orders, while grasshoppers are members of a phylogenetically more primitive order. Thus, one interpretation of the available data is that pair-rule pre-patterning evolved sometime during the evolution of the phylogenetically more advanced insects. However, it is possible that evidence for pair-rule pre-patterning will emerge when additional *Schistocerca* homologs of *Drosophila* pair-rule genes are characterized. Certainly the results from *Tribolium* prove that development in a cellular environment and the sequential appearance of *engrailed* stripes do not necessarily rule out the presence of pair-rule pre-patterning. Alternatively, as Sander (1988) pointed out, pair-rule pre-patterning may be an ancient property of insects that has simply been lost in the evolution of the lineage leading to *Schistocerca*. The validity of each of these hypotheses can be tested by looking at the expression of additional pair-rule gene homologs in *Schistocerca* as well as by expanding these sorts of studies to additional phylogenetically primitive insects and to other arthropods outside of the insects.

CONCLUSIONS

The molecular comparisons made between various organisms have given us important insights into the evolution of insect segmentation. Parasegments are almost certainly the fundamental units of development in all insects and crustaceans. Short germ insects generate segments sequentially at both the molecular and morphological levels. Pair-rule patterning is evident in several orders of phylogenetically advanced insects but may be absent from phylogenetically primitive insect orders. Many questions that were posed in the 1988 Development Supplement remain unanswered, of course, and what new data we have also raises a number of new questions.

For example, work in *Tribolium* points to the clear involvement of gap gene patterning in the development of this embryo (Sommer and Tautz, 1993); will gap gene patterning also participate in the development of embryos from phylogenetically more primitive insects? In *Drosophila*, gap gene products presumably diffuse in the syncytial blastoderm to generate local gradients (reviewed by Pankratz and Jäckle, 1993). Can similar gradients form in the cellular environment of the growing *Tribolium* embryo? If pair-rule pre-patterning is not at work in *Schistocerca* development, how are segment polarity gene expression patterns initiated? While both *Drosophila* and *Tribolium* display pair-rule prepatterning, some details of pair-rule gene expression are different between the two insects. What is responsible for these slightly different pair-rule patterns and how do these differences influence subsequent development? Evidence for the maternal establishment of embryo polarity has been obtained for many insects, but is the maternal information always transmitted by the same genetic system that has been characterized in *Drosophila* (reviewed by St. Johnston and Nüsslein-Volhard, 1992)? Does the mode of oogenesis (i.e., with or without nurse cells attached to the anterior end of the developing oocyte) limit the extent to which maternal components can be used to establish the body plan of the embryo (French, 1988; Patel, 1993)? Most analyses have been confined to insects and crustaceans, but what sorts of patterning systems will be seen in myriapods and chelicerates?

While continued molecular comparisons will provide some answers to these questions, additional approaches are needed. In particular, gene expression patterns can be quite suggestive, but it will be important to devise ways to eliminate the function of particular genes in various arthropods in order to critically access their developmental roles. Moreover, although identifying homologs of *Drosophila* segmentation genes has been a useful approach, it will not identify novel patterning mechanisms that might be at work in other arthropods. Thus, additional organisms amenable to genetic analysis must be sought out. In another five or six years, significant progress will certainly be made in our continuing efforts to understand the processes that have guided the evolution of pattern formation in the arthropods. The results obtained will also help us to understand how complex developmental patterning systems in other phyla may have evolved.

REFERENCES

Akam, M. (1987). The molecular basis for metameric pattern in the *Drosophila* embryo. *Development* **101**, 1-22.

Brown, S. J., Patel, N. H. and Denell, R. E. (1994). Embryonic expression of the single *Tribolium engrailed* homolog. *Dev Genet.* **15**, 7-18.

Dawes, R., Dawson, I., Falciani, F., Tear, G. and Akam, M. (1994). *Dax*, a locust Hox gene related to *fushi-tarazu* but showing no pair-rule expression. *Development* **120**, 1561-1572.

Dohle, W. (1976). Die Bildung und Differenzierung des postnauplialen Keimstreifs von *Diastylis rathkei* (Crustacea, Cumacea). II. Die Differenzierung und Musterbildung des Ektoderms. *Zoomorphologie* **84**, 235-277.

Dohle, W. and Scholtz, G. (1988). Clonal analysis of the crustacean segment: the discordance between genealogical and segmental borders. *Development* **104 Supplement,** 147-160.

Fleig, R. (1990). *Engrailed* expression and body segmentation in the honeybee *Apis millifera*. *Roux's Arch. Dev. Biol.* **198**, 467-473.

French, V. (1988). Gradients and insect segmentation. *Development* **104 Supplement**, 3-16.

Ingham, P. W. (1988). The molecular genetics of embryonic pattern formation in *Drosophila*. *Nature* **335**, 25-34.

Kelsh, R., Dawson, I. and Akam, M. (1993). An analysis of *Abdominal-B* expression in the locust *Schistocerca gregaria*. *Development* **117**, 293-305.

Kelsh, R., Weinzierl, R. O. J., White, R. A. H. and Akam, M. (1994). Homeotic gene expression in the locust *Schistocerca*: An antibody that detects conserved epitopes in Ultrabithorax and abdominal-A proteins. *Dev. Genet.* **15**, 19-31.

Lawrence, P. (1988). The present status of the parasegment. *Development* **104 Supplement**, 61-65.

Manzanares, M., Marco, R. and Garesse, R. (1993). Genomic organization and developmental pattern of expression of the *engrailed* gene from the brine shrimp *Artemia*. *Development* **118**, 1209-1219.

Martinez Arias, A. and Lawrence, P. A. (1985). Parasegments and compartments in the *Drosophila* embryo. *Nature* **313**, 639-642.

Martinez Arias, A. (1993). Development and patterning of the larval epidermis of *Drosophila*. In *The Development of Drosophila melanogaster* (ed. M. Bate and A. Martinez Arias), pp. 517-608. New York: Cold Spring Harbor Laboratory Press.

Mee, J. and French, V. (1986). Disruption of segmentation in a short germ insect embryo I. The localization of segmental abnormalities induced by heat shock. *J. Embryol. exp. Morph.* **96**, 245-266.

Nagy, L. M., Booker, R. and Riddiford, L. M. (1991). Isolation and embryonic expression of a *abdominal-A*-like gene from the lepidopteran, *Manduca sexta*. *Development* **112**, 119-129.

Nagy, L. M. and Carroll, S. (1994). Conservation of wingless patterning functions in the short-germ embryos of *Tribolium castaneum*. *Nature* **367**, 460-463.

Pankratz, M. J. and Jäckle, H. (1993). Blastoderm segmentation. In *The Development of Drosophila melanogaster* (ed. M. Bate and A. Martinez Arias), pp. 467-516. New York: Cold Spring Harbor Laboratory Press.

Patel, N. H., Martin-Blanco, E., Coleman, K. G., Poole, S. J., Ellis, M. C., Kornberg, T. B. and Goodman, C. S. (1989a). Expression of *engrailed* proteins in arthropods, annelids, and chordates. *Cell* **58**, 955-968.

Patel, N. H., Kornberg, T. B. and Goodman, C. S. (1989b). Expression of *engrailed* during segmentation in grasshopper and crayfish. *Development* **107**, 201-212.

Patel, N. H., Ball, E. E. and Goodman, C. S. (1992). Changing role of *even-skipped* during the evolution of insect pattern formation. *Nature* **357**, 339-342.

Patel, N. H. (1993). Evolution of insect pattern formation: a molecular analysis of short germband segmentation. In *Evolutionary Conservation of Developmental Mechanisms* (ed. A. C. Spradling), pp. 85-110. New York: Wiley-Liss, Inc.

Patel, N. H., Condron, B. G. and Zinn, K. (1994). Pair-rule expression patterns of *even-skipped* are found in both short and long germ beetles. *Nature* **367**, 429-434.

Sander, K. (1976). Specification of the basic body pattern in insect embryogenesis. *Adv. Insect. Physiol.* **12**, 125-238.

Sander, K., Gutzeit, H. O. and Jäckle, H. (1985). Insect embryogenesis: morphology, physiology, genetical, and molecular aspects. In *Comprehensive Insect Physiology, Biochemistry, and Pharmacology* (eds. G. A. Kerkut and L. I. Gilbert), pp. 319-385. New York: Pergamon Press.

Sander, K. (1988). Studies in insect segmentation: from teratology to phenogenetics. *Development* **104 Supplement**, 112-121.

Scholtz, G. (1992). Cell lineage studies in the crayfish *Cherax destructor* (Crustacea, Decapoda): germ band formation, segmentation, and early neurogenesis. *Roux's Arch. Dev. Biol.* **202**, 36-48.

Scholtz, G. Dohle, W., Sandeman, R. and Richter, S. (1993). Expression of engrailed can be lost and regained in cells of one clone in crustacean embryos. *Int. J. Dev. Biol.* **37**, 299-304.

Sommer, R. J. and Tautz, D. (1993). Involvement of an orthologue of the *Drosophila* pair-rule gene *hairy* in segment formation of the short germ band embryo of *Tribolium* (Coleoptera). *Nature* **361**, 448-450.

St Johnston, D. and Nüsslein-Volhard, C. (1992). The origin of pattern and polarity in the *Drosophila* embryo. *Cell* **68**, 201-220.

Stuart, J. J., Brown, S. J., Beeman, R. W. and Denell, R. E. (1993). The *Tribolium* homeotic gene *Abdominal* is homologous to *abdominal-A* of the *Drosophila* bithorax complex. *Development* **117**, 233-243.

Tear, G., Bate, C. M. and Martinez Arias, A. (1988). A phylogenetic interpretation of the patterns of gene expression in *Drosophila* embryos. *Development* **104 Supplement**, 135-146.

Tear, G., Akam, M. and Martinez Arias, A. (1990). Isolation of an *abdominal-A* gene from the locust *Schistocerca gregaria* and its expression during embryogenesis. *Development* **110**, 915-925.

Vincent, J. P. and O'Farrell, P. H. (1992). The state of *engrailed* expression is not clonally transmitted during early *Drosophila* development. *Cell* **69**, 923-931.

Development 1994 Supplement, 209-215 (1994)
Printed in Great Britain © The Company of Biologists Limited 1994

The evolving role of Hox genes in arthropods

Michael Akam, Michalis Averof, James Castelli-Gair, Rachel Dawes, Francesco Falciani and David Ferrier

Wellcome/CRC Institute and Department of Genetics, Tennis Court Road, Cambridge, CB2 1QR, UK

SUMMARY

Comparisons between Hox genes in different arthropods suggest that the diversity of Antennapedia-class homeotic genes present in modern insects had already arisen before the divergence of insects and crustaceans, probably during the Cambrian. Hox gene duplications are therefore unlikely to have occurred concomitantly with trunk segment diversification in the lineage leading to insects. Available data suggest that domains of homeotic gene expression are also generally conserved among insects, but changes in Hox gene regulation may have played a significant role in segment diversification. Differences that have been documented alter specific aspects of Hox gene regulation within segments and correlate with alterations in segment morphology rather than overt homeotic transformations.

The *Drosophila* Hox cluster contains several homeobox genes that are not homeotic genes – *bicoid*, *fushi-tarazu* and *zen*. The role of these genes during early development has been studied in some detail. It appears to be without parallel among the vertebrate Hox genes. No well conserved homologues of these genes have been found in other taxa, suggesting that they are evolving faster than the homeotic genes. Relatively divergent Antp-class genes isolated from other insects are probably homologues of *fushi-tarazu*, but these are almost unrecognisable outside of their homeodomains, and have accumulated approximately 10 times as many changes in their homeodomains as have homeotic genes in the same comparisons. They show conserved patterns of expression in the nervous system, but not during early development.

Key words: homeobox, homeotic gene, tagmosis, insect development evolution

INTRODUCTION

In *Drosophila*, the differences between the trunk segments are controlled by the homeotic genes of the Antennapedia and Bithorax gene complexes (Lewis, 1978; Sánchez-Herrero et al., 1985; Kaufman et al., 1990). This paper is concerned with the extent to which changes affecting these Hox genes may have contributed to the evolution of body form and developmental mechanism in the arthropods, and in particular the insects. We begin by considering the timing of gene duplications that gave rise to the Insect Hox cluster*, in relation to the origin of invertebrate phyla and arthropod classes. Then we consider how changes in the regulation of Hox genes may have contributed to the diversification of the insects. Finally, we present evidence that the selective constraints on some genes in the Hox clusters appear to have been relaxed in some insect lineages, allowing them to diverge relatively rapidly, both in terms of sequence and role in development.

*Here we shall use the term Hox genes to refer generically to the homeobox genes contained within the arthropod homeotic gene clusters, as well as their vertebrate homologues. The term Hom/Hox, never euphonic, is now redundant. The epithet Hox should no longer be applied to other classes of homeobox genes (Scott, 1992). We include within the 'Hox' designation the non-homeotic genes of the *Drosophila* Hox clusters (*bcd*, *ftz*, *zen*). The final section of this manuscript provides some justification for this usage. We use the term Antennapedia-class (Antp-class) Hox genes more restrictively, to refer to that subset of the Hox genes that have a homeobox conforming to the Antp-class consensus defined by Scott et al (1989). This includes all the central genes of the Hox clusters, but excludes the labial, proboscipedia, Deformed and Abdominal-B classes (vertebrate paralogy groups 1-4, 9-13).

HOX GENE DUPLICATIONS AND THE ORIGIN OF ARTHROPOD BODY PLANS

An explicit hypothesis relating Hox genes to evolutionary change was articulated by E. B. Lewis in the opening paragraph of his classic paper (Lewis, 1978):

> "During the evolution of the fly, two major groups of genes must have evolved: 'leg suppressing' genes which removed legs from abdominal segments of millipede like ancestors, followed by 'haltere-promoting' genes which suppressed the second pair of wings of four-winged ancestors...
> During evolution a tandem array of redundant genes presumably diversified by mutation to produce this complex." (the Bithorax Complex).

In essence, Lewis proposed that the evolutionary history of the homeotic gene clusters would be related to the evolution of morphology in the arthropod lineage, with increasing complexity of the gene cluster underlying increasing complexity of the body plan. No doubt he had in mind a scheme proposed by Snodgrass (1935), where the archetypal insect is derived from a superficially myriapod-like ancestor in which all the segments between the mouthparts and the anal segment were similar. In the parlance of arthropod morphologists, such an animal is homonomous; all the segments of the trunk form a single tagma, or body region. This contrasts with the regional specialisation of segment form and function that constitutes tagmosis, and characterises most arthropod classes.

Snodgrass' scheme assumes that in the proposed insect/myriapod lineage, a homonomous trunk is the primitive state, and tagmosis a derived character. At first sight this conflicts with the observation that most representatives of the other major group of mandibulate arthropods – the Crustacea – also display clear trunk tagmosis, and typically have distinct thorax and abdomen (Schram, 1986). However, conventional phylogenies argue that these analogous states have been acquired independently in the two lineages (Brusca and Brusca, 1990): homonomous Crustacea exist (the remipedes), and are assumed to reflect the condition ancestral to this whole class (Fig. 1).

It is not obvious why a homonomous, myriapod-like animal would need the array of homeotic genes that specify trunk segment diversity in *Drosophila*, most particularly the set of distinct Antp-class Hox genes that specify the thoracic and abdominal segments to be different from one another (*Antennapedia* (*Antp*), *Ultrabithorax* (*Ubx*) and *abdominal-A* (*abd-A*)). Indeed, we previously argued that the duplication leading to these particular genes may have occurred relatively late, in association with the origin of the thorax/abdomen distinction in insects (Akam et al., 1988).

One test of this model is to compare the structure and expression of Antp-class Hox genes in crustaceans and insects. If the mechanism to specify thorax and abdomen arose independently in these two lineages, we should see signs of this when we examine the molecular machinery. The first step in this comparison has been achieved (Averof and Akam, 1993). In a study of Hox gene sequences in the branchiopod crustacean *Artemia*, we have shown that specific homologues of *Antp*, *Ubx* and *abd-A*, as well as *Sex Combs Reduced* (*Scr*) and *Deformed* (*Dfd*) are present in Crustacea. Other, more divergent Hox gene classes (labial, proboscipedia, Abdominal-B) are known to be of even more ancient origin (Fig. 2; Schubert et al., 1993). We conclude that the existing diversity of homeotic genes in insects (or at least *Drosophila*) derives from gene duplications that occurred prior to the insect/crustacean split.

Arthropods recognisable as crustaceans are well documented among Cambrian fossils (Robison and Kaesler, 1987), including probable Branchiopods from the lower Cambrian, 550 Myr before present (Butterfield, 1994). In contrast, the earliest insect fossils appear significantly later, during the Devonian, 400 Myr before present (Jeram et al., 1990). These insects are likely to derive from an early crustacean-mandibulate clade, in which case the final diversification of the Antp-class Hox genes may have occurred within this arthropodan lineage. However, at least one of the relevant gene duplication events occurred earlier, before the separation of annelid and arthropod lineages: leeches possess a *Ubx/abd-A* type gene (*Lox2*), distinct from other Antp class sequences (*Lox 1*, *Lox5*; Wysocka-Diller et al., 1989; Nardelli-Haefliger and Shankland, 1992).

These findings argue against the strong version of the Lewis hypothesis – that the acquisition of trunk tagmosis in insects was directly correlated with the origin of new Hox genes. However, they do not rule out the possibility that significant changes in the regulation of pre-existing Hox genes may have played a crucial part in this transition. Further work will be required to establish if and how the Antp class Hox genes are used to define trunk segment identities in Crustacea, and in homonomous arthropods (e.g. myriapods).

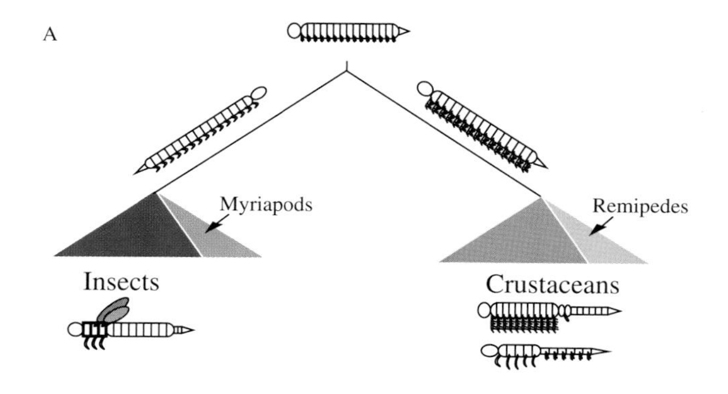

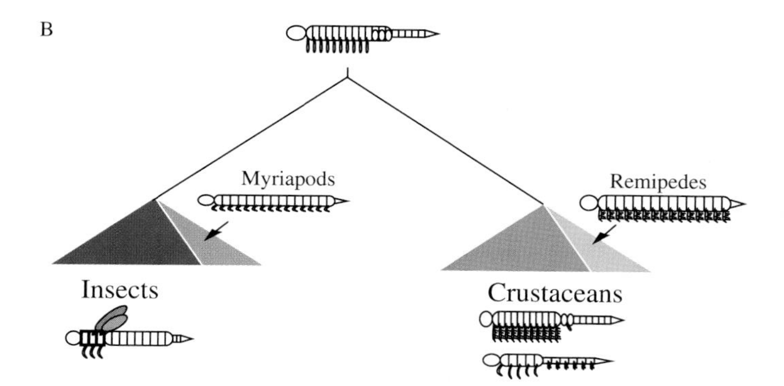

Fig. 1. 'Conventional' and 'alternative' hypotheses concerning the origin of trunk tagmosis in relation to arthropod phylogeny. (A) Conventional view depicting the common ancestor of insects and crustaceans with a homonomous trunk. Thorax/abdomen tagmosis is assumed to have arisen independently in the two lineages. Myriapods, assumed to constitute a sister group to the insects, and remipedes within the Crustacea, retain a homonomous trunk reflecting the primitive condition for this whole lineage (B) An alternative scenario in which insects and Crustacea derive from a common ancestor that already displays well differentiated trunk segments (perhaps controlled by the differential expression of multiple Antp-class Hox genes, which we infer to have been present in this ancestor). Those myriapods (some or all) that constitute a sister group to the insects, and remipedes among the Crustacea, would then show a secondarily simplified trunk. Note that the monophyly of the myriapods is uncertain, and the position of some myriapods within this lineage has recently been challenged: Sequence data (Turbeville et al., 1991; Ballard et al., 1992; Averof, 1994) and patterns of neural development (Whitington et al., 1991) suggest that many myriapods may form an outgroup to the whole lineage depicted here.

Even before these data are available, it is perhaps worth questioning the validity of one assumption on which the Lewis model is based, for the argument illuminates one contribution that developmental data may make to evolutionary discussions. This is the assumption that homonomy is always primitive. As developmental geneticists, we would emphasise that there is an enormous asymmetry between the invention of a complex feature (e.g. the transition from homonomy to heteronomy) and its loss (heteronomy→homonomy). An animal that has no

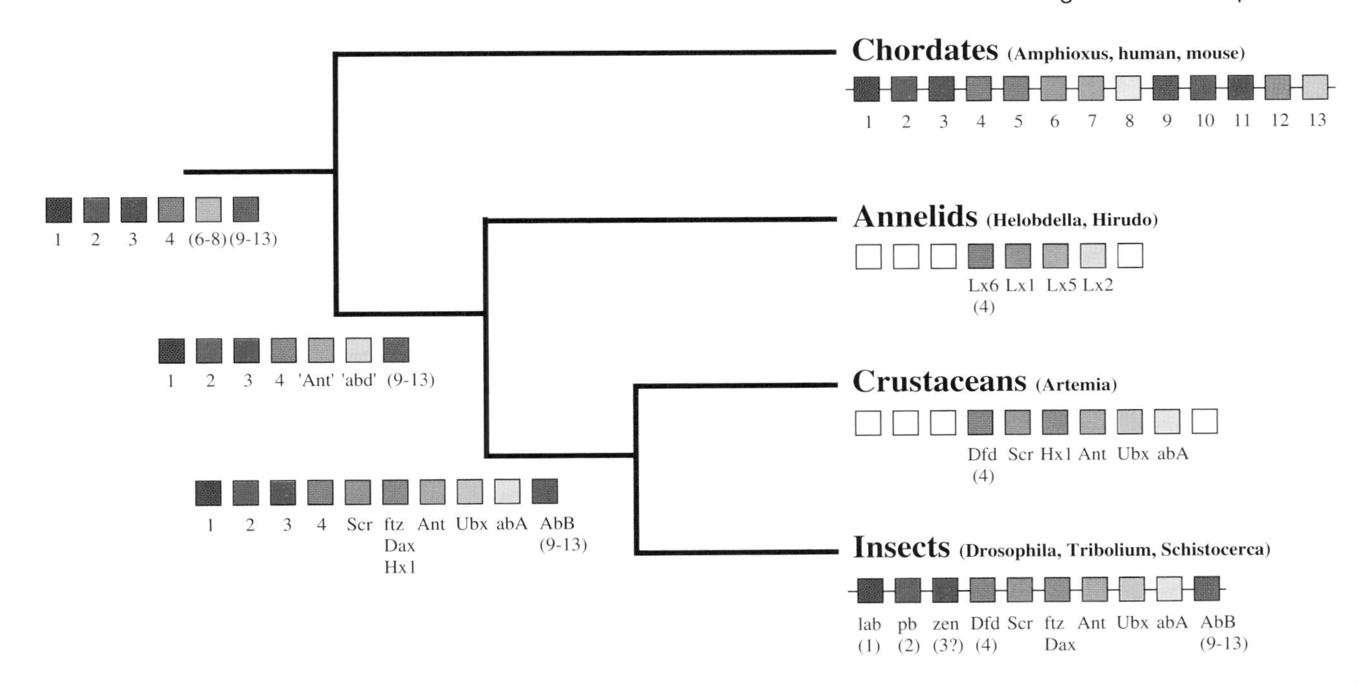

Fig. 2. The minimal inferred complexity of Hox gene clusters in the lineage leading to the insects. Panels on the right summarise the diversity of Hox genes described from insects, Crustacea, annelids (Class Hirudinea, leeches) and chordates (*Amphioxus*, and several vertebrates; a single 'complete' chordate cluster is illustrated, based on Amphioxus data; Garcia-Fernandez and Holland, 1994). Genes characterised by at least the full homeobox sequence are shown as coloured boxes. Open boxes in the annelid and crustacean panels are used for genes that are assumed to exist in these taxa, but have not yet been characterised adequately. Identical colours are used for genes that can be assigned unique homologues (orthologues) within the clusters of different taxa. Similar but non-identical colours are used for genes where orthology relationships are unclear. Boxes are joined only where linkage has been established in some member of the taxon. The assumed phylogenetic relationships of these four taxa are not disputed. Panels on the left show the minimal complexity of the Hox cluster in each of the stem groups that is implied by considering this phylogeny together with the assumed orthology relationships. A single Antp-class ancestor is shown for the basal stem group (though *Scr* and the genes of vertebrate paralogy group 5 may derive from a second ancestral Antp-class sequence present at the base of this lineage. The origin of these genes is not clear; Bürglin, 1994). Distinct *Antp*-like and *Ubx/abd-A*-like genes are shared by arthropods and annelids. All homeotic gene classes are shared by insects and Crustacea. The inclusion of a group 3 homologue in the arthropod lineages is based on unpublished data (see text), and on the isolation of a homeobox PCR fragment that has residues diagnostic for class 3 Hox genes from *Limulus*, a chelicerate arthropod (Cartwright et al., 1993). Gene symbols have been abbreviated as follows: Lx, *Lox* (leech Hox) genes; Ant, *Antp*; abd, *Ubx/abd-A* related; ab-A, *abd-A*; Ab-B, *Abd-B*. (References: Chordates: Bürglin, 1994; Garcia-Fernandez and Holland, 1994. Annelids: *Lox1*, Aisemberg and Macagno, 1994); *Lox2*,Wysocka-Diller et al., 1989; Nardelli-Haefliger and Shankland, 1992; Lox 5,6,7, Shankland et al., 1991. Crustaceans: Averof and Akam, 1993; see also legend to Fig. 4. Insects: For a summary of *Drosophila* gene sequences, see Bürglin, 1994; for other insects, see legend to Fig. 4.)

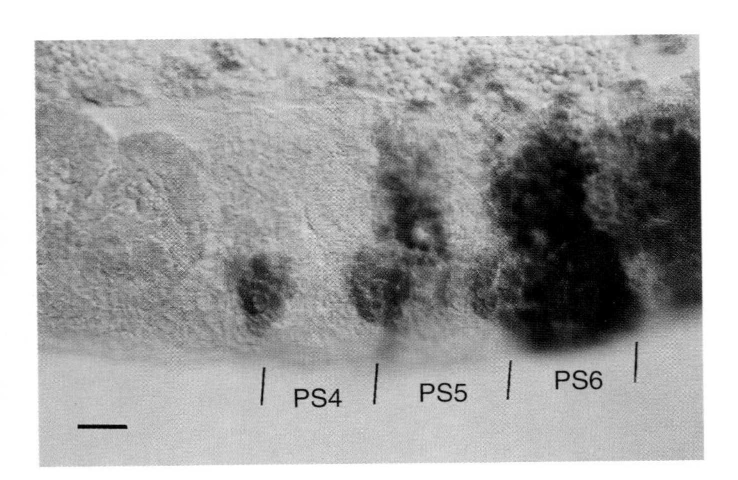

Fig. 3. Segment specification involves precise spatial regulation of homeotic genes in *Drosophila.* A region of the *Drosophila* embryo at extended germ band stage (maxillary to second abdominal segments) has been stained for protein produced by the *Ubx* gene (black). At this stage, homeotic genes are regulated in parasegmental domains.*Ubx* plays no role in segment specification in parasegment 4 (PS4; posterior first thoracic segment/anterior second). Near ubiquitous expression of *Ubx* characterises parasegment 6 (PS6, T3p/A1a), and can be shown to specify PS6 by experimental manipulation (Gonzalez-Reyes et al., 1990; Mann and Hogness, 1990). Spatially regulated expression of *Ubx* characterises parasegment 5 (PS5, T2p/T3A), and is necessary for its correct specification. In this preparation, counter staining for the product of a *distal-less* reporter gene (Vachon et al., 1992) marks the developing leg discs of the thoracic segments (brown). Scale, 10 μm.

developmental mechanism to make segments different must invent both the mechanism to specify the differences, and also the machinery to elaborate distinct, adaptive segment types. Loss of heteronomy can be achieved by what seem to us to be much simpler processes – by turning off the mechanism that specifies segments to be different, or simply by eliminating some tagma from the adult body plan altogether.

We do not seriously doubt the hypothesis that the first metamerically segmented ancestors of the arthropods were homonomous – though there is at present no way of knowing where to place that ancestor on the phylogenetic tree. We do question whether homonomy is generally a primitive trait among extant arthropods. In an environment where it is useful for all segments to have legs, this adaptive change may be relatively easy to achieve. A single Bx-C mutation will make a 'myriapod' of a fly (albeit not a viable one; Lewis, 1978), but we cannot conceive of the reverse transition occurring in a single step.

An analogous situation pertains to the origin and loss of insect wings. It takes many genes to make a wing (Williams and Carroll, 1993) yet, in *Drosophila*, we know of several single gene mutations that will suppress wing formation in an otherwise viable fly (*wingless, apterous, vestigial*; Lindsley and Zimm, 1992). Among the insects, where the phylogenetic context is much less ambiguous, wings are believed to have evolved only once; we do not hesitate to invoke convergence repeatedly to explain the origin of wingless taxa at many phylogenetic levels among the pterygotes, from whole orders (e.g. fleas, lice) to individual species (Kristensen, 1991).

If heteronomy is hard to achieve de novo, but useful and easy to modify once invented, we should expect to see two very different modes of evolution: clades that generate a diversity of taxa which retain essentially homonomous body plans, and clades that exploit heteronomy to fill many niches. Trilobites may exemplify the first mode; despite their abundance and taxonomic diversity throughout the Palaeozoic, virtually all trilobite families retain largely homonomous trunk segments (Eldridge, 1977). The Crustacea present a very different picture, with a Palaeozoic radiation giving rise to a huge diversity of body plans (Cisne, 1974; Schram, 1986; Briggs et al., 1992).

HOMEOTIC GENE REGULATION AND EVOLUTIONARY CHANGE WITHIN INSECTS

Hox gene duplications may have been a necessary precondition for segment diversification within the insects, but the results summarised above suggest that they were not an immediate driving force for its appearance. There is some evidence, however, that changes in the regulation of the Hox genes have occurred within insects, and may have contributed to morphological evolution. These changes are to be seen against a broadly conserved role for many Hox genes, at least in so far as can be inferred from the phenotypes of mutations in the Hox clusters of beetles and moths (Beeman et al., 1993; Tazima, 1964; Booker and Truman, 1989; Ueno et al., 1992), and from the patterns of expression of Hox genes in these and other insects (see below).

Most data are available for *abd-A*. In each of four species where the expression of *abd-A* has been examined (*Drosophila*, Karch et al., 1990; Macias et al., 1990; *Manduca*, Nagy et al., 1991; *Tribolium*, Stuart et al., 1993; *Schistocerca*, Tear et al., 1990), the gene is expressed throughout much of the abdomen, with no evidence for expression in the head or thorax. Moreover, in each case, the gene shows an abrupt anterior boundary of expression within the first abdominal segment. Where tested, this lies at the parasegmental boundary defined by *engrailed* expression, suggesting that, at least in this case, the precise co-ordination of segmentation and homeotic gene expression is maintained in insects as diverse as Orthoptera and Diptera (Tear et al., 1990).

In some other cases, the domains of homeotic gene expression are broadly conserved, but the precise limits of expression are not the same as those in *Drosophila*. For example, the *Abd-B* genes in *Tribolium* and *Schistocerca* appear to be expressed in the epidermis of only the most posterior abdominal segments (from A8p back; Kelsh et al., 1993; J. He & R. Denell, personal communication), whereas in *Drosophila Abd-B* is also expressed in the more anterior abdominal segments, A5p-A8a. This anterior expression of the gene, which is necessary for normal development, is regulated quite independently of that in the more posterior abdomen: It derives from a separate promoter and is activated through different regulatory elements (Boulet et al., 1991). Altered regulation of the *Abd-B* gene may have played a role in the evolution of the modified abdomen that is characteristic of the higher Diptera.

Comparisons between *Schistocerca* and *Drosophila* reveal one clear case of 'molecular heterochrony'. In *Drosophila*, *abd-A* is normally never expressed in the ninth or more posterior abdominal segments (parasegment 14). All of these segments fuse with A8 to form the 'terminalia'. In *Schistocerca, abd-A* is expressed at high levels in A9 and A10 during early embryogenesis, but this expression is rapidly lost, at the same time that *Abd-B* expression extends anteriorly (Tear et al., 1990; Kelsh et al., 1994). It seems most likely that this is due to the direct repression of the *abd-A* gene by Abd-B protein, for this same interaction has been demonstrated in *Drosophila*. In *Abd-B* mutants of *Drosophila*, *abd-A* expression extends posteriorly into A9 (Macias et al., 1990), and this segment now develops characteristics of the more anterior abdomen, including a denticle belt (Sánchez-Herrero et al., 1985). Thus a regulatory interaction that occurs as a temporal sequence after segment formation in *Schistocerca*, appears to be 'hard wired' at an earlier stage in *Drosophila* development, and is only apparent in mutants.

The two cases above suggest that alterations in the timing or extent of homeotic gene expression may be used to modify segment development – even though they do not result in striking 'homeotic' transformations. However, there are a number of observations in the entomological literature suggestive of homeotic transformations that have been fixed in the course of evolution. One that we are unlikely to be able to study in detail concerns the Monura – an extinct group of insects that flourished in the Carboniferous (Kukalova-Peck, 1991). These insects, in common with a number of other early groups, had reduced but well-formed legs (with terminal claws) on each abdominal segment A1 to A10: thorax-abdomen tagmosis was not complete. However, in contrast with all other known insects, in the Monura, these legs were also present on the terminal abdominal segment, in place of the filiform cerci that are the usual appendages of A11. The presumed sister group (Paleodictyoptera) and outgroups (Thysanura, Symphyla) to the Monura had cerci, so it is possible that a homeotic transformation of cerci to leglets occurred in this particular order. Alternatively, of course, this 'simple' character state may represent a retained primitive feature that has been lost several times in other groups.

The case of the Strepsipteran haltere is more amenable to experimental analysis. The Strepsiptera are a small order of parasitic insects. The adult females are neotenic, retaining larval morphology. It is the males that are interesting for our purposes (Whiting and Wheeler, 1994). They have only a single pair of wings, but these are developed from the third thoracic segment, not the second, as in the Diptera. The dorsal appendages of the second thoracic segment are reduced to halteres, which in many details resemble the halteres derived from the third thoracic segment of Diptera. Until recently, this resemblance has been ascribed to convergence, for the Strepsiptera have been assumed to be only distantly related to the Diptera. Ribosomal DNA sequence data now provides strong evidence that the Strepsiptera are in fact the sister group to the Diptera. This prompted Whiting and Wheeler to suggest that the Strepsipteran haltere is indeed homologous to the haltere of Diptera, and that this developmental pathway has been adopted in T2 of the Strepsiptera, presumably by a change in the segmental regulation of *Ubx*. This hypothesis is open to test, using a recently described antibody that cross-reacts with Ubx protein in a wide range of insect species (Kelsh et al., 1994).

However, we should be cautious in making such predictions. This dramatic modification of adult segment morphology may depend entirely on the regulation of genes acting downstream of the homeotics. Even if it does involve modification of homeotic gene expression, it may involve only their regulation in a subset of tissues at a particular stage in development (e.g. in the developing imaginal discs). It is not necessary to assume that evolutionary change affecting the homeotic genes must involve major, saltatory change in the morphology of whole segments. It may much more frequently proceed by incremental changes affecting levels of expression, or by the alteration of those enhancer elements within their promoters that control activity in particular structures within a segment.

This view is supported by careful studies of the role of one homeotic gene, *Ubx*, in *Drosophila* (J. Castelli-Gair and M. Akam, unpublished data). The *Ubx* gene is responsible for defining the identities of two quite different (para)segments, parasegment 5 (T3) and parasegment 6 (A1). This differential function does not appear to depend on the segment-specific expression of different protein variants (Busturia et al., 1990). Rather it seems to involve the precise temporal and spatial regulation of the gene in parasegment 5 (Fig. 3). Examples of how changes in the precise pattern of within-segment regulation may have contributed to insect segment diversity are provided by the work of Carroll et al. (this volume). They show, for example, that the loss of expression of *abd-A* specifically within some limb buds may allow the development of larval abdominal prolegs in Lepidoptera.

HOX GENES THAT GOT AWAY

Up to this point, we have taken for granted that Hox genes cloned from any insect can be identified without question as the specific homologue (orthologue) of one of the Hox cluster genes defined in *Drosophila*. In general, this is the case. With few exceptions, the homeodomains are almost identical, and gene-specific sequences around the homeodomain and extending upstream to the hexapeptide motif (Bürglin, 1994) are sufficiently similar to make identification unambiguous.

This is true, however, only for homologues of the 'homeotic' genes within the *Drosophila* clusters (*labial, proboscipedia, Dfd, Scr, Antp, Ubx, abd-A* and *Abd-B*). Well-conserved homologues have not been isolated for the several homeobox genes of the Antp complex that have 'anomalous' roles in development. There are four such genes in *Drosophila melanogaster* – *bicoid, fushi-taruzu* and the two *zen* genes.

Bicoid is only expressed maternally. It encodes a gradient morphogen in the early (syncytial) embryo (Driever and Nüsslein-Volhard, 1988). *Fushi-taruzu* (*ftz*) is a 'pair-rule' segmentation gene (Kaufman et al., 1990). It is re-used during neurogenesis to specify the identity of certain neurons in each trunk segment (Doe et al., 1988). *Zen1* is required very early for the specification of the dorsal pattern elements in the embryo (Ferguson and Anderson, 1991). None of these roles is analogous to that of the typical homeotic genes, or to those of the Hox genes in vertebrates, so far as we know. The existence of these genes within the insect Hox clusters has been something of an enigma, but isolation of several 'divergent' Hox genes from other insects throws some light on their origin.

The best studied of these are putative homologues of *fushi-tarazu*, isolated from *Tribolium* (Brown et al., 1994) and *Schistocerca* (Dawes et al., 1994). Neither of these genes shows extensive similarity to *ftz* (or any other Hox gene) outside of the homeobox, and even within the homeobox they are much more divergent than is typical for the other Hox genes (Fig. 4). However, both of these genes are closely linked to the Hox cluster, and both show a pattern of expression in the nervous system very similar to that of the *fushi-tarazu* gene in *Drosophila*.

Outside of the homeodomain, short conserved peptides are shared uniquely between *Drosophila ftz* and the *Tribolium ftz*-like gene, and between the *Tribolium* and *Schistocerca* genes. In the latter case, the conserved region is centred on a YPWM sequence just upstream of the homeodomain. This so-called hexapeptide motif is characteristic of the homeotic Hox genes, and generally well conserved between distant species, often showing extended similarity between insect and vertebrate genes of the same class. Its presence in the *Schistocerca* and *Tribolium* genes strongly suggests that, on structural grounds, these should be regarded as 'typical' Hox genes. However, note that no such motif is conserved in the *Drosophila ftz* gene.

In contrast to the conserved expression of *ftz* in the nervous system, these putative *ftz* homologues show different patterns of expression during early development. *Drosophila ftz* is expressed in parasegmental stripes at two-segment intervals, and is needed for the formation of alternate segment boundaries (Lawrence and Johnston, 1989). In *Tribolium*, the *ftz* gene is expressed in less well resolved stripes, and is apparently not needed for making the correct number of segments (Stuart et al., 1991; Brown et al., 1994). The *Schistocerca* gene is expressed in a posterior domain of the early embryo, but never in stripes (Dawes et al., 1994); its function in early development is unknown.

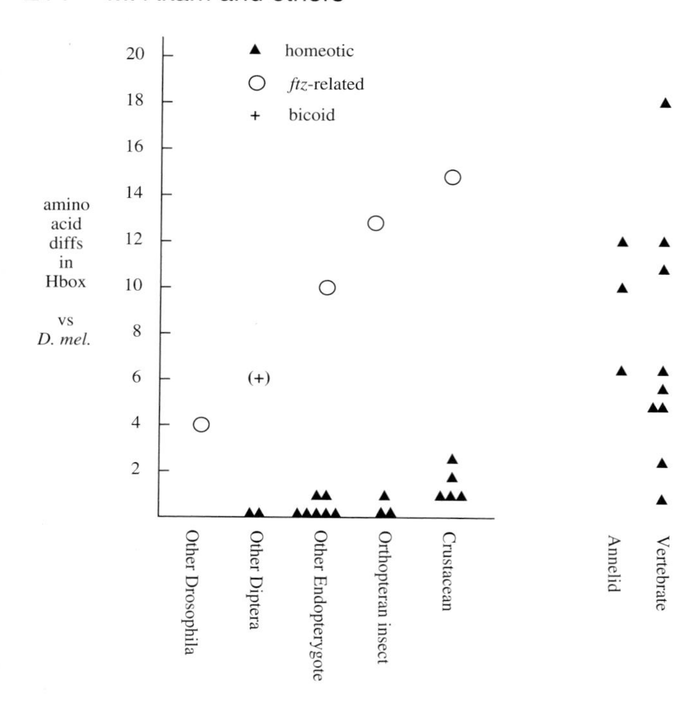

Fig. 4. Relative divergence of homeotic and other Hox cluster homeodomains within Arthropods. Species for which Hox gene sequences are available have been grouped according to approximate phylogenetic distance from *Drosophila melanogaster*. Rough estimates of divergence times are given below, based on immunological similarity (within Diptera; Beverley and Wilson, 1984)) or on the first appearance of other taxonomic groups in the fossil record (Kukalova-Peck, 1991). Other *Drosophila* subgenera, (40-60 My): *D. hydei ftz* (Jost and D. Maier, personal communication; EMBL accession X79494); other Dipteran family/suborder, (100-200My): *Musca bicoid* (n.b. 6 changes in 44 residues of homeodomain sequenced (Sommer and Tautz, 1991); *Aedes abd-A* (Eggleston, 1992); other Endopterygote insects (250My): Bombyx (Lepidoptera) *Ubx, abd-A* (Ueno et al., 1992); *Manduca* (Lepidoptera) *abd-A* (Nagy et al., 1991); *Apis* (Hymenoptera) *Antp, Scr, Dfd* (Fleig et al., 1988; Walldorf et al., 1989); *Tribolium* (Coleoptera) *abd-A* (Stuart et al., 1993), *ftz* (Brown et al., 1994). *Schistocerca* (Orthoptera, approx. 300My) *abd-A* (Tear et al., 1990), *Abd-B* (Kelsh et al., 1993), *Scr* (Akam et al., 1988), *Dax* (*ftz*-related gene (Dawes et al., 1994); crustacean: *Artemia* (Anostraca, 550 My; Averof and Akam, 1993). The assignment of the divergent Artemia gene *AfHx1* to the *ftz* homology group is tentative, but is consistent with the fast divergence of these genes. For comparison, the divergence between some annelid or vertebrate Hox sequences and their closest *Drosophila* match are shown on the right.

We believe that all of these *ftz*-related genes derive from a single *Antp*-class Hox gene that acquired novel functions in arthropod development. They are not involved in the maintenance of segmental identity in the mesoderm or the epidermis. They may have a role in very early patterning of the embryo in all insects, but this role appears to be different in different insects. In *Schistocerca* and *Tribolium*, whatever constraints maintain the integrity of the YPWM motif and its flanking sequences are retained, but the function of *ftz* in *Drosophila* appears no longer to impose the same sequence requirement. Because the changes in the homeobox are confined to regions other than those that contact DNA, we suggest that it may be the lower complexity of protein-protein interactions that allows *ftz* to diverge more rapidly than the canonical homeotic genes.

The other 'non-homeotic' genes of the *Drosophila* Ant-C are in some ways analogous to *ftz*. Their homeoboxes are very divergent, with respect to all other non-Dipteran homeobox genes; they lack any recognisable hexapeptide motif; comparison between Dipteran species shows that the *bicoid* homeobox is diverging rapidly (Sommer and Tautz, 1991; Schröder and Sander, 1993). Activity capable of rescuing bicoid mutants can be recovered from the eggs of other cyclorapphan Diptera, but was not detected in honeybee, or even lower Diptera (Schröder and Sander, 1993). Searches for *bicoid* genes in other insects have been unsuccessful (I. Dawson, S. Frenk and M. A., unpublished results).

We think that these genes may be derived, like *ftz*, from Hox genes which, in the lineage leading to *Drosophila*, have escaped from the conservative selection that characterises homeotic genes. We have recently isolated from *Schistocerca* a cDNA that lends some support to this hypothesis. It encodes a homeodomain protein that shows some similarities to the *proboscipedia* and *Zen* genes, but has no close homologue in *Drosophila*. It resembles most closely Hox class 3 vertebrate sequences (F. F., unpublished results). As yet we know nothing of the function or expression of this gene in *Schistocerca*, but we guess that it may derive from an ancestral sequence that gave rise, after considerable divergence, to the *zen* and/or *bicoid* genes of *Drosophila*.

Work from the authors laboratory was supported by the Wellcome Trust. We thank Nick Butterfield, Dieter Maier, Sue Brown and Rob Denell for communicating results prior to publication; Geoffrey Fryer and Hans-Georg Frohnhöfer for their advice at the early stages of our Crustacean work, Peter Holland and David Stern for valuable comments on the manuscript.

REFERENCES

Aisemberg, G. O. and Macagno, E. R. (1994). *Lox1*, an Antennapedia-Class homeobox gene is expressed during leech gangliogenesis in both transient and stable central neurons. *Dev. Biol.* **161**, 455-465.

Akam, M., Dawson, I. and Tear, G. (1988). Homeotic genes and the control of segment identity. *Development* **104 Supplement**, 123-133.

Averof, M. (1994). HOM/Hox Genes of a Crustacean; Evolutionary Implications. Ph.D. thesis, University of Cambridge.

Averof, M. and Akam, M. (1993). HOM/Hox genes of *Artemia*: implications for the origin of insect and crustacean body plans. *Current Biology* **3**, 73-78.

Ballard, J. W. O., Olsen, G. J., Faith, D. P., Odgers, W. A., Rowell, D. M. and Atkinson, P. W. (1992). Evidence from 12S ribosomal RNA sequences that onychophorans are modified arthropods. *Science* **258**, 1345-1348.

Beeman, R. W., Stuart, J. J., Brown, S. J. and Denell, R. E. (1993). Structure and function of the homeotic gene complex (HOM-C) in the beetle *Tribolium castaneum*. *BioEssays* **15**, 439-444.

Beverley, S. M. and Wilson, A. C. (1984). Molecular evolution in *Drosophila* and the higher Diptera.2. A time scale for fly evolution. *J. Mol. Evol.* **21**, 1-13.

Booker, R. B. and Truman, J. W. (1989). *Octopod*, a homeotic mutation of the moth *Manduca sexta*, influences the fate of identifiable pattern elements within the CNS. *Development* **105**, 621-629.

Boulet, A. M., Lloyd, A. and Sakonju, S. (1991). Molecular definition of the morphogenetic and regulatory functions and the *cis*-regulatory elements of the *Drosophila Abd*-B homeotic gene. *Development* **111**, 393-406.

Briggs, D. E. G., Fortey, R. A. and Wills, M. A. (1992). Morphological disparity in the Cambrian. *Science* **256**, 1670-1673.

Brown, S. J., Hilgenfeld, R. B. and Denell, R. E. (1994). The beetle *Tribolium castaneum* has a *fushi tarazu* homolog expressed in stripes during segmentation. *Proc. Nat. Acad. Sci. USA* (in press).

Brusca, R. C. and Brusca, G. J. (1990). *Invertebrates*. Massachusetts: Sinauer.

Bürglin, T. R. (1994). A comprehensive classification of homeobox genes. In *Guidebook to the Homeobox Genes* (ed. D. Duboule), pp. 25-71. Oxford: Oxford University Press.

Busturia, A., Vernos, I., Macias, A., Casanova, J. A. and Morata, G. (1990). Different forms of Ultrabithorax proteins generated by alternative splicing are functionally equivalent. *EMBO J.* **9**, 3551-3555.

Butterfield, N. J. (1994). Burgess Shale-type fossils from an early Cambrian shallow-shelf sequence, northwestern Canada. *Nature* **369**, 477-479.

Carroll, S. B. (1994). Developmental regulatory mechanisms in the evolution of insect diversity. *Development* **1994 Supplement**,

Cartwright, P., Dick, M. and Buss, L. W. (1993). Hom/Hox type homeoboxes in the Chelicerate *Limulus polyphemus*. *Mol. Phylogenet. Evol.* **2**, 185-192.

Cisne, J. L. (1974). Evolution of the world fauna of aquatic free-living arthropods. *Evolution* **28**, 337-366.

Dawes, R., Dawson, I., Falciani, F., Tear, G. and Akam, M. (1994). *Dax*, a Locust Hox gene related to *fushi-tarazu* but showing no pair-rule expression. *Development* **120**, 1561-1572.

Doe, C. Q., Hiromi, Y., Gehring, W. J. and Goodman, C. S. (1988). Expression and function of the segmentation gene *fushi tarazu* during *Drosophila* neurogenesis. *Science* **239**, 170-175.

Driever, W. and Nüsslein-Volhard, C. (1988). The bicoid protein determines position in the *Drosophila* embryo in a concentration-dependent manner. *Cell* **54**, 95-104.

Eggleston, P. (1992). Identification of the *abdominal-A* homologue from *Aedes aegypti* and structural comparisons among related genes. *Nucleic Acids Research* **20**, 4095-4096.

Eldridge, N. (1977). Trilobites and evolutionary patterns. In *Patterns of evolution, as illustrated by the fossil record* (ed. A. Hallam), pp. 305-332. Amsterdam: Elsevier.

Ferguson, E. L. and Anderson, K. Y. (1991). Dorsal-ventral pattern formation in the *Drosophila* embryo – the role of zygotically active genes. *Current Topics Dev. Biol.* **25**, 17-43.

Fleig, R., Walldorf, U., Gehring, W. J. and Sander, K. (1988). In situ localisation of the transcripts of a homeobox gene in the honeybee *Apis mellifora* L. (Hymenoptera). *Roux's Arch. Dev. Biol.* **197**, 269-276.

Garcia-Fernandez, J. and Holland, P. W. H. (1994). Archetypal organization of the *Amphioxus Hox* gene cluster. *Nature* **370**, 563-566.

Gonzalez-Reyes, A., Urquia, N., Gehring, W., Struhl, G. and Morata, G. (1990). Are cross regulatory interactions between homeotic genes functionally significant? *Nature* **344**, 78-80.

Jeram, A. J., Selden, P. A. and Edwards, D. (1990). Land animals in the Silurian: arachnids and myriapods from Shropshire, England. *Science* 658-669.

Karch, F., Bender, W. and Weiffenbach, B. (1990). *abdominal-A* expression in *Drosophila* embryos. *Genes Dev.* **4**, 1573-1588.

Kaufman, T. C., Seeger, M. A. and Olsen, G. (1990). Molecular and genetic organisation of the Antennapedia complex of *Drosophila melanogaster*. In *Genetic Regulatory Hierarchies in Development* (ed. T. R. F. Wright), pp. Academic Press.

Kelsh, R., Dawson, I. A. and Akam, M. (1993). An analysis of *Abdominal-B* expression in the locust *Schistocerca gregaria*. *Development* **117**, 293-305.

Kelsh, R., Weinzierl, R. O. J., White, R. A. H. and Akam, M. (1994). Homeotic gene expression in the Locust *Schistocerca*: an antibody that detects conserved epitopes in Ultrabithorax and abdominal-A proteins. *Developmental Genetics* **15**, 19-31.

Kristensen, N. P. (1991). Phylogeny of extant hexapods. In *The Insects of Australia* (ed. CSIRO. Div. of Entomology), pp. 125-140. Melbourne: Melbourne University Press.

Kukalova-Peck, J. (1991). Fossil history and the evolution of hexapod structures. In *The Insects of Australia* (ed. CSIRO Div. of Entomology), pp. 141-179. Melbourne: Melbourne Univ. Press.

Lawrence, P. A. and Johnston, P. (1989). Pattern formation in the *Drosophila* embryo: allocation of cells to parasegments by *even-skipped* and *fushi tarazu*. *Development* **105**, 761-769.

Lewis, E. B. (1978). A gene complex controlling segmentation in *Drosophila*. *Nature* **276**, 565-570.

Lindsley, D. L. and Zimm, G. G. (1992). *The genome of Drosophila melanogaster*. San Diego: Academic Press.

Macias, A., Casanova, J. and Morata, G. (1990). Expression and regulation of the *abd-A* gene of *Drosophila*. *Development* **110**, 1197-1207.

Mann, R. S. and Hogness, D. S. (1990). Functional dissection of Ultrabithorax proteins in *D. melanogaster*. *Cell* **60**, 597-610.

Nagy, L. M., Booker, R. and Riddiford, L. M. (1991). Isolation and embryonic expression of an *abdominal-A*-like gene from the lepidopteran, *Manduca sexta*. *Development* **112**, 119-131.

Nardelli-Haefliger, D. and Shankland, M. (1992). *Lox2*, a putative leech segment identity gene, is expressed in the same segmental domain in different stem cell lineages. *Development* **116**, 697-710.

Robison, R. A. and Kaesler, R. L. (1987). Phylum Arthropoda. In *Fossil Invertebrates* (ed. R. S. Boardman, A. H. Cheetham and A. J. Rowell), pp. 205-269. Palo Alto, CA: Blackwell Scientific.

Sánchez-Herrero, E., Vernós, I., Marco, R. and Morata, G. (1985). Genetic organisation of the *Drosophila* bithorax complex. *Nature* **313**, 108-113.

Schram, F. R. (1986). *Crustacea*. Oxford: University Press.

Schröder, R. and Sander, K. (1993). A comparison of transplantable *bicoid* activity and partial *bicoid* homeobox sequences in several *Drosophila* and blowfly species (Calliphoridae). *Roux's Arch. Dev. Biol.* **203**, 34-43.

Schubert, F. R., Nieselt-Struwe, K. and Gruss, P. (1993). The Antennapedia-type homeobox genes have evolved from three precursors separated early in metazoan evolution. *PNAS* **90**, 143-147.

Scott, M. P. (1992). Vertebrate Homeobox gene nomenclature. *Cell* **71**, 551-553.

Scott, M. P., Tamkun, J. W. and Hartzell, G. W. I. (1989). The structure and function of the homeodomain. *Biochim. Biophys. Acta* **989**, 25-48.

Shankland, M., Martindale, M. Q., Nardelli-Haefliger, D., Baxter, E. and Price, D. J. (1991). Origin of segmental identity in the leech nervous system. *Development* **Supplement 2**, 29-38.

Snodgrass, R. E. (1935). *Principles of Insect Morphology*. London and New York: McGraw-Hill.

Sommer, R. and Tautz, D. (1991). Segmentation gene expression in the housefly *Musca domestica*. *Development* **113**, 419-30.

Stuart, J. J., Brown, S. J., Beeman, R. W. and Denell, R. E. (1991). A deficiency of the homeotic complex of the beetle *Tribolium*. *Nature* **350**, 72-74.

Stuart, J. J., Brown, S. J., Beeman, R. W. and Denell, R. E. (1993). The *Tribolium* homeotic gene *Abdominal* is homologous to *abdominal-A* of the *Drosophila* bithorax complex. *Development* **117**, 233-243.

Tazima, Y. (1964). *The Genetics of the Silkworm*. London: Logos Press.

Tear, G., Akam, M. and Martinez-Arias, A. (1990). Isolation of an *abdominal-A* gene from the locust *Schistocerca gregaria* and its expression during early embryogenesis. *Development* **110**, 915-926.

Turbeville, J. M., Pfeifer, D. M., Field, K. G. and Raff, R. A. (1991). The phylogenetic status of arthropods as inferred from 18S rRNA sequences. *Mol. Biol. Evol.* **8**, 669-686.

Ueno, K., Hui, C.-C., Fukuta, M. and Suzuki, Y. (1992). Molecular analysis of the deletion mutants in the E homeotic complex of the silkworm *Bombyx mori*. *Development* **114**, 555-563.

Vachon, G., Cohen, B., Pfeifle, C., McGuffin, M. E., Botas, J. and Cohen, S. M. (1992). Homeotic genes of the bithorax complex repress limb development in the abdomen of the *Drosophila* embryo through target gene *Distal-less*. *Cell* **71**, 437-450.

Walldorf, U., Fleig, R. and Gehring, W. J. (1989). Comparison of homeobox-containing genes of the honeybee and *Drosophila*. *Proc. Natl. Acad. Sci. USA* **86**, 9971-9975.

Whiting, M. F. and Wheeler, W. C. (1994). Insect homeotic transformation. *Nature* **368**, 696.

Whitington, P. M., Meier, T. and King, P. (1991). Segmentation, neurogenesis and formation of early axonal pathways in the centipede, *Ethmostigmus rubripes* (Brandt). *Roux's Arch Dev Biol* **199**, 349-363.

Williams, J. A. and Carroll, S. B. (1993). The origin, patterning and evolution of insect appendages. *BioEssays* **15**, 567-577.

Wysocka-Diller, J. W., Aisemberg, G. O., Baumgarten, M., Levine, M. and Macagno, E. R. (1989). Characterization of a homologue of bithorax complex genes in the leech *Hirudo medicinalis*. *Nature* **341**, 760-763.

Development 1994 Supplement, 217-223 (1994)
Printed in Great Britain © The Company of Biologists Limited 1994

Developmental regulatory mechanisms in the evolution of insect diversity

Sean B. Carroll

Howard Hughes Medical Institute, and Laboratory of Molecular Biology, University of Wisconsin-Madison, 1525 Linden Drive, Madison, WI 53706, USA

SUMMARY

The major architectural differences between most Arthropod classes and orders involve variations in the number, type and pattern of body appendages. We have utilized the emerging knowledge of appendage formation in fruit flies to begin to address the developmental and genetic basis of morphological diversity among insects. Butterflies, for example, differ from fruit flies in possessing larval abdominal limbs, two pairs of adult wings, and a sophisticated system of wing color pattern formation. We have found that the genetic bases for these three major morphological features involve differences between flies and butterflies at three levels of genetic regulation during development.

First, we show that the presence of abdominal limbs in butterflies is associated with striking changes in the regulation of specific homeotic genes in the abdominal segments of the butterfly embryo. Second, we suggest that the two-winged state of the fruit fly and the distinct pattern of the butterfly hindwing are the consequence of many accumulated changes in the target genes regulated by the *Ultrabithorax* homeotic gene. And finally, we demonstrate that a new genetic program, involving many of the same genes that specify the conserved global patterning coordinates of fruit fly and butterfly wings, has been superimposed onto the butterfly wing to create their unique color patterning system. These findings demonstrate how morphological diversity arises from the different ways in which conserved sets of regulatory genes are deployed during development.

Key words: appendages, homeotic genes, butterflies, color patterns

> "...remember that we are still at the beginning, that the complexity of the problem of Specific Difference is hardly less now than it was when Darwin first shewed that Natural History is a problem and no vain riddle."
>
> William Bateson, preface to *Materials for the Study of Variation* (1894)

William Bateson's book, in which he catalogued and interpreted a dazzling variety of nature's oddities and coined the term "homoeosis", marked one of the first post-Darwinian inquiries into the relationship between heredity, development, and morphological evolution. Bateson, and later biologists such as Goldschmidt (1940) who attempted similar syntheses, were lacking crucial information about mechanisms of inheritance and animal development. Now, a century later, with our mastery of genetics and rapidly expanding knowledge of embryology, the long anticipated integration of these disciplines with evolution appears within reach. Molecular genetics has provided the means to understand both gene function during development and to explore evolutionary relationships (Raff, 1992). We are now in a position to expand beyond the standard model organisms, which represent just a few phyla, in order to ask specific questions about the developmental mechanisms underlying morphological evolution.

The past decade's progress in dissecting animal development in genetically or experimentally favorable model species and recent forays into comparative molecular embryology have uncovered two general trends. First, the types of molecules involved in cell interactions, signal transduction, and gene regulation have been conserved among most metazoans. And second, many of the molecular pathways connecting these processes during pattern development in diverse organisms have been remarkably conserved. These sometimes startling similarities between animals are satisfying, in terms of identifying common elements of animal development, but also a great challenge in that we now must ask, 'how are these common batteries of gene families and signalling pathways used to create animal diversity?' To Allan Wilson, the answer was plainly evident two decades ago. Namely, that the essence of morphological diversity is based on changes in the *mechanisms controlling the expression of genes* rather than sequence changes in proteins (Wilson et al., 1974; King and Wilson, 1975; Wilson, 1985). Therefore, not only should we identify which genes function in morphogenesis, but also decipher how their regulation and function contributes to diversity.

The fruit fly *Drosophila melanogaster* has yielded critical information about the genetic basis of morphogenesis. The past decade's work on this animal has allowed us to draw the crucial connections between phenotype and gene function and regulation during development. The visualization of the chemical changes (i.e. gene expression) in cells that precede the overt formation of pattern has revolutionized our perspective and approach to embryogenesis. While the discovery of particular conserved sequences (e.g. homeoboxes) has provided an important dimension to understanding animal relationships, the direct inspection and perturbation of gene function during development has painted a rich picture that forms the foundation of a new comparative embryology. Now that we have considerable genetic, molecular and developmental understanding of oogenesis, segmentation, neurogenesis, homeosis, organogenesis, appendage formation and cell differentiation from fruit flies, evolutionary aspects of these processes can be addressed.

However, given the emerging picture of molecular conservation, the tricky question becomes how to study *diversity*. For instance, we know a fair amount about bristle formation and patterning in flies. However, the Dipteran bristle pattern is so static it offers little hope of finding any meaningful mechanisms of diversity. The same may be said largely of the insect eye, the development of which in *Drosophila* is becoming increasingly well understood, but what little variation there may be in insect retinal cell architecture is of dubious significance. It follows then that we need some criteria for choosing topics and organisms to study. First, the specific difference in question must be of obvious functional consequence. Second, we should have a reasonably deep grasp of the formal genetic logic and identity of the molecules involved in the formation of the structure, trait, or pattern in a reference species (for our purposes here - fruit flies). Finally, we should focus on animals for comparison whose molecular genetics and embryology present no obvious barriers or better yet, afford distinct experimental advantages (e.g. husbandry, some genetics, generation time, all relevant stages of development easy to obtain, etc.).

We have chosen to work with butterflies for precisely these reasons. Several major body characters differ in obvious ways between flies and butterflies, particularly with respect to their appendages and the cell types found in them (Table 1). Compared with other insects, they are fairly closely related to flies, and therefore gene homologs may be readily obtained. Finally, there is a reasonable knowledge of Lepidopteran embryology (Nagy et al., 1994).

We were also drawn to butterflies out of interest in the development and evolution of appendages, especially insect wings. The analysis of wing morphogenesis in *Drosophila* is one of the few genetically accessible models for understanding pattern formation in a growing cellular field and is providing a general framework for the analysis of wing development and evolution (Williams and Carroll, 1993). Given that the wing evolved in the course of insect evolution (see Kukalova-Peck, 1978 for a fascinating discussion), comparative analysis of wing patterning genes will allow us to explore the origin of this adaptation. In addition, butterflies differ greatly in their wing patterns. Analysis of the development of these wing color patterns may both shed light on an important but unsolved dimension of pattern formation (coloration) and create the opportunity to investigate patterning differences at the species level. Here, I will review the results of our initial investigation of butterfly appendage formation and patterning, in which we have found that three order-specific differences between butterflies and fruit flies can be correlated with three types of genetic regulatory differences in development.

Table 1. Order-specific differences between fruit flies and butterflies

Character	*Drosophila*	*Precis coenia*
Limb number	No abdominal limbs	4 pairs of larval abdominal prolegs, anal prolegs
Wing number, type	Two, halteres on T3	Four, hindwing large
Wing pattern	Unpigmented	Elaborate color pattern
Wing cell types	Bristles, epithelia	Scales (modified bristle?), epithelia

GENERAL APPROACH

To investigate specific differences between fruit flies and butterflies, we had to choose a suitable butterfly species for initial study. We decided on the Buckeye (*Precis coenia*: Nymphalidae) because there has been considerable investigation into the development of its wing color pattern (Nijhout, 1980a,b) and this species can be mass-reared in population cages on a simple diet. As a start, we constructed wing and embryonic cDNA libraries and developed embryonic and imaginal in situ hybridization and immunohistochemical procedures for *Precis coenia*. It is expected that most butterfly groups that will be of comparative interest in the future have been separated by less than 100-150 Myr, with some genera such as the tropical *Heliconius* representing much more recent radiations (Sheppard et al., 1985). Based upon the experiences of many investigators, including ourselves, in analyzing gene expression in different Diptera that are 80-100 Myr divergent (often utilizing antibodies to *D. melanogaster* proteins), we are hopeful that nucleic acid probes and/or antibodies for *P. coenia* gene products will provide tools for exploring butterfly relationships.

With libraries constructed, we cloned various butterfly homologs of *Drosophila* genes involved in appendage formation, segment identity, segment polarity, and intercellular signalling (Table 2). Of the genes shown in Table 2, all but *wingless* (cloned by PCR) were cloned by cross-hybridization with *Drosophila* probes. In only one instance (*rhomboid*), have we not yet been able to clone a *P. coenia* homolog, however this protein lacks homology to other known proteins or motifs that would help to target our approach. In the next three sections, I will describe what the analysis of the expression of these genes has taught us about the formation of the butterfly body pattern relative to that of *Drosophila*.

DIFFERENCES IN APPENDAGE NUMBER: REGULATION OF HOMEOTIC GENES

P. coenia larvae possess prolegs on the third through sixth and terminal abdominal segments (Fig. 1A). There is not full agreement as to the evolutionary relationships between prolegs and other appendages or indeed their status as limbs (Birket-Smith, 1984). Since it has been established in *Drosophila* that the *Distal-less* (*Dll*) homeobox gene is the earliest marker of the limb primordia (Fig. 1B; Cohen et al., 1989) and is required for the elaboration of the proximodistal axis of adult limbs (Cohen and Jürgens, 1989), we reasoned that the *P. coenia Dll* gene could reveal new information about proleg formation. The *P. coenia Dll* gene is expressed in all appendages (except the mandible), including the abdominal and anal prolegs (Fig. 1C; Panganiban et al., 1994). This suggests that the proximodistal axis of the prolegs is elaborated in a fashion similar to cephalic and thoracic limbs.

Unlike *P. coenia*, *D. melanogaster* lacks abdominal *Dll* expression and larval limbs. The lack of abdominal *Dll* expression has been shown to be due to the direct repressive action of the bithorax complex (BX-C) homeotic gene products on the *Dll* gene (Vachon et al., 1992). The abdominal expression of *Dll* in *P. coenia* indicates that either the *P. coenia* BX-C gene products do not repress *Dll* in the same way or that they are deployed differently during embryonic development. To address this question we cloned *Sex combs reduced* (*Scr*), *Antennapedia* (*Antp*), *Ultrabithorax* (*Ubx*), and *abdominal-A* (*abd-A*) cDNAs from *P. coenia* and examined their expression during embryogenesis (Warren et al., unpublished data). These genes are initially expressed in similar segmental registers as their *Drosophila* counterparts (Warren et al., unpublished data). However, after the initial deployment of *Ubx* and *abd-A* in the abdomen, these two genes are selectively shut off in the abdominal proleg primordia (Fig. 1D; Warren et al., unpublished data).

This observation that abdominal proleg formation is enabled by a striking alteration in homeotic gene regulation was unexpected. *Dll* de-repression in the abdomen could have easily occurred through alteration of *cis*-regulatory elements of the *Dll* gene without any changes in homeotic genes. There are many potential molecular mechanisms for the selective repression of BX-C genes in the prolegs (Warren et al., unpublished data), however, the fact that this feature, specific to the Lepidopteran order, is developmentally regulated at the homeotic level indicates that limb number can be regulated differently in the presence of the same complement of homeotic genes. It is possible that the suppression of limbs in insect ancestors was due to the evolution of a regulatory link between BX-C genes and limb-patterning genes, and not the evolution of BX-C genes.

DIFFERENCES IN APPENDAGE TYPE: REGULATION OF HOMEOTIC TARGET GENES

The flight appendages of flies and butterflies differ in that the metathoracic (T3) segment in flies bears very reduced structures, the halteres (Fig. 2A), which are used in balancing the animal during flight. Butterflies possess two well-developed pairs of wings (Fig. 2B) and pattern formation on the two wing pairs is often coordinated so as to present a continuous overall pattern when the animal's wings are at rest. The existence of mutations in the *Ubx* gene that revert the fly to its ancestral four-winged state has led to the suggestion that haltere formation was related to the evolution of the *Ubx* gene (Lewis, 1978).

It is possible, then, that if the presence of Ubx is sufficient to suppress the posterior flight appendages, the developing hindwings of butterflies might lack Ubx expression (to permit their growth as full-sized wings). To investigate this possibility, we examined Ubx expression in *P. coenia* wing imaginal discs. Like the *Drosophila* metathoracic haltere disc (Fig. 2C), the *P. coenia* hindwing imaginal disc expresses high levels of Ubx protein (Fig. 2D). This demonstrates that the presence of *Ubx is* not sufficient to promote reduction of wing size. The deployment of *Ubx* in the hindwings of a four-winged insect and the identification of the *Ubx* gene in other insects (Ueno et al., 1992), crustacea (Averof and Akam, 1993) and even annelids (Wysocka-Diller et al., 1989) demon-

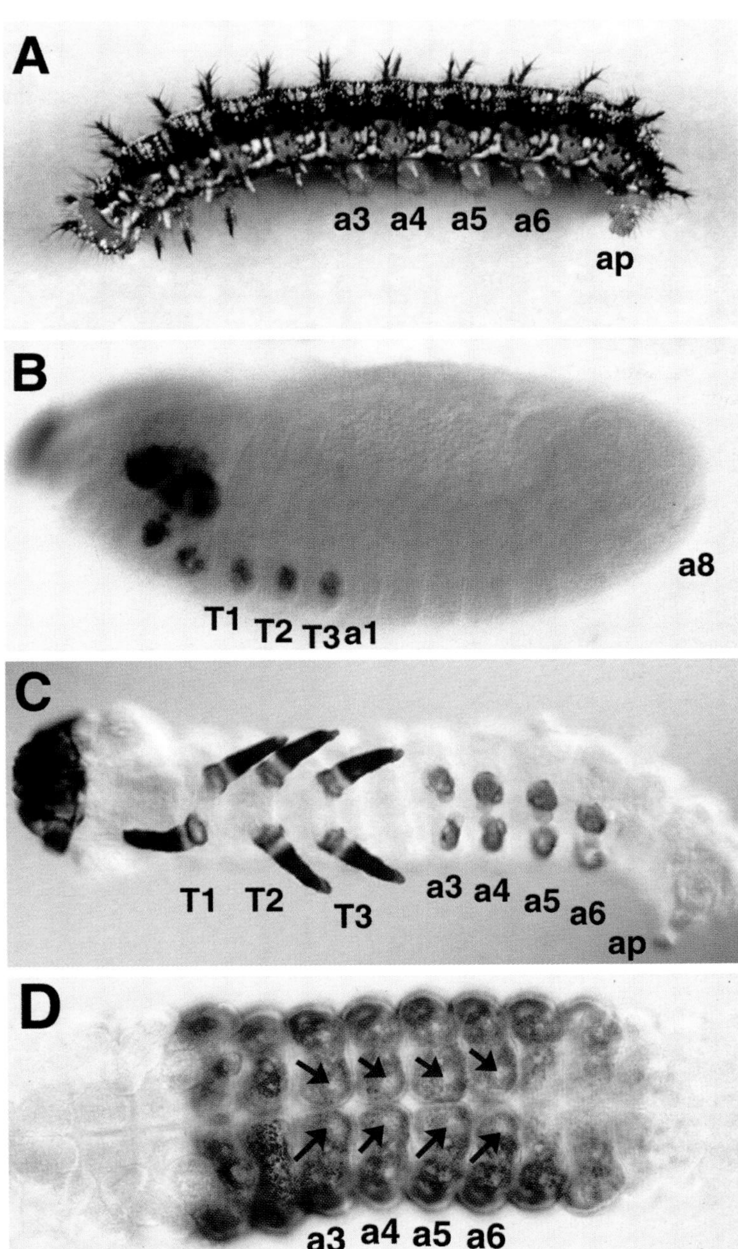

Fig. 1. Insect limb number and the regulation of *Dll* and homeotic gene expression (A) *P. coenia* larva displays four pairs of abdominal prolegs (a3-a6) and one pair of anal prolegs (ap). (B) *Dll* expression in a *Drosophila* embryo beginning germband retraction. The antennal, maxillary, labial, and thoracic (t1-t3) limb primordia express *Dll* (C). *P. coenia* embryo at 40% of embryogenesis, *Dll* RNA is expressed in a proximal ring and large distal section of each thoracic limb and in the abdominal (a3-a6) and anal prolegs (ap). (D) *Ubx* and *abd-A* protein expression in *P. coenia* embryo at about 25% of development, detected with a monoclonal antibody that recognizes both proteins (Kelsh et al., 1994). Both proteins are selectively shut off in the abdominal proleg primordia.

Table 2. *P. coenia* homologs of *D. melanogaster* genes

P. coenia gene	Protein function	Comments
Homeotic		
*Sex combs reduced**	Homeodomain, DNA-binding	Initial segmental register conserved
*Antennapedia**	Homeodomain, DNA-binding	High level expression in thorax
*Ultrabithorax**	Homeodomain, DNA-binding	Repressed in prolegs
*abdominal-A**	Homeodomain, DNA-binding	Repressed in prolegs
Appendage formation		
Distal-less†	Homeodomain, DNA-binding	Expressed in all limb primordia except mandible, also in wing eyespots
apterous†	LIM-type homeodomain	Dorsal only wing expression conserved
scalloped‡	TEA family transcription factor	In all cells of the wing
engrailed§	Homeodomain, DNA-binding	Marks anterior/posterior boundary
invected‡	Homeodomain, DNA-binding	Marks anterior/posterior boundary
wingless‡	Secreted signalling molecule	Marks wing margin
decapentaplegic‡	TGF-β type signalling molecule	Deployed in wing cell
wnt-3§	*Wingless* related protein	Expression pattern unknown
patched¶	Transmembrane receptor?	A/P boundary conserved
Others		
6OA§	TGF-β type protein	Pattern unknown
extramacrochaetae§	Helix-loop-helix negative regulatory protein	Pattern unknown

References: *Warren et al., unpublished data. †Panganiban et al. (1994). ‡Carroll et al. (1994). §Williams, Keys, Selegue and Carroll, unpublished. ¶Goodrich, L. and M.P. Scott, unpublished.

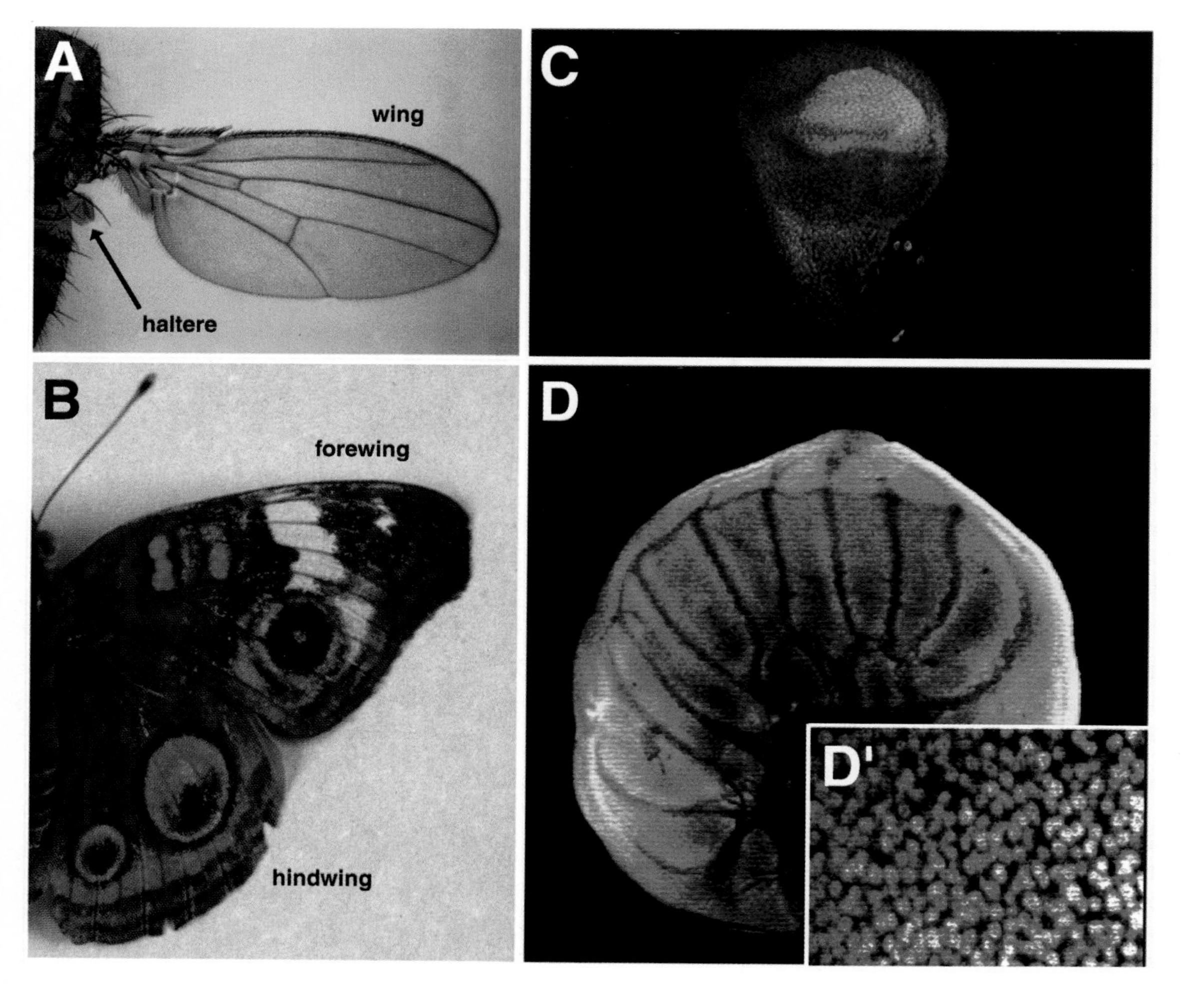

Fig. 2. The role of the *Ubx* gene in determining the type of flight appendage. (A) The adult *Drosophila* wing and haltere. Photo courtesy of Juan Pablo Couso. (B) The adult *P. coenia* forewing and hindwing. (C) The Ubx protein is expressed throughout the third instar *Drosophila* haltere imaginal disc. (D) Ubx protein is also expressed throughout the fifth instar *P. coenia* hindwing imaginal disc. Ubx expression in both halteres and hindwings indicates that the mere presence of Ubx is not sufficient to reduce the posterior flight appendages.

strates that the *Ubx* gene existed well before the evolution of the haltere.

Since no mutations are known outside of the BX-C and its regulators that transform halteres to wings, it is likely that the haltere is the product of multigenic regulation by *Ubx*. As flies evolved from a four-winged ancestor (Riek, 1977),

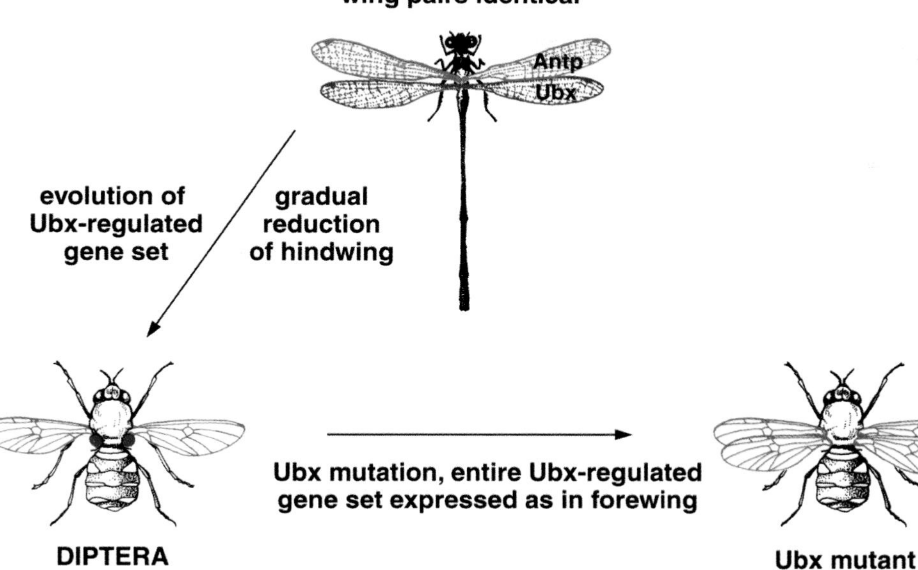

Fig. 3. The role of the *Ubx* gene in the evolution of the haltere in Diptera. *Ubx* was probably expressed in the developing hindwings of the four-winged ancestor of the Diptera (as it is in Lepidopteran hindwings). The constellation of *Ubx*-regulated genes evolved such that the hindwing was gradually reduced. Loss of *Ubx* by mutations does not simply reverse the course of evolution, it eliminates the *Ubx*-dependent regulation of a large gene set in the posterior flight appendages.

numerous genes have probably fallen under control of the *Ubx* gene. The regulation of this network of genes by *Ubx* during imaginal disc formation and morphogenesis results in the transformation of the potential wing into a haltere (Fig. 3). The reduction of the posterior wings may have been a gradual process, since fossil ancestors of modern Diptera exhibited reduced hindwings (Riek, 1977). The dramatic four-winged phenotype brought about by mutations in the *Ubx* region (Lewis, 1978) is not the simple reverse of the pathway followed by evolution; it represents a complete loss of the *Ubx*-regulated gene network in the posterior flight appendages (Fig. 3).

DIFFERENCES IN APPENDAGE PATTERN: CREATING NEW PATTERNS FROM OLD CIRCUITS

Wings of different butterfly species sport a dazzling variety of colors and patterns, organized on a similar framework. Nevertheless, all patterns can be interpreted as variations on a basic groundplan (Nijhout, 1991). The evolution of the patterning system that gives rise to this groundplan appears to be unique to butterflies (see Nijhout, this volume), while other insects and Lepidoptera have evolved pigmentation, none possess a system as robust as that found in butterflies. The fundamental units of pattern formation in the butterfly wing are the subdivisions bounded by veins and the wing margin, termed "wing cells". Pattern formation within each wing cell is largely independent of that in other wing cells. The overall wing pattern consists of a serial repetition of the pattern elements of each wing cell, although not all elements present in one wing cell are necessarily present in the others. Among the more striking pattern elements are the eyespots, concentric rings of pigmentation surrounding a center or focus. Transplantation and cautery experiments suggest that the focus is an organizer capable of inducing surrounding cells into eyespot formation (Nijhout, 1980b, 1985; French and Brakefield, 1992). Determination of the eyespot takes place in the late larval and early pupal imaginal wing disc. This is at a time well before the scale cells differentiate and produce pigment. Therefore, we have sought to identify molecules that might be involved in a prepattern in the imaginal wings.

Our bias was that genes involved in appendage pattern formation and cell-cell signalling might play a role in butterfly wing patterns. We cloned butterfly homologs of the *Drosophila* genes that play key roles in dorsal/ventral (*apterous*), anterior/posterior (*invected/ engrailed/ decapentaplegic*), and proximal/distal (*scalloped, Distal-less*) organization of the wing and in formation of the wing margin (*wingless*; Carroll et al., 1994). The expression of each of these genes reflects a common organization of the wings of *P. coenia* and *Drosophila.* Specifically, the spatial coordinates in the *Drosophila* wing are marked by the expression of particular pattern-regulating genes. Comparable coordinates in the butterfly wing are marked by homologs to the *Drosophila* pattern-regulating genes. Unexpectedly, however, we found that most of the *P. coenia* genes displayed a second pattern of transcription during the key fifth larval instar that was reiterated within each wing cell. These patterns included rays of *wingless* transcription in cells flanking the midline at the distal end of each wing cell (Fig. 4B), striped accumulations of *scalloped* transcripts along the veins (Fig. 4C), and chevrons of *apterous* transcription at the distal end of dorsal wing cells (Fig. 4D). While none of these patterns neatly correspond to color pattern elements in *P. coenia*, midline stripes, chevrons, and venous stripes occur in many butterflies (Nijhout, 1991). We suggest that these transcription patterns reflect a fundamental dynamic patterning system within each wing cell, and that the interpretation of this landscape is species-specific (Carroll et al., 1994).

The best evidence to date for a dynamic butterfly wing cell patterning system comes from analysis of *P. coenia Dll* expression. In *P. coenia, Dll* is initially transcribed in a distal band along the wing with higher levels of transcript accumulating along the midline of each wing cell. At the proximal

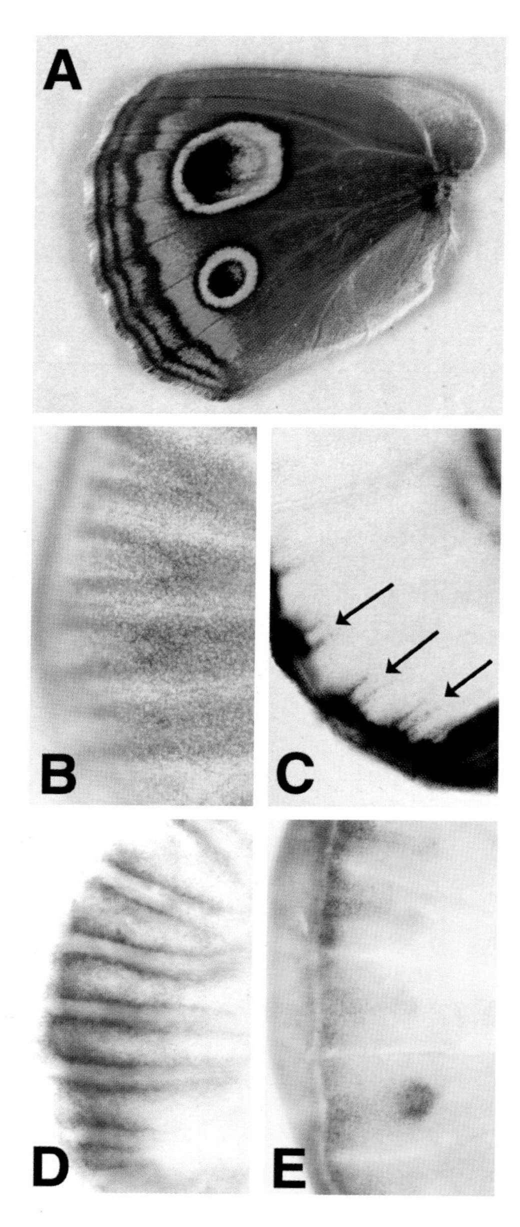

Fig. 4. Utilization of appendage-patterning genes in novel wing patterning circuit in the butterfly. (A-E) In all panels, anterior is toward the top and the distal wing margin to the left. (A) *P. coenia* hindwing displaying one anterior and one posterior eyespot. (B-E) In situ hybridizations of *P. coenia* gene probes to fifth instar hindwing imaginal discs. (B) The *P. coenia wg* gene is transcribed in pairs of rays emanating from the margin of each wing cell. (C) The *P. coenia sd* gene is transcribed at higher levels in cells lying along the veins of each wing cell. (D) The *P. coenia ap* gene is modulated in a chevron-like pattern at the distal end of each wing cell. (E) The *P. coenia Dll* gene is transcribed at high levels in cells that will give rise to the posterior eyespot. The wing cell transcription patterns in B-E suggest that a dynamic patterning system, involving some of the same global wing patterning genes, operates in each wing cell.

boundary of these midline rays, *Dll* transcription expands transiently into small circles, and in the posterior hindwing and forewing cells that develop eyespots, the *Dll* circles expand further to form spots comprising about 200 cells by the late fifth instar whereas *Dll* expression is lost from the centers of other wing cells (Fig. 4E). These features of *Dll* expression suggest that a wing cell patterning system creates the initial proximodistal restriction, the midline rays, the proximal enlargements and finally the posterior eyespot prepattern. Since *Dll* plays a crucial role in appendage formation and its expression is spatially regulated along the primary body axis and within limb fields, we propose that this appendage-patterning gene has been recruited into another regulatory circuit that is devoted to generating patterns within the wing cells. The specific transcription patterns of the other appendage-patterning genes noted above support the idea that many elements of a proximodistal patterning circuit are being deployed to pattern these tertiary wing cell fields.

THREE LEVELS OF REGULATORY EVOLUTION

Our studies have illustrated how order-specific differences in appendage number, type, and pattern involve the evolution of regulatory systems at three levels. First, we showed that abdominal limb formation is enabled by a major difference in the way BX-C homeotic genes are regulated in the abdomen. Second, we have deduced that haltere formation in the Diptera is probably the product of many accumulated changes in the network of genes regulated by the *Ubx* protein. And third, we've presented evidence that a second regulatory circuit, involving many of the same genes required for global organization of insect wings, operates specifically within butterfly wings to generate their unique groundplan.

The clear implication of this work is that much morphological diversity is derived from the way conserved batteries of developmental genes are regulated and does not require the expansion of existing gene families or the invention of new ones. While the latter have certainly occurred, the chemical evolution of animals has not been nearly as great as their morphological evolution. Changes in the number, type, position, and pattern of structures, as illustrated here, are likely to involve changes in the timing and spatial regulation of existing genes. It is, as Francois Jacob put it, a matter of "tinkering" - using the same elements and adjusting, altering, and arranging combinations to produce new morphologies (Jacob, 1977). The challenge then for comparative embryology is to decipher where and what kind of tinkering has taken place.

Further studies of the evolution of appendage number, type, and pattern may help to address the developmental and genetic roots of morphological evolution at both greater and lesser taxonomic distances than those illustrated here with flies and butterflies. For example, since wings arose within the insects, it would be interesting to look at the deployment of wing patterning genes in primitive wingless insects. What role do *apterous*, *vestigial*, *scalloped*, etc. play in an animal that never had wings? We would predict that these genes exist in all insects, but their deployment could shed light into the origins of the genetic circuitry guiding wing formation or to a morphogenetic program that may have been co-opted into wing formation.

Second, it may be possible to study morphological evolution within an order. Clearly, some of the features that distinguish butterflies from fruit flies are the product of regulatory evolution; what about family, genus, or species-level differences? Of the 15,000 or so butterfly species, at what genetic and developmental level may we expect to explain differences in wing patterns or other traits? The simplest model is that more related species possess similar wing ground plans and developmental prepatterns and that the interpretation of the patterning landscape by downstream genes, such as the pigmentation genes, may differ. It is also possible that elements of the prepattern differ according to which pattern elements are present. For example, the number as well as the coloration of eyespots differs between species. While a mechanistic grasp of color pattern formation is still just a distant hope, we can use molecular probes to ask at what level of butterflies tinker, i.e. how the patterning landscape of the butterfly wing and/or its interpretation differs between species.

The goal of this comparative molecular embryology, as with most new areas of investigation, is to identify a few first principles about the developmental and genetic mechanisms of morphological change. As Garcia-Bellido (1993) points out, we are going to have to trade the precision of mechanistic details in individual model species for deeper understanding of animal relationships and the process of change. Our understanding of and appreciation for the functions and regulation of *Drosophila* genes that have been at the center of attention for decades, such as the homeotic genes, is greatly enriched by this comparative approach. The shut-off of BX-C genes in specific butterfly limb primordia and the role (or more accurately, lack thereof) for *Ubx* gene evolution in the origin of the haltere are new chapters in the homeotic story.

There are important lessons in the haltere example in that while genetics can demonstrate a pathway back to a wing from a haltere via a simple mutation, this is clearly not the reverse of the course followed by evolution. Neither genetics nor comparative embryology will tell us the exact sequence of steps in evolution. Nevertheless, we have, in Jacob's (1982) phraseology, distinguished the possible (the evolution of the *Ubx* gene driving haltere formation) from the actual (genes downstream of *Ubx* determine haltere character). The evolution of the *Ubx* gene provided the opportunity for distinguishing insect flight appendages but the course of fly evolution is only one of many paths by which this was achieved. Butterflies, on the other hand, probably use this same gene to distinguish hindwing from forewing patterns. Our *Drosophila*-centric view might have led us to believe that *Ubx* was invented by the Diptera to make halteres. The deeper understanding is available only through a comparative lens.

I thank J. Williams, J. Selegue, L. Nagy, J. Gates, B. Warren, D. Keys, G. Panganiban, and S. Paddock who have gotten the embryology and molecular biology of *P. coenia* off the ground and who have provided the data for the illustrations in this review; Steve Paddock and Leanne Olds for preparation of the figures; Juan Pablo Couso for Figure 2A; Peter Carroll, Steve Paddock, Lisa Nagy, Bob Warren and Grace Panganiban for comments on the manuscript; and Jamie Wilson for its preparation. I am very grateful to the NSF, the Shaw Scientist Program of the Milwaukee Foundation, and the Howard Hughes Medical Institute whose generous and flexible support made these studies possible.

REFERENCES

Averof, M. and Akam, M. (1993). HOM/Hox genes of *Artemia*: implications for the origin of insect and crustacean body plans. *Current Biology* **3,** 73-78.

Bateson, W. (1894). *Materials for the Study of Variation.* Macmillan and Co., London.

Birket-Smith, S. J. R. (1984). *Prolegs, Legs and Wings of Insects.* Copenhagen: Scandinavian Science Press Ltd.

Carroll, S. B., Gates, J., Keys, D., Paddock, S. W., Panganiban, G. F., Selegue, J. and Williams, J. A. (1994). Pattern formation and eyespot determination in butterfly wings. *Science* **265**, 109-114.

Cohen, S. M. and Jürgens, G. (1989). Proximal-distal pattern formation in *Drosophila*: cell autonomous requirement for *Distal-less* gene activity in limb development. *EMBO J.* **8,** 2045-2055.

Cohen, S. M., Brönner, G., Küttner, F., Jürgens, G. and Jackle, H. (1989). *Distal-less* encodes a homoeodomain protein required for limb development in *Drosophila. Nature* **338,** 432-434.

French, V. and Brakefield, P. M. (1992). The development of eyespot patterns on butterfly wings: morphogen sources or sinks? *Development* **116,** 103-109.

García-Bellido, A. (1993). Coming of age. *Trends Genet.* **100,** 102-103.

Goldschmidt, R. (1940). *The Material Basis of Evolution.* New Haven: Yale University Press.

Jacob, F. (1977). Evolution and tinkering. *Science* **196,** 1161-1166.

Jacob, F. (1982). *The Possible and the Actual.* New York: Pantheon Books.

King, M.-C. and Wilson, A. C. (1975). Evolution at two levels in humans and chimpanzees. *Science* **188,** 107-116.

Kelsh, R., Weinzierl, R. O. J., White, R. A. H., Akam, M. (1994). Homeotic gene expression in the locust *Schistocerca*: An antibody that detects conserved epitopes in Ultrabithorax and abdominal-A genes. *Dev. Genetics* **15**, 19-31.

Kukalova-Peck, J. (1978). Origin and evolution of insect wings and their relation to metamorphosis, as documented by the fossil record. *J. Morphol.* **156**, 53-126.

Lewis, E. B. (1978). A gene complex controlling segmentation in *Drosophila. Nature* **276,** 565-570.

Nagy, L., Riddiford, L. and Kiguchi, K. (1994). Morphogenesis in the early embryo of the Lepidopteran *Bombyx mori. Dev. Biol.* in press.

Nijhout, H. F. (1980a). Ontogeny of the color pattern on the wings of *Precis coenia* (Lepidoptera: Nymphalidae). *Dev. Biol.* **80,** 275-288.

Nijhout, H. F. (1980b). Pattern formation on lepidopteran wings: Determination of an eyespot. *Dev. Biol.* **80,** 267-274.

Nijhout, H. F. (1985). Cautery-induced colour patterns in *Precis coenia* (Lepidoptera:Nymphalidae). *J. Embryol. exp. Morphol.* **86,**191-203.

Nijhout, H. F. (1991). *The Development and Evolution of Butterfly Wing Patterns.* Smithsonian: Washington.

Nijhout, H. F. (1994). Symmetry systems and compartments in Lepidopteran wings: the evolution of a patterning mechanism. *Development* **120**, **Supplement**, 225-233.

Panganiban, G., Nagy, L. and Carroll, S. B. (1994). The development and evolution of insect limb types. *Current Biology* **4**, 671-675.

Raff, R. A. (1992). Evolution of developmental decisions and morphogenesis: the view from two camps. *Development* **Supplement,** 15-22.

Riek, E. F. (1977). Four-winged Diptera from the upper permian of Australia. *Proc. Linn. Soc. New South Wales* **101,** 250-255.

Sheppard, P. M., Turner, J. R. G., Brown, K. S., Benson, W. W. and Singer, M. C. (1985). Genetics and the evolution of muellerian mimicry in *Heliconius* butterflies. *Phil. Trans. Royal Soc. London. B.* **308,** 433-613.

Ueno, K., Hui, C.-C., Fukukta, M. and Suzuki, Y. (1992). Molecular analysis of the deletion mutants in the E homeotic complex of the silkworm *Bombyx mori. Development* **114,** 555-563.

Vachon, G., Cohen, B., Pfeifle, C., McGuffin, M. E., Botas, J. and Cohen, S. M. (1992). Homeotic genes of the Bithorax complex repress limb development in the abdomen of the Drosophila embryo through the target gene *Distal-less. Cell* **71,** 437-450.

Williams, J. A. and Carroll, S. B. (1993). The origin, patterning and evolution of insect appendages. *BioEssays* **15,** 567-578.

Wilson, A. C., Sarich, V. M. and Maxson, L. R. (1974). The importance of gene rearrangement in evolution: evidence from studies on rates of chromosomal, protein, and anatomical evolution. *Proc. Nat. Acad. Sci. USA* **71,** 3028-3030.

Wilson, A. C. (1985). The molecular basis of evolution. *Scientific American* **253,** 164-173.

Wysocka-Diller, J. W., Aisemberg, G. O., Baumgarten, M., Levine, M. and Macagno, E. R. (1989). Characterization of a homologue of bithorax-complex genes in the leech *Hirudo medicinalis. Nature* **341,** 760-763.

Development 1994 Supplement, 225-233 (1994)
Printed in Great Britain © The Company of Biologists Limited 1994

Symmetry systems and compartments in Lepidopteran wings: the evolution of a patterning mechanism

H. Frederik Nijhout

Department of Zoology, Duke University, Durham, North Carolina 27708-0325, USA

SUMMARY

The wing patterns of butterflies are made up of an array of discrete pattern elements. Wing patterns evolve through changes in the size, shape and color of these pattern elements. The pattern elements are arranged in several parallel symmetry systems that develop independently from one another. The wing is further compartmentalized for color pattern formation by the wing veins. Pattern development in these compartments is largely independent from that in adjacent compartments. This two-fold compartmentalization of the color pattern (by symmetry systems and wing veins) has resulted in an extremely flexible developmental system that allows each pattern element to vary and evolve independently, without the burden of correlated evolution in other elements. The lack of developmental constraints on pattern evolution may explain why butterflies have diverged so dramatically in their color patterns, and why accurate mimicry has evolved so frequently.

This flexible developmental system appears to have evolved from the convergence of two ancient patterning systems that the butterflies inherited from their ancestors. Mapping of various pattern types onto a phylogeny of the Lepidoptera indicates that symmetry systems evolved in several steps from simple spotting patterns. Initially all such patterns were developmentally identical but each became individuated in the immediate ancestors of the butterflies. Compartmentalization by wing veins is found in all Lepidoptera and their sister group the Trichoptera, but affects primarily the ripple patterns that form the background upon which spotting patterns and symmetry systems develop. These background pattern are determined earlier in ontogeny than are the symmetry systems, and the compartmentalization mechanism is presumably no longer active when the latter develop. It appears that both individuation of symmetry systems and compartmentalization by the wing veins began at or near the wing margin. Only the butterflies and their immediate ancestors evolved a pattern formation mechanism that combines the development of a regular array of well-differentiated symmetry systems with the mechanism that compartmentalizes the wing with respect to color pattern formation. The result was an uncoupling of symmetry system development in each wing cell. This, together with the individuation of symmetry systems, yielded an essentially mosaic developmental system of unprecedented permutational flexibility that enabled the great radiation of butterfly wing patterns.

Key words: symmetry, compartment, Lepidoptera, wing, pattern formation, butterfly

INTRODUCTION

The color patterns on the wings of butterflies are examples of pure pattern formation without morphogenesis. Cell movement and cell division play no role pattern determination and differentiation in this system. Each surface of a wing is a monolayer of epidermal cells. The color pattern is a product of the interactions among intercellular signalling mechanism that determine the spatially patterned synthesis of pigments (Nijhout, 1990, 1991). The color patterns that result are highly organized systems of discrete pattern elements, each with a species-characteristic position, size, shape and color. Not only is the arrangement and morphology of the pattern elements identical among the individuals of a species, but individual pattern elements can be traced from species to species, across genera, and often across families. Thus in the course of pattern evolution, each of the pattern elements that makes up the overall wing pattern has become individuated and the elements of the color pattern adhere to a system of homologies that is every bit as consistent as that observed in the bones of tetrapod limbs and vertebrate skulls.

The system of homologies among pattern elements in butterflies is called the nymphalid groundplan (Schwanwitsch, 1924; Süffert, 1927; Nijhout, 1991). The nymphalid groundplan consists of three sets of paired bands called symmetry systems. Butterflies have three such symmetry systems on their wings, the basal symmetry system, central symmetry system and border symmetry system. In a few species, the bands of these symmetry systems run uninterrupted from the anterior to the posterior margin of the wing, but in the vast majority of species the bands are interrupted and their position is dislocated wherever they cross a wing vein. Each band is thus broken up into a series of short segments, or pattern elements, by the venation system of the wing.

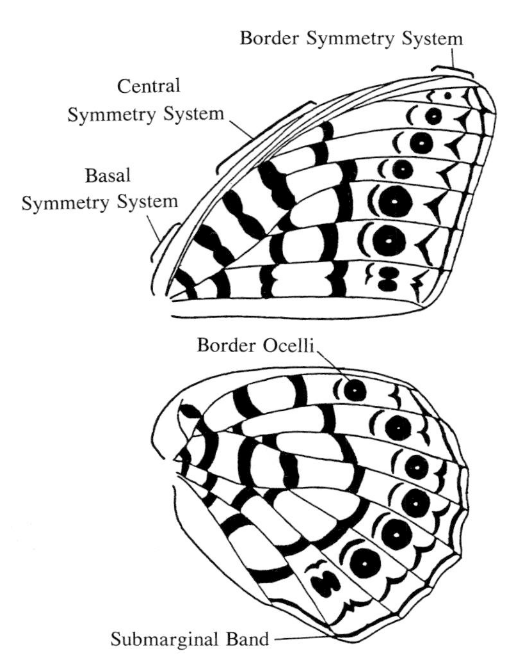

Fig. 1. The nymphalid groundplan. The butterfly wing pattern is compartmentalized into three developmentally independent symmetry systems of pigmented bands. The border symmetry system is usually the most elaborate with eyespot patterns developed along its midline; there is characteristically one such eyespot in each wing-cell. The wing pattern is also compartmentalized by the wing veins so that the elements of each symmetry system in a given wing cell develop largely independently from their homologs in adjacent wing-cells.

In butterflies, the wing veins in effect compartmentalize the color pattern, and the overall wing pattern is a serial repetition of fundamentally similar pattern elements in each wing-cell (the area between wing veins). Fig. 1 shows a version of the nymphalid groundplan that emphasizes this compartmentalization of the wing pattern by the veins. Each pattern element, then, can be viewed as a member of a rank of serial homologues that corresponds to one of the bands of the classical nymphalid groundplan.

Pattern evolution and diversification in butterflies has occurred via the modification of individual pattern elements. While in some species all members of a homologous series are identical and can be readily recognized as belonging to a single system, in most species individual pattern elements in one or more wing cells have become greatly modified in color, shape or size, and frequently even lost (see Nijhout, 1991, for illustrations). Morphometric and quantitative genetic analyses of pattern variation have revealed that there are no phenotypic or genetic correlations between pattern elements that belong to different homologous series, even when they are adjacent to each other in a wing-cell. (Paulsen and Nijhout, 1993; Paulsen, 1994). These findings suggest that adjacent symmetry systems share few if any genetic determinants. There are small to moderate phenotypic and genetic correlations among the members of a homologous series; these correlations are strongest among elements that have not diverged in morphology and weakest among elements that have diverged greatly (Paulsen and Nijhout, 1993; Paulsen, 1994). Experimental perturbation studies likewise indicate that the development of each pattern element is uncoupled from that of other elements on the wing, insofar as surgical perturbations of the wing typically only affect the morphology of a single pattern element, leaving the morphology of its immediate neighbors unaffected (Nijhout, 1981, 1991). The genetic and phenotypic evidence thus indicates that the wing pattern is composed of semi-independent pattern elements whose position, size, shape and color can be individually modified to achieve a particular optical effect.

It appears at present that there are few genetic constraints on the independent evolution of pattern elements. The developmental basis of this lack of internal constraint is that the position and morphology of each pattern element is determined by signalling sources whose effects extend only over short distances, and whose signal does not appear to pass across wing veins except in a few instances (Nijhout, 1990, 1991). The evolutionary consequence of this lack of constraint is that it enables natural selection to act on small portions of the pattern, and the response of those parts to selection is not constrained by the correlated evolution of other parts. Pattern formation on butterfly wings apparently exist at or near one of the extremes of a spectrum of developmental constraints. It is a system with relatively little internal fabricational constraint, and this lack of constraint has enabled not only the vast morphological radiation of color patterns we see today, but also the ability of butterfly patterns to easily evolve camouflage and mimicry of great accuracy and detail (Nijhout, 1994a).

The reason this developmental system is so highly flexible is that the wing pattern is compartmentalized in two ways. First, there is a proximodistal compartmentalization by independent symmetry systems, each organized around a discrete set of signalling sources. Second, there is an anteroposterior compartmentalization of the wing surface by the wing veins into compartments, the wing-cells, in which pattern development proceeds autonomously or nearly so. This highly compartmentalized, essentially mosaic, developmental system is to my knowledge unique, and it is therefore of interest to investigate how it might have arisen in evolution.

THE EVOLUTION OF A DEVELOPMENTAL SYSTEM

Molecular evidence

The molecular biology of wing pattern development is still in its infancy. There is preliminary evidence from Sean B. Carroll and his associates (Carroll et al., 1994; Nijhout, 1994b) that the organizing center of developing eyespots in *Precis coenia* is marked by the expression of the *Distal-less* gene. Carroll et al. (1994) hypothesize that the gradients that give rise to eyespots (and by analogy, to the bands of symmetry systems) may be generated by a process similar to, and perhaps evolved from, the proximodistal pattern formation system in insect appendages (Williams and Carroll, 1993). If this hypothesis is correct, then the multiple organizing centers for the color pattern on a butterfly wing may all be derived from the replication and patterned re-expression of the proximodistal axis determining system later in development.

Phylogenetic background

Evidence about the evolutionary origin of a particular feature, when this feature may have originated only once and in the

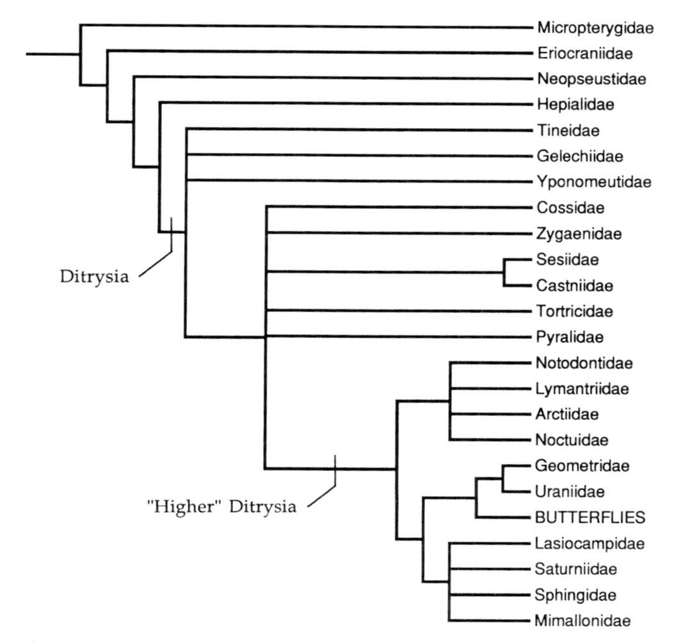

Fig. 2. Phylogenetic relationships among the major families of Lepidoptera. The position of the butterflies (Superfamily Papilionoidea) is indicated. (Based on Nielsen, 1989; Scoble, 1992; Kristensen, 1984).

distant past, must come from comparative studies. The steps involved in the evolution of a particular character from an ancestor that did not possess that character can often be reconstructed by examining how different states of the character are distributed in sister taxa of the group of interest. In the case of butterfly wing patterns, then, we need to ask what evidence we can find for the evolution of this double compartmentalized system from the way in which wing patterns are organized in the moths that constitute the sister taxa to the butterflies.

The basis of all comparative studies is a good phylogeny of the group in question. Whether one interprets a particular character state as primitive or derived in a particular taxonomic group depends on whether or not that character state occurs in the sister group (the out-group rule: Hennig, 1965; Wiley, 1981). Constructing a phylogeny is itself a major task and, while our understanding of the detailed phylogeny of the Lepidoptera is still incomplete, systematists are gradually converging on a common interpretation of its broad outlines. It is fortunate that lepidopteran systematists have not used color pattern characters in their phylogenetic reconstruction, because this means that the existing phylogenies are not loaded, or biased, in favor of a particular model of color pattern evolution.

Fig. 2 shows the phylogenetic relationships among the major families of moths and the placement of the butterflies within the moths. There is a gradually building recognition that the butterflies may share a most recent common ancestor with the geometrid and uraniid moths (Nielsen, 1989), and not with a group of day-flying moths, the Castniidae, as was once thought (Brock, 1971; Common, 1975). The relationships among many of the major groups and families of moths remains unresolved, as can be seen from the many polychotomies in Fig. 2, and lepidopteran systematists continue to disagree on which characters contain useful phylogenetic information (Nielsen, 1989; Scoble, 1992). The major groups of Lepidoptera belong to the Ditrysia (having two genital openings). Within this group, most authors isolate a large lineage variously called the Higher Ditrysia (shown in Fig. 2) or Obtectomera (Minet, 1991; Nielsen, 1989; Scoble, 1992). There is some disagreement on the placement of the Pyralidae. They are here placed outside the Higher Ditrysia, as suggested by Scoble (1992). The sister group to the order Lepidoptera is the Trichoptera (caddis flies).

Below, I will take note of the distribution of various features of the color pattern in the families shown in this phylogeny. My conclusions are based on many years of comparative studies involving thousands of species of moths. While I have examined species in all families, it is also clear that my studies have not taken in all the species that exist and, insofar as I may have missed some key species, some of the conclusions presented below may be subject to revision. In general, I will assume that if a character is found in a single species of a family, that character is at least potentially present in all members of that family and in all families more distal in the phylogeny, even if it is not expressed in any of them. Many wing pattern characters are expressed sporadically (and apparently erratically) in single species or genera of distantly related families. This observation suggests that the development and expression of many color pattern characters can be readily turned on and off, much in the way a threshold character can disappear and reappear by successively raising and lowering the threshold for its expression. Such evolutionary reversals can be accomplished relatively easily and rapidly.

Not all moths have symmetry systems, nor do all moths have their color pattern compartmentalized by wing veins, and the occurrence of these two features is not correlated in the phylogeny. The question of the evolutionary origin of the butterfly patterning system can therefore be divided into two separate questions: when do symmetry systems first appear, and when does compartmentalization by wing veins first appear? As we will see below, neither of these two simple questions has a simple answer.

The evolution of symmetry systems

Clear examples of multiple, parallel symmetry systems, that run from the anterior to the posterior margin of the wing, are found in all the Higher Ditrysia and in the Pyralidae. The remaining families have either only a single symmetry system or have a foreground pattern of multiple irregular spots or circles. These spots, in turn, may remain isolated or may be fused with one another to various degrees. Families that are most basal in the phylogeny, such as the Micropterygidae (mandibled moths) and many of the 'lower' Ditrysia, have a much simplified banding pattern. In some species there is a single color partition across the wing so that the proximal half is a different color from the distal half. Other species have an irregular band that is either Y-shaped, or broad near both the anterior and posterior margins and constricted or broken near the middle of the wing.

There is no single pattern type that is unambiguously basal or primitive for the Lepidoptera as a whole. There is, not unexpectedly, an overall tendency for patterns to become more complex (i.e. to be composed of more parts, with more morphological diversity among those parts) as one goes up the phylogeny. There is also a general tendency for the number of parallel bands that make up the pattern to increase in number, and a trend from irregular spotting patterns to regular banding patterns. But there are exceptions (reversals) to all these trends in many families. The detection of trends or transition series is complicated by the fact that the pattern is scale dependent. Small moths have simpler wing patterns than large ones, composed of fewer parts and with less detail in the differentiation of those parts. Some apparent reversals of evolutionary trends in the color pattern are no doubt due to the fact that some families contain many species of small body size. Conversely, in some families with relatively 'primitive' patterns such as the Hepialidae (swift moths), the patterns of species with unusually large body sizes can be surprisingly complex. In order to correctly interpret this diversity of patterns it is necessary to understand the developmental origin of symmetry systems.

Symmetry systems develop around discrete organizing centers. This has been experimentally demonstrated by the work of Kühn and Von Engelhardt (1933), Henke (1933), Wehrmaker (1959), Schwartz (1962), and Toussaint and French (1988). These organizing centers act as sources that establish a gradient whose local value determines the synthesis of different pigments within its field. A point source then gives rise to a pattern of concentric circular bands such as is seen in the eyespots of butterflies; around a row of closely spaced point sources the bands fuse, producing the pattern of parallel bands that we recognize as a symmetry system. The position of a symmetry system is therefore specified by the positions of these organizing centers: in species with a single symmetry system there is a single array of organizing centers; in species with three symmetry systems there are three parallel rows of organizing centers (Nijhout, 1991). In the butterflies there is potentially one such organizing center per wing-cell, but this is not true in the moths. In Pyralidae there are two or three organizing centers for the central symmetry system (Kühn and Von Engelhardt, 1933; Schwartz, 1962; Nijhout, 1991). In the Saturniidae and Lymantriidae too, there appear to be at least two organizing centers (Henke, 1933; Nijhout, 1978). In the Geometridae there appear many centers of origin for the central symmetry system. In some species there is possibly one per wing-cell at the center of a well-developed symmetry system (Fig. 3A), while in others there is no clear symmetry system and the pattern appears to originate from a handful of source scattered across the wing (Fig. 3B). The developmental physiology of these geometrid patterns has not yet been investigated experimentally.

In the ditrysian moths that have well-developed patterns but do not have symmetry systems, such as the Zygaenidae, the wing pattern is composed of roughly circular patterns, often with concentric bands of different colors. These spots may fuse with each other to different degrees and in different combinations, depending on the species. When two spots merge their bands become smoothly continuous. Individual variability and species diversity show that any pair of neighboring spots can fuse (Fig. 4), which implies that all the spots are developmentally equivalent. It is reasonable to assume that these circular patterns represent the precursors (or ancestors) of symmetry systems and that each is organized around a point source of pattern determination, much like the eyespots of butterflies (a proposition that can be tested by ablating the centers of these spots early in development). If this interpretation is correct, then the fact that any pair of spots can fuse implies that the primitive wing pattern develops around an assortment of identical signalling sources. I assume here that these spots develop around organizing centers similar to those described by Kühn and Von Engelhardt (1933), though this proposition has yet to be tested experimentally in the Zygaenidae and most other families of moths.

Families of moths differ in the number and distribution of organizing centers on their wings. The more basal families of moths have relatively few such centers. Most patterns in the Micropterygidae, Gelechiidae and related families, for instance, appear to be derived from a small number of such centers spaced around the periphery of the wing (Fig. 5). These centers can expand to form spots or elongate to form bands, and the spots and bands can fuse in various ways, producing large areas of contrasting color, interdigitating bands and Y-shaped patterns. Each family of these primitive moths tends to have a characteristic pattern of organizing centers and a limited diversity of ways in which the spots and bands that these centers produce expand and fuse with one another. In the Hepialidae, for instance, there are a larger number of sources distributed in what appears to be an haphazard arrangement (Fig. 6), although this arrangement is very consistent throughout the family and produces an array of color patterns that is highly distinctive of this family. In the Hepialidae we also see incipient symmetry systems made up by the fusion of rings that develop around rows of adjoining sources.

The colorful and complex wing patterns of the Arctiidae (tiger moths) are likewise composed of irregular spots and bands that fuse in various patterns. In the Arctiidae we also see the first appearance of a multiple parallel banding pattern (Fig. 7). These bands are derived from the irregular banding pattern, presumably by some process that causes the sources to become aligned in relatively straight and parallel rows. There can be as many as six parallel bands evenly spaced on the wing, all self-symmetrical, and all identical in color. Each of these bands can now be seen as a symmetry system. In the species with such parallel bands there is often a partial local fusion between

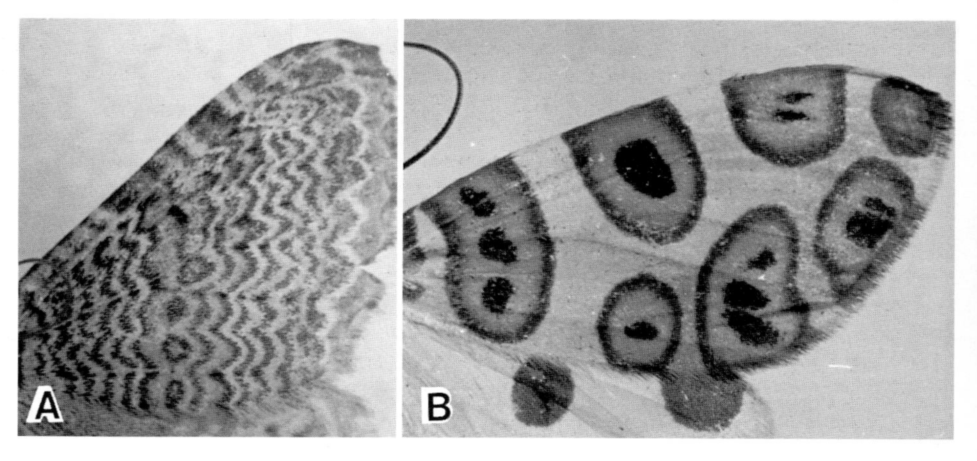

Fig. 3. Wing patterns of two moths in the family Geometridae. (A) *Hydria undulata;* (B) *Pantheroides pardaclis.*

3/94 LIB/LEND/001

Robert Scott Library

St. Peter's Square
Wolverhampton WV1 1RH

Wolverhampton (01902) 322305

This item may be recalled at any time. Keeping it after it has been recalled or beyond the date stamped may result in a fine.
See tariff of fines displayed at the counter.

-9 FEB 1996
-9 FEB 1996
30 JAN 1997
28 MAY 1998
26 OCT 1998
25 NOV 1999
14 JAN 2000
-5 APR 2000

WP 0866174 X